STRUCTURE REPORTS
for 1965
Vol. 30 B

Structure Reports is prepared under the guidance of a Commission of the International Union of Crystallography. The members of the Commission sometime concerned with the preparation of this volume are listed below.

STRUCTURE REPORTS

FOR 1965

Volume 30 B

GENERAL EDITOR

W. B. PEARSON

SECTION EDITOR

A. W. HANSON

Springer Science+Business Media, B.V.

First published in 1974

ISBN 978-94-017-3108-9   ISBN 978-94-017-3106-5 (eBook)
DOI 10.1007/978-94-017-3106-5

Koninklijke Drukkerij Van de Garde B.V., Zaltbommel

TABLE OF CONTENTS

SYMBOLS

The letters a, b, c; α, β, γ are used consistently for the edges and angles of the unit cell. Other letters used consistently are as follows.

U	Volume of unit cell
D_m	Measured density in g/cm³ or specific gravity
D_x	Density in g/cm³ calculated from cell volume and contents
Z	Number of times the formula quoted is repeated in the unit cell (Number of atoms per unit cell in alloys of simple structure)
x, y, z	Atomic coordinates as fractions of cell edge (Occasionally u, v, w or other letters are used)
X, Y, Z	Atomic coordinates in Ångström units
X', Y', Z'	Atomic coordinates in Ångström units, referred to orthogonal axes (Used only in the Organic Section)
F.W.	Formula weight
A, B, C	Types of layer in layer structures
M, A, B	Variable metal atom(s) in a sequence of related structures
X, H	Variable non-metals, usually halogen, in a sequence of related structures
R	Variable organic radical, or reliability index
s, m, w, v b	Strong, medium, weak, very, broad
TSNR	Taylor-Sinclair-Nelson-Riley (extrapolation function)

LIMITS OF ERROR

Errors are generally quoted in units in the last place. Thus $4 \cdot 8754 \pm 3$ means $4 \cdot 8754 \pm 0 \cdot 0003$, $4 \cdot 87 \pm 3$ means $4 \cdot 87 \pm 0 \cdot 03$, and $4 \cdot 875 \pm 15$ means $4 \cdot 875 \pm 0 \cdot 015$. Occasionally a very doubtful last digit is given below line.

TRANSLITERATION OF RUSSIAN

а	a	и	i	р	r	ш	š
б	b	й	j	с	s	щ	šč
в	v	к	k	т	t	ы	y
г	g	л	l	у	u	ъ	”
д	d	м	m	ф	f	ь	’
е	e	н	n	х	kh	э	ė
ж	ž	о	o	ц	c	ю	ju
з	z	п	p	ч	č	я	ja

INTRODUCTION

Starting with Volume **30**, 1965, *Structure Reports* is produced in a new format by photo-offset printing from typed manuscript with unjustified lines. At the time when the decision for this change was taken, the cost of setting the manuscript in type was becoming so high as to render the cost of individual subscription prohibitive. At that time automatic typing methods giving justified lines, etc. for photo-offset reproduction did not offer any saving over type setting, but hand typing of the manuscripts could give a considerable saving in production costs. In the belief that a publication that is too expensive to buy is of little value, the format has been changed sacrificing elegance to availability.

The new format does not lead to increased length of the volumes since the information content of the typed and typeset pages is practically identical. However, the amount of work to be reported demands the eventual separation of *Structure Reports* into two volumes, A. *Metals and Inorganic* and B. *Organic*. It was convenient to introduce this change also at Volume **30**, and with Volume **31** further to restrict the publication of crystal data, so that from 1966 onwards the reports deal almost entirely with complete structure determinations only.

In the past the aim of *Structure Reports* has been to present critical reports on all work of crystallographic structural interest, whether it is derived directly from X-ray, electron or neutron diffraction, or even indirectly from other experiments. The reports were intended to be critical and not mere abstracts, except in some cases when a brief indication of the content of a paper of related interest was included in the form of an abstract. In selecting topics for reporting, the criterion "of structural interest" was freely interpreted in terms of what was topically interesting. However, the amount of literature covering matters of structural interest has become so large that this policy can no longer be followed. From Volume **28** onwards, critical reports are only given on actual structure determinations. That is to say reports are only written on papers recording the determination of the positions of the atoms in structures. Nevertheless, Volumes **28** to **30** still do record much information on lattice parameters and space groups of the structures of alloys and compounds, although in the form of tables. From Volume **31** onwards, even this information is generally omitted and only full structure determinations are reported. One of the reasons for this omission is that such data are obtainable from the continuing volumes of *Crystal Data*.† Only in this way is it possible to keep yearly volumes of *Structure Reports* to a fairly uniform and usable size in view of the ever increasing amount of published material.

Ideally, the reports have been prepared in such a way that no further structural information would be gained by consulting the original paper itself, although from Volume **21** onwards, atomic parameters are not generally reproduced for structures containing more than about 30 independent atoms. The main reason for this is

† *Crystal Data* Third Edition, Ed. J. D. H. DONNAY and H. M. ONDIK; U.S. Department of Commerce and JCPDS.

that the chance of including typographical errors in reproducing extensive tables of data is such, that anybody wishing to make detailed use of them would in any case consult the primary references.

Although the data in *Structure Reports* must be presented as briefly as possible, every effort is made to avoid jargon, so that the information is readily understandable by the non-crystallographer as well as the crystallographer. Nevertheless it is assumed that the reader has available Volume I of the *International Tables**, wherein he can obtain details of space group settings and equivalent positions of the atoms.

The arrangements in individual reports is generally: Name, Formula, Papers reported, Unit cell, Space group, Atomic positions, Interatomic and intermolecular distances, Material, Discussion, Details of analysis and References. Editorial comments are enclosed in square brackets, and it may be assumed that material not distinguished in this manner is based directly on the papers reported. The volumes are divided into three main sections: *Metals, Inorganic Compounds* and *Organic Compounds*. In the earlier volumes of *Structure Reports* the arrangement of the *Metals* section was strictly alphabetical, with crossreferences given in the text, so that the preparation and use of an index to find data on alloys was unnecessary. In order to save space this practice was discontinued in Volume **18** and subsequently. Although the arrangement of the *Metals* section remains roughly alphabetical, *it is now essential to use the Subject and Formula indexes when seeking work on elemental metals, intermetallic compounds and alloys*. The arrangement of the *Inorganic* and *Organic* sections is roughly in the order of increasing complexity of composition, related substances and related structures being kept together as far as possible. Inorganic and organic compounds should therefore also be sought in the Subject or Formula indexes. The Subject index is arranged alphabetically by the names printed as the headings of the reports, and it also includes other common names and information. The Formula index in the A volumes, *Metals and Inorganic Compounds*, is arranged in alphabetical order of the chemical symbols. In the B volumes, *Organic Compounds*, the classification is by the number of carbon atoms and secondary classification by the number of hydrogen atoms; other constituents then follow alphabetically. In the formula indexes solvents of crystallization follow at the end of the formulae, but double entries are frequently made, especially for inorganic compounds containing water, possibly or certainly as OH groups.

The scheme generally employed for the transliteration of Russian is reproduced on p. VI, and the usual abbreviations of journal titles are listed in earlier volumes. Transliteration is in accordance with draft recommendation no. 6 of the International Organization for Standardization, and the abbreviations are based on the World List of Scientific Periodicals.

W. B. PEARSON

University of Waterloo
Waterloo, Ont., Canada.
24 April 1974

---

* *International Tables for X-ray Crystallography*, Vol. I, *Symmetry Groups*. International Union of Crystallography, Kynoch Press, Birmingham, 1952.

STRUCTURE REPORTS

SECTION III

ORGANIC COMPOUNDS

EDITED BY
A. W. HANSON

WITH THE ASSISTANCE OF THE FOLLOWING REPORTERS
D. R. POLLARD
E. W. MACAULAY

ARRANGEMENT

The order of entries is that used in the Bibliography (1935-69) "Molecular Structures and Dimensions", Vols. I and II, 1970, Oosthoek, Utrecht.

TARTRONIC ACID

$C_3H_4O_5$ F.W. = 120.1

I. The crystal structure of tartronic acid. B.P. VAN EIJCK, J.A. KANTERS and J. KROON, 1965. *Acta Cryst.*, 19, 435-439.

Orthorhombic, $a = 4.485\pm1$, $b = 8.813\pm2$, $c = 10.895\pm3\text{Å}$, $[U = 430.6\text{Å}^3]$, $D_m = 1.83$, $z = 4$, $D_x = 1.84$ (CuK_α, $\lambda = 1.5418\text{Å}$).

Space group $P2_12_12_1$ (D_2^4)

Atomic positions

	x	y	z
O(1)	0.1305	−0.1396	0.3601
O(2)	−0.0271	0.0927	0.3069
O(3)	0.0153	0.1156	0.6625
O(4)	0.1314	−0.1279	0.6278
O(5)	−0.4277	0.0993	0.4889
C(1)	−0.0229	−0.0152	0.3753
C(2)	−0.2129	−0.0155	0.4921
C(3)	−0.0066	−0.0010	0.6043
H(1)	−0.314	−0.122	0.498
H(2)	−0.348	0.180	0.446
H(3)	0.195	−0.141	0.292
H(4)	0.221	−0.115	0.698

Standard deviations are: O, 0.002; C, 0.003; H, 0.04Å. Anisotropic temperature factors are given for the non-hydrogen atoms.

Structure

The central carbon atom is essentially coplanar with each of the carboxyl groups. With respect to the central C-O bond, these groups are rotated about C-C bonds by 15.0° and 18.5°. The sense of the rotations is such as to preserve approximate mirror symmetry. Some mean bond lengths and angles are:

a = 1.210Å	ac = 121.4°
b = 1.304	bc = 112.6
c = 1.535	cd = 111.6
d = 1.397	cc' = 109.0

(The standard deviations are 0.005Å and 0.2°.)

Details of analysis

The intensity data were measured with a four-circle diffractometer and scintillation counter, using CuK_α radiation. 534 reflexions were observed and measured. Refinement was by block-diagonal least squares to a final R index of 0.037.

SUCCINIC ANHYDRIDE

$C_4H_4O_3$ F.W. = 100.1

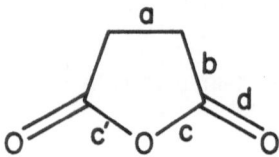

I. The crystal structure of succinic anhydride. M. EHRENBERG, 1965. *Acta Cryst.*, **19**, 698-703.

II. Struttura cristallina e molecolare dell'anidride succinica. S. BIAGINI and M. CANNAS, 1965. *Ric. Sci.*, **2**, A, *Ital.*, B, 1518-1526.

I. *Orthorhombic*, a = 6.963±15, b = 11.71±4, c = 5.402±10Å, [U = 440.5Å^3], D_m = 1.504, z = 4, D_x = 1.509.
Space group $P2_12_12_1$ (D_2^4)

Atomic positions

	x	y	z
C(1)	0.1655±10	0.2119± 6	0.3422±12
C(2)	0.0300±10	0.1786± 6	0.1475±12
C(3)	-0.0391± 9	0.0608± 6	0.2196±12
C(4)	0.0603±10	0.0373± 6	0.4572±12
O(5)	0.1768± 6	0.1272± 4	0.5204± 8
O(6)	0.2596± 8	0.2963± 4	0.3643±11
O(7)	0.0527± 9	-0.0436± 5	0.5900±10

Anisotropic temperature factors are given and are analysed in terms of rigid-body vibrations. Assumed hydrogen positions are given also.

Structure

The molecule is slightly non-planar. [The confidence level for this statement is about 1%]. Mean bond lengths and angles are:

a = 1.51±1Å	ab = 104.6°
b = 1.48±1	bc = 110.3
c = 1.38±1	cc' = 110.1
d = 1.19±1	bd = 130.4
	cd = 119.4

Intermolecular distances are consistent with van der Waals interaction.

Details of analysis

Intensity data were recorded on equi-inclination Weissenberg photographs ($hk0$ to $hk3$; $h0l$ to $h8l$) using CuK_α radiation. 390 reflexions (about 65% of those within the CuK_α sphere) were observed and measured visually. The structure was refined by least squares to a final R index of 0.076.

II. This is an independent study of the same crystal structure, but of somewhat lower accuracy. The dimensions reported are in acceptable agreement with those of I.

ADIPID ACID

$C_6H_{10}O_4$ $COOH(CH_2)_4COOH$ F.W. = 146.1

I. Localisation des atomes d'hydrogéne dans l'acide adipique $COOH[CH_2]_4COOH$.
J. HOUSTY and M. HOSPITAL, 1965. *Acta Cryst.*, 18, 693–697.
(For earlier studies see 1).

Monoclinic, a = 10.0±1, b = 5.15±1, c = 10.06±1Å, β = 136°45', $[U$ = 355.3Å³$]$,
D_m = 1.36, Z = 2, D_x = 1.355 [1.365], $(CuK_\alpha$, λ = 1.542Å).

Space group $P2_1/c$ (C_{2h}^5)
Molecular symmetry, centre

Atomic positions

	x	y	z
C(1)	0.0520	0.0392	-0.0252
C(2)	0.2056	-0.1624	0.0493
C(3)	0.3310	-0.0907	0.0236
O(1)	0.2961	0.1013	-0.0707
O(2)	0.4746	-0.2450	0.1023
H(11)	0.090	0.210	0.010
H(12)	-0.055	0.080	-0.175
H(21)	0.140	-0.325	-0.030
H(22)	0.300	-0.200	0.195
H(4)	0.580	-0.150	0.100

Isotropic temperature factors are given also.

Structure

The structure is essentially as reported in 1, except that the hydrogen atoms have been located [but without indication of accuracy]. The bond lengths are given in Fig. 1; bond angles are given also.

Details of analysis

The intensity data were recorded with the DeJong retigraph, using CuK_α radiation. 380 independent reflexions were observed and estimated visually. Refinement was by least squares to a final R index of 0.08.

1.   *Structure Reports*, 12, 321.

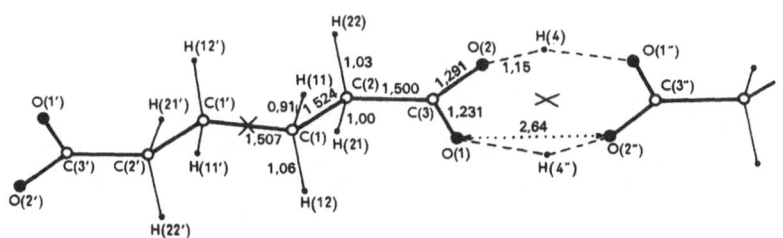

Fig. 1.   Bond lengths in adipic acid. Standard deviations, 0.005Å for C-C, C-O, O-H..O; [not given for C-H or O-H.]

SUBERIC ACID

$C_8H_{14}O_4$ F.W. = 174.2

COOH$(CH_2)_6$COOH

I. Localisation des atomes d'hydrogene dans l'acide suberique COOH$(CH_2)_6$ COOH.
J. HOUSTY and M. HOSPITAL, 1965. *Acta Cryst.*, 18, 753-755.
(For a preliminary determination from projections, see 1.)

Monoclinic, a = 8.98±1, b = 5.06±1, c = 10.12 1Å, β = 97° 50' [U = 455.5 Å³],
D_m = 1.270, Z = 2, D_x = 1.262.

Space group $P2_1/c$ (C_{2h}^5) Molecular symmetry: centre

Atomic positions

	x	y	z
C(1)	0.0622	0.0917	0.0282
C(2)	0.1487	0.0035	0.1612
C(3)	0.2760	0.1879	0.2122
C(4)	0.3701	0.1009	0.3384
O(1)	0.3414	-0.0953	0.4016
O(2)	0.4840	0.2550	0.3801
H(11)	0.1250	0.0700	-0.0310
H(12)	0.0120	0.2280	0.0700
H(21)	0.0880	0.0040	0.2290
H(22)	0.1970	-0.1480	0.1200
H(31)	0.2450	0.3550	0.2430
H(32)	0.3350	0.2040	0.1460
H(2)	0.5470	0.1800	0.4845

Temperature factors (anisotropic for oxygen) are given for all atoms.

Structure

The structure is essentially as reported in 1, but the results are more
accurate, and the positions of the hydrogen atoms have been determined. [There
is no indication of the accuracy of the hydrogen positions, however]. The bond
lengths are given in Fig. 2.

Details of analysis

435 measured reflexions were used in a least-squares refinement. The final
R index is 0.13.

1.   *Structure Reports*, 29, 470.

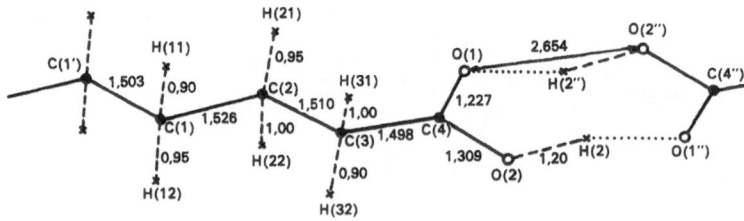

Fig. 2.   Interatomic distances in suberic acid. Standard deviations are ±0.005Å
for distances not involving hydrogen atoms.

PEROXYPELARGONIC ACID

$C_9H_{18}O_3$ F.W. = 174.2

$$CH_3(CH_2)_7COOOH$$

I. The crystal structure of peroxypelargonic acid. D. BELITSKUS and G.A. JEFFREY,
1965. *Acta Cryst.*, <u>18</u>, 458–463.

Monoclinic, a = 23.49±5, b = 4.80±5, c = 9.6±5Å, β = 106.0±5°, U = 1044Å, D_m = 1.11,
Z = 4, D_x = 1.11.

[Temperature not explicitly stated, but is probably -30°C.]

Space group $P2_1/c$ (C_{2h}^5)

Atomic positions

	x	y	z		x	y	z
O(1)	-0.0172	-0.108	-0.190	C(4)	0.2095	-0.275	0.302
O(2)	0.0225	-0.250	-0.069	C(5)	0.2556	-0.095	0.409
O(3)	0.0865	0.090	-0.084	C(6)	0.3031	-0.277	0.515
C(1)	0.0759	-0.125	-0.027	C(7)	0.3510	-0.096	0.623
C(2)	0.1158	-0.282	0.096	C(8)	0.3967	-0.288	0.728
C(3)	0.1636	-0.095	0.192	C(9)	0.4437	-0.115	0.839

Standard deviations are given in full, and range from 0.008 to 0.021Å.
Temperature factors are given, as are assumed positions and temperature factors
for the hydrogen atoms.

Structure

The structure is illustrated in Fig. 3. Adjacent molecules related by the
screw diad axis are linked by hydrogen bonds (O(1)...O(3'), 2.74±1Å) to form an
infinite helical chain parallel to b. The carbon backbone of the molecule from
C(3) to C(9) is planar, as is the peroxyacid group COOOH. Bond lengths and
angles are given in full; those of the peroxyacid group are given in Fig. 4.

Details of analysis

Three-dimensional intensity data were recorded on equi-inclination Weissenberg
photographs ($h0l$ to $h2l$, and $hk0$ to $hk5$) at -30°C. 886 reflexions (about 31%
of those in the CuK$_\alpha$ sphere) were accessible on the films, and of these only 197
were observed above background. Refinement was by least-squares to a final R index
of 0.18 for observed reflexions.

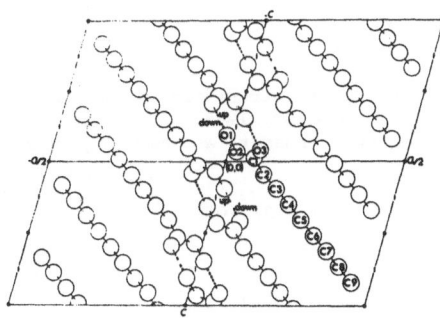

Fig. 3.   The peroxypelargonic acid structure, viewed along b.

Fig. 4.   Bond lengths and angles in the peroxyacid group.   Standard deviations
          are 0.01 to 0.02Å, and 1°.

METHYL *m*-BROMOCINNAMATE

METHYL *p*-BROMOCINNAMATE

$C_{10}H_9O_2Br$ F.W. = 241.1

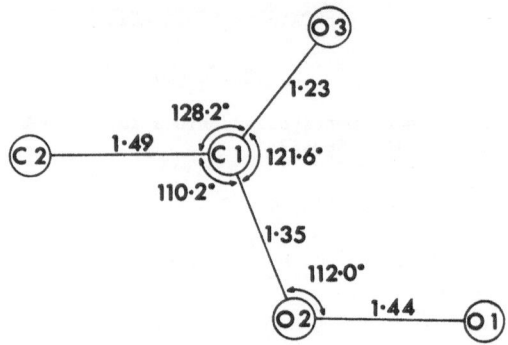

I. Topochemisty. XI. The crystal structures of methyl *m*- and *p*-bromocinnamates.
   L. LEISEROWITZ and G.M.J. SCHMIDT, 1965. *Acta Cryst.*, <u>18</u>, 1058-1067.
   (For crystal data of methyl cinnamate and methyl *p*-chlorocinnamate, see <u>1</u>).

Meta derivative:

Monoclinic, a = 7.830, *b* = 5.976, *c* = 21.208Å, β = 99°31', [*U* = 978.7Å^3], *z* = 4,
D_x = 1.64.
Space group $P2_1/a$ (C_{2h}^5)

Para derivative:

Monoclinic, a = 8.485, *b* = 20.703, *c* = 5.764Å, β = 92.2°, [*U* = 1012Å^3], *z* = 4,
D_x = 1.58.
Space group $P2_1/n$ (C_{2h}^5)

Atomic positions

Meta derivative:

	x	y	z
Br	0.1969	0.1898	0.0496
O(1)	0.5888	−0.0924	0.3823
O(2)	0.5685	0.2615	0.4225
C(1)	0.6866	−0.1596	0.4421
C(2)	0.5370	0.1332	0.3793
C(3)	0.4502	0.1831	0.3121
C(4)	0.3905	0.3795	0.2982
C(5)	0.3005	0.4460	0.2328
C(6)	0.2119	0.6578	0.2268
C(7)	0.1249	0.7380	0.1692
C(8)	0.1186	0.5900	0.1131
C(9)	0.2024	0.3787	0.1196
C(10)	0.2968	0.3148	0.1795

(Original in Å). Also given are anisotropic temperature factors for all except C(1) of the atoms above, and assumed positions for all the hydrogen atoms except those of the methyl group. There is no estimate of error.

Para derivative:

	x	y	z
Br	−0.0913	0.2353	0.0186
O(1)	0.1347	0.5592	1.0461
O(2)	0.4218	0.5562	0.7647
C(1)	0.3559	0.6077	1.1542
C(2)	0.3031	0.5381	0.8468
C(3)	0.1961	0.4874	0.7327
C(4)	0.2359	0.4582	0.5498
C(5)	0.1517	0.4068	0.4304
C(6)	0.2084	0.3820	0.2146
C(7)	0.1410	0.3315	0.0918
C(8)	0.0052	0.3053	0.1836
C(9)	−0.0576	0.3278	0.3806
C(10)	0.0143	0.3775	0.5007

(Original in Å). Also given are anisotropic temperature factors for all atoms. There is no estimate of error.

Structure

Bond lengths and angles are given in full, and are consistent with expectation. Both molecules are roughly planar, with the C=O bond *cis* to the C=C bond. Intermolecular distances are consistent with van der Waals interaction.

Details of analysis

Meta derivative:

The intensity data were recorded for zones normal to a and b (no details). Refinement was by diagonal least squares to a final R index of 0.066.

Para derivative:

The intensity data for the principal zones were recorded photographically, using CuK_{α} radiation, and measured visually. Refinement was by full-matrix least squares to a final R index of 0.101.

1.   *This volume*, p. 423.

TRILAURIN (β-FORM)

$C_{39}H_{74}O_6$ F.W. = 639

$(C_{11}H_{23}COO)_3C_3H_5$

I. The crystal structure of the β-form of trilaurin. K. LARSSON, 1965. *Ark.
Kemi., 23, 1–15.
(For a preliminary report, see 1.)

Triclinic, a = 12.35±8, b = 5.44±4, c = 31.75±10Å, α = 94.0±5, β = 96.7±5,
γ = 99.2±5°, [U = 2083Å^3], z = 2, [D_x = 1.02].
(Crystal data for the isomorphous triglyceride of 11-bromoundecanoic acid are
given also, and are reported in 2.)

Space group   P1 (C_1^1)   or   P$\bar{1}$ (C_i^1)   P$\bar{1}$ is confirmed by analysis.

Atomic positions of the 45 non-hydrogen atoms are given in the original.

Structure

    Bond lengths and angles are given in full. The molecular packing is illus-
trated in Fig. 5. The C–COO groups are planar. The configuration of the molecule
is compared to that of a tuning fork.

Details of analysis

    Three-dimensional intensity data were recorded on Weissenberg photographs,
using CuK_α radiation, and estimated visually. Refinement was by block-diagonal
least squares to a final R index of 0.20.

1.   K. LARSSON, 1963. *Proc. Chem. Soc.*, 87.
2.   *This volume*, p. 436.

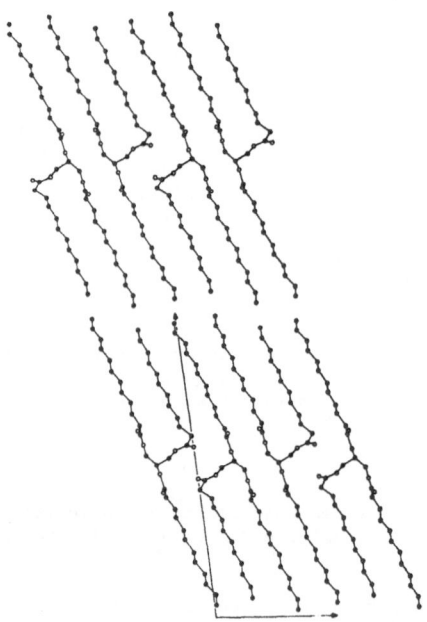

Fig. 5.   The structure of the β form of trilaurin projected along *b*.

$C_{15}H_{30}O_4$ 2-MONOLAURIN   F.W. = 274.4

$C_{11}H_{23}COOCH(CH_2OH)_2$

I. On the crystal structure of 2-monolaurin. K. LARSSON, 1965. *Ark. Kemi*, <u>23</u>, 23–27.

Triclinic, a = 5.90±4, b = 4.97±3, c = 32.9±2Å, α = 91.8±8, β = 91.8±8, γ = 118.5±10°, $[U = 846Å^3]$, z = 2, $[D_x = 1.08]$.

(Crystal data for the isomorphous monoglyceride of 11-bromoundecanoic acid are given also, and are reported in <u>1</u>.)

Space group   P1 (C_1^1) or P$\bar{1}$ (C_i^1) P$\bar{1}$ is confirmed by analysis.

Atomic positions

	x	z		x	z
C(1)	0.242	0.0417	C(11)	0.955	0.3220
C(2)	0.484	0.0570	C(12)	0.122	0.3585
C(3)	0.602	0.0962	C(13)	0.418	0.4212
C(4)	0.827	0.1182	C(14)	0.664	0.4253
C(5)	0.916	0.1583	C(15)	0.287	0.4580
C(6)	0.190	0.1791	O(1)	0.212	0.4770
C(7)	0.266	0.2165	O(2)	0.807	0.4512
C(8)	0.493	0.2365	O(3)	0.253	0.3524
C(9)	0.611	0.2708	O(4)	0.178	0.3935
C(10)	0.799	0.2936			

Isotropic temperature factors are given also. The standard deviations are estimated to be 0.04Å and 0.02Å for carbon and oxygen respectively.

Structure

The molecular arrangement is illustrated in Fig. 6. The molecules are linked by O–H..O bonds into infinite chains as shown; the hydrogen atoms must depart from P$\bar{1}$ symmetry, and are probably disordered.

Details of analysis

The intensity data for the $h0l$ zone were recorded on Weissenberg photographs, using CuK$_\alpha$ radiation, and measured visually. Refinement was by block-diagonal least squares to a final R index of 0.20.

1.   *This volume*, p. 429.

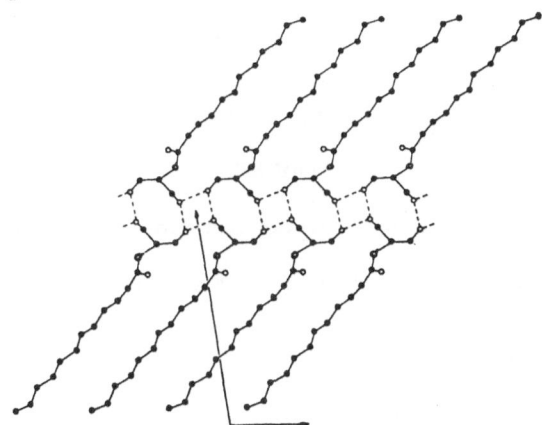

Fig. 6.   The structure of 2-monolaurin viewed along b. Carbon atoms are shown as solid, and oxygen atoms as open circles. [Note that one of the oxygen atoms is incorrectly represented in the original.]

1-MONOSTEARIN (β_1 AND β_2 FORMS)

$C_{21}H_{42}O_4$ $C_{17}H_{35}COOCH_2CHOHCH_2OH$ F.W. = 358.6

I. On the crystal structure of the forms β_1 and β_2 of racemic 1-monoglycerides.
K. LARSSON, 1965.   *Ark. Kemi*, <u>23</u>, 29-33.

Monoclinic, a = 9.20±6, b = 100.5±6, c = 5.04±3Å, β = 100.2±6°, [U = 4586Å^3],
D_m = 1.057, Z = 8, D_x = 1.04   (same parameters for β_1 and β_2).

Space group β_1 form, $P2_1$ (C_2^2) or $P2_1/m$ (C_{2h}^2). The analysis confirms $P2_1/m$.

β_2 form, Am   (C_s^3) [The absences given are consistent with Aa (C_s^4).]

Atomic positions are not given.

Structure

The similarities and the differences in the molecular packing are illustrated
in Fig. 7. The direction of the chain tilt alternates in successive double layers
for the β_1 form, and in successive single layers for the β_2 form.

Details of analysis

There are few details. The structures were partially solved by study of the
three-dimensional Patterson function, and refined by means of *c*-axis Fourier
projections. The polar end groups were not resolved.

Fig. 7.   Schematic representation of the chain packing of 1-monostearin.
Upper, β_1; lower, β_2.

ACETAMIDE (ORTHORHOMBIC FORM)

C_2H_5NO CH_3CONH_2 F.W. = 59.07

I. The crystal structure of orthorhombic acetamide. W.C. HAMILTON, 1965. *Acta
Cryst.*, <u>18</u>, 866-870.

Orthorhombic, a = 7.76±1, b = 19.00±4, c = 9.51±2Å, [U = 1402Å^3], z = 16,
D_x = 1.119.

Space group Pccn (D_{2h}^{10})

Atomic positions

	x	y	z
C(1)	0.4890±18	0.7399±7	0.8578±20
Me(1)	0.5389±21	0.8150±6	0.8372±18
N(1)	0.4640±15	0.7022±5	0.7391±14
O(1)	0.4769±15	0.7144±4	0.9787±14
C(2)	0.2998±19	0.5340±8	0.9117±21
Me(2)	0.2214±19	0.4612±6	0.9289±17
N(2)	0.3165±16	0.5743±6	1.0235±13
O(2)	0.3496±12	0.5528±4	0.7901±13

Anisotropic temperature factors are given also.

Structure

The bond lengths and angles are given in full for the two independent molecules. Mean values are: C=O, 1.26±11; C–N, 1.334±17; C–C, 1.505±13Å; C–C–N,
117.2±1.5; C–C–O, 119.6±1.1; N–C–O, 123.1±0.5°. Each molecule is planar, and
the two are nearly coplanar. The two independent molecules in the asymmetric
unit are joined by a pair of N–H..O bonds. Moreover each molecule is hydrogen-
bonded to two of its own kind in chains parallel to *c*. The N...O distances
range from 2.873 to 3.014Å. The hydrogen-bonding scheme is shown in Fig. 8.

Details of analysis

The intensity data were recorded on Weissenberg photographs (*hk*0 to *hk*6)
using CuK_α radiation. The intensities of 459 independent reflexions were estimated
visually. Refinement was by least squares to a final *R* index of 0.101.

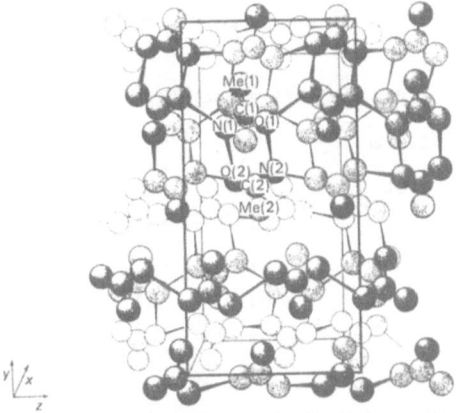

Fig. 8.   A view of the structure of acetamide down *a*. The hydrogen bonds are
indicated by heavy black lines. The atoms in one of the dimers are
labelled. (Me stands for methyl).

POTASSIUM HYDROGEN DI-TRIFLUOROACETATE
RUBIDIUM HYDROGEN DI-TRIFLUOROACETATE
CAESIUM HYDROGEN DI-TRIFLUOROACETATE

$C_4HF_6O_4K$ $MH(CF_3CO_2)_2$ F.W. = 266.1
$C_4HF_6O_4Rb$ F.W. = 312.5
$C_4HF_6O_4Cs$ F.W. = 359.9

I. The crystal structures of the acid salts of some monobasic acids.   Part X.
Potassium, rubidium, and caesium hydrogen di-trifluoroacetates.   L. GOLIC
and J.C. SPEAKMAN, 1965.   *J. Chem. Soc.*, 2530–2542.

POTASSIUM SALT

Monoclinic, a = 8.773±5, b = 10.169±6, c = 9.255±6Å, β = 99.85±7°, U = 813.5Å^3,
D_m = 2.085, z = 4, D_X = 2.171,
(calibration with aluminium, a = 4.04907Å).

Space group   Ia (C_s^4) or   I2/a (C_{2h}^6)

I2/a is confirmed by the analysis.   Molecular symmetry, two-fold axis for potassium
atom, centre for hydrogen atom.

Atomic positions

	x	y	z
K	1/2	0.21876	1/4
F(1)	0.06204	0.32461	0.59619
F(2)	0.23590	0.38638	0.77440
F(3)	0.10544	0.53053	0.63729
O(1)	0.27638	0.33947	0.43080
O(2)	0.39977	0.49506	0.57579
C(1)	0.17507	0.41230	0.63804
C(2)	0.29307	0.41198	0.53617
H	0	0	0

Standard deviations (in Å) are given in full, and for F, O and C range from
0.003 to 0.005Å.   Anisotropic temperature factors are given for the non-hydrogen
atoms.

Structure

Bond lengths and angles for the trifluoroacetate residue are given below.

C(2)	–	O(1)	1.212Å	O(1)	–	C(2)	–	O(2)	128.4°
C(2)	–	O(2)	1.268	O(1)	–	C(2)	–	C(1)	120.1
C(2)	–	C(1)	1.515	O(2)	–	C(2)	–	C(1)	111.6
C(1)	–	F(1)	1.341					
C(1)	–	F(2)	1.310	F(1)	–	C(1)	–	C(2)	111.7
C(1)	–	F(3)	1.348	F(2)	–	C(1)	–	C(2)	113.0
				F(3)	–	C(1)	–	C(2)	110.9
				F(1)	–	C(1)	–	F(2)	107.7
				F(2)	–	C(1)	–	F(3)	107.1
				F(1)	–	C(1)	–	F(3)	106.3

Standard deviations are 0.006 to 0.007Å and 0.4°.

Consideration of the dimensions of the carboxyl group (C(2), O(1), O(2))
suggests that the acid residue is intermediate between anionic and neutral states.
As illustrated in Fig. 9, the carboxyl groups of adjacent residues are joined
across a centre of symmetry by a hydrogen bond, of length 2.435±7Å.   The hydrogen
atom is required by symmetry to lie on the centre, and is thus shared equally by
the two residues.   This situation could be simulated by a disordered structure,
but the short O...O distance suggests that the hydrogen bond is truly symmetrical.
Potassium hydrogen di-trifluoroacetate is therefore an acid salt of class A.   The
non-hydrogen atoms of the acid residue are coplanar.   The potassium atom is coordinated to eight oxygen atoms at distances of 2.84 to 2.97Å.

Details of analysis

The intensity data were recorded on Weissenberg photographs (principally hk0
to hk4) using CuK$_\alpha$ radiation.   601 reflexions (about 64% of those accessible) were
observed, and measured visually.   Refinement was by least-squares to a final R
index of 0.068.

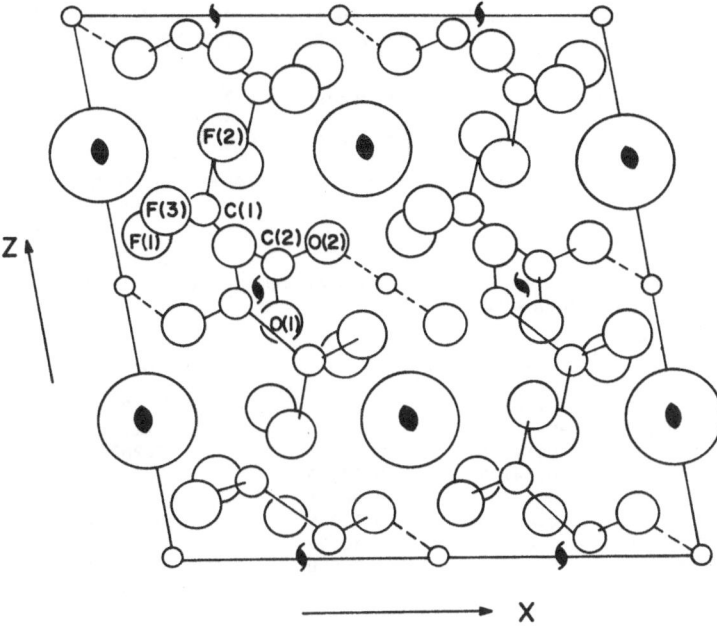

Fig. 9. The structure of potassium hydrogen di-trifluoroacetate viewed along *b*.

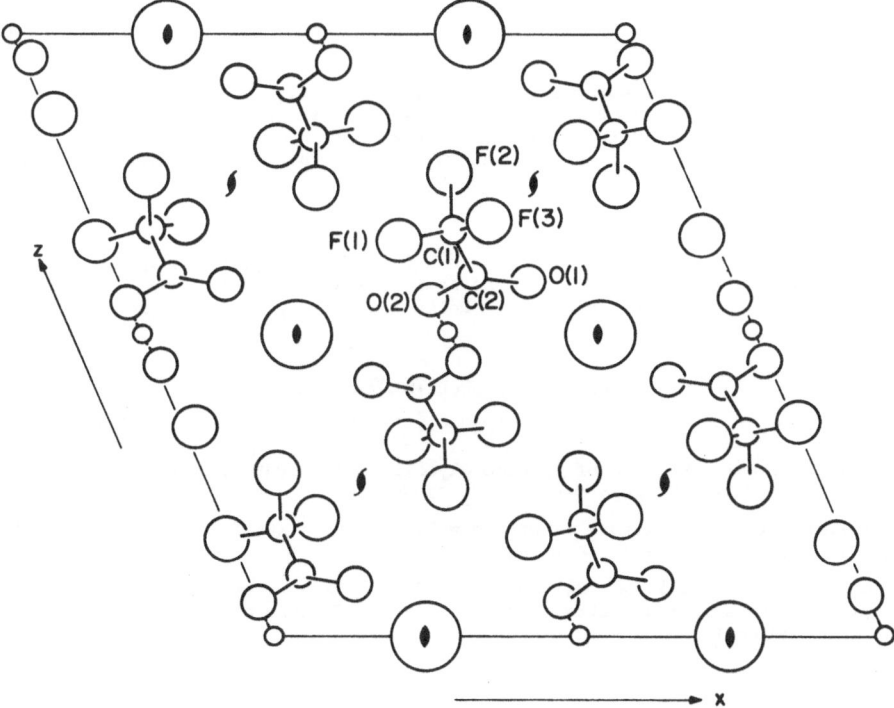

Fig. 10. The structure of caesium hydrogen di-trifluoroacetate viewed along *b*.

RUBIDIUM SALT

Monoclinic, a = 8.813±3, [18.813 in original], *b* = 10.550±4, *c* = 9.546±6Å,
β = 100.66±6°, *U* = 872.2Å^3, *Z* = 4, D_x = 2.380, (calibration with aluminum as
for potassium salt).

The compound is isotypic with the potassium salt.

CAESIUM SALT

Monoclinic, a = 13.623±3, *b* = 5.033±2, *c* = 14.741±4Å, β = 112.76±4°, *U* = 932.0 Å^3,
D_m = 2.61, *Z* = 4, D_x = 2.565, (calibration with aluminum   as for potassium salt).
Space group   Aa (C_s^4) or   A2/a (C_{2h}^6)

A2/a is confirmed by the analysis. Molecular symmetry, two-fold axis for caesium
atom, centre for hydrogen atom.

Atomic positions

	x	y	z
Cs	-1/4	0.21074	0
F(1)	-0.01397	0.59125	0.15638
F(2)	0.11625	0.38021	0.26241
F(3)	0.14726	0.70498	0.18430
O(1)	0.16498	0.28404	0.08000
O(2)	-0.00479	0.17766	0.05131
C(1)	0.08426	0.50222	0.17385
C(2)	0.08574	0.30548	0.09601
H	0	0	0

Standard deviations (in Å) are given in full, and for F, O and C range
from 0.01 to 0.02Å. Anisotropic temperature factors are given for the non-
hydrogen atoms.

Structure

The structure, illustrated in Fig. 10, is formally similar to that of the
potassium salt. Bond lengths and angles are given in full, but are of lower
accuracy. The length of the apparently symmetrical hydrogen bond which links
the adjacent acid residues is 2.38±3Å. The caesium atom is coordinated to six
oxygen atoms at distances of 3.13 to 3.19Å.

Details of analysis

The intensity data were recorded on Weissenberg photographs (principally h0l
to h3l) using CuK$_α$ radiation. 767 reflexions (about 72% of those accessible)
were observed and measured visually. Refinement was by least squares to a final
R index of 0.103.

LITHIUM GLYCOLLATE MONOHYDRATE

$C_2H_3LiO_3 \cdot H_2O$ [HOCH$_2$COO]$^-$.Li$^+$.H$_2$O   F.W. = 100.0

I. The crystal structure of lithium glycollate monohydrate. R.H. COLTON and
   D.E. HENN, 1965. *Acta Cryst.*, 18, 820-822.

Monoclinic, a = 11.21±1, *b* = 6.80±1, *c* = 5.97±1Å, β = 108.2±1°, [U = 432.2 Å^3],
D_m = 1.542, *Z* = 4, D_x = 1.546 [1.536].
Space group   C2 (C_m^3), Cm (C_5^3), or C2/m (C_{2h}^3) C2/m is consistent with deduced
structure. Molecular symmetry, mirror plane for lithium glycollate, two-fold axis
for water molecule.

Atomic positions

	x	y	z
C(1)	0.2378	0	0.6921
C(2)	0.3040	0	0.4955
O(1)	0.1235	0	0.6347
O(2)	0.3095	0	0.9078
O(3)	0.2119	0	0.2698
OH$_2$	0	0.2206	0
Li+	0.0310	0	0.2595

Anistropic temperature factors are given for the above atoms.

Structure

Interatomic distances and the structural arrangement are given in Fig. 11.

$$O(3)C(2)C(1) = 107° 52'$$
$$O(1)C(1)C(2) = 120° 21'$$
$$O(2)C(1)C(2) = 115° 20'$$
$$O(1)C(1)O(2) = 124° 19'$$

(All angles ± 35'.)

The coordination of the lithium ion is slightly distorted trigonal bipyramidal.

Details of analysis

Intensity data were recorded on equi-inclination Weissenberg photographs (*h0l* to *h6l*) using CuK$_\alpha$ radiation. 385 independent reflexions were measured photometrically. Refinement was by least squares to a final R index of 0.11.

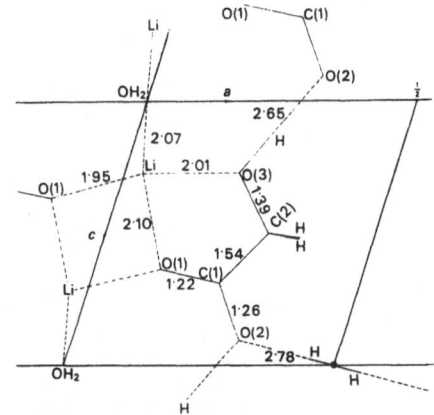

Fig. 11. The structure of lithium glycollate monohydrate projected along *b*. Standard deviations of the distances shown range from 0.005 to 0.015 Å.

RUBIDIUM HYDROGEN BISGLYCOLLATE

C$_4$H$_7$O$_3$Rb   RbH(C$_2$H$_3$O$_3$)$_2$ F.W. = 236.6

I. The crystal structures of the acid salts of some monobasic acids. Part IX. Rubidium hydrogen bisglycollate. L. GOLIC and J.C. SPEAKMAN, 1965. *J. Chem. Soc.*, 2521–2530.

Monoclinic, $a = 4.052\pm4$, $b = 17.91\pm1$, $c = 10.52\pm1\overset{\circ}{A}$, $\beta = 98°26\pm4'$, $U = 755\overset{\circ}{A}^3$, $D_m = 2.05$, $Z = 4$, $D_x = 2.08$. (Crystal data are given also for the isotypic potassium salt. They are: $a = 3.97$, $b = 17.62$, $c = 10.40\overset{\circ}{A}$, $\beta = 100°24'$, $U = 716\overset{\circ}{A}^3$, $D_m = 1.7$, $D_x = 1.76$.)

Space group $P2_1/n$ (C_{2h}^5)

Atomic positions

	x	y	z
Rb	0.77710	0.38262	0.52374
O(1)	0.33101	0.48240	0.64689
O(2)	0.59049	0.53550	0.82392
O(3)	0.08162	0.37120	0.78801
O(4)	-0.03249	0.25416	0.32791
O(5)	0.21459	0.36313	0.31510
O(6)	-0.19675	0.24133	0.07192
C(1)	0.40759	0.48293	0.76264
C(2)	0.31371	0.42504	0.84986
C(3)	0.06182	0.30649	0.26472
C(4)	-0.02435	0.30737	0.11923

Standard deviations (in $\overset{\circ}{A}$) are given in full, and for the carbon and oxygen atoms they range from 0.007 to 0.016$\overset{\circ}{A}$. Anisotropic temperature factors are given. The hydrogen atoms have been located, and somewhat idealized positions are given for them.

Structure

The bond lengths and angles in the two glycollate residues are as follows:

	(1)		(2)
C(1) – O(1)	$1.212\pm14\overset{\circ}{A}$	C(3) – O(4)	$1.241\pm14\overset{\circ}{A}$
C(1) – O(2)	1.308 ± 14	C(3) – O(5)	1.264 ± 14
C(2) – O(3)	0.435 ± 14	C(4) – O(6)	1.426 ± 14
C(1) – C(2)	1.471 ± 16	C(3) – C(4)	1.519 ± 16
O(1) – C(1) – O(2)	$123.0\pm1.0°$	O(4) – C(3) – O(5)	$123.5\pm1.0°$
O(1) – C(1) – C(2)	124.8 ± 1.0	O(4) – C(3) – C(4)	120.3 ± 1.0
O(2) – C(1) – C(2)	112.3 ± 1.0	O(5) – C(3) – C(4)	116.1 ± 1.0
O(3) – C(2) – C(1)	113.8 ± 1.0	O(6) – C(4) – C(3)	111.8 ± 1.0

Consideration of the dimensions of the carboxylate groups (C(1), O(1), O(2) and C(3), O(4), O(5)) indicates that (1) is neutral ($CH_2OH.COOH$), and that (2) is anionic ($CH_2OH.COO^-$). This conclusion is supported by the position of the relevant hydrogen atom, which shows O(2) to be hydroxylic. Rubidium hydrogen bisglycollate is therefore an acid salt of class *B*. The non-hydrogen atoms of the anionic residue are coplanar, but those of the other deviate from coplanarity by a twist of 7° about the C–C bond. A view of the structure is given in Fig. 12. The lengths of the O–H..O bonds shown are: O(2)...O(5'), 2.53$\overset{\circ}{A}$; O(3)...O(4'), 2.73$\overset{\circ}{A}$; O(4)...O(6'), 2.72$\overset{\circ}{A}$. The rubidium atom is coordinated to eight oxygen atoms at distances of 2.88 to 3.07$\overset{\circ}{A}$.

Details of analysis

The intensity data were recorded on Weissenberg photographs (principally $0kl$ to $2kl$) using CuK_α radiation. 1039 reflexions (constituting about 61% of those accessible) were observed and measured visually. Refinement was by least squares to a final R index of 0.093.

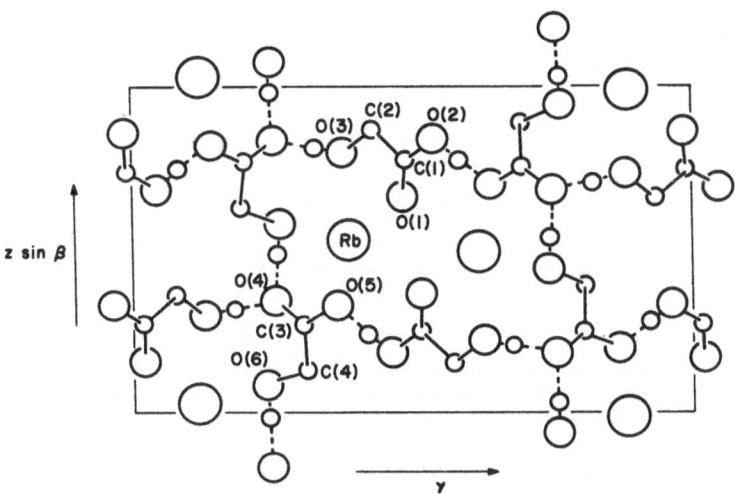

Fig. 12. The structure of rubidium hydrogen bisglycollate viewed along a.

POTASSIUM PALMITATE

$C_{16}H_{31}KO_2$ $CH_3(CH_2)_{14}CO_2K$ F.W. = 294.5

I. The crystal structure of potassium palmitate (form B). J.H. DUMBLETON and T.R. LOMER, 1965. *Acta Cryst.*, 19, 301–307.

Triclinic, a = 4.15±1, b = 5.60±2, c = 37.82±4Å, α = 93.0±3, β = 91.4±3, γ = 92.4±3°, [U = 876.7Å^3], D_m = 1.117, Z = 2, D_x = 1.109.

Space group P1 (C_1^1) or P$\bar{1}$ (C_i^1)

P$\bar{1}$ is consistent with the deduced structure.

Atomic positions

	x	y	z
K	0.23761	0.25687	0.02755
O(1)	0.76645	0.57339	0.04122
O(2)	0.72621	0.96148	0.04114
C(1)	0.71214	0.77592	0.05632
C(2)	0.61077	0.77670	0.09534
C(3)	0.50343	0.54146	0.11026
C(4)	0.43396	0.56098	0.14913
C(5)	0.31384	0.32458	0.16358
C(6)	0.23401	0.33782	0.20269
C(7)	0.12604	0.10154	0.21710
C(8)	0.03194	0.11470	0.25574
C(9)	0.93194	0.88061	0.27044
C(10)	0.84033	0.89108	0.30903
C(11)	0.73586	0.65645	0.32386
C(12)	0.64259	0.67219	0.36263
C(13)	0.53943	0.43008	0.37654
C(14)	0.44560	0.44556	0.41574
C(15)	0.34819	0.20426	0.42890
C(16)	0.25107	0.21069	0.46828

Standard deviations are given in full. Values are 0.0038Å for potassium, and from 0.010 to 0.026Å for carbon and oxygen. Also given are assumed hydrogen positions and isotropic temperature factors for all atoms.

Structure

The carbon chain of the palmitate ion is planar, and the carboxyl group is rotated about the first C–C bond by 16.4° from the plane of the chain. Bond lengths and angles are given in full. Mean values are:

C–C	=	1.52±1Å	C–C–C	=	114±1°
C–O	=	1.24±3	O–C–O	=	123±1

The structural organization is indicated in Fig. 13 and 14. The potassium and oxygen atoms form double layers, parallel to (001), which are separated from each other by the hydrocarbon chains. Each potassium ion is coordinated to four oxygen atoms in the same half of the double layer, and to two more in the other half. K–O distances range from 2.71 to 2.82Å.

Details of analysis

Three-dimensional intensity data were recorded on Weissenberg photographs. [No details]. About 850 reflexions were observed. Refinement was by least squares to a final R index of 0.116.

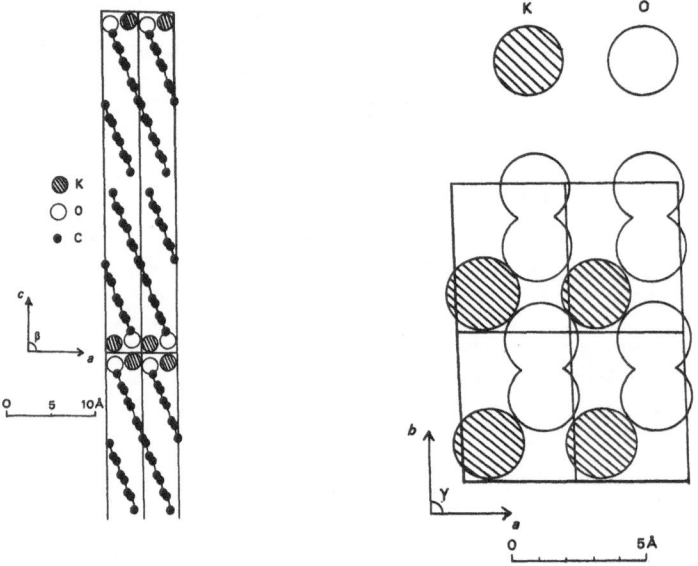

Fig. 13 (left). Projection of the structure of potassium palmitate along *b*.

Fig. 14 (right). Arrangement of the atoms in one half of the ionic double layer in potassium palmitate.

AMMONIUM OXALATE MONOHYDRATE

$C_2H_8N_2O_4 \cdot H_2O$ $(NH_4)_2(COO)_2 \cdot H_2O$ F.W. = 142.1

I. Ammonium oxalate monohydrate: structure refinement at 30°K. J.H. ROBERTSON, 1965. *Acta Cryst.*, 18, 410–417.

II. Enantiomorphism of the oxalate ion in ammonium oxalate, *Idem*, 1965. *Ibid.*, 18, 417–419.

III. Neutron diffraction study of ammonium oxalate monohydrate, $(NH_4)_2C_2O_4 \cdot H_2O$. V.M. PADMANABHAN, S. SRIKANTHA and S. MEDHI ALI, 1965. *Ibid.*, 18, 567–568.

  (For earlier work, see 1, 2.)

I. *Orthorhombic*, $a = 8.017 \pm 4$, $b = 10.309 \pm 4$, $c = 3.735 \pm 2$Å, $U = 308.7$Å^3, $Z = 2$, $D_x = 1.526$, (all at 30°K). Corresponding room-temperature values are: $a = 8.035 \pm 4$, $b = 10.309 \pm 4$, $c = 3.795 \pm 2$Å, $U = 314.4$Å^3, $D_m = 1.50$, $D_x = 1.500$.

Space group $P2_12_12$ (D_2^3) Molecular symmetry, two-fold axis.

Atomic positions

	x	y	z
C	0.0934	0.0227	0.0693
O(1)	0.2028	−0.0604	0.1398
O(2)	0.1192	0.1411	0.0013
N	0.3897	0.2256	0.4286
O(w)	0.4999	0	−0.1839

  In the original, these are given in Å. Standard deviations range from 0.003 to 0.005Å. Anisotropic temperature factors are given also; the thermal motion has been analysed in terms of rigid-body modes, and the positions of the atoms of the oxalate ion, above, have been appropriately corrected. (Uncorrected positions are not given.) The positions of the hydrogen atoms are also given.

Structure

  The structure is essentially as reported earlier, but has been determined more accurately. The angle of twist of the oxalate ion about the central bond is 26.6°. The bond lengths and angles in the oxalate ion are: C–C, 1.569±8Å; C–O, 1.257±6 Å; O–C–O, 126.0±5°. The hydrogen bonding environment is shown in Fig. 15. In II it is pointed out that, because of the twist of the oxalate ion, there must be enantiomorphic forms of ammonium oxalate monohydrate; it is predicted that optical rotation will be observed.

Details of analysis

  The intensity data were recorded at 30°K (liquid hydrogen temperature) on oscillation and Weissenberg photographs about a, b, c and [110], using CuK_α radiation. 432 of 456 accessible reflexions were observed. Refinement was by least squares to a final R index of 0.080.

III. The crystal data are those given in 2. Those reported in I, above, are probably more accurate.

Atomic positions

	x	y	z
C	0.092	0.027	0.068
O(1)	C.200	−0.056	0.140
O(2)	0.118	0.142	0.001
N	0.386	0.229	0.427
O(w)	0	0.500	0.198
H(w)	0.094	0.476	0.053
H(1)	0.482	0.268	0.287
H(2)	0.430	0.151	0.568
H(3)	0.288	0.194	0.291
H(4)	0.347	0.306	0.582

Standard deviations are 0.018Å for hydrogen, and 0.010Å for other atoms.

Structure

The structure is as reported in I, above, but the hydrogen nuclei have been located with significant precision. Some average bond lengths involving the hydrogen atoms are: N–H, 1.02±3Å, O–H, 0.97±3Å; H–O–H, 106±2°; H–N–H, 104° to 117°

Details of analysis

Neutron diffraction intensities were measured for 125 *hk*0 and 45 0*kl* reflexions with a single-crystal diffractometer, with λ = 1.029Å. Refinement was by Fourier and least-squares methods.

1.   *Structure Reports,* <u>16</u>, 429.
2.   *Strukturbericht,* <u>4</u>, 291.

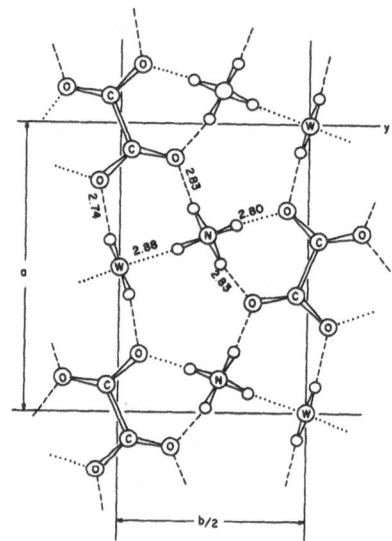

Fig. 15.   A view of one layer of the ammonium oxalate monohydrate structure, viewed along *c*, showing the hydrogen bonds. Dashed lines show intra-layer bonds; dotted lines involve a molecular lying in the layer above or below the one shown. Distances shown are for O...O and N...O, and standard deviations are about 0.006Å.

CALCIUM OXALATE POLYHYDRATE (WEDDELLITE)
STRONTIUM OXALATE POLYHYDRATE

$C_2CaO_4 \cdot 2.17H_2O$

$C_2SrO_4 \cdot 2.17H_2O$ $M \cdot C_2O_4 \cdot 2.17H_2O$ (M = Ca, Sr)

F.W. = 167

F.W. = 215

I. Crystal structure analysis of weddellite, $CaC_2O_4 \cdot (2+x)H_2O$. C. STERLING, 1965. *Acta Cryst.,* <u>18</u>, 917-921. (For earlier work, see <u>1</u>).

II. Crystal structure of tetragonal strontium oxalate. *Idem,* 1965. *Nature, Lond.,* <u>205</u>, 588-589.

I.   CALCIUM SALT

Tetragonal, $a = 12.30\pm2$, $c = 7.34\pm2$Å, $[U = 1110$Å$^3]$, $D_m = 2.000$, $Z = 8$, $D_x = 2.000$.

Space group $I4$ (C_4^5), $I\bar{4}$ (S_4^2) or $I4/m$ (C_{4h}^5)

$I4/m$ is confirmed by analysis. Molecular symmetry, four-fold axis for one water molecule, mirror plane for remaining water molecules and for the oxalate and calcium ions.

Atomic positions

	x	y	z
Ca	0.200	0.300	0
O(1)	0.356	0.246	0.181
O(2)	0.236	0.463	0.181
O(5)	0.144	0.114	0
O(6)	0.020	0.386	0
O(7)	0.500	0.500	0.210
C	0.445	0.242	0.105

Isotropic temperature factors are given also.

Structure

Within experimental error, the oxalate ion has *mmm* symmetry with C-C = 1.55 ± 2Å, C–O = 1.25 ± 1Å, and O–C–O = $127\pm1°$. The projection along c is shown in Fig. 16. The calcium ion is coordinated to two oxygen atoms belonging to water molecules, and to six oxygen atoms belonging to oxalate ions, at distances ranging from 2.40 to 2.50Å. The hydrogen atoms have not been located, but a hydrogen bonding system is inferred from OH--O distances which range from 2.87 to 2.93Å. The fractional (zeolitic) water molecule does not approach closer than 3.11Å to any other atom, and is therefore probably quite free in the structure.

Details of analysis

The intensity data were recorded on equi-inclination Weissenberg photographs ($0kl$ to $8kl$) using CuK_α radiation. 636 of 751 accessible reflexions were observed. Refinement was by least squares to a final R index of 0.13.

II.   STRONTIUM SALT

Tetragonal, $a = 12.82\pm3$, $c = 7.50\pm2$Å, $[U = 1233$Å$^3]$, $D_m = 2.36$, $Z = 8$, $D_x = 2.36$.

Space group $I4/m$ (C_{4h}^5) (as for I).

Atomic positions

	x	y	z
Sr	0.196	0.301	0
O(1)	0.361	0.248	0.180
O(2)	0.231	0.469	0.179
O(5)	0.149	0.106	0
O(6)	0.013	0.390	0
O(7)	0.500	0.500	0.210
C	0.445	0.241	0.101

(The numbering system has been altered to correspond to that of I.) Isotropic temperature factors are given also.

Structure

The structure is isotypic with that of I. Bond lengths and angles are given, but with less accuracy than in I. The strontium atom is eight-coordinated, with Sr-O distances ranging from 2.56 to 2.61Å.

Details of analysis

The intensity data were recorded with a diffractometer and scintillation counter, using CuK$_\alpha$ radiation. 607 independent reflexions were observed. Refinement was by least squares to a final R index of 0.11.

1. *Strukturbericht*, 5, 159.

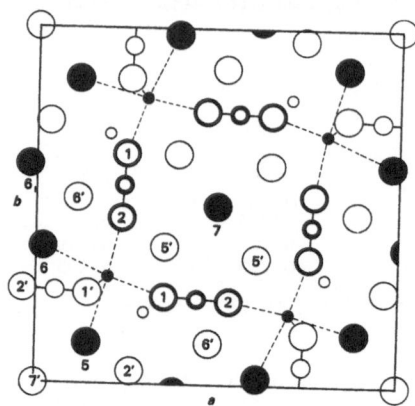

Fig. 16. The structure of weddellite viewed along *c*.

RUBIDIUM OXALATE MONOHYDRATE

C$_2$O$_4$Rb$_2$.H$_2$O   F.W. = 277.0

I. The crystal structure of rubidium oxalate monohydrate, Rb$_2$C$_2$O$_4$.H$_2$O. B.F. PEDERSEN, 1965. *Acta Chem. Scand.*, 19, 1815-1818.

Monoclinic, a = 9.662, b = 6.350, c = 11.088Å, β = 109.4°, [U = 641.7Å^3], z = 4, [D_x = 2.85].

Space group   Cc   (C$_s^4$)   (C$_c^6$)   (C$_{2h}^6$)   C2/c confirmed by structure analysis. Molecular symmetry, centre for rubidium oxalate, two-fold axis for water molecule.

Atomic positions

	x	y	z
Rb	0.1290±1	0.8156±5	0.1297±1
O(1)	0.1335±11	0.2748±37	0.0933±13
O(2)	0.3262±15	0.4768±27	0.0940±11
O(w)	0	0.4733±40	1/4
C	0.2396±23	0.3216±38	0.0546±17

Anisotropic temperature factors are given also.

Structure

The structure is isotypic with potassium oxalate monohydrate (1). The Rb-O distances range from 2.946 to 3.215Å, with standard deviation 0.016Å. The oxalate ion is planar.

Details of analysis

The intensity data were recorded on equi-inclination Weissenberg photographs (*h0l* to *h4l*) using CuK$_\alpha$ radiation and measured visually. Refinement was by full-matrix least squares to a final R index of 0.102.

1. *Structure Reports*, 29, 463.

POTASSIUM HYDROGEN CHLOROMALEATE

$C_4H_2ClKO_4$ F.W. = 188.6

I. A centered hydrogen bond in potassium hydrogen chloromaleate: a neutron
diffraction structure determination. R.D. ELLISON and H.A. LEVY, 1965.
Acta Cryst., <u>19</u>, 260-268.

*Orthorhombic, a = 15.815±15, b = 10.928±6, c = 7.707±5Å, [U = 1332 Å^3], D_m = 1.868,
Z = 8, D_x = 1.881.*

Space group Pbcn (D_{2h}^{14})

(Unit cell and space group from <u>1</u>.)

Atomic positions

	x	y	z
Cl	0.1161±1	-0.0197±1	-0.0534±2
C(1)	0.1394±1	0.1988±1	0.1041±1
C(2)	0.1850±1	0.0860±1	0.0381±2
C(3)	0.2677±1	0.0575±1	0.0387±2
C(4)	0.3414±1	0.1281±2	0.1108±2
O(1)	0.0638±1	0.2100±2	0.0727±3
O(2)	0.1817±1	0.2767±2	0.1918±3
O(3)	0.3290±1	0.2234±2	0.2039±4
O(4)	0.4131±1	0.0880±2	0.0787±3
H(1)	0.2558±3	0.2522±4	0.2032±6
H(2)	0.2863±2	-0.0289±3	-0.0203±6
K(1)	0.5000	-0.0946±4	0.2500
K(2)	0.5000	0.2973±4	0.2500

Neutron scattering amplitudes and anisotropic temperature factors are given
also.

Structure

Bond lengths and angles for the chloromaleate ion are given in Fig. 17. The
ion deviates appreciably from planarity only by rotation of the carboxyl groups
about the adjacent C-C bonds, 8.5±3° for one, 9.0±3° for the other. The sense of
the rotations is such as to displace O(2) and O(3) to the same side of the mean
plane. It is to be noted that the O..H..O distance is unusually short, and that
the O..H distances are, within experimental error, equal. The possibility is
considered that the central position of the hydrogen atom occurs only for a time-
space average of two or more equilibrium positions. It is considered that if two
equilibrium positions are present, each must be less than 0.1Å from the mid-point,
and the barrier between them must be extremely small. The coordination of the
potassium ions is described in detail. One is surrounded by six oxygen atoms,
and the other by six oxygen and four chlorine atoms.

Details of analysis

The intensity data were measured with a four-circle diffractometer, using a
neutron wave length of 1.078Å. 1959 independent reflexions were measured in the
range $\sin\theta/\lambda \leqslant 0.71$Å^{-1}. Absorption corrections were applied. Refinement was by
full-matrix least squares to a final R index of 0.079. An analysis of the thermal
motion has been used to correct the bond lengths.

1. *Structure Reports*, <u>26</u>, 554.

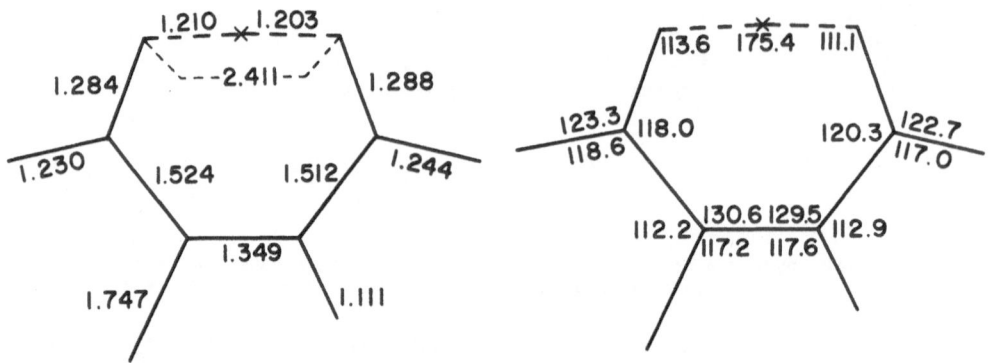

Fig. 17. Bond lengths and angles in the chloromaleate ion. The standard
          deviations are 0.001 to 0.005Å and 0.1 to 0.3°. The bond lengths
          have been corrected for thermal motion.

POTASSIUM MESOTARTRATE DIHYDRATE
RUBIDIUM MESOTARTRATE DIHYDRATE

$C_4H_4O_6K_2 \cdot 2H_2O$ F.W. = 262.3

$C_4H_4O_6Rb_2 \cdot 2H_2O$ F.W. = 355.0

$$\left[\begin{array}{c} O \\ \| \\ C \\ \| \\ O \end{array} - CH - \overset{\displaystyle OH}{\underset{\displaystyle OH}{\,}} CH - \begin{array}{c} O \\ \| \\ C \\ \| \\ O \end{array} \right]^{2-} , K_2^+ , 2H_2O$$

I. The crystal structure of potassium mesotartrate dihydrate and the isomorphous
   rubidium salt. J. KROON, A.F. PEERDEMAN and J.M. BIJVOET, 1965. *Acta Cryst.*,
   19, 293–297.

POTASSIUM SALT

Triclinic, a = 7.05, b = 6.87, c = 11.16Å, α = 95.87, β = 103.27, γ = 62.22°,
[U = 465.4Å^3], D_m = 1.89, Z = 2, D_x = 1.89.

RUBIDIUM SALT

Triclinic, a = 7.11, b = 6.93, c = 11.5Å, α = 96.0, β = 101.6, γ = 61.8°, [U = 489.1Å^3]
D_m = 2.38, Z = 2, D_x = 2.41.

Space group   P1 (C_1^1) or P$\bar{1}$ (C_i^1) P$\bar{1}$ confirmed by analysis.

Atomic positions

	Potassium mesotartrate			Rubidium mesotartrate		
	x	y	z	x	y	z
(K/Rb)(1)	0.9983	0.2063	0.3604	0.9979	0.2069	0.3604
(K/Rb)(2)	0.2945	0.4498	0.1379	0.2941	0.4608	0.1375
O(1)	0.8852	0.7420	0.0706	0.8434	0.7560	0.0683
O(2)	0.6928	1.0775	0.1534	0.7126	1.0753	0.1653
O(3)	0.5727	0.6216	0.1381	0.5794	0.6438	0.1384
O(4)	0.8534	0.6560	0.3667	0.8346	0.6697	0.3550
O(5)	0.3836	1.2004	0.3529	0.3867	1.2123	0.3534

	Potassium mesotartrate			Rubidium mesotratrate		
	x	y	z	x	y	z
O(6)	0.7401	1.0669	0.4439	0.7289	1.0727	0.4447
O(7)	0.9545	0.2306	0.1063	0.9533	0.2423	0.1038
O(8)	0.2358	0.6482	0.3777	0.2360	0.6548	0.3852
C(1)	0.7092	0.8817	0.1309	0.7270	0.8911	0.1358
C(2)	0.5669	0.8214	0.1845	0.5754	0.8370	0.1848
C(3)	0.6204	0.8240	0.3232	0.6245	0.8333	0.3169
C(4)	0.5800	1.0463	0.3775	0.5795	1.0556	0.3735

Isotropic temperature factors are given. For the potassium salt the mean standard deviations are: K, 0.003Å; C and O, 0.015Å. Hydrogen positions are given also. For the rubidium salt the standard deviations are: Rb, 0.002Å; C and O, 0.02Å.

Structure

The mesotartrate ion has the ethane conformation, with two nearly-planar halves inclined to each other at 57°. (C(1), C(2), O(1), O(2) are coplanar, as are C(3), C(4), O(5), O(6); O(3) and O(4) lie 0.22Å and 0.17Å respectively from the mean planes of these groupings.) The carbon atoms (unlike those of the tartrate ion) are not coplanar. Instead, C(2), C(3), and C(4) are essentially coplanar with O(3). The conformation is asymmetric, and because of the operation of the centre of symmetry, both antipodes, or conformational isomers, are present.

Bond lengths and angles are given in full, and are consistent with expectation.

The view of the structure given in Fig. 18 illustrates the linking by water molecules of successive mesotartrate ions into infinite chains in the direction -a+b+c. Both conformational isomers are present in a given chain. The coordination of the metal ions, and the hydrogen-bonding arrangement are discussed. It is noted that, whereas the equivalent non-bonded intermolecular distances in the two isomorphs differ appreciably, the hydrogen-bonded distances are remarkably equal.

Details of analysis

The intensities for the three principal zones were recorded on integrated Weissenberg photographs, using CuK_α radiation, and measured photometrically. Refinement was by least squares to a final R index of 0.080 for the potassium salt, and 0.074 for the rubidium salt.

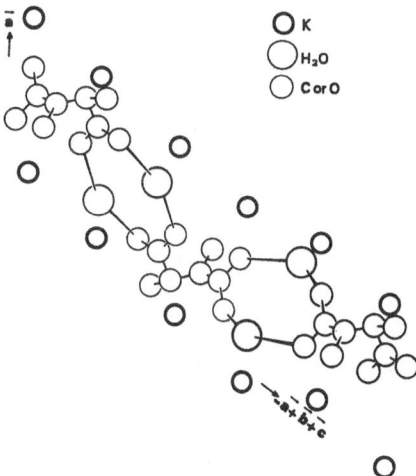

Fig. 18. A mesotartrate chain viewed along the *b* axis.

SODIUM DIHYDROGEN CITRATE

LITHIUM DIHYDROGEN CITRATE

$C_6H_7NaO_7$ F.W. = 214.1

$C_6H_7LiO_7$ F.W. = 198.0

$$\begin{array}{c} CH_2COOH \\ | \\ HO-C-COO^- \qquad M^+ \\ | \\ CH_2COOH \end{array}$$

(M = Na, Li)

I. X-ray crystal analysis of the substrates of aconitase. VI. The structures of sodium and lithium dihydrogen citrates. J.P. GLUSKER, D. VAN DER HELM, W.E. LOVE, M.L. DORNBERG, J.A. MINKIN, C.K. JOHNSON and A.L. PATTERSON, 1965. *Acta Cryst.*, 19, 561–572. (For a preliminary report, see 1).

SODIUM SALT

Monoclinic, a = 9.668, b = 11.682, c = 7.484Å (all ± 0.13%), β = 105°20+6', $[U = 815.2\text{Å}^3]$, D_m = 1.747, Z = 4, D_x = 1.744.

Space group $P2_1/a$ (C_{2h}^5)

Atomic positions

	x	y	z
Na	0.0903±2	0.1251±1	0.1024±2
O(1)	0.3549±4	0.3212±4	0.4777±4
O(2)	0.2033±5	0.1735±3	0.4092±5
O(3)	−0.0391±4	0.5487±3	0.7588±5
O(4)	0.1835±4	0.5741±3	0.9328±5
O(5)	0.1312±3	0.3123±2	1.0186±4
O(6)	0.3637±4	0.3495±3	1.1414±4
O(7)	0.4005±3	0.4197±2	0.8221±4
C(1)	0.2714±5	0.2414±4	0.5185±6
C(2)	0.2725±5	0.2405±3	0.7218±5
C(3)	0.2691±4	0.3594±3	0.8087±5
C(4)	0.1393±4	0.4278±3	0.6979±5
C(5)	0.0984±4	0.5246±3	0.8079±5
C(6)	0.2544±4	0.3390±3	1.0083±5
H(1)	0.361 ±8	0.195 ±6	0.783 ±11
H(2)	0.192 ±8	0.194 ±7	0.737 ±10
H(3)	0.064 ±9	0.371 ±6	0.662 ±11
H(4)	0.156 ±8	0.460 ±6	0.580 ±10
H(5)	0.423 ±8	0.420 ±7	0.721 ±10
H(6)	0.335 ±10	0.328 ±8	0.358 ±12
H(7)	−0.074 ±10	0.610 ±8	0.828 ±13

Temperature factors (anisotropic for the non-hydrogen atoms) are given also.

Structure

Bond lengths and angles for the dihydrogen citrate ion are given in Fig. 19. The C-O distances identify the central carboxyl group as the one which is ionized. (C-O distances corrected for thermal motion are given also; the corrections are rather small.) A view of the ion, with some O-O and O-H distances, is given in Fig. 20.

The sodium ion is coordinated to six oxygen atoms at distances of 2.305 to 2.461Å. The coordination octahedra occur in centrosymmetrically related pairs with one edge shared. Each octahedron is made up of two oxygen atoms from each of two dihydrogen citrate ions, and one from each of two more.

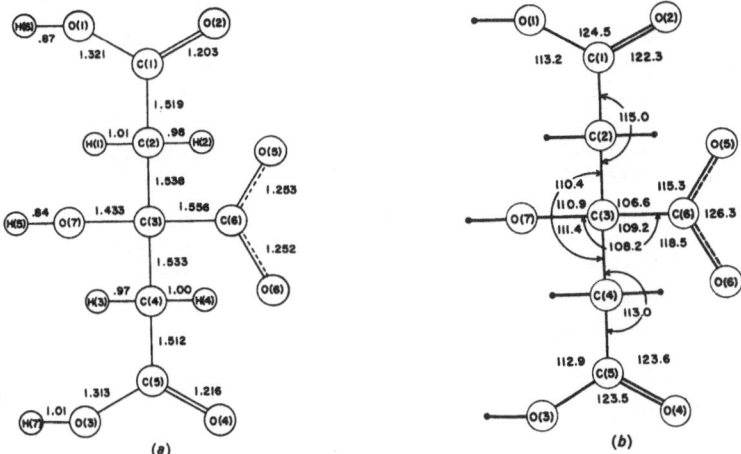

Fig. 19. (a) Covalent bond lengths and (b) angles for the dihydrogen citrate ion in the sodium salt.

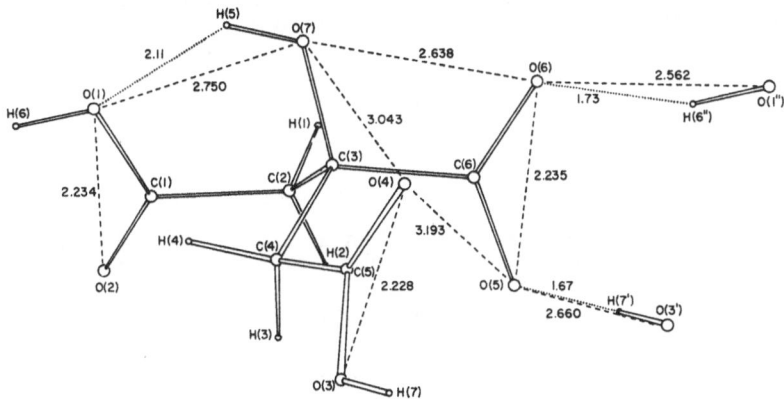

Fig. 20. Some oxygen-oxygen and hydrogen-oxygen distances (Å) in sodium dihydrogen citrate.

Details of analysis

Intensity data were recorded on integrated precession photographs (MoK_α) and integrated Weissenberg photographs (CuK_α), and measured photometrically and visually. Some were remeasured with a four-circle diffractometer, using CuK_α radiation. 1790 of a possible 2004 reflexions were observed above threshold. Refinement was by block-diagonal least squares to a final R index of 0.066.

LITHIUM SALT

Monoclinic, a = 9.143, b = 11.354, c = 7.475 Å (all ±0.13%), β = 107°45±6', $[U = 739.0 \text{ Å}^3]$, D_m = 1.783, z = 4, D_x = 1.780.

Space group $P2_1/a$ (C_{2h}^5)

Atomic positions

	x	y	z
Li	0.0894±23	0.1314±19	0.1053±32
O(1)	0.3685±10	0.3229±8	0.4752±11
O(2)	0.2007±10	0.1741±7	0.3889±12

ORGANIC COMPOUNDS

O(3)	−0.0393±9	0.5532±7	0.7637±12
O(4)	0.2045±9	0.5759±8	0.9381±10
O(5)	0.1359±8	0.3051±7	1.0129±11
O(6)	0.3885±10	0.3389±7	1.1408±10
O(7)	0.4229±8	0.4169±7	0.8305±11
C(1)	0.2763±13	0.2415±17	0.5027±14
C(2)	0.2764±14	0.2339±12	0.7100±13
C(3)	0.2796±12	0.3555±10	0.8063±14
C(4)	0.1461±11	0.4321±9	0.6951±13
C(5)	0.1091±11	0.5285±9	0.8166±14
C(6)	0.2680±14	0.3337±10	1.0056±19

Isotropic temperature factors are given also.

Structure

The material is nearly isomorphous and isostructural with the sodium salt reported above.

Details of analysis

Intensity data for the three principal zones were recorded on integrated precession (MoK_α) and Weissenberg (CuK_α) photographs and measured photometrically. 261 of a possible 345 reflexions were observed above threshold. Refinement was by full-matrix least squares to a final R index of 0.084.

1. *Idem*, 1964. *J. Amer. Chem. Soc.*, <u>82</u>, 2964.

MAGNESIUM CITRATE DECAHYDRATE

$C_{12}H_{10}Mg_3O_{14}.10H_2O$ F.W. = 631.3

$$Mg(H_2O)_6.[Mg(COO.CH_2\overset{\overset{\displaystyle OH}{|}}{\underset{\underset{\displaystyle COO}{|}}{C}}.CH_2.COO).H_2O]_2.2H_2O$$

I. X-ray crystal analysis of the substrates of aconitase. V. Magnesium citrate decahydrate [Mg(H₂O)₆][MgC₆H₅O₇(H₂O)]₂.2H₂O. C.K. JOHNSON, 1965. *Acta Cryst.*, <u>18</u>, 1004–1018.

Monoclinic, $a = 20.222$, $b = 6.686$, $c = 9.135$ Å (all ±0.02%), $\beta = 96.86±3°$, $[U = 1226$ Å³], $D_m = 1.71$, $z = 2$, $D_x = 1.709$.

Space group $P2_1/n$ (C_{2h}^5) Molecular symmetry, centre.

Atomic positions

	x	y	z
Mg(1)	0.00000	0.00000	0.00000
Mg(2)	0.21788±3	0.50189±8	0.53903±6
O(1)	0.19066±6	−0.0273 ±2	0.8340 ±1
O(2)	0.20762±7	−0.2632 ±2	0.6734 ±1
O(3)	0.02617±7	0.4971 ±2	0.2942 ±2
O(4)	0.12989±6	0.5318 ±2	0.3982 ±1
O(5)	0.23356±6	0.2266 ±2	0.4425 ±1
O(6)	0.16724±6	−0.0037 ±2	0.3287 ±1
O(7)	0.16006±6	0.3011 ±2	0.6502 ±1
O(8)	0.26173±6	0.6641 ±2	0.3893 ±1
O(9)	0.03990±6	−0.1129 ±2	0.2039 ±1
O(10)	0.08787±6	0.1547 ±2	−0.0182 ±2
O(11)	−0.04031±7	0.2409 ±2	0.0987 ±2
O(12)	0.13837±7	0.5501 ±2	0.0763 ±2
C(1)	0.17481±7	−0.1204 ±2	0.7143 ±2
C(2)	0.11485±8	−0.0411 ±2	0.6162 ±2
C(3)	0.13136±7	0.1575 ±2	0.5431 ±2
C(4)	0.06670±7	0.2499 ±2	0.4681 ±2
C(5)	0.07520±8	0.4397 ±2	0.3811 ±2
C(6)	0.18128±7	0.1223 ±2	0.4288 ±2

The positions of the hydrogen atoms are given also. Anisotropic termperature factors are given for the non-hydrogen atoms, and the corresponding thermal ellipsoids are characterized.

Structure

Bond lengths and angles are given in full. For the citrate ion (Fig. 21) the bond lengths are as follows:

C(1)	-	C(2)	=	1.515 Å	C(1) - O(1)	=	1.270
C(4)	-	C(5)	=	1.518	C(1) - O(2)	=	1.261
C(3)	-	C(6)	=	1.555	C(5) - O(3)	=	1.266
C(2)	-	C(3)	=	1.541	C(5) - O(4)	=	1.268
C(3)	-	C(4)	=	1.532	C(6) - O(5)	=	1.270
C(3)	-	O(7)	=	1.443	C(6) - O(6)	=	1.261

(All ±0.002 Å)

The C-O distances other than C(3)-O(7) have been corrected for thermal motion. The carbon backbone of the ion is essentially planar, and one terminal carboxyl group lies approximately in this plane. The other deviates from it by a rotation about the adjacent C-C bond of approximately 65°. The remaining carboxyl group, the hydroxyl oxygen atom, and the central carbon atom are practically coplanar. The angle between their plane and that of the carbon backbone is 86.4°.

Each magnesium ion is coordinated to six oxygen atoms, and there are no shared edges or corners between coordination octahedra. Mg-O distances range from 2.02 to 2.12 Å. Mg(1) is coordinated only to water molecules. Mg(2) is coordinated to a hydroxyl and two carboxyl oxygen atoms (O(7), O(4), O(5)) from one citrate ion, a carboxyl oxygen atom from each of two other citrate ions (O(2') and O(2'')), and a water molecule. The coordination octahedra are shown in Fig. 22.

Details of analysis

The intensity data were measured with a four-circle diffractometer and scintillation counter, using CuK_α radiation. 2484 of a possible 2725 independent reflexions were observed above background. Refinement was by full-matrix least squares to a final R index of 0.028.

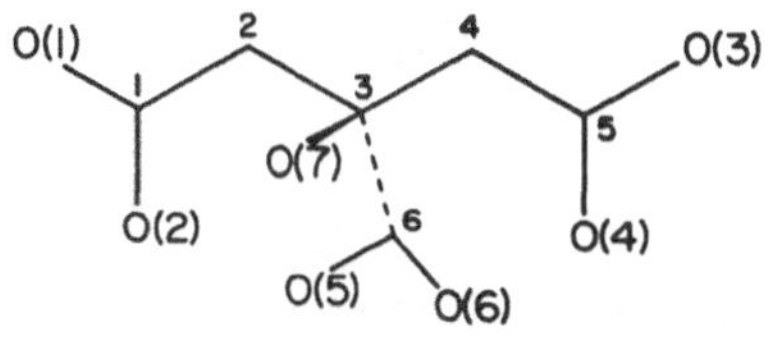

Fig. 21.   Numbering for the citrate ion in magnesium citrate decahydrate.

DIPOTASSIUM ETHYLENETETRACARBOXYLATE

$C_6H_2K_2O_8$ F.W. = 280.3

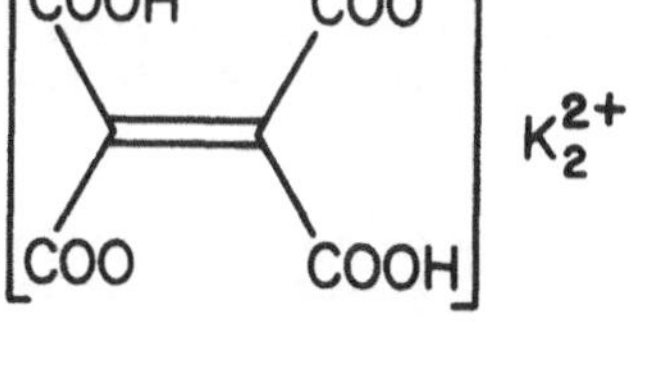

I.   The crystal structure of dipotassium ethylenetetracarboxylate. S.K. KUMRA and S.F. DARLOW, 1965. *Acta Cryst.*, 18, 98-104.

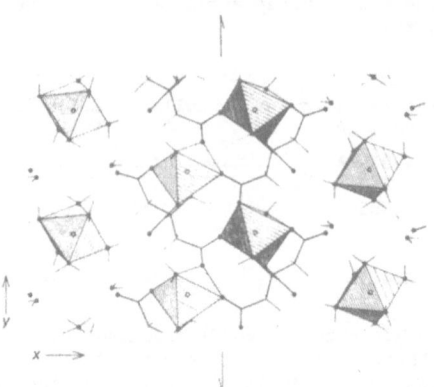

Fig. 22.   Coordination octahedra and hydrogen bonding in the (30$\bar{1}$) sheet of the
            magnesium citrate decahydrate structure.

Orthorhombic, a = 9.724±15, b = 6.440±2, c = 14.175±7 Å, [U = 887.7 Å^3], D_m =
2.095, Z = 4, D_x = 2.10.

Space group   *Pbca* (D_{2h}^{15})   Molecular symmetry, centre

Atomic positions

	x	y	z
K	0.1061	-0.1690	0.2951
O(1)	-0.0128	0.3262	-0.1185
O(2)	0.1297	0.0889	-0.1712
O(3)	0.2897	0.1080	0.0081
O(4)	0.1599	0.1037	0.1373
C	0.0548	0.0603	-0.0127
C(1)	0.0594	0.1658	-0.1099
C(2)	0.1740	0.0911	0.0510

        Standard deviations for K, O and C are 0.004, 0.009 and 0.015 Å respectively
for the x coordinates, and about half these values for the y and z coordinates.
Also given are anisotropic temperature factors for the above atoms, and an assumed
position for the hydrogen atom.

Structure

        The six carbon atoms of the ethylenetetracarboxylate anion are coplanar, and
the carboxylic groups deviate from this plane by rotations of 72° and 34° about
their C-C bonds.   Bond lengths and angles are given in full. A short (2.51 Å)
intermolecular distance between O(1) and O(3') is presumably a hydrogen bond.
Each potassium ion is coordinated to eight oxygen atoms at distances ranging from
2.66 to 3.12 Å. A view of the structure is given in Fig. 23.

Details of analysis

        The intensity data were recorded on equi-inclination Weissenberg photographs
(0kl to 6kl) using CuK_α radiation. Refinement was by least squares to a final
R index of 0.104.

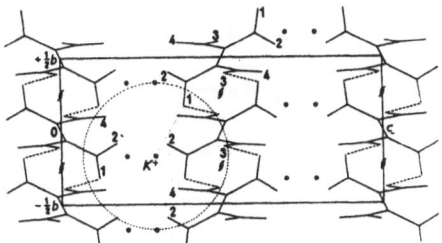

Fig. 23.   A view of the dipotassium ethylenetetracarboxylate structure along *a*.
The dotted lines are K–O contacts, and the dashed line is believed to
be a hydrogen bond.

TRIETHYLAMINE HYDROCHLORIDE

$C_6H_{15}N.HCl$ F.W. = 137.7

$$
\begin{array}{c}
CH_3 \\
| \\
CH_2 \qquad CH_2 \\
\diagdown \; N \; \diagup \quad CH_3 \qquad HCl\\
| \\
CH_2 \\
| \\
CH_3
\end{array}
$$

I. Contribution à l'étude de la liaison hydrogène dans quelques chlorhydrates
d'amines. F. GENET, 1965. *Bull. Soc. Fr. Minér. Crist.*, <u>88</u>, 463–482.
(For earlier work, see <u>1</u>).

Hexagonal, $a = 8.38\pm2$, $c = 7.07\pm2$ Å, $U = 429$ Å^3, $D_m = 1.01$, $Z = 2$, $D_x = 1.06$.

Space group $P6_3mc$ (C_{6v}^4) $P\bar{6}2c$ (D_{3h}^4) or $P6_3/mmc$ (D_{6h}^4)

$P6_3mc$ is consistent with analysis. Molecular symmetry, $3m$ for a disordered
molecule.

Atomic positions

	x	y	z
Cl	1/3	−1/3	0
N	1/3	−1/3	0.560
C(1)	0.180	−0.305	0.512
C(2)	0.165	−0.135	0.599

Anisotropic temperature factors are given also.

Structure

The triethylamine molecule has the conformation indicated above, with C–N–C
= 114°. The chlorine atom is hydrogen bonded to the nitrogen atom (N...Cl = 3.11
Å), completing an approximately tetrahedral environment. The complex molecule thus
has the appearance of a three-bladed propeller, of which the blades are the ethyl
groups. The shafts (N...Cl) coincide with the three-fold axes. Both right- and
left-handed forms occur randomly, giving rise to crystalline disorder. Other bond
lengths and angles are given, and are consistent with expectation, but there is no
estimate of error.

Details of analysis

The intensity data were measured with a four-circle diffractometer. Refinement was by full-matrix least squares to a final R index of 0.10.

1.   *Strukturbericht*, <u>1</u>, 635.

TETRAETHYLAMMONIUM OXOTETRABROMOAQUORHENATE

$C_8H_{22}O_2NBr_4Re$ F.W. = 670.1

$$[(C_2H_5)_4N][ReBr_4O(H_2O)]$$

I. Chemical and structural studies of rhenium(V) oxyhalide complexes. I.
Complexes from rhenium(III) bromide. F.A. COTTON and S.J. LIPPARD, 1965.
Inorg. Chem., **4**, 1621-1629.

Orthorhombic, $a = 11.14\pm2$, $b = 11.54\pm2$, $c = 12.90\pm2$ Å, $[U = 1658$ Å$^3]$, $D_m = 2.7$,
$Z = 4$, $D_x = 2.67$.

Space group $Pna2_1$ (C_{2v}^9) or $Pnam$ (D_{2h}^{16}) $Pnam$ is consistent with the analysis,
assuming a disordered tetraethylammonium ion. Molecular symmetry, mirror plane.

Atomic positions

	x	y	z
Re	0.2299±2	0.0953±2	1/4
Br(1)	0.1433±4	0.2277±4	0.1148±7
Br(2)	0.3617±5	0.0065±4	0.1134±8
O(1)	0.119 ±3	−0.008 ±3	1/4
O(2)	0.385 ±3	0.229 ±3	1/4
N	0.260 ±4	0.119 ±3	3/4
C(1)	0.370 ±6	0.234 ±6	0.608 ±9
C(2)	0.138 ±7	0.006 ±6	0.610 ±9
C(3)	0.134 ±7	0.125 ±6	0.683 ±8
C(4)	0.262 ±9	0.007 ±8	0.692 ±9
C(5)	0.253 ±7	0.218 ±7	0.678 ±9
C(6)	0.370 ±9	0.110 ±8	0.683 ±10

C(3), C(4) and C(5), C(6) represent alternative sites for the disordered
methylenic carbon atoms, and are assigned occupancy factors of 0.5. Isotropic
temperature factors are given also.

Structure

The structure consists of monomeric $[ReBr_4O(H_2O)]^-$ units, and $[(C_2H_5)_4N]^+$
ions. The latter are disordered, with each of the inner (methylenic) carbon atoms
adopting two alternative positions. The coordination of the rhenium atom is
probably best described as square pyramidal. The rhenium atom is displaced from
the basal plane defined by the bromine atoms, towards the apical oxygen atom, by
a distance of 0.32 Å. However, distorted octahedral coordination is completed
by a weakly-bonded water molecule. Relevant distances are: Re–Br, 2.51±1 Å;
Re–O, 1.71±4 Å; Re–O(H$_2$O), 2.32±4 Å. Other distances and angles are given, and
appear to be normal.

Details of analysis

The intensity data were measured with a four-circle diffractometer, using
CuK_α radiation. About 700 reflexions with $l \leqslant 5$ were observed. Refinement was
by full-matrix least squares to a final R index of 0.135.

HEXAMETHYLENEBIS [TRIMETHYLAMMONIUM BROMIDE] DIHYDRATE
DECAMETHYLENEBIS [TRIMETHYLAMMONIUM BROMIDE] DIHYDRATE

$C_{12}H_{30}Br_2N_2.2H_2O$ F.W. = 398.1

$C_{16}H_{38}Br_2N_2.2H_2O$ F.W. = 454.2

$$(CH_3)_3N(CH_2)_nN(CH_3)_3.2Br.2H_2O$$

$$(n = 6, 10)$$

I. Structure studies of hexa-and deca-methonium bromides, $(CH_3)_3N(CH_2)_nN(CH_3)_3$.
2Br.2H$_2$O, in relation to their pharmacological action. K. LONSDALE, H.J.
MILLEDGE and L.M. PANT, 1965. *Acta Cryst.*, **19**, 827-840.

HEXA–SALT

Monoclinic, $a = 7.41$, $b = 18.40$, $c = 7.20$ Å, $\beta = 109°$, $[U = 928$ Å$^3]$, $D_m = 1.406$, $Z = 2$, $D_x = 1.415$.

Space group $P2_1/c$ (C_{2h}^5) Molecular symmetry, centre.

Atomic positions

	x	y	z
Br	0.3182	0.3677	0.1657
C(H_2)(1)	−0.0552	0.0363	−0.0042
C(H_2)(2)	−0.1167	0.0496	0.1691
C(H_2)(3)	−0.2293	0.1173	0.1464
N	−0.3500	0.1233	0.2808
C(H_3)(1)	−0.4232	0.2012	0.2705
C(H_3)(2)	−0.5105	0.0719	0.2348
C(H_3)(3)	−0.2285	0.1118	0.4919
(H_2)O	0.0545	0.2723	0.4010

Temperature factors (anisotropic for the bromine atom) are given also.

DECA–SALT

Monoclinic, $a = 13.49$, $b = 10.07$, $c = 9.04$ Å, $\beta = 109°$, $[U = 1161$ Å$^3]$, $D_m = 1.305$, $Z = 2$, $D_x = 1.299$.

Space group $P2_1/a$ (C_{2h}^5) Molecular symmetry, centre.

Atomic positions

	x	y	z
Br	0.5943	0.1459	0.7454
C(H_2)(1)	0.0323	0.0484	0.0697
C(H_2)(2)	0.0936	−0.0292	0.2061
C(H_2)(3)	0.1496	0.0605	0.3553
C(H_2)(4)	0.2021	−0.0210	0.5105
C(H_2)(5)	0.2584	0.0787	0.6384
N	0.2954	0.0170	0.7968
C(H_3)(1)	0.3495	0.1203	0.9179
C(H_3)(2)	0.2090	−0.0344	0.8534
C(H_3)(3)	0.3724	−0.0953	0.8115
(H_2)O	0.4219	0.1823	0.3729

Temperature factors (anisotropic for the bromine atom) are given also.

Structure

The organic ions in both structures are in the extended chain conformation, with no unusual bond lengths or angles. There appears to be no interaction between the nitrogen atoms and any other atoms to which they are not directly bonded. Views of the structures are given in Figs. 24 and 25.

Details of analysis

The intensity data for both structures were recorded on equi-inclination Weissenberg photographs, using CuK_α radiation, and measured visually. For the *hexa*-salt, 502 reflexions were observed in the range sin $\leqslant 0.65$ on the levels $0kl$ to $5kl$. The crystals were invariably twinned, so that not all the accessible reflexions could be measured. For the *deca*-salt, 673 reflexions were observed on the levels $h0l$ to h, 11, l. Refinement of both structures was by least squares to final R indices of about 0.10.

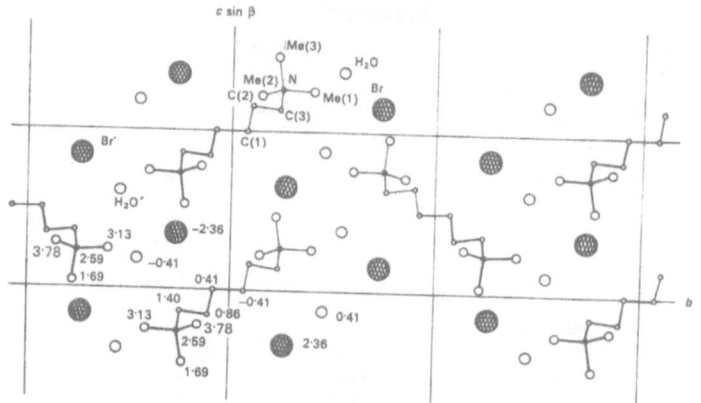

Fig. 24. The structure of the *hexa*-salt, projected along *a*. The numbers show the distances of atoms above or below the plane $x = 0$.

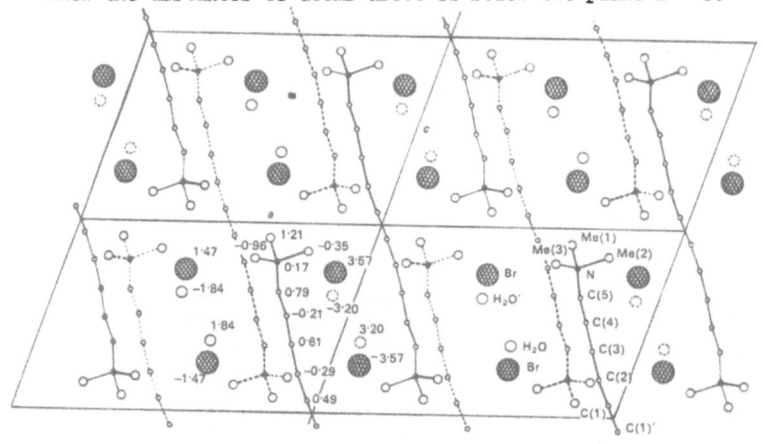

Fig. 25. The structure of the *deca*-salt, projected along *b*. The numbers show the distances of atoms above or below the plane $y = 0$. The molecule is reflected in the plane at $y = b/4$ to give atoms at $\frac{1}{2} + x$, $\frac{1}{2} - y$, z.

DITHIO-OXAMIDE (RUBEANIC ACID)

$C_2H_4N_2S_2$ F.W. = 120.2

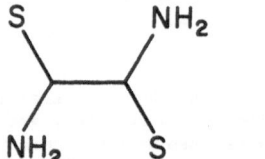

I. The crystal and molecular structure of dithio-oxamide (rubeanic acid). P.J. WHEATLEY, 1965. *J. Chem. Soc.*, 396–402. (For a preliminary report, see 1.)

Triclinic, $a = 5.858\pm3$, $b = 10.757\pm5$, $c = 3.936\pm3$ Å, $\alpha = 92°30.0\pm5'$, $\beta = 102°45.0\pm10'$, $\gamma = 92°19.0\pm5'$, $U = 241.7$ Å³, $D_m = 1.66$, $Z = 2$, $D_x = 1.651$, CuK_α, $\gamma = 1.54050$ Å). [The accuracy claimed for the cell angles (one minute of arc, or better) seems questionable.]

Space group $P1$ (C_1^1) or $P\bar{1}$ (C_i^1) $P\bar{1}$ confirmed by analysis. Molecular symmetry, centre.

Atomic positions

	x	y	z
S(1)	-0.2987	-0.1302	-0.1082
N(1)	-0.1849	0.0950	-0.2997
C(1)	-0.1249	-0.0043	-0.1141
S(2)	-0.2279	0.3641	0.1681
N(2)	-0.2496	0.5917	0.4129
C(2)	-0.1275	0.4932	0.3950

Coordinates in the original are given in Å. Standard deviations are given in full. Mean values are: S, 0.001; N and C, 0.005 Å. Anisotropic temperature factors are given also.

Structure

There are two independent molecules in the structure, each occupying a centre of symmetry, and accurately planar. There are however inexplicable differences in their bond distances. Mean bond lengths and angles are:

C − S = 1.649±5 Å	S − C − N = 125.0±4°
C − N = 1.324±7	S − C − C = 120.0±4
C − C = 1.537±14	N − C − C = 114.9±4

N-H--S bonding is suggested by intermolecular N-S distances of 3.432 and 3.456 Å.

Details of analysis

Intensity data were recorded on equi-inclination Weissenberg photographs about the three principal axes, using CuK_{α} radiation. 832 of a possible 1070 reflexions were observed and estimated visually. Refinement was by block-diagonal least squares to a final R index of 0.071.

1.   *Structure Reports*, 18, 654.

cis, cis-1,2,3,4-TETRAPHENYLBUTADIENE

$C_{28}H_{22}$ F.W. = 358.5

$$
\begin{array}{ccc}
 & & C_6H_5 \\
 & H & | \\
H_5C_6-C=C-C=C-C_6H_5 \\
 & | & \\
 & C_6H_5 &
\end{array}
$$

I. The crystal and molecular structure of *cis, cis*-1,2,3,4-tetraphenylbutadiene.
   I.L. KARLE and K.S. DRAGONETTE, 1965. *Acta Cryst.*, 19, 500-503.

Monoclinic, $a = 5.87±2$, $b = 21.31±2$ $c = 8.13±2$ Å, $\beta = 97°\ 5±10'$, [$U = 1009$ Å³],
$D_m = 1.181$, $Z = 2$, $D_x = 1.191$.
Space group $P2_1/c$ (C_{2h}^5) Molecular symmetry, centre.

Atomic positions

	x	y	z
C(1)	0.2372	0.0117	0.6706
C(2)	0.0725	0.0252	0.5434
C(3)	0.0309	0.0918	0.4804
C(4)	0.1955	0.1188	0.3914
C(5)	0.1603	0.1828	0.3379
C(6)	-0.0276	0.2160	0.3769
C(7)	-0.1864	0.1889	0.4657
C(8)	-0.1614	0.1251	0.5181
C(9)	0.3908	0.0547	0.7746
C(10)	0.3254	0.1164	0.8168
C(11)	0.4766	0.1522	0.9254
C(12)	0.6947	0.1304	0.9906
C(13)	0.7616	0.0701	0.9465
C(14)	0.6071	0.0328	0.8416

The mean standard error is 0.007 Å. Anisotropic temperature factors are given.

Structure

The bond lengths and angles are given in Fig. 26. The phenyl rings are planar, as is the group consisting of the butadiene chain and its four attached carbon atoms. The angle between the ring C(3) to C(8) and the chain group is 75°; the corresponding angle for the ring C(9) to C(14) is 34°. The sense of these rotations is such that both C(8) and C(10) in Fig. 26 lie on the same side of the plane of the chain group. Intermolecular distances are consistent with van der Waals interaction.

Details of analysis

Intensity data were recorded on equi-inclination Weissenberg photographs (0kl to 4kl) using CuK$_\alpha$ radiation. 1100 of a possible 1400 reflexions were observed and estimated visually. Refinement was by full-matrix least squares to a final R index of 0.12.

Fig. 26.   Bond distances and angles in *cis, cis*-1,2,3,4-tetraphenylbutadiene. The standard deviation for the bond lengths ranges between 0.009 and 0.012 Å. The standard error for the angles is approximately 0.5°.

p-IODOBENZONITRILE

C$_7$H$_4$NI

I —a— (benzene ring: b, c, d) —e— C $\equiv$ N (f)

F.W. = 229.0

I. The crystal structure of *p*-iodobenzonitrile. E.O. SCHLEMPER and D. BRITTON, 1965. *Acta Cryst.*, <u>18</u>, 419–424.

Monoclinic, $a = 10.36\pm5$, $b = 10.63\pm5$, $c = 9.10\pm5$ Å, $\beta = 133.1\pm2°$, $[U = 732$ Å$^3]$, $Z = 4$, $D_x = 2.08$, (CuK$_\alpha$, $\lambda = 1.5418$ Å; MoK$_\alpha$, $\lambda = 0.7107$ Å).

Space group   *Cc* (C_s^4) or *C2/c* (C_{2h}^6) *C2/c* confirmed by analysis. Molecular symmetry, two-fold axis.

Atomic positions   Anisotropic temperature factors are given also

	x	y	z
I	0	1.02650±4	1/4
C(1)	0	0.8324 ±7	1/4
C(2)	0.0926±5	0.7642 ±6	0.2141±6
C(3)	0.0945±5	0.6339 ±6	0.2162±6
C(4)	0	0.5675 ±8	1/4
C(5)	0	0.4386 ±8	1/4
N	0	0.3258 ±8	1/4

Structure

The molecule is planar, and the bonds *a*, *e*, and *f* coincide with a crystallographic two-fold axis. The intermolecular distance I–N (along the two-fold axis) is 3.18 Å, (the sum of nominal van der Waals radii is 3.65 Å). Other intermolecular distances are consistent with van der Waals interaction. Bond lengths and angles in the molecule are:

a = 2.064±8 Å	a̅b̅ = 121.0±5'
b = 1.407±9	b̅c̅ = 121.2±5
c = 1.385±8	c̅d̅ = 120.1±5
d = 1.401±10	d̅e̅ = 120.2±5
e = 1.370±12	
f = 1.199±12	

(It is suggested that these standard deviations may be too small by a factor of two.) [Moreover, *f* may be appreciably underestimated because of severe thermal motion, normal to the bond, of the nitrogen atom.]

Details of analysis

Intensity data were recorded on Weissenberg photographs (*hk*0 to *hk*8) using MoK_α radiation. 883 of 1189 possible reflexions were observed and measured visually. Refinement was by least squares to a final *R* index of 0.117.

AMMONIUM TRICYANOMETHIDE

$C_4H_4N_4$ $NH_4C(CN)_3$ F.W. = 108.1

I. The crystal structure of ammonium tricyanomethide, $NH_4C(CN)_3$. R. DESIDERATO and R.L. SASS, 1965. *Acta Cryst.*, 18, 1–4.

Monoclinic, *a* = 9.055±7, *b* = 3.87±1, *c* = 17.325±14 Å, β = 104.6±2°, [*U* = 587.5 Å³], *z* = 4, D_x = 1.18 [1.22].

Space group $P2_1/c$ (C_{2h}^5)

Atomic positions

	x	*y*	*z*
C(1)	0.2976±9	0.1451±30	0.3609±4
C(2)	0.2462±9	0.3264±34	0.2891±5
C(3)	0.1928±9	0.0308±31	0.4021±4
C(4)	0.4490±9	0.0274±33	0.3835±4
N(1)	0.1718±7	0.5405±24	0.5945±3
N(2)	0.2055±9	0.4798±29	0.2305±4
N(3)	0.1079±8	0.9407±26	0.4366±4
N(4)	0.5741±9	0.9300±29	0.4024±4

Isotropic temperature factors are given also.

Structure

The tricyanomethide ion has non-crystallographic 3*m* symmetry. It is nonplanar, with the linear C–C–N linkages making an angle of about 87° to the three-fold axis. The mean bond lengths are: C–C, 1.40±1 Å; C–N, 1.15±1 Å. The structure is illustrated in Fig. 27; it consists of stacks of tricyanomethide ions, extended in the *b* direction, and held together by ammonium ions. The environment of the ammonium ion is distorted tetragonal prismatic, with eight nitrogen atoms at distances ranging from 2.95 to 3.52 Å.

Details of analysis

The intensity data were recorded on Weissenberg photographs (*h*0*l* to *h*2*l*) using CuK_α radiation. 429 reflexions were observed and measured visually. The refinement was by full-matrix least squares to a final *R* index of 0.104.

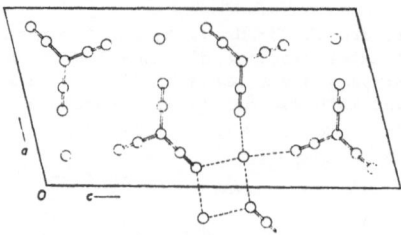

Fig. 27.   The structure of ammonium tricyanomethide viewed along b.

7,7,8,8-TETRACYANOQUINODIMETHANE

$C_{12}H_4N_4$ F.W. = 204.2

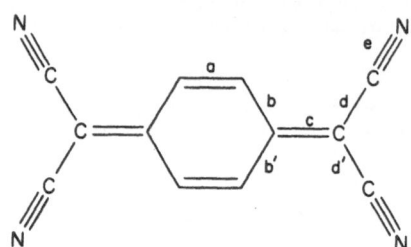

I. The crystal and molecular structure of 7,7,8,8-tetracyanoquinodimethane.
   R.E. LONG, R.A. SPARKS and K.N. TRUEBLOOD, 1965. *Acta Cryst.*, <u>18</u>, 932–939.

Monoclinic, a = 8.90±6, b = 7.060±4, c = 16.395±5 Å, β = 98.54±4°, [U = 1019.4 Å³],
D_m = 1.315, Z = 4, D_x = 1.329, (calibration with quartz crystal, a = 4.9131 Å).

Space group Cc (C_S^4) or $C2/c$ (C_{2h}^6) $C2/c$ is confirmed by the analysis.
Molecular symmetry, centre.

Atomic positions

	x	y	z
C(1)	−0.0344±2	0.0781±3	0.0743±1
C(2)	0.1502±2	−0.0255±3	−0.0164±1
C(3)	0.1215±2	0.0550±2	0.0612±1
C(4)	0.2396±2	0.1079±3	0.1205±1
C(5)	0.2146±2	0.1882±3	0.1982±1
C(6)	0.3957±2	0.0839±3	0.1090±1
N(1)	0.1957±2	0.2529±4	0.2596±1
N(2)	0.5188±2	0.0646±3	0.0994±1
H(1)	−0.0581±34	0.1315±45	0.1218±20
H(2)	0.2452±31	−0.0296±37	−0.0255±16

    Anisotropic temperature factors for the non-hydrogen atoms are given, and
have been analyzed in terms of rigid-body motion. The positions given here have
been corrected for libration.

Structure

    The molecule is very slightly, but significantly, non-planar. The mean bond
lengths and angles (assuming *mmm* symmetry) are:

    a = 1.346 Å   bb' = 118.3°
    b = 1.448   dd' = 116.1
    c = 1.374   de = 179.5
    d = 1.440
    e = 1.140

    (The standard deviations are 0.003 Å and 0.2°.)   The intermolecular distances
are consistent with van der Waals interaction.

Details of analysis

The intensity data were recorded on integrated Weissenberg photographs (six self-correlating levels for a crystal mounted about [110]) using CuK_α radiation. 976 of 1120 accessible reflexions were observed and measured with a densitometer. (The weaker reflexions were measured visually.) Refinement was by full-matrix least squares to a final R index of 0.081.

TRIURET

(CARBONYL DIUREA)

$C_3H_6N_4O_3$ F.W. = 146.1

I. The molecular and crystal structure of triuret. D. CARLSTRÖM and H. RINGERTZ, 1965. *Acta Cryst.*, 18, 307–313.

Monoclinic, $a = 7.209\pm2$, $b = 7.143\pm2$, $c = 24.836\pm3$ Å, $\beta = 119.83\pm1°$, $U = 1109.5$ Å^3, $D_m = 1.745$, $Z = 8$, $D_x = 1.749$, (calibration with silicon powder, $a = 5.4306$ Å).

Space group Cc (C_s^4) or $C2/c$ (C_{2h}^6) $C2/c$ is confirmed by the structure analysis.

Atomic positions

	x	y	z
C(1)	0.2186±6	0.5906±6	0.0633±2
C(2)	0.2461±6	0.3086±5	0.1266±2
C(3)	0.2443±6	0.0988±6	0.2059±2
O(1)	0.1633±5	0.7563±5	0.0560±2
O(2)	0.3062±5	0.1963±5	0.1015±1
O(3)	0.2498±5	0.0993±5	0.2565±2
N(1)	0.2988±6	0.5027±6	0.0329±2
N(2)	0.1918±5	0.4898±5	0.1077±2
N(3)	0.2222±5	0.2706±6	0.1770±2
N(4)	0.2553±6	−0.0568±6	0.1785±2

Temperature factors are given. The positions of the six hydrogen atoms are given also, with standard deviations of about 0.08 Å.

Structure

The molecule is not planar, but has a slight helical twist resulting from rotations about the C-N bonds. However, the configuration of bonds to each carbon atom is planar; the dihedral angles between adjacent planes so defined are 7.4° and 16.0°. Some mean bond lengths and angles are:

a = 1.22 Å	bb' = 111°
b = 1.37	bc = 128
c = 1.40	cd = 120
d = 1.33	ce = 117
e = 1.24	

(Standard deviations are 0.01 Å and 0.6°.) The structure is stabilized by an extensive system of N-H..O bonds.

Details of analysis

The intensity data were recorded on equi-inclination Weissenberg photographs (h0l to h6l), using CuK_α radiation. 1063 reflexions (86.4% of those accessible) were observed and measured photometrically. Refinement was by block-diagonal least squares to a final R index of 0.126.

ALLYLTHIOUREA

$C_4H_8N_2S$ F.W. = 116.2

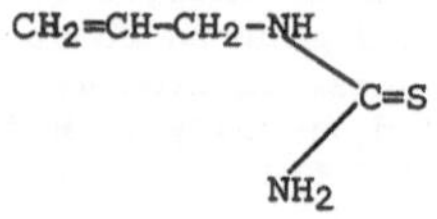

$$CH_2=CH-CH_2-NH$$
$$C=S$$
$$NH_2$$

I. The crystal and molecular structure of allylthiourea. K.S. DRAGONETTE and
   I.L. KARLE, 1965. *Acta Cryst.*, <u>19</u>, 978–983.

Monoclinic, a = 8.39±2, b = 8.58±1, c = 9.77±1 Å, β = 119°45±10', [U = 610.6 Å^3],
D_m = 1.247, Z = 4, D_x = 1.264.
Space group $P2_1/c$ (C_{2h}^5)

Atomic positions

	x	y	z
S	0.5510±3	0.5601±4	0.2311±3
N(1)	0.3344±9	0.4239±12	0.3162±8
C(2)	0.3975±11	0.5613±16	0.2912±11
N(3)	0.3335±9	0.6978±12	0.3148±8
C(4)	0.2110±11	0.7064±16	0.3830±11
C(5)	0.0103±11	0.6674±16	0.2644±11
C(6)	−0.0641±11	0.6512±16	0.1160±11

Anisotropic temperature factors, and positions of hydrogen atoms are given.

Structure

     The bond lengths and angles are given in Fig. 28. (These values are not cor-
rected for thermal motion, which could result in underestimation of some bond lengths
by as much as 0.02 Å). The thiourea group (S,N(1), C(2), N(3)) is planar. The plane
of the allyl group (C(4), C(5), C(6)) makes an angle of 99.6° with that of the
thiourea group. Each sulphur atom is hydrogen bonded to two nitrogen atoms, each
in a different adjacent molecule. The structure thus consists of endless chains
parallel to b, as shown in Fig. 29.

Details of analysis

     Intensity data were recorded on equi-inclination Weissenberg photographs (h0l
to h6l) using CuKα radiation. 760 of a possible 910 reflexions were observed and
measured visually. Refinement was by full-matrix least squares to a final R index
of 0.101.

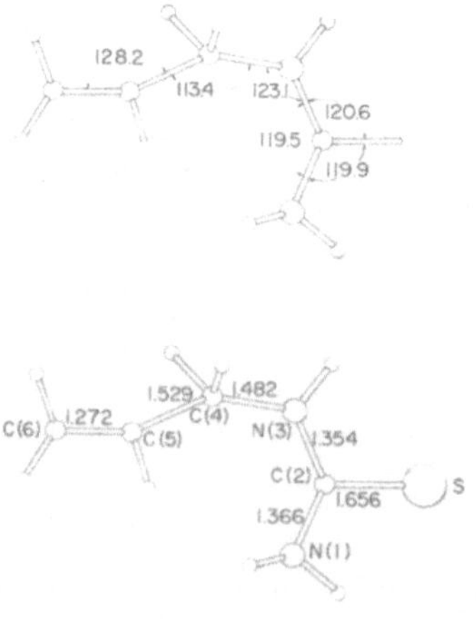

Fig. 28.   Bond lengths and angles in allylthiourea. Standard deviations are
            0.010 to 0.014 Å, and .1°.

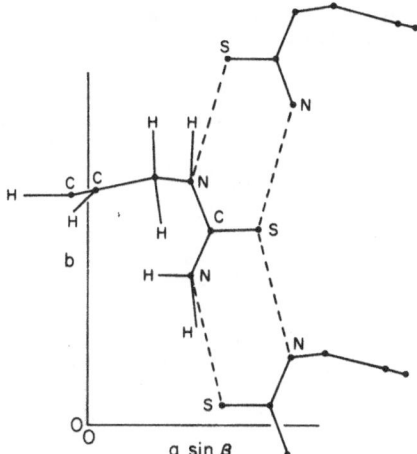

Fig. 29.   The structure of allylthiourea viewed along c, showing hydrogen bonds.

N-PHENYL-N'-BENZOYLSELENOUREA

$C_{14}H_{12}N_2OSe$ F.W. = 303.2

I. The crystal structure of N-phenyl-N'-benzoyl-selenourea. H. HOPE, 1965.
   Acta Cryst., <u>18</u>, 259-264.

Monoclinic, a = 13.160±5, b = 5.064±3, c = 19.940±5 Å, β = 103.65±3°, [U = 1291 Å³],
D_m = 1.60, z = 4, D_x = 1.56.
Space group $P2_1/c$ (C_{2h}^5)

Atomic positions

	x	y	z
Se(1)	0.1077±1	0.2808±2	0.5847±5
C(2)	0.2096±7	0.3771±17	0.5416±5
N(3)	0.3059±6	0.2864±16	0.5513±4
C(4)	0.3630±7	0.0872±19	0.5952±5
C(5)	0.3250±9	−0.0692±19	0.6406±6
C(6)	0.3916±10	−0.2584±25	0.6793±6
C(7)	0.4901±9	−0.2926±24	0.6721±6
C(8)	0.5279±11	−0.1365±28	0.6253±7
C(9)	0.4632±9	0.0529±24	0.5876±6
N(10)	0.1818±6	0.5676±15	0.4897±4
C(11)	0.2410±9	0.6517±19	0.4443±5
C(12)	0.1880±7	0.8354±19	0.3893±5
C(13)	0.2188±9	0.8229±23	0.3270±6
C(14)	0.1765±9	0.9909±24	0.2741±6
C(15)	0.1011±9	1.1731±24	0.2814±6
C(16)	0.0688±9	1.1817±22	0.3427±5
C(17)	0.1127±7	1.0136±20	0.3966±5
O(18)	0.3295±6	0.5724±17	0.4485±4

Also given are temperature factors (some anisotropic) for the above atoms, and the positions of the hydrogen atoms.

Structure

Bond lengths and angles are given in full. Some of the bond lengths are:

$$a = 1.820 \pm 10 \text{ Å}$$
$$b = 1.320 \pm 13$$
$$c = 1.430 \pm 12$$
$$d = 1.400 \pm 14$$
$$e = 1.390 \pm 14$$
$$f = 1.480 \pm 14$$
$$g = 1.220 \pm 14$$

The molecule deviates from planarity chiefly by rotation about the bonds d (7.7°), e (4.2°) and f (30.5°). There is an intramolecular N-H..O bond of length 2.590 ±12 Å. The molecular shape and packing are further illustrated in Fig. 30. There is an intermolecular selenium-nitrogen distance of 3.83 Å. The corresponding selenium-hydrogen distance is 2.94 Å.

Details of analysis

The intensity data were recorded on integrated Weissenberg photographs ($h0l$ to $h4l$) using CuK$_\alpha$ radiation. Interlayer scaling was accomplished with the aid of some diffractometer measurements. The 1432 observed reflexions were measured photometrically. Refinement was by full-matrix least squares to a final R index of 0.071.

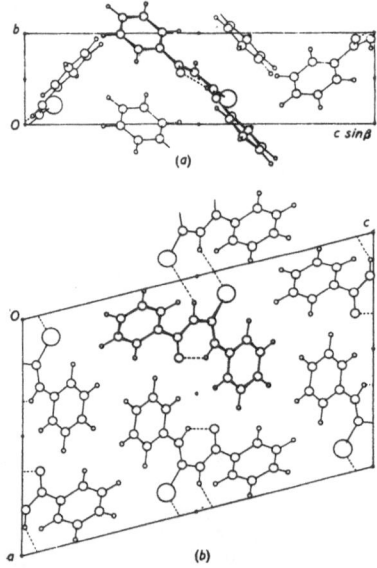

Fig. 30.   (a) Projection of part of the structure of N-phenyl-N'-benzoylselenourea
                 along a.
            (b) Projection of the structure along b. The intramolecular hydrogen
                 bond and a short selenium-hydrogen distance are indicated.

GUANIDINIUM CHLORIDE

CH_6ClN_3 $C(NH_2)_3^+\ Cl^-$ F.W. = 95.7

I. The crystal structure of guanidinium chloride. D.J. HAAS, D.R. HARRIS and
   H.H. MILLS, 1965. *Acta Cryst.*, <u>19</u>, 676–679.

Orthorhombic, $a = 9.184\pm5$, $b = 13.039\pm5$, $c = 7.765\pm5$ Å, $[U = 929.9$ Å$^3]$, $D_m = 1.35$, $z = 8$, $D_x = 1.365$.

Space group $Pbca$ (D_{2h}^{15})

Atomic positions

	x	y	z
Cl	0.0203	0.1187	0.1823
C	0.3457	0.3750	0.3885
N(1)	0.2882	0.2901	0.4505
N(2)	0.2879	0.4657	0.4240
N(3)	0.4619	0.3700	0.2878
H(1)	0.3308	0.2398	0.3998
H(2)	0.2089	0.3000	0.5290
H(3)	0.2221	0.4610	0.4757
H(4)	0.3397	0.5147	0.3787
H(5)	0.4910	0.4504	0.2324
H(6)	0.4907	0.2860	0.2472

[In the original these values are stated to be Å. They are fractional, however.] Standard deviations are given in Å. Mean values are Cl, 0.0014; C and N, 0.0055; H, 0.1 Å. Temperature factors (anisotropic for non-hydrogen atoms) are given also.

Structure

To the accuracy of the analysis the guanidinium ion is planar, with trigonal symmetry. The mean C–N distance is 1.323±3 Å. Details of this packing are given in Fig. 31 and Fig. 32. The chlorine is coordinated to three guanidinium ions, two of which are nearly coplanar. The mean NH–Cl distance is 3.303 Å. Each nitrogen atom is coordinated to two chlorine atoms, *via* hydrogen bonds.

Details of analysis

Intensity data were measured with a four–circle diffractometer, in the range $2\theta < 150°$, using CuK$_\alpha$ radiation. 673 of a possible 960 reflexions were observed above background. Refinement was block–diagonal least squares to a final R index of 0.062.

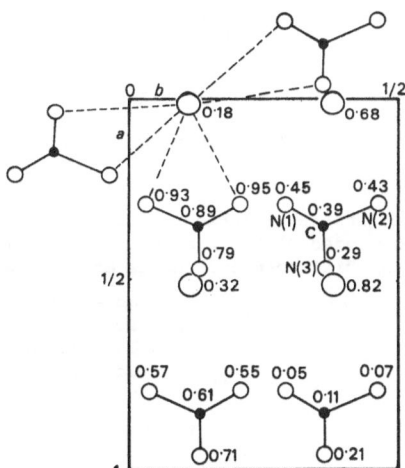

Fig. 31.   The structure of guanidinium chloride projected along c. Numerical values of z are shown.

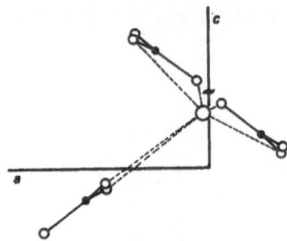

Fig. 32.   Part of the structure projected along b, showing coordination of the chloride ion.

SEMICARBAZIDE HYDROCHLORIDE

CH_6ClN_3O F.W. = 111.5

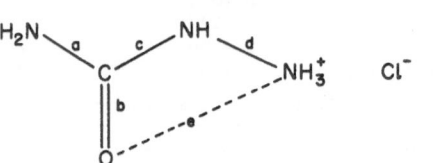

I. The crystal and molecular structure of semicarbazide hydrochloride. M. NARDELLI, G. FAVA and G. GIRALDI, 1965. *Acta Cryst.*, <u>19</u>, 1038-1042. (For earlier work on this compound, see <u>1</u>. Many of the results of <u>1</u> are given in I.)

Orthorhombic, a = 7.51±1, b = 13.13±1, c = 4.64±1 Å, U = 457.8 Å³, z = 4, D_x = 1.618.

Space group $P2_12_12_1$ (D_2^4)

Atomic positions

	x	y	z
Cl	0.2543	0.1399	0.9636
O	0.7587	0.0874	0.8233
N(1)	0.9267	0.1605	0.4787
N(2)	0.6335	0.1320	0.3930
N(3)	0.4909	0.0655	0.4551
C	0.7767	0.1236	0.5787

Standard deviations (in Å) are given in full. Mean values are: Cl, 0.003; O, 0.007; C and N, 0.008 to 0.012 Å. Also given are anisotropic temperature factors for these atoms, and assumed hydrogen positions. The atomic positions from <u>1</u> are given also, and there is good agreement between the two sets.

Structure

A view of the structure, showing interionic distances, is given in Fig. 33. The carbon atom and the atoms adjacent to it are coplanar, and the NH^+ group lies 0.44 Å from their plane. Some distances and angles are as follows:

a = 1.311±18 Å	ab = 124±1
b = 1.238±16	ac = 115±1
c = 1.382±16	bc = 121±1
d = 1.411±14	cd = 114±1
e = 2.655±16	

Details of analysis

The intensity data were recorded on integrated and non-integrated Weissenberg photographs ($hk0$ to $hk4$; $0kl$ to $1kl1$ using CuK_α radiation. 513 of 588 accessible reflexions were observed and measured photometrically. Refinement was by differential syntheses to a final R index of 0.128.

1.   Q.C. JOHNSON, 1960. *Ph.D. Thesis, University of California.*

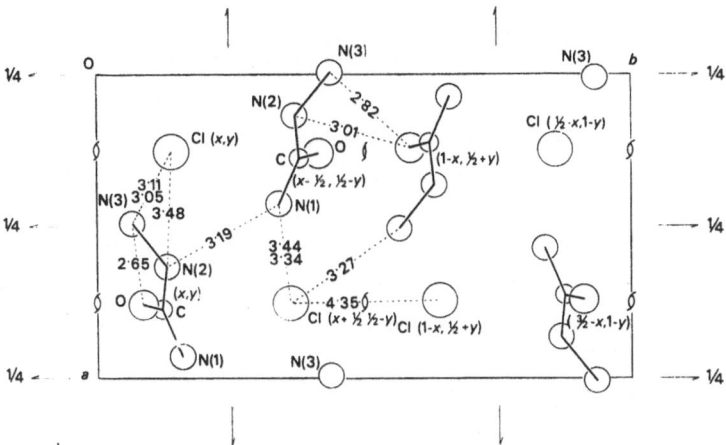

Fig. 33.   The structure of semicarbazide hydrochloride viewed along c.
Hydrogen bonds and other short interatomic distances are indicated
by broken lines.

PHENYLHYDRAZINE HYDROCHLORIDE

$C_6H_9ClN_2$ F.W. = 144.6

I. The crystal structure of monoclinic phenylhydrazine hydrochloride, $C_6H_5NHNH_3Cl$.
   CHUNG HOE KOO, 1965. *Bull. Chem. Soc. Japan*, <u>38</u>, 286–290.

Monoclinic, $a = 7.82\pm3$, $b = 30.80\pm6$, $c = 3.87\pm2$ Å, $\beta = 130.3\pm5°$, [$U = 711$ Å³],
$z = 4$, $D_x = 1.344$.
Space group $P2_1/a$ (C_{2h}^5)

Atomic positions

	x	y	z
C(1)	0.7608	0.1225	−0.0233
C(2)	0.9400	0.1429	0.0433
C(3)	0.9367	0.1888	0.0650
C(4)	0.7575	0.2092	0.0083
C(5)	0.5800	0.1883	−0.0483
C(6)	0.5833	0.1425	−0.0700
N(1)	0.7633	0.0754	−0.0500
N(2)	0.6667	0.0492	0.0900
Cl	0.2433	0.0480	0.0167

     Standard deviations of x and y are: Cl, 0.002; N, 0.005; C, 0.006 Å. Standard
deviations of z are about twice as large.

Structure

     Bond lengths and angles are given in full. The distance C–N is 1.455±8 Å,
and the distance N–N is 1.432±12 Å. The molecule deviates from planarity by a
rotation of 67° of the N–N bond about the C–N bond. The terminal nitrogen atom
has three chlorine atoms as nearest neighbors at 3.13, 3.16, and 3.22 Å;   the
other nitrogen atom has two at 3.36 and 3.43 Å. There is an NH...NH$_3^+$ contact of
3.01 Å.

Details of analysis

The intensity data were recorded photographically and measured visually. The structure was refined with the aid of Fourier projections along a and c. Final R indices were 0.10 and 0.12.

POTASSIUM *syn*-METHYLDIAZOTATE

CH_3N_2OK $(H_3CN_2O)^- K^+$ F.W. = 98.15

I. Kristal-und Molekülstruktur des *syn*-methyldiazotatkaliums CH_3N_2OK. R. HUBER, R. SANGER and W. HOPPE, 1965. *Acta Cryst.*, 18, 467–473. (For a preliminary account, see 1).

Monoclinic, $a = 12.49\pm2$, $b = 9.97\pm2$, $c = 6.61\pm2$ Å, $\beta = 98°\ 30\pm5'$, $[U = 814.1\ \text{Å}^3]$, $D_m = 1.60$, $Z = 8$, $D_x = 1.61$. [Original gives $\beta = 81°\ 30'$; the sense of the c axis has been reversed in order to make β obtuse.].

Space group Cc (C_s^4) or $C2/c$ (C_{2h}^6) $C2/c$ is confirmed by analysis.

Atomic positions

	x	y	z
N(1)	0.15595±42	0.28068±53	0.22847±69
N(2)	0.13919±37	0.16349±51	0.16610±64
O	0.09060±32	0.14199±46	-0.01991±52
C	0.11941±65	0.38816±74	0.08010±110
K	0.12956±9	-0.12392±13	0.09767±16
H(1)	0.100	0.346	-0.022
H(2)	0.046	0.392	0.058
H(3)	0.154	0.396	-0.033

Anisotropic temperature factors are given for the non-hydrogen atoms. [The signs of the z coordinates above have been reversed].

Structure

The non-hydrogen atoms of the anion are essentially coplanar. Bond lengths and angles are N–O, 1.306±7; N–N, 1.246±8; C–N, 1.477±10 Å; O–N–N, 119.8±5°; N–N–C, 116.2±5°. Bond lengths and angles involving the hydrogen atoms are given also. The C..O distance within the anion is 2.55 10 Å. Moreover, one of the hydrogen atoms is *cis* to the N–N linkage, and the H–O distance is 2.03 Å. These distances suggest C–H..O bonding. A view of the structure is given in Fig. 34. Potassium-oxygen distances are 2.650, 2.728 and 2.785 Å, while potassium-nitrogen distances range from 2.900 to 2.957 Å.

Details of analysis

Intensity data were recorded on precession and Weissenberg photographs using MoK_α radiation. 1230 independent reflexions were measured photometrically. Refinement was by full-matrix least squares, and the hydrogen atoms were located in a difference Fourier synthesis. The final R index was 0.084.

1.   E. Müller *et al.*, 1963. *Chem. Ber.*, 96, 1712.

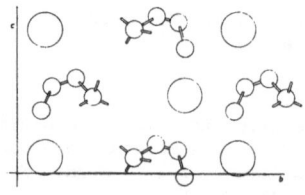

Fig. 34.   The structure of potassium *syn*-methyldiazotate projected along *a*.

2,4-DIBROMODIAZOAMINOBENZENE

$C_{12}H_9Br_2N_3$ F.W. = 355.0

I. [Crystal and molecular structures of 2,4-dibromo-diazo-aminobenzene.]   J.A.
   OMEL'ČENKO and J.D. KONDRAŠEV, 1965. *Kristallografija*, 10, 822–827 [*Soviet
   Physics – Crystallography*, 10, 690–697].

Monoclinic, a = 12.05±1, b = 23.88±2, c = 4.48±1 Å, [β = 95°47±15', (84°13' in
original)], [U = 1283 Å³], D_m = 1.82, z = 4, D_x = 1.84.

Space group $P2_1/n$ (C_{2h}^5)

Atomic positions

	x	y	z
Br(1)	0.662	0.983	0.219
Br(2)	0.078	0.082	0.714
C(1)	0.800	0.131	0.143
C(2)	0.903	0.132	0.033
C(3)	0.935	0.087	0.861
C(4)	0.865	0.013	0.790
C(5)	0.761	0.013	0.897
C(6)	0.729	0.087	0.070
C(7)	0.468	0.162	0.640
C(8)	0.360	0.157	0.720
C(9)	0.322	0.194	0.923
C(10)	0.390	0.237	0.046
C(11)	0.499	0.210	0.963
C(12)	0.536	0.204	0.755
N(1)	0.620	0.083	0.174
N(2)	0.595	0.126	0.382
N(3)	0.500	0.117	0.463

The original z coordinates are here given as 1-z in conformity with the conventional choice of β as obtuse. The mean standard deviations are: Br, 0.002 Å;
C, 0.017 Å; N, 0.012 Å. [These estimates seem optimistic.]

Structure

Bond lengths and angles are given in full, and are consistent with the formulation indicated above. The dihedral angle between the aromatic rings is 9°.
Pairs of molecules are held together as centrosymmetrical, approximately flat,
dimers by charge-transfer bonds Br(1) – N(2') of length 3.22 Å.

Details of analysis

Intensity data were variously recorded on Weissenberg photographs, and with
a diffractometer, using CuK_α radiation, for levels hk0, hk1, and hk3. 1095 reflexions were observed above background. Refinement was by Fourier methods.

ORGANIC COMPOUNDS

p-NITROPHENYL AZIDE

$C_6H_4N_4O_2$ F.W. = 164.1

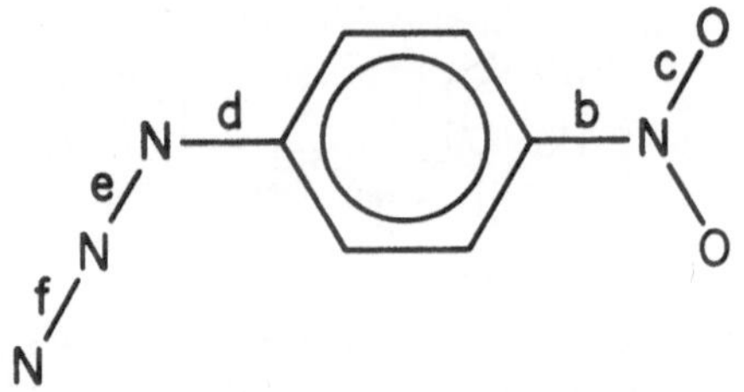

I. Crystal, molecular, and electronic structure of p-nitrophenyl azide.
A. MUGNOLI, C. MARIANI and M. SIMONETTA, 1965. *Acta Cryst.*, 19, 367–372.
(For reports of preliminary work, see 1).

Orthorhombic, a = 18.05±2, b = 10.29±1, c = 3.73±1 Å, $[U$ = 692.8 Å$^3]$, D_{m_o} = 1.50, Z = 4, D_x = 1.57, (Cell and intensity data at 100±5°K; CuK_α, λ = 1.5418 Å).

Space group $P2_12_12_1$ (D_2^4)

Atomic positions

	x	y	z
C(1)	0.4659±8	0.1729±13	0.5377±49
C(2)	0.4199±8	0.0647±13	0.5190±50
C(3)	0.3487±7	0.0776±11	0.6566±47
C(4)	0.3247±7	0.1949±12	0.7886±46
C(5)	0.3715±7	0.3016±12	0.7950±47
C(6)	0.4429±7	0.2907±12	0.6681±51
N(1)	0.5397±6	0.1614±10	0.3893±42
N(2)	0.2507±6	0.1959±10	0.9153±40
N(3)	0.2308±6	0.3006±10	1.0675±41
N(4)	0.2068±7	0.3878±11	1.2085±45
O(1)	0.5607±5	0.0565±10	0.2688±37
O(2)	0.5801±5	0.2601±9	0.3767±37

Also given are isotropic temperature factors for the above atoms, and the positions of the hydrogen atoms.

Structure

The molecule is essentially planar, except that the NNN chain is displaced from the ring plane by a rotation of 6° about the C–N bond. Bond lengths and angles are given in full; selected average values are:

C–C = 1.38±2 Å de = 115±1°
   b = 1.45±2 ef = 173±1
   c = 1.24±2 cc' = 121±1
   d = 1.42±2
   e = 1.27±2
   f = 1.13±2

Intermolecular distances are consistent with van der Waals interaction.

Details of analysis

Intensity data were recorded on equi-inclination Weissenberg photographs ($hk0$ to $hk2$), using CuK_α radiation at 100±5°K. 586 of a possible 722 reflexions were observed and measured with a microdensitometer. Refinement was by successive differential syntheses to a final R index of 0.087.

1. A. MUGNOLI and C. MARIANI, 1964. *Gazz. Chim. Ital.*, 94, 665.

FERRIC CHLORIDE AND o-METHOXYPHENYL DIAZONIUM
CHLORIDE (DOUBLE SALT)

$C_7H_7Cl_4FeN_2O$ F.W. = 332.8

I. [Structures of double diazonium salt crystals 1. Structure of the double
salt of ferric chloride and o-methoxyphenyl diazonium chloride.] T.N. POLYNOVA,
N.G. BOKIJ and M.A. PORAJ-KOŠIC, 1965. Ž. Strukt. Khim. SSSR, 6, 878-887 [J.
Struct. Chem., 6, 841-849]. (For crystal data of corresponding salt of antimony
pentachloride, see 1.)

Orthorhombic, $a = 11.16\pm2$, $b = 16.36\pm3$, $c = 7.18\pm2$ Å, [$U = 1311$ Å^3], $D_m = 1.71$,
$Z = 4$, $D_x = 1.69$.

Space group Pbn2$_1$ (C_{2v}^9) or Pbnm (D_{2h}^{16}) Pbnm from statistical analysis, and
confirmed by structure analysis. Molecular symmetry, mirror plane.

Atomic positions

	x	y	z
Fe	0.1521	0.2354	1/4
Cl(1)	0.0379	0.2319	0.4955
Cl(2)	0.2727	0.1318	1/4
Cl(3)	0.2468	0.3554	1/4
O	0.592	0.537	1/4
N(1)	0.749	0.662	1/4
N(2)	0.714	0.725	1/4
C(1)	0.793	0.577	1/4
C(2)	0.710	0.515	1/4
C(3)	0.751	0.434	1/4
C(4)	0.877	0.427	1/4
C(5)	0.960	0.492	1/4
C(6)	0.926	0.580	1/4
C(7)	0.500	0.474	1/4

Standard deviations are estimated to be: Fe, 0.002; Cl, 0.003; N and O,
0.01; C, 0.015 Å. However, these estimates are regarded as optimistic, and it
is suggested that they should be increased by a factor of three or four. Isotropic temperature factors are given also.

Structure

Bond lengths and angles are given, but are not regarded as accurate. The
analysis confirms the ionic formulation given above. The FeCl$_4$ anion is roughly
tetrahedral, with Fe-Cl distances ranging from 2.16 to 2.23 Å. The coordination
of FeCl$_4$ anions about the diazonium group is illustrated in Fig. 35. The N-Cl
distances suggest that the positive charge is concentrated on the terminal nitrogen
atom.

Details of analysis

Intensity data were recorded photographically (hk0 to hk3; 0kl to 2kl; h0l)
using MoK$_\alpha$ radiation, and estimated visually. Specimens were found to decompose
on exposure to X-rays, to the detriment of the accuracy of the data. Refinement
was by least squares to a final R index of 0.133.

1. This volume, p. 418.

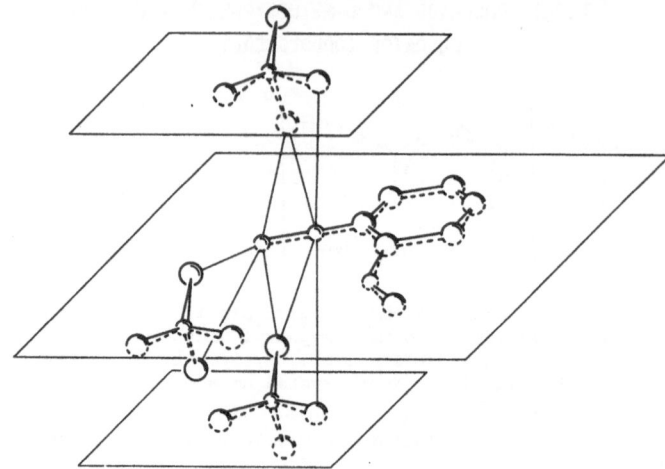

Fig. 35. Coordination of anions about the diazonium group.

FORMAMIDOXIME

CH$_4$N$_2$O   NH$_2$CHNOH   F.W. = 60.1

I. A refinement of the structure of formamidoxime. D. HALL, 1965. *Acta Cryst.*,
18, 955–958.

This is a refinement of a structure previously reported. For details not
given here see 1.

Atomic positions

	x	y	z
C	0.2930	0.2944	0.2432
N(1)	0.4243	0.4001	0.2223
N(2)	0.2554	0.1603	0.0828
O	0.3778	0.1360	-0.1221
H(1)	0.473	0.467	0.370
H(2)	0.506	0.380	0.086
H(3)	0.215	0.310	0.386
H(4)	0.341	0.60	-0.214

Standard deviations are given for the non-hydrogen atoms, and range from
0.002 to 0.004 Å. Anisotropic temperature factors for these atoms are given also.

Structure

The structure is essentially as described in 1. New bond lengths and angles
are:

C–N(1)	1.334 Å	N(1)–C–N(2)	126.7°
C–N(2)	1.288	C–N(2)–O	109.7°
O–N(2)	1.415		

Standard deviations of bond lengths are 0.005 Å.

The absence of a hydrogen atom bonded to N(2) confirms the predicted amidoxime
formulation. For a view of the structure, see 1, Fig. 1. The short intermolecular
separations indicated there are here confirmed as hydrogen bonds.

Details of analysis

Refinement of original data was by successive differential syntheses, to a
final *R* index of 0.079.

1.   *Structure Reports*, 20, 495.

TRITHIOCARBONIC ACID

CH_2S_3 $SC(SH)_2$ F.W. = 110.2

I. Das Kohlenstoffsulfid–bis–(hydrogensulfid) $SC(SH)_2$ und das System H_2S–CS_2. 6. Die Kristallstruktur der Trithio–Kohlensäure bei –100°. B. KREBS and G. GATTOW, 1965. Z. anorg. Chem., 340, 294–311.

Monoclinic, a = 13.12±4, b = 22.58±6, c = 5.88±2 Å, β = 90.0±3°, [U = 1742 Å^3], Z = 16, D_x = 1.68 (all at –100°C).

Space group $P2_1/a$ (C_{2h}^5)

Atomic positions

	x	y	z
C(1)	0.044	0.145	0.447
C(2)	0.219	0.986	0.935
C(3)	0.455	0.105	0.479
C(4)	0.282	0.262	0.918
S(1)	0.990	0.083	0.333
S(2)	0.004	0.169	0.718
S(3)	0.135	0.182	0.299
S(4)	0.170	0.048	0.816
S(5)	0.316	0.948	0.805
S(6)	0.171	0.960	0.195
S(7)	0.512	0.165	0.349
S(8)	0.501	0.081	0.746
S(9)	0.356	0.068	0.353
S(10)	0.195	0.298	0.752
S(11)	0.329	0.195	0.840
S(12)	0.325	0.296	0.171

Structure

Within the accuracy of the analysis, the molecules have 3m symmetry. The mean C–S distance is 1.72±3 Å. The molecules are linked by S–H..S bonds (of length 3.50 to 3.70 Å) to form folded chains extended in the c direction.

Details of analysis

The intensity data were recorded at –100°C on zero–level precession (MoK_α: $hk0$) and Weissenberg (CuK_α; $0kl$) photographs. The intensities were measured visually. Refinement was by Fourier methods to final R indices of 0.09 for both zones.

TRIS(METHYLSULPHONYL)METHANE

$C_4H_{10}O_6S_3$ $CH(SO_2CH_3)_3$ F.W. = 250.3

I. The disordered crystal and molecular structure of tris(methylsulfonyl)methane. J.V. SILVERTON, D.T. GIBSON and S.C. ABRAHAMS, 1965. *Acta Cryst.*, 19, 651–657. (For earlier work, see 1)

Trigonal, a = 12.89±2, c = 9.53±2 Å^3, U = 1370.3 Å^3, D_m = 1.83, Z = 6, D_x = 1.82.

Space group $R3c$ (C_{3v}^6) or $R\bar{3}c$ (D_{3d}^6) $R3c$ is consistent with analysis. Molecular symmetry, three–fold axis.

Atomic positions

	x	y	z
S	0.1525±4	0.1040±3	0.3291±6
O(1)	0.2217±14	0.0469±14	0.3072±14
O(2)	0.1415±12	0.1401±13	0.4690±12
C(1)	0.2061±16	0.2230±15	0.2132±14
C(2)	0	0	0.2694±32

The occupancy factor of the molecule defined by these coordinates is about 0.61. A second molecule, with occupancy 0.39, is defined by x, y, $1/2-z$. Isotropic temperature factors are given also.

Structure

The molecule is oriented with the C-H bond lying along the three-fold axis. The configuration of bonds at the central carbon atom is tetrahedral. Bond lengths and angles (averaged where appropriate) are:

S-CH	= 1.83±1 Å	S-CH-S	= 110.8±4°
S-CH$_3$	= 1.73±2	O-S-O	= 119±1
S-O	= 1.43±1	O-S-CH$_3$	= 109±1
		O-S-CH	= 107±1
		CH-S-CH$_3$	= 105±1

The structure consists of stacks of molecules, extended in the z direction. All molecules in a given stack point in the same direction; that is, their C-H vectors all have the same sense. The molecules in an adjacent stack may all point in this, or in the opposite direction. It is found that about 61% of all stacks point in one direction, and the remaining 39% in the other. It is shown that crystalline disorder of this type should give rise to the temperature-independent diffuse scattering characteristic of the material.

Details of analysis

The intensity data were recorded on Weissenberg photographs, using MoK_α radiation. About 263 of 337 accessible reflexions were observed and measured visually. Refinement was by least squares to a final R index of 0.091.

1.   *Structure Reports*, <u>20</u>, 521.

TRIPHENYLMETHYL PERCHLORATE

C$_{19}$H$_{15}$ClO$_4$ F.W. = 342.8

Cl O$_4^-$

I. The structure of triphenylmethyl perchlorate at 85°C.   A.H. GOMES DE MESQUITA, C.H. MacGILLAVRY and K. ERIKS, 1965. *Acta Cryst.*, <u>18</u>, 437-443.

Cubic, a = 18.91±2 Å, [U = 6762 Å^3], D_m = 1.37, Z = 16, D_x = 1.346, (cell and intensity data at 85°C).

Space group $F4_132$ (O^4)   Molecular symmetry, 32 for triphenylcarbonium ion.

Atomic positions

	x	y	z
C(c)	0.1250	0.1250	0.1250
C(1)	0.1791±7	0.1250	0.0709±7
C(2)	0.2482±6	0.1012±6	0.0838±6
C(3)	0.2997±6	0.0993±7	0.0335±6
C(4)	0.2816±19	0.1250	0.9684±19
ClO$_4$(a)	0.0000	0.0000	0.0000
Cl(b)	0.5000	0.5000	0.5000
O(b)	0.5419±9	0.5419±9	0.5419±9

The distance Cl-O in the disordered ion ClO$_4$(a) is assumed to be 1.45 Å. Also given are assumed hydrogen positions and anisotropic temperature factors.

Structure

The triphenylcarbonium ion is propeller–shaped, with three coplanar central bonds. Each aromatic ring is rotated out of the plane of the central bonds by $31.8\pm6°$. Bond lengths and angles (corrected for thermal libration) are given in full, and appear to be normal. One perchlorate ion is ordered. The Cl–O distance is given as 1.37 ± 3 Å, but this value is recognized to be unreliable. The other perchlorate ion is disordered, and no really satisfactory model was found. The packing of the ions is discussed in detail.

Details of analysis

Intensity data were recorded on equi–inclination Weissenberg photographs, using CuK_{α} radiation. 186 of a possible 420 reflexions were observed and estimated visually. Refinement was by block–diagonal least squares. Final R indices were 0.091 and 0.084, depending on the model used to describe the disorder.

TRI-p-METHOXYPHENYLMETHYL HYDROGEN DICHLORIDE TETRAHYDRATE
TRI-p-METHOXYPHENYLMETHYL HYDROGEN DIBROMIDE TETRAHYDRATE

$C_{22}H_{22}Cl_2O_3 \cdot 4H_2O$ F.W. = 477.4

$$C(p-C_6H_5OCH_3)_3HX_2 \cdot 4H_2O$$

$C_{22}H_{22}Br_2O_3 \cdot 4H_2O$ F.W. = 566.3

I. Structure investigation of tri-p-methoxyphenylmethyl carbonium ion. P. ANDERSEN and B. KLEWE, 1965. *Acta Chem. Scand.*, 19, 791–796.

Rhombohedral a = 13.24Å, α = 118.9°, U = 614.3 Å^3, D_m (dichloride) = 1.26 – 1.27, D_m (dibromide) = 1.50 – 1.52, Z = 1, D_x (dichloride) = 1.29, D_x (dibromide) = 1.53.
(Crystal data of the anhydrous compounds, and of the corresponding perchlorate are also given in this paper.)

Space group $R3$ (C_3^4) ($R\bar{3}$ is rejected because the carbonium ion has no centre of symmetry.) Molecular symmetry, three–fold axis.

Structure

The compounds are isotypic. The analysis indicates that the central bonds of the carbonium ion are coplanar, and that the phenyl groups are twisted by about 30° out of this plane. The hydrogen dihalide anion and the water molecules are disordered, and cannot be located.

Details of analysis

The intensity data were recorded photographically by various techniques, using CuK_{α} and MoK_{α} radiation. The information reported was obtained from a c–axis Fourier projection of the chlorine derivative. Disorder of the anions and water molecules prevented significant refinement.

p-AMINOBENZOIC ACID

$C_7H_7NO_2$ F.W. = 137.2

$$NH_2C_6H_4COOH$$

I. Twinning in p-aminobenzoic acid. R.C.G. KILLEAN, 1965. *Acta Cryst.*, 19, 482–483.

Three crystalline modifications are described, all twinned, and all having approximate *mmm* symmetry of the reciprocal lattice. The crystal class could not be determined for any of them. One modification had axial lengths a = 25.50, b = 27.16, c = 3.85 Å. The (001) projection (with plane group *Pgg*) was solved. From the proximity of symmetry–related carboxyl groups it was concluded that the molecules are associated as hydrogen–bonded dimeric pairs. (For an independent study of this compound, see 1.)

Details of analysis

The intensity data for the projection were recorded on Weissenberg photographs. Refinement was by least squares to a final R index of 0.10.

1. T.F. LAI and RICHARD E. MARSH, 1967. *Acta Cryst.*, **22**, 885.

SALICYLIC ACID

$C_7H_6O_3$ F.W. = 138.1

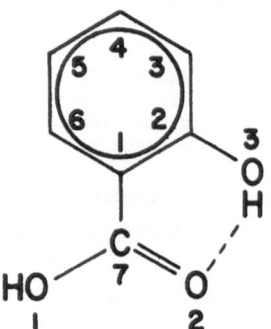

I. Refinement of the structure of salicylic acid. M. SUNDARALINGAM and L.H. JENSEN, 1965. *Acta Cryst.*, **18**, 1053-1058. [This is a redetermination. For earlier work, see 1.]

Monoclinic, $a = 11.52\pm12$, $b = 11.21\pm11$, $c = 4.920\pm5$ Å, $\beta = 90°\ 50\pm2'$, $U = 635$ Å³, $Z = 4$, $D_x = 1.44$.

Space group $P2_1/a$ (C_{2h}^5) (Cell and space group data from 1).

Atomic positions

	x	y	z
O(1)	-0.05050±22	0.13729±21	0.83797±46
O(2)	0.09881±20	0.01676±22	0.77027±43
O(3)	0.25092±23	0.07103±25	0.39921±52
C(1)	0.08361±27	0.19261±29	0.50601±59
C(2)	0.18470±29	0.16977±32	0.35871±62
C(3)	0.22036±33	0.25058±35	0.16466±70
C(4)	0.15721±36	0.35264±34	0.11374±75
C(5)	0.05729±40	0.37610±33	0.25742±75
C(6)	0.02141±33	0.29689±32	0.45012±69
C(7)	0.04504±28	0.10881±27	0.71260±60
H(1)	-0.0692 ±34	0.0781 ±34	0.9828 ±78
H(2)	0.2096 ±38	0.0213 ±39	0.5449 ±73
H(3)	0.2936 ±35	0.2380 ±36	0.0674 ±81
H(4)	0.1801 ±41	0.4123 ±40	-0.0338 ±85
H(5)	0.0085 ±37	0.4365 ±36	0.2139 ±77
H(6)	-0.0511 ±33	0.3094 ±35	0.5489 ±69

Temperature factors (anisotropic for the non-hydrogen atoms) are given also.

Structure

The molecule is essentially planar, although O(2) lies 0.028 Å from the ring plane. Bond lengths and angles are given in full; the C-C and C-O distances are:

C(1)—C(2)	= 1.404 Å		C(6)—C(1)	= 1.394 Å
C(2)—C(3)	= 1.381		C(1)—C(7)	= 1.457
C(3)—C(4)	= 1.379		C(7)—O(1)	= 1.307
C(4)—C(5)	= 1.384		C(7)—O(2)	= 1.234
C(5)—C(6)	= 1.365		C(2)—O(3)	= 1.358

[Standard deviations are not given explicitly, but are presumably about 0.005 Å.]

The hydrogen-bonding arrangement is as described in 1; the length of the intramolecular bond O(2)...O(3) is 2.620 Å, and that of the intra-dimer bond O(1)...O(2') is 2.653 Å.

Details of analysis

Intensity data were recorded on uni-dimensionally integrated equi-inclination Weissenberg photographs (*hk*0 to *hk*4). 732 of 1460 reflexions accessible within the CuK$_\alpha$ sphere were observed and measured photometrically. Refinement was by full-matrix least squares to a final *R* index of 0.059.

1. *Structure Reports*, 17, 702.

3,5-DIBROMO-*p*-AMINOBENZOIC ACID

C$_7$H$_5$NO$_2$Br$_2$ F.W. = 295.0

I. The crystal structure of 3,5-dibromo-*p*-aminobenzoic acid at room temperature (25°C approx.) and at -150°C. A.K. PANT, 1965. *Acta Cryst.*, 19, 440-448. (For a preliminary report, see 1).

Orthorhombic, *a* = 22.46±2, *b* = 19.47±2, *c* = 3.94±1 Å, [*U* = 1723 Å^3], *D$_m$* = 2.26, *Z* = 8, *D$_x$* = 2.27. (All at 25°C. Cell dimensions at -150°C are: *a* = 22.38±2, *b* = 19.46±2, *c* = 3.88±1 Å).

Space group P2an (*C$_{2v}^6$*) or *Pman* (*D$_{2h}^7$*) *Pman* suggested by statistics, and confirmed by analysis. Molecular symmetry, mirror plane for one, diad axis for the other of two independent molecules.

Atomic positions	*x*	*y*	*z*
Br	0.1280	0.2828	0.172
N	0	0.3380	0.023
C(1)	0.	0.2766	0.213
C(2)	0.0537	0.2430	0.294
C(3)	0.0548	0.1805	0.481
C(4)	0	0.1490	0.570
C(5)	0	0.0850	0.753
O	0.0495	0.0573	0.837
Br'	0.2152	0.1328	0.816
N'	0.1609	0	0.500
C(1')	0.2222	0	0.500
C(2')	0.2546	0.0560	0.649
C(3')	0.3173	0.0557	0.649
C(4')	0.3499	0	0.500
C(5')	0.4133	0	0.500
O	0.4428	0.0506	0.638

Standard deviations are not given, but are presumably rather high. Anisotropic temperature factors are given for the bromine atoms. The positions given here are for the structure at -150°C. Positions are given also for the structure at 25°C.

Structure

Bond lengths and angles for the structure at −150°C are given in full, but are rather inaccurate. Both molecules are planar within experimental error. A view of the structure is given in Fig. 36. Each molecule forms a dimer with a symmetry-related neighbour by hydrogen bonding between adjacent carboxyl groups. The symmetry elements involved (mirror plane or diad axis) imply symmetrical hydrogen bonding. However, disorder or a wrong space group assignment are possible explanations. Inter-dimer N–O distances of 2.84 and 3.03 Å are presumably N–H..O bonds.

Details of analysis

Intensity data were recorded on Weissenberg photographs of the three principal zones, both at 25°C and at −150°C, using CuK_{α} radiation. Intensities were measured photometrically. Refinement was by a variety of techniques. The R index for the low-temperature data was 0.086.

1.   *Structure Reports*, 27, 898.

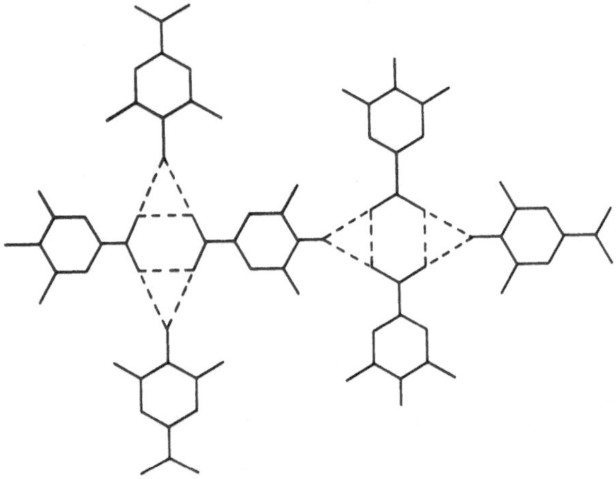

Fig. 36.   A view of the 3,5-dibromo-*p*-aminobenzoic acid structure, showing hydrogen bonding.

o-NITROPEROXYBENZOIC ACID

$C_2H_5NO_5$ F.W. = 183.1

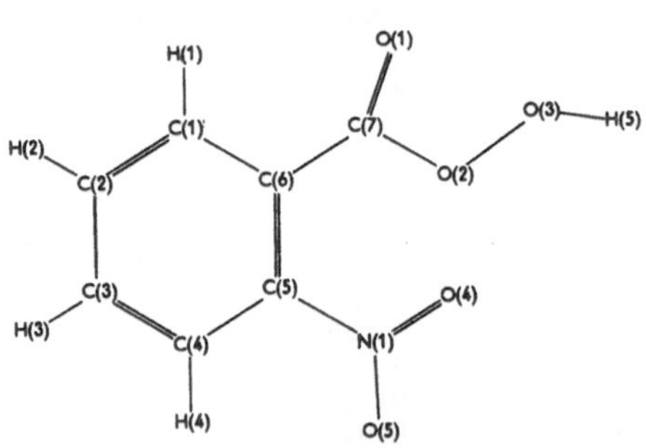

I. The crystal structure of *o*-nitroperoxybenzoic acid. M. SAX, P. BEURSKENS and S. CHU, 1965. *Acta Cryst.*, 18, 252–258.

Monoclinic, $a = 13.75\pm2$, $b = 7.95\pm1$, $c = 7.47\pm1$ Å, $\beta = 112°40\pm10'$, $U = 753.5$ Å^3, $Z = 4$, $D_X = 1.614$. (Cell and intensity data at -15°C. Corresponding values at 25°C are given also. They are: $a = 13.84\pm2$, $b = 8.03\pm1$, $c = 7.51\pm1$ Å, $\beta = 112°0\pm30'$, $U = 773.8$ Å^3, $D_m = 1.576$, $D_X = 1.572$.)

Space group $P2_1/c$ (C_{2h}^5)

Atomic positions

	x	y	z
C(1)	0.1784	−0.3018	0.8097
C(2)	0.2448	−0.4384	0.8218
C(3)	0.3455	−0.4132	0.8290
C(4)	0.3838	−0.2504	0.8256
C(5)	0.3172	−0.1152	0.8173
C(6)	0.2146	−0.1376	0.8056
C(7)	0.1382	0.0024	0.7822
O(1)	0.0906	0.0272	0.8870
O(2)	0.1232	0.0882	0.6197
O(3)	0.0476	0.2260	0.6000
O(4)	0.3258	0.1648	0.9070
O(5)	0.4364	0.0759	0.7845
N(1)	0.3630	0.0551	0.8345

The mean standard deviation for the atoms above is 0.005 Å. Also given are anisotropic temperature factors, and the positions of the hydrogen atoms. (Two sets of parameters are given, each refined by a different technique. Values given above are the arithmetic mean of those given in the paper.)

Structure

The bond lengths and angles are as follows:

C(1) − C(2)	1.399±7 Å		C(6)C(1)C(2)	119.7±4°
C(2) − C(3)	1.379±8		C(1)C(2)C(3)	120.7±5
C(3) − C(4)	1.401±7		C(2)C(3)C(4)	120.8±5
C(4) − C(5)	1.397±7		C(3)C(4)C(5)	117.8±5
C(5) − C(6)	0.390±7		C(4)C(5)C(6)	122.4±5
C(6) − C(1)	1.401±7		C(5)C(6)C(1)	118.6±4
C(6) − C(7)	1.495±7		C(1)C(6)C(7)	117.1±4
C(7) − O(1)	1.214±7		C(5)C(6)C(7)	124.2±4
C(7) − O(2)	1.337±6		C(6)C(7)O(1)	125.1±5
O(2) − O(3)	1.478±7		C(6)C(7)O(2)	109.9±4
O(5) − N(1)	1.478±7		O(1)C(7)O(2)	124.7±5
N(1) − O(4)	1.236±7		C(7)O(2)O(3)	108.9±4
N(1) − O(5)	1.215±7		C(5)N(1)O(4)	116.8±5
			C(5)N(1)O(5)	118.3±5
			C(6)C(5)N(1)	120.7±4
			C(4)C(5)N(1)	116.8±4
			O(4)N(1)O(5)	124.8±5

In the benzene ring the atoms other than C(5) are coplanar; C(5) lies 0.025 Å from their mean plane. The nitrogen atom lies 0.154 Å from the ring plane, and the O−N−O plane makes an angles of 28° with this plane. The peroxycarboxyl carbon atom lies 0.074 Å from the ring plane, on the opposite side to N(1); the angle between the O−C−O plane and the ring plane is 58°. A perspective view of the structure is given in Fig. 37. Glide-related molecules are hydrogen-bonded into an infinite chain.

Details of analysis

Intensity data were recorded on equi-inclination Weissenberg photographs taken about the three principal axes at -15°C, using CuK$_\alpha$ radiation. 1277 of a possible 1450 reflexions were observed and measured visually. The structure was refined both by differential syntheses and full-matrix least squares to a final R index of 0.12.

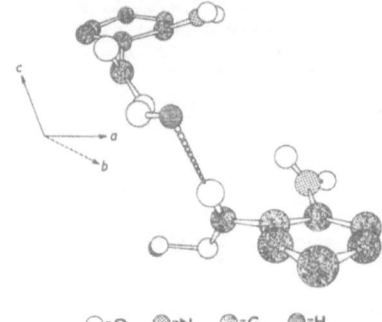

O=O.   ●=N.   ⦿=C.   ◉=H

Fig. 37.   Perspective view of glide-related molecules in a hydrogen-bonded chain
of *o*-nitroperoxybenzoic acid.

p-FLUOROBENZAMIDE

C_7H_6FNO F.W. = 139.1

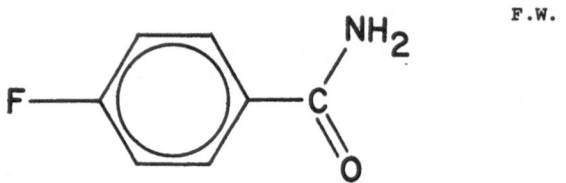

I. The crystal structure of *p*-fluorobenzamide, C_7H_6NOF. Y. TAKAKI, T. TANIGUCHI
and K. SAKURAI, 1965. *Mem. Osaka Gakugei Univ.*, B14, 48-55.

Monoclinic, a = 9.66±2, *b* = 7.51±1, *c* = 9.34±1 Å, β = 92.8±2°, [*U* = 677 Å³], *z* = 4,
[D_x = 1.36].

Space group $P2_1/a$ (C_{2h}^5)

Atomic positions

	x	*y*	*z*
C(1)	0.026	0.397	0.214
C(2)	0.132	0.410	0.318
C(3)	0.147	0.566	0.393
C(4)	0.059	0.709	0.364
C(5)	-0.046	0.696	0.260
C(6)	-0.062	0.540	0.185
C(7)	0.016	0.227	0.128
N	-0.110	0.181	0.069
O	0.121	0.138	0.106
F	0.079	0.862	0.441

The positions of the hydrogen atoms are given also, as are isotropic temperature factors for all atoms.

Structure

Bond lengths and angles are given in full, and are consistent with expectation.
The fluorine atom is coplanar with the benzene ring, but the amide group is rotated
out of the plane by about 29°. The hydrogen-bonding arrangement is illustrated in
Fig. 38. Adjacent molecules are joined by N-H..O bonds across a centre of symmetry,
and adjacent pairs related by the *a* glide plane are linked to form endless chains
extended along *a*.

Details of analysis

The intensity data for the principal zones were recorded on Weissenberg photographs, using CuK$_\alpha$ radiation. Refinement was by Fourier methods to final R indices as follows: (0*kl*), 0.18; (*h0l*), 0.14; (*hk0*), 0.12.

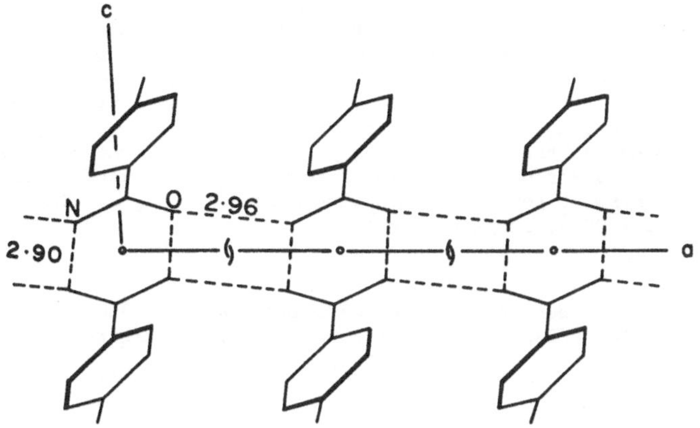

Fig. 38. A part of the *p*-fluorobenzamide structure, viewed along *b*.

m-FLUOROBENZAMIDE

C$_7$H$_6$FNO   F.W. = 139.1

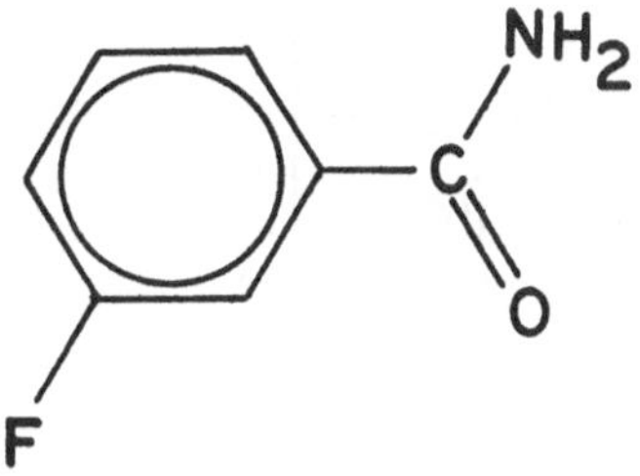

I. The crystal structure of *m*-fluorobenzamide, C$_7$H$_6$NOF. T. TANIGUCHI, Y. KATO, Y. TAKAKI and K. SAKATA, 1965. *Mem. Osaka Gakugei Univ.*, B14, 56–64.

Monoclinic, a = 23.78±3, b = 5.56±1, c = 5.04±1 Å, β = 100.3±5°, U = 656 Å^3, Z = 4, D_x = 1.41.

Space group P2$_1$/a (C_{2h}^5)

Atomic positions

	x	y	z
C(1)	0.105	−0.459	0.024
C(2)	0.150	−0.457	0.235
C(3)	0.193	−0.630	0.246
C(4)	0.189	−0.805	0.048
C(5)	0.144	−0.803	−0.163
C(6)	0.102	−0.633	−0.175
C(7)	0.058	−0.284	0.026
N	0.028	−0.226	−0.217
O	0.049	−0.187	0.232
F	0.239	−0.623	0.431
F'	0.151	−0.960	−0.352

ORGANIC COMPOUNDS

F and F' define fluorine positions with occupancy factors of 2/3 and 1/3 respectively. The positions of the hydrogen atoms are given also, as are isotropic temperature factors for all atoms.

Structure

Two molecular species are present, one related to the other by a 180° rotation of the fluorobenzene moiety about the amide linkage. The two species are present in the ratio 2:1, the predominant conformation being that indicated above. Bond lengths and angles are given in full, and are consistent with expectation. The amide group is rotated out of the ring plane by about 26°. The hydrogen-bonding arrangement resembles that found for the corresponding *para* isomer (1). Adjacent molecules are joined by N-H..O bonds across one centre of symmetry, and adjacent pairs are joined across a second centre to form endless chains extended along *a*. The lengths of the bonds are 2.93 and 2.88 Å, respectively.

Details of analysis

The intensity data were recorded on zero-level Weissenberg photographs (*hk*0, 129 of 169 accessible reflexions; *h*0*l*, 145 of 156) using CuK_α radiation. Refinement was by Fourier methods to a final *R* index of 0.16.

1.   This volume, p. 60.

POTASSIUM ACID PHTHALATE

$C_8H_5O_4K$ F.W. = 204.2

I. The crystal structure of potassium acid phthalate, $KC_6H_4COOH \cdot COO$. Y. OKAYA, 1965. <u>19</u>, 879–882.

Orthorhombic, *a* = 6.46 ± 6, *b* = 9.60 ± 9, *c* = 13.85 ± 7 Å, [*U* = 861 Å³], *z* = 4, [D_x = 1.57].

Space group $P2_1ab$ (C_{2v}^5) or $Pmab$ (D_{2h}^{11}) $P2_1ab$ is confirmed by analysis

Atomic positions

	x	*y*	*z*
K	0.25000	0.09898	0.03878
C(1)	0.00561	−0.17780	0.21761
O(1)	0.07871	−0.29831	0.26258
O(1')	0.01748	−0.16109	0.12732
C(I)	−0.23842	0.06887	0.15685
O(II)	−0.15861	0.14522	0.09314
O(II')	−0.40385	0.00006	0.14401
C(1)	−0.02519	−0.05921	0.28937
C(2)	0.06263	−0.06367	0.38466
C(3)	0.03824	0.04731	0.45050
C(4)	−0.07350	0.16280	0.42058
C(5)	−0.15805	0.16928	0.32513
C(6)	−0.13588	0.05799	0.25841
H(0)	0.038	−0.380	0.235
H(2)	0.140	−0.146	0.400
H(3)	0.096	0.048	0.521
H(4)	−0.091	0.237	0.464
H(5)	−0.231	0.252	0.304

Standard deviations are given in full for the non-hydrogen atoms. Mean values are: K, 0.0007 Å; C and O, 0.0023 to 0.0036 Å. Temperature factors (aniso-tropic for the non-hydrogen atoms) are given also.

Structure

Within experimental error the phenyl ring is a regular hexagon of side 1.388±6 Å. Some bond lengths and angles are:

a	=	1.498±7 Å	ab	=	113.2°
b	=	1.305±6	ac	=	122.4
c	=	1.210±7	bc	=	124.4
d	=	1.504±7	de	=	117.1
e	=	1.269±6	df	=	118.1
f	=	1.232±6	ef	=	124.8

These values identify *ef* as the ionized carboxyl group. Both groups are rotated out of the ring plane, *bc* by 31.7°, and *ef* by 75.4°. The rotations are in the same sense, and too close O...O contact is thereby avoided. A view of the structure is given in Fig. 39. The potassium ion is coordinated to six oxygen atoms, as shown, and adjacent phthalate ions are joined by a hydrogen bond of length 2.546 Å. It is to be noted that *e*, the bond adjacent to the hydrogen-bond acceptor, is significantly longer than *f*.

Details of analysis

The intensity data were measured with a four-circle diffractometer, using MoK$_\alpha$ radiation. 1478 of 1490 accessible reflexions were observed in the range sinθ/λ ⩽ 0.81. Refinement was by full-matrix least squares to a final R index of 0.044.

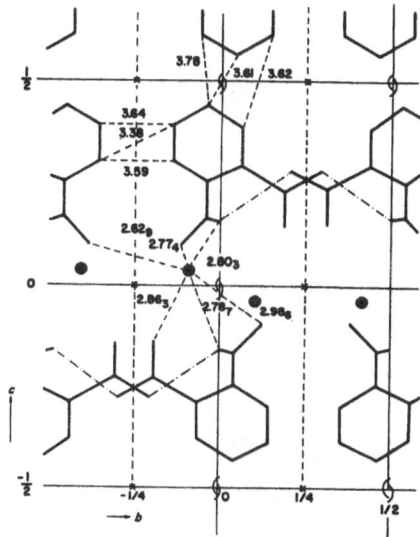

Fig. 39.   Schematic drawing of the potassium acid phthalate structure viewed along *a*. Short O-H...O hydrogen bonds are indicated by chain lines.

m-CHLORONITROBENZENE

C$_6$H$_4$ClO$_2$N   F.W. = 157.6

I.   X-ray structure analysis of *meta* chloronitrobenzene. E.M. GOPALAKRISHNA, 1965. *Z. Kristallogr.*, <u>121</u>, 378-384. (For a preliminary study, see <u>1</u>.)

Orthorhombic, $a = 6.00$, $b = 21.40$, $c = 5.35$ Å, $[U = 686.9$ Å$^3]$, $D_m = 1.555$, $z = 4$, $D_x = 1.523$.

Space group $Pbn2_1$ (C_{2v}^9) or $Pbnm$ (D_{2h}^{16}) $Pbn2_1$ is confirmed by the analysis.

Atomic positions

	x	y	z
Cl	0.070	0.237	0.000
C(1)	0.187	0.172	0.163
C(2)	0.076	0.146	0.366
C(3)	0.181	0.094	0.477
C(4)	0.398	0.072	0.414
C(5)	0.498	0.099	0.208
C(6)	0.401	0.151	0.099
N	0.064	0.064	0.688
O(1)	-0.120	0.084	0.761
O(2)	0.150	0.022	0.809

Structure

The structure is isotypic with the corresponding bromine derivative (2). The molecule appears to be planar. Bond lengths and angles are given in full, and are consistent with expectation.

Details of analysis

The intensity data were recorded on Weissenberg photographs, using CuK_α radiation. 110 of a possible 160 $hk0$ reflexions, and 77 of a possible 80 $0kl$ reflexions, were observed and measured visually. Refinement was by Fourier methods to final R indices of 0.16 and 0.13 for the two zones.

1.   *Structure Reports*, 27, 893.
2.   T.L. CHARLTON and J. TROTTER, 1963. *Acta Cryst.*, 16, 313.

ANILINE *p*-THIOCYANATE

$C_7H_6N_2S$ F.W. = 150.2

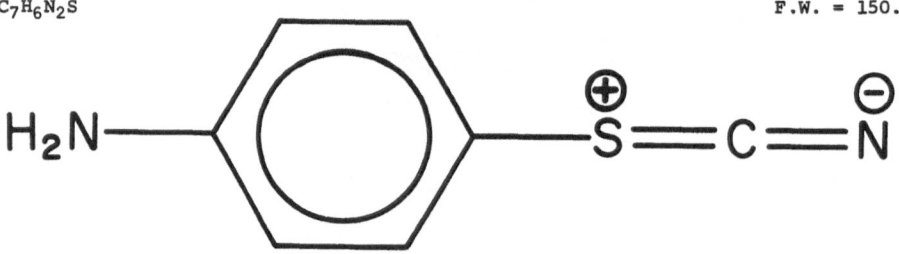

I. [The crystal structure of aniline *p*-thiocyanate.]   I.V. ISAKOV and Z.V. ZVONKOVA, 1965. *Kristallografija*, 10, 194–198 [*Soviet Physics - Crystallography*, 10, 144–147].   (This appears to be a preliminary report).

Monoclinic, $a = 12.35$, $b = 4.40$, $c = 18.94$ Å, $\beta = 135°8'$, $[U = 726$ Å$^3]$, $D_m = 1.37$, $z = 4$, $D_x = 1.37$.

Space group $P2_1/c$ (C_{2h}^5)

Atomic positions

	x	y	z
S	0.995	0.55	0.341
C(1)	0.130	0.76	0.434
C(2)	0.132	0.85	0.505

C(3)	0.247	0.03	0.587
C(4)	0.359	0.11	0.594
C(5)	0.365	0.03	0.528
C(6)	0.252	0.85	0.447
C(7)	0.850	0.34	0.267
N(1)	0.738	0.17	0.209
N(2)	0.472	0.29	0.674

Structure

It is reasonably concluded that the molecule has the formulation given above.

Details of analysis

Intensity data were recorded photographically (*h0l* and *0kl*). The structure was refined by Fourier projections along *b*, and the *y* coordinates were deduced from known bond lengths and stereochemical considerations. The proposed structure was confirmed by a Fourier projection along *a*. The final *R* index for *h0l* reflexions was 0.26.

METANILIC ACID

$C_6H_7NO_3S$ F.W. = 173.1

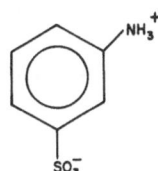

I. The crystal structure of metanilic acid.   S.R. HALL and E.N. MASLEN, 1965.
    Acta Cryst., <u>18</u>, 301-306.

Orthorhombic, a = 8.500±1, b = 11.944±1, c = 6.756±5 Å, U = 685.9 Å^3, D_m = 1.69, Z = 4, D_x = 1.677.

Space group $Pna2_1$ (C_{2v}^9) or $Pnam$ (D_{2h}^{16}) $Pnam$ is consistent with the refined structure. Molecular symmetry, mirror plane.

Atomic positions

	x	y	z
S	0.3114	0.6226	0.2500
O(1)	0.2400	0.7320	0.2500
O(2)	0.4000	0.5990	0.4270
N	-0.2880	0.5300	0.2500
C(1)	0.1460	0.5280	0.2500
C(2)	0.1790	0.4120	0.2500
C(3)	0.0560	0.3380	0.2500
C(4)	-0.0970	0.3760	0.2500
C(5)	-0.1270	0.4920	0.2500
C(6)	-0.0060	0.5670	0.2500

Temperature factors (some anisotropic) are given for each of the two projections from which the coordinates were derived. Assumed hydrogen positions are given also.

Structure

Bond lengths and angles are given in full, and are consistent with expectation. The structure is held together by strong N–H..O bonds, with lengths of 2.84 to 2.85 Å. The hydrogen-bonding arrangement is consistent with the zwitterionic configuration depected above.

Details of analysis

The intensity data for the [001] and [100] zones were recorded on Weissenberg photographs, using CuK$_\alpha$ radiation. 156 of 192 accessible reflexions were observed and measured visually. Refinement was by Fourier methods to final R indices of 0.059 for [001], and 0.065 for [100].

SULPHANILAMIDE

$C_6H_8N_2O_2S$ F.W. = 172.2

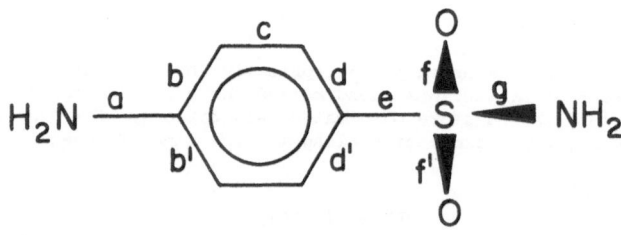

I. The crystal structure of α–sulphanilamide. B.H. O'CONNOR and E.N. MASLEN, 1965. *Acta Cryst.*, <u>18</u>, 363–366.

II. Affinement tridimensionnel du sulfanilamide β. M. ALLEAUME and J. DECAP, 1965. *Acta Cryst.*, <u>18</u>, 731–736.

III. Affinement tridimensionnel du sulfanilamide γ. *Idem*, 1965. *Ibid.*, <u>19</u>, 934–938.

I. *Orthorhombic*, a = 5.65±5, b = 18.509±5, c = 14.794±4 Å, U = 1560 Å³, D_m = 1.47, z = 8, D_x = 1.479.

Space group Pbca (D_{2h}^{15})

Atomic positions

	x	y	z
S	0.2484	0.0602	0.1054
O(1)	0.2518	0.0449	0.0117
O(2)	0.4580	0.0444	0.1603
N(1)	0.0505	0.3741	0.1327
N(2)	0.0355	0.0129	0.1470
C(1)	0.0972	0.3000	0.1250
C(2)	0.2974	0.2719	0.1657
C(3)	0.3400	0.1979	0.1631
C(4)	0.1819	0.1515	0.1178
C(5)	−0.0236	0.1801	0.0788
C(6)	−0.0737	0.2530	0.0838

Temperature factors, (some anisotropic) are given also.

II. *Monoclinic*, a = 9.000±5, b = 9.015±5, c = 10.05±1 Å, β = 111°30±10', [U = 758.7 Å³], D_m = 1.5, z = 4, D_x = 1.506.

Space group P2$_1$/c (C_{2h}^5)

Atomic positions

	x	y	z
S	0.0822	0.8531	0.2875
O(1)	0.0081	0.8602	0.1336
O(2)	0.1179	0.9906	0.3666
N(1)	0.9595	0.7601	0.3417
N(2)	0.6926	0.5318	0.4090
C(1)	0.2624	0.7570	0.3276
C(2)	0.2654	0.6273	0.2509
C(3)	0.4082	0.5528	0.2804
C(4)	0.5507	0.6058	0.3841
C(5)	0.5451	0.7351	0.4609
C(6)	0.4022	0.8092	0.4333

The positions of the hydrogen atoms are given also. Temperature factors, (ansiotropic for sulphur) are given for all atoms.

III. *Monoclinic*, $a = 7.95 \pm 1$, $b = 12.945 \pm 5$, $c = 7.79 \pm 1$ Å, $\beta = 106°30 \pm 10'$, $[U = 768.7$ Å$^3]$, $D_m = 1.5$, $Z = 4$, $D_x = 1.486$.

Space group $P2_1/c$ (C_{2h}^5)

Atomic positions

	x	y	z
S	0.3916	0.1619	0.9627
O(1)	0.3968	0.2732	0.9876
O(2)	0.4354	0.0983	0.1207
N(1)	0.5340	0.1342	0.8483
N(2)	0.6856	0.0476	0.5110
C(1)	0.1824	0.1296	0.8311
C(2)	0.0762	0.2019	0.7261
C(3)	0.9135	0.1754	0.6166
C(4)	0.8494	0.0765	0.6182
C(5)	0.9545	0.0034	0.7259
C(6)	0.1183	0.0304	0.8367

The positions of the ring hydrogen atoms are given also. Anisotropic temperature factors are given for the non—hydrogen atoms.

Structure

I, II, and III report molecular dimensions which are in good agreement. Mean values (assuming the chemical equivalence implied above) are:

a = 1.39 Å		bb' = 121°	
b = 1.40		bc = 119	
c = 1.38		cd = 120	
d = 1.40		dd' = 120	
e = 1.74		ef = 108	
f = 1.44		ff' = 118	
g = 1.65		eg = 109	
		fg = 107	

(distances ±0.01 to 0.02 Å; angles ±1 to 2°.)

It appears that the sulphonamide group can, by rotation about C–S, adopt the conformation best suited to specific packing requirements. Each crystalline form is characterized by an extensive system of N–H..O bonds, for details of which the original papers should be consulted.

Details of analysis

I. Intensity data for the zone 0kl were measured with a diffractometer and Geiger counter. Further data (1kl and 4kl) were recorded on equi-inclination Weissenberg photographs, and measured visually. CuK_α radiation was used throughout. 481 of a possible 660 reflexions were observed. Refinement was by block–diagonal least squares to a final R index of 0.158.

II. Intensity data were recorded on De Jong retigrams, using CuK_α radiation, and measured visually. 913 of a possible 1031 reflexions were observed. Refinement was by least squares to a final R index of 0.10.

III. Intensity data were recorded on De Jong retigrams, using CuK_α radiation, and measured visually. 703 of a possible 876 reflexions were observed. Refinement was by least squares to a final R index of 0.092.

2-CHLORO-4-NITROANILINE

$C_6H_5ClN_2O_2$

F.W. = 172.6

M.P. = 105–106°C.

I. X-ray studies of molecular overcrowding. Part V. The crystal and molecular structure of 2-chloro-4-nitroaniline. A.T. McPHAIL and G.A. SIM, 1965. J. Chem. Soc., 227–236.

Orthorhombic, a = 11.25, b = 16.85, c = 3.87 Å, U = 734.0 Å³, D_m = 1.545, z = 4, D_x = 1.562.

Space group Pna2₁ (C_{2v}^9) or Pnam (D_{2h}^{16}) Pna2₁ confirmed by analysis.

Atomic positions

	x	y	z		x	y	z
C(1)	0.1849	0.4534	0.1159	N(1)	0.1101	0.5147	0.0218
C(2)	0.2947	0.4641	0.2762	N(2)	0.4019	0.2573	0.4090
C(3)	0.3676	0.4014	0.3793	O(1)	0.4937	0.2714	0.5727
C(4)	0.3295	0.3247	0.2942	O(2)	0.3707	0.1895	0.3287
C(5)	0.2217	0.3095	0.1339	Cl	0.3396	0.5614	0.3790
C(6)	0.1513	0.3740	0.0414				

Standard deviations are given in Å; mean values are: Cl, $\sigma(x) = \sigma(y) = 0.002$ Å, $\sigma(z) = 0.007$ Å; other atoms, $\sigma(x) = \sigma(y) = 0.009$ Å, $\sigma(z) = 0.015$ Å. Anisotropic temperature factors are given also. These have been analyzed and the bond lengths corrected appropriately.

Structure

Except for the NO_2 group, the molecule is approximately planar. This group is rotated by 3.4° about the bond $C–NO_2$, which is itself bent about 3° out of the ring plane. Bond lengths and angles are given in full. The structure is held together by a system of N–H··O bonds of length 3.05 Å.

Details of analysis

Partial three-dimensional intensity data were recorded on equi-inclination Weissenberg photographs (hk0 to hk2) using CuK_α radiation. 614 reflexions were observed and measured visually. Refinement was by least squares to a final R index of 0.115, the scale factors for individual layers being adjusted by comparison of F_O and F_C. The thermal parameters were analyzed in terms of rigid-body motion, and appropriate corrections were applied to the bond lengths. [In view of the restriction on the number of data collected, and of the method used for interlayer scaling, the thermal parameters must be regarded as unreliable, and their analysis as questionable.]

N,N-DIMETHYL-p-NITROANILINE

$C_8H_{10}N_2O_2$

F.W. = 166.2

I. The crystal and molecular structure of N,N-dimethyl-p-nitroaniline. T.C.W. MAK and J. TROTTER, 1965. Acta Cryst., 18, 68–74.

Monoclinic, $a = 9.73 \pm 1$, $b = 10.56 \pm 1$, $c = 3.964 \pm 5$ Å, $\beta = 91°28 \pm 5'$, $U = 407.2$ Å^3, $D_m = 1.35$, $Z = 2$, $D_x = 1.355$, (CuK$_\alpha$, $\lambda = 1.5418$ Å, MoK$_\alpha$, $\lambda = 0.7107$ Å).

Space group $P2_1$ (C_2^2) or $P2_1/m$ (C_{2h}^2) $P2_1$ is confirmed by the analysis.

Atomic positions

	x	y	z
C(1)	0.3159	0.2490	0.1321
C(2)	0.3744	0.1463	0.3106
C(3)	0.5095	0.1466	0.4127
C(4)	0.5944	0.2473	0.3351
C(5)	0.5429	0.3535	0.1580
C(6)	0.4056	0.3558	0.0625
C(7)	0.0944	0.1354	0.0886
C(8)	0.1153	0.3644	-0.1057
N(9)	0.1815	0.2527	0.0369
N(10)	0.7324	0.2517	0.4429
O(11)	0.7827	0.1562	0.5924
O(12)	0.8014	0.3431	0.3853

Anisotropic temperature factors are given also, as are the positions of the hydrogen atoms.

Structure

Bond lengths and angles (the former corrected for the rather intense thermal motion) are given in full, and are consistent with expectation. The aromatic ring is planar, but the substituent groups are twisted slightly out of the ring plane, the methylamine group by 7°, and the nitro group by 3°. The intermolecular distances are consistent with van der Waals interaction.

Details of analysis

Intensity data were collected with a four-circle diffractometer and scintillation counter. CuK$_\alpha$ radiation was used, and the range $2\theta < 102°$ was scanned. 383 reflexions were observed, constituting 84% of those in the range examined. Refinement was by block-diagonal least squares to a final R index of 0.116.

1,3,5-TRIAMINO-2,4,6-TRINITROBENZENE

$C_6H_6N_6O_6$ F.W. = 258.2

I. The crystal structure of 1,3,5-triamino-2,4,6-trinitrobenzene. H.H. CADY and A.C. LARSON, 1965. *Acta Cryst.*, **18**, 485–496.

Triclinic, $a = 9.010 \pm 3$, $b = 9.028 \pm 3$, $c = 6.812 \pm 3$ Å, $\alpha = 108.59 \pm 2$, $\beta = 91.82 \pm 3$, $\gamma = 119.97 \pm 1°$, [$U = 442.5$ Å^3], $D_m = 1.93$, $Z = 2$, $D_x = 1.937$, (MoK$_\alpha$, $\lambda = 0.70926$ Å).

Space group $P1$ (C_1^1) or $P\bar{1}$ (C_i^1) $P\bar{1}$ is confirmed by the analysis.

Atomic positions

	x	y	z
C(1)	0.5332±7	0.1651±8	0.2568±10
C(2)	0.3733±7	0.0026±7	0.2487±9
C(3)	0.2150±7	0.0073±7	0.2487±11
C(4)	0.2144±7	0.1667±7	0.2511±10
C(5)	0.3760±7	0.3218+8	0.3443±11

	x	y	z
C(6)	0.5379±7	0.3268±7	0.2520±10
N(1)	0.6921±6	0.1683±6	0.2695±9
N(2)	0.3693±7	−0.1466±7	0.2389±11
N(3)	0.0565±6	−0.1493±6	0.2467±9
N(4)	0.0710±7	0.1722±7	0.2578±11
N(5)	0.3747±6	0.4763±6	0.2350±8
N(6)	0.6842±6	0.4709±7	0.2517±10
O(1)	0.8335±5	0.3069±6	0.2820±10
O(2)	0.6927±6	0.0297±6	0.2641±9
O(3)	0.0512±6	−0.2919±6	0.2377±10
O(4)	−0.0811±5	−0.1499±6	0.2525±9
O(5)	0.2380±6	0.4789±6	0.2304±9
O(6)	0.5131±5	0.6175±6	0.2387±8
H(1)	0.467 ±9	−0.143 ±9	0.238 ±1
H(2)	0.280 ±9	−0.232 ±9	0.260 ±11
H(3)	−0.017 ±10	0.091 ±10	0.254 ±12
H(4)	0.071 ±11	0.288 ±12	0.254 ±14
H(5)	0.682 ±10	0.564 ±11	0.244 ±12
H(6)	0.779 ±12	0.454 ±12	0.224 ±14

Temperature factors (anisotropic for the non-hydrogen atoms) are given also.

Structure

The molecule is very neárly planar, and has reasonably exact 3m symmetry. The greatest distance of any non-hydrogen atom from the ring plane is 0.15 Å. The structure consists of sheets of coplanar molecules lying parallel to the (110) plane. This arrangement is stabilized by a system of bifurcated hydrogen bonds; each amino group participates in two such N–H..O bonds, each of which has an intra- and an inter-molecular branch. The bond distances and the hydrogen bonding scheme are indicated in Fig. 40. (Bond angles also are given in full.) The bond lengths shown have not been corrected for thermal motion; it is suggested that the N–O distances should be increased by about 0.023 Å. The C–C and C–N distances are believed to indicate the predominance of unusual resonance forms of the molecule. Non-bonded intermolecular distances have normal van der Waals values.

Details of analysis

Intensity data were collected with a four-circle diffractometer and scintilla-tion counter. MoK_α radiation was used, and the sphere 2θ<60° was scanned. 928 of a possible 2591 reflexions were observed above background. Refinement was by full-matrix least squares, with unit weights, and non-spherical atomic scattering factors to a final R index of 0.056. A parallel refinement using spherical scattering factors is described also.

o-AMINOPHENOL HYDROCHLORIDE

C_6H_8NOCl $C_6H_4OHNH_3^+ .Cl^-$ F.W. = 145.6

I. Hydrogen bonding in o-aminophenol hydrochloride. A.F. CESUR and J.P.G. RICHARDS, 1965. Z. *Kristallogr.*, 122, 283–297.

Monoclinic, a = 10.280±6, b = 4.938±2, c = 28.010±6 Å, β = 92.12±8°, [U = 1421 Å^3], D_m = 1.37, Z = 8, D_x = 1.36.

Space group P2$_1$/c   (C_{2h}^5)

Atomic positions

	x	y	z
Cl(1)	0.1579±2	0.0184±4	−0.0882±1
Cl(2)	0.1271±2	0.2112±4	0.1935±1
O(1)	0.1778±5	0.3800±12	−0.1739±2
O(2)	0.0945±5	0.2374±11	0.0109±2
N(1)	0.1600±6	0.7840±13	−0.2371±2
N(2)	0.0374±6	0.4803±13	0.0936±2

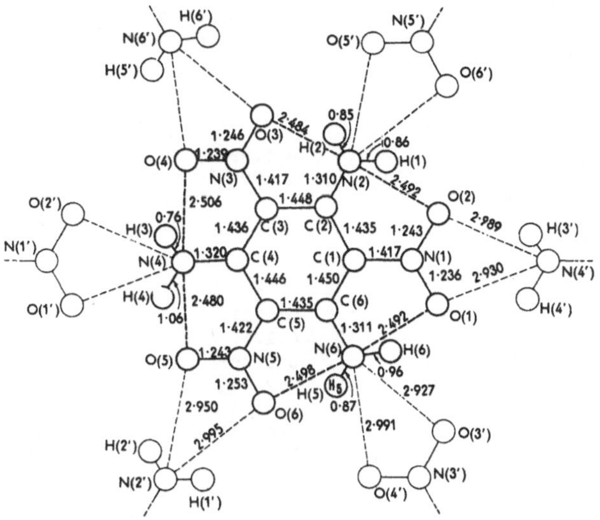

Fig. 40.   Bond distances and hydrogen bonding in 1,3,5-**triamino-2,4,6-trinitroben-**
zene. Standard deviations are 0.007 Å for bonds not involving hydrogen
atoms, 0.08 Å for those that do.

	x	y	z
C(1)	0.2830±6	0.7435±15	−0.2092±2
C(2)	0.3861±7	0.9062±18	−0.2164±3
C(3)	0.5017±8	0.8633±20	−0.1902±3
C(4)	0.5065±9	0.6576±20	−0.1564±3
C(5)	0.4004±8	0.4874±17	−0.1484±3
C(6)	0.2867±7	0.5353±15	−0.1763±2
C(7)	0.1558±6	0.5625±15	0.0693±2
C(8)	0.1807±6	0.4328±15	0.0270±2
C(9)	0.2895±8	0.5094±16	0.0019±3
C(10)	0.3713±9	0.7130±19	0.0210±3
C(11)	0.3435±9	0.8378±20	0.0632±3
C(12)	0.2375±8	0.7639±18	0.0882±3

Temperature factors (anisotropic for chlorine) are given for the above atoms.
The positions of the hydrogen atoms are given also.

Structure

The analysis shows that the amino group is protonated, as indicated above.
Bond lengths and angles are given in full, and are consistent with expectation.
The structure is held together by N–H..Cl and O–H..Cl bonds, each chlorine atom
forming four such bonds with different o–aminophenol ions. For further details
of this complex bonding system the original should be consulted.

Details of analysis

The intensity data were recorded on equi-inclination Weissenberg photographs
($h0l$ to $h4l$) using CuK$_\alpha$ radiation. 2070 of 2830 accessible reflexions were observed. Refinement was by least squares to a final R index of 0.11. The phenolic
and amino hydrogen atoms were observed in a difference Fourier synthesis.

p-NITROPHENOL

$C_6H_5NO_3$ F.W. = 139.1

I. The crystal structure of the α modification of *p*-nitrophenol near 90°K.
P. COPPENS and G.M.J. SCHMIDT, 1965. *Acta Cryst.*, 18, 62–67.

II. The crystal structure of the metastable (β) modification of *p*-nitrophenol.
Idem, 1965. *Ibid.*, 18, 654–663. (For earlier work on the β modification see
1).

I. *Monoclinic, a* = 11.66, *b* = 8.78, *c* = 6.098Å, (all ±0.1%), β = 107°32±6',
$[U = 595.3 \text{ Å}^3]$, $Z = 4$, $D_x = 1.551$, (all at 90°K).

Space group $P2_1/n$ (C_{2h}^5)

Atomic positions

	x	y	z
O(1)	−0.0586±3	0.2931±4	0.8930±6
O(2)	−0.0864±3	0.1267±4	0.6215±6
O(3)	0.3188±3	0.5238±4	0.4370±6
N	−0.0350±3	0.2409±5	0.7240±7
C(1)	0.0554±3	0.3144±5	0.6431±8
C(2)	0.1138±3	0.4413±5	0.7612±7
C(3)	0.2013±4	0.5104±5	0.6860±8
C(4)	0.2298±4	0.4526±5	0.4971±8
C(5)	0.1689±3	0.3267±5	0.3777±7
C(6)	0.0800±3	0.2580±5	0.4503±9

Also given are anisotropic temperature factors for the above atoms, and the
positions and isotropic temperature factors of the hydrogen atoms.

Structure

Bond lengths and angles are given in full, and do not differ significantly
from the more accurate values reported for II. The benzene ring is planar, but
the nitrogen and oxygen atoms are displaced from the benzene plane by 0.03 to
0.07 Å. As illustrated in Fig. 41, the structure consists of endless chains of
glide-related molecules linked by O–H··O bonds of length 2.82 Å.

Details of analysis

Partial three-dimensional intensity data were recorded on Weissenberg photographs (hk0, hk3, h0l), using CuK$_\alpha$ radiation. 358 of a possible 393 reflexions
were observed and measured visually. Refinement was by diagonal least squares to
a final *R* index of 0.083.

II. *Monoclinic, a* = 15.403±3, *b* = 11.117±2, *c* = 3.785±1 Å, β = 107°4±2',
$[U = 619.6 \text{ Å}^3]$, $Z = 4$, $D_x = 1.491$. (Cell constants at 90°K are given also. They
are: *a* = 15.21±2, *b* = 11.04±2, *c* = 3.622±6 Å, β = 106°49±10'.)

Space group $P2_1/a$ (C_{2h}^5)

Atomic positions

	x	y	z		x	y	z
O(1)	−0.0469	0.3229	0.3131	H(O)	0.377	0.414	0.205
O(2)	−0.0117	0.1575	0.0983	H(2)	0.063	0.478	0.340
O(3)	0.3389	0.4605	0.2571	H(3)	0.206	0.563	0.376
N	0.0063	0.2609	0.2067	H(5)	0.286	0.237	0.109
C(1)	0.0933	0.3109	0.2114	H(6)	0.141	0.159	0.076
C(2)	0.1096	0.4314	0.3008				
C(3)	0.1926	0.4792	0.3132				
C(4)	0.2590	0.4069	0.2393				
C(5)	0.2417	0.2863	0.1508				
C(6)	0.1582	0.2378	0.1331				

The standard deviations range from 0.015 to 0.030 Å for the hydrogen atoms,
and from 0.0015 to 0.0020 Å for the others. Temperature factors (isotropic for
the hydrogen, anisotropic for the other atoms) are given also. The thermal motion
has been analyzed in terms of rigid-body modes, and a set of corrected positions
is given for the non-hydrogen atoms.

Structure

The bond lengths and angles are given in Fig. 42. The benzene ring is planar, but the nitrogen and oxygen atoms are displaced from the benzene plane by as much as 0.18 Å. Like the α-modification, and as illustrated in Fig. 43, the β-structure consists of infinite chains of glide-related molecules, linked by O–H..O bonds of length 2.84 Å. Intermolecular contacts of the type CH...ON are discussed in relation to the photochemical reactivities of the α and β structures. It is pointed out that for the α-structure, O...C vectors are inclined to the benzene ring by as much as 73°, while for the β-structure, the corresponding inclination is only 18.5°. In the latter case, a hydrogen atom is thus interposed between the carbon and oxygen atoms, and photochemically-induced interaction is inhibited. This is not the case for the α-modification, which is photochemically unstable. The two types of C...O contact are illustrated in Fig. 44.

Details of analysis

Three-dimensional intensity data were measured with a four-circle diffratometer and proportional counter, using CuK$_\alpha$ radiation. Refinement was by diagonal least squares to a final R index of 0.069.

1.   *Structure Reports*, 18, 701.

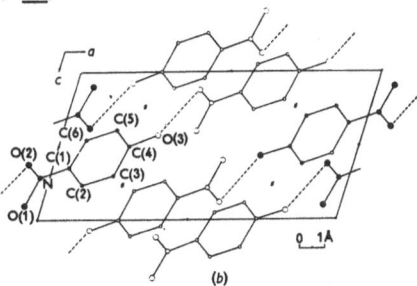

(b)

Fig. 41.   The α-structure of *p*-nitrophenol viewed along *b*.

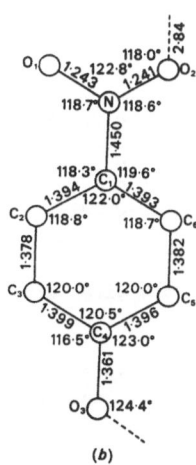

(b)

Fig. 42.   Bond lengths and angles for the β-modification of *p*-nitrophenol, after correction for thermal motion. The standard deviations of distances are 0.003 to 0.005 Å.

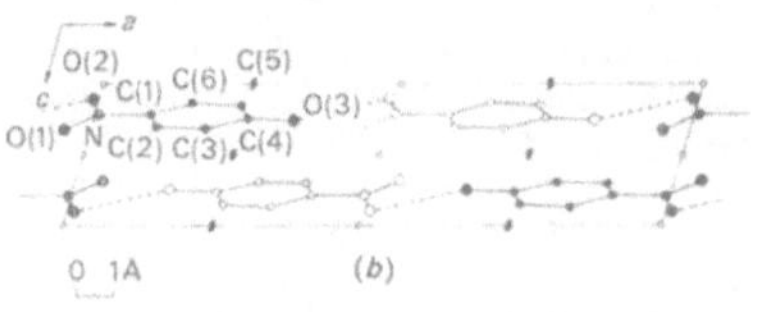

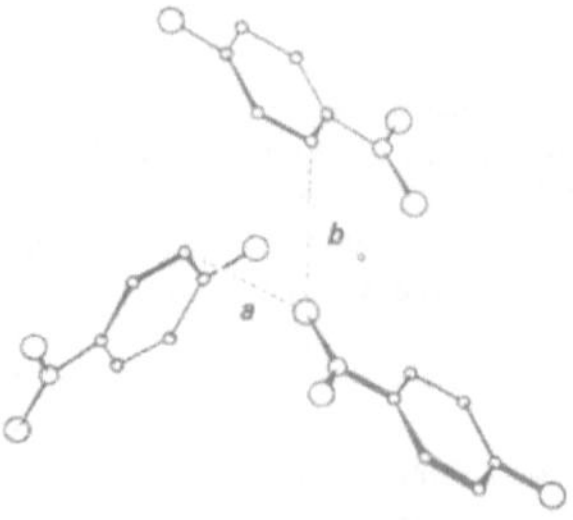

Fig. 43 (left). The β-structure of *p*-nitrophenol viewed along *b*.
Fig. 44 (right). A perspective view of the packing of three molecules in α-*p*-
                  nitrophenol, to illustrate two types of C-H...O-N contacts.
                  a) nearly linear contact common to the α and β forms; b) near-
                  perpendicular contact characteristic of the α form only and
                  regarded as responsible for the redox reaction.

2,5-DIMETHYL PHENOL

$C_8H_{10}O$ F.W. = 122.2

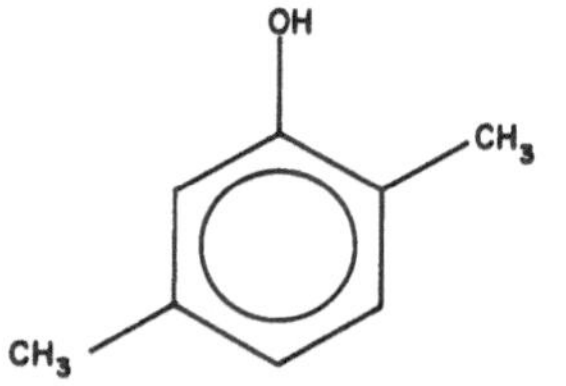

I. Structure cristalline du dimethl-2,5 phenol. H. GILLIER-PANDRAUD, 1965.
   Bull. Soc. Chim. Fr., 3267-3270.

Monoclinic, a = 5.94±2, b = 4.91±2, c = 12.48±3 Å, β = 109.7±5°, U = 342 Å^3,
D_m = 1.13, z = 2, D_x = 1.18.

Space group $P2_1$ (C_2^2) or $P2_1/m$ (C_{2h}^2) $P2_1$ is confirmed by analysis

Atomic positions

	x	y	z
C(1)	0.469±2	0.093±4	0.154±1
C(2)	0.303±2	−0.030±5	0.198±1
C(3)	0.313±3	0.060±5	0.305±1
C(4)	0.481±3	0.247±5	0.364±1
C(5)	0.652±2	0.356±4	0.322±1
C(6)	0.640±2	0.271±5	0.214±1
C(7)	0.113±3	−0.227±5	0.130±1
C(8)	0.838±3	0.562±5	0.387±1
O	0.454±2	0	0.0429±8

Isotropic temperature factors are given also.

Structure

    Bond lengths and angles are given in full, and are consistent with expecta-
tion. The molecule is planar. The structure consists of chains of molecules
linked by hydrogen bonds as shown in Fig. 45, and extended along *b*.

Details of analysis

    Three-dimensional intensity data were recorded on Weissenberg photographs,
using CuK$_\alpha$ radiation, and measured with a microdensitometer. Refinement was by full-
matrix least squares.

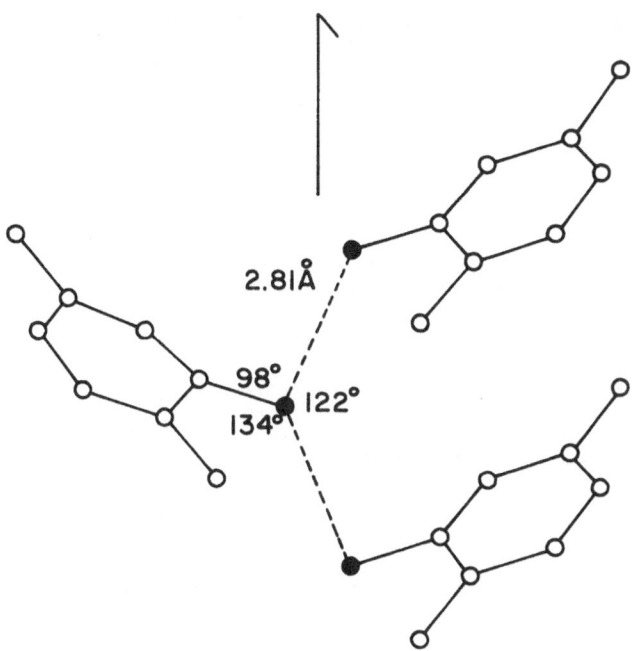

Fig. 45.   Hydrogen bonding in 2,5-dimethylphenol.

QUINHYDRONE

$C_{12}H_{10}O_4$

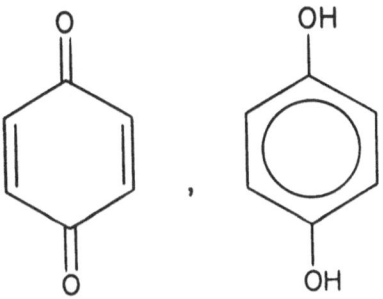

F.W. = 218.2

I. The crystal structure of the triclinic modification of quinhydrone. T. SAKURAI, 1965. *Acta Cryst.*, <u>19</u>, 320-330. (For a report on the monoclinic modification of quinhydrone see <u>1.</u>)

Triclinic, a = 7.652±22, b = 5.95±13, c = 6.770±20 Å, α = 107°37±7, β = 121°56±3, γ = 90°17±19', U = 244.3 Å³, D_m = 1.45, Z = 1, D_x = 1.48.

Space group P1 (C_1^1) or   PĪ (C_i^1) PĪ confirmed by analysis. Molecular symmetry, centre.

Atomic positions

	x	y	z		x	y	z
O(1)	0.1302	0.2796	−0.1756	O(2)	0.6115	0.2834	−0.1665
C(1)	0.0659	0.1419	−0.0846	C(4)	0.5610	0.1490	−0.0897
C(2)	0.1239	0.2278	0.1623	C(5)	0.6279	0.2327	0.1751
C(3)	0.0560	0.0818	0.2441	C(6)	0.5666	0.0897	0.2551
H(1)	0.214	0.419	−0.040	H(4)	0.718	0.410	0.278
H(2)	0.213	0.405	0.270	H(5)	0.588	0.144	0.412
H(3)	0.080	0.138	0.412				

Mean standard deviations are: O, 0.004 Å; C, 0.005 Å; H, 0.05 Å. Anisotropic temperature factors are given for carbon and oxygen, and are analyzed in terms of rigid body vibration.

Structure

Bond lengths and angles of the two molecules are given in Fig. 46. Both molecules are planar, but the deviations from strict *mmm* symmetry are considered to be real. The structure consists of endless chains of alternating molecules, linked as in Fig. 47, extended in the direction [120]. Adjacent chains overlap (with an interplanar spacing of about 3.2 Å) in such a way that the C=O double bond of the quinone lies over the ring of the hydroquinone molecule. Integration of the charge density over both molecules suggests that as much as 0.21 electron units of charge may be transferred from the hydroquinone to the quinone molecule.

Details of analysis

Intensity data were recorded on equi-inclination Weissenberg photographs (0*kl* to 6*kl*; layers around [120] up to the tenth) using CuK_α radiation. Of 1050 possible reflexions 848 were observed and measured visually. Refinement was by full-matrix least squares to a final *R* index of 0.101.

1.   *Structure Reports*, <u>22</u>, 694.

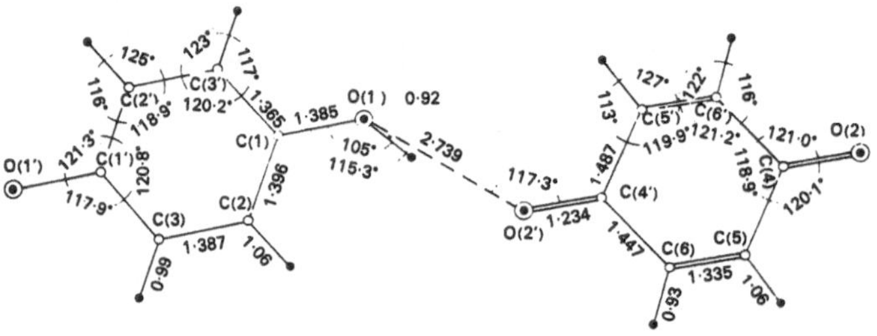

Fig. 46.   Bond lengths and angles in quinhydrone. Standard deviations: 0.008 Å and 0.5° where hydrogen is not involved; otherwise 0.06 Å and 3°.

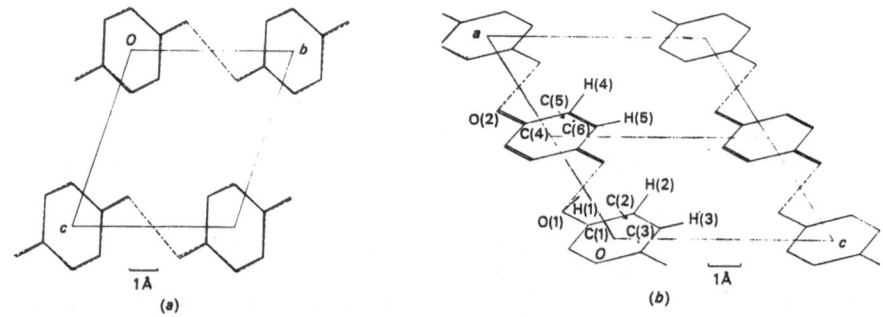

Fig. 47.   Configuration of molecules in the quinhydrone crystal. a) projection along *a*. b) projection along *b*.

PHLOROGLUCINOL
(1,3,5-TRIHYDROXYBENZENE)

$C_6H_6O_3$ F.W. = 126.1

I. The crystal and molecular structure of phloroglucinol. K. MAARTMANN-MOE,
1965. *Acta Cryst.*, <u>19</u>, 155-157.

Orthorhombic, a = 4.83±1, *b* = 9.37±2, *c* = 12.56±3 Å, [*U* = 568.4 Å^3], D_m = 1.46,
z = 4, D_x = 1.47.

Space group $P2_12_12_1$ (D_2^4)

Atomic positions

	x	*y*	*z*
C(1)	0.4082	0.4929	0.4815
C(2)	0.5106	0.3693	0.4379
C(3)	0.6993	0.3786	0.3556
C(4)	0.7758	0.5097	0.3133
C(5)	0.6660	0.6318	0.3578
C(6)	0.4792	0.6255	0.4421
O(7)	0.2251	0.4810	0.5659
O(8)	0.8150	0.2582	0.3121
O(9)	0.7424	0.7647	0.3230

   Hydrogen positions and temperature factors (anisotropic for carbon and oxygen,
isotropic for hydrogen) are given. Average standard deviations are: C, 0.0053;
O, 0038 Å.

Structure

   The benzene ring is planar, and the oxygen atoms are displaced from the ring
plane by 0.024, 0.039, and 0.045 Å, all in the same direction. The mean bond
lengths are: C—C, 1.38 Å, and C—O, 1.37 Å. The mean C-C-C angle is 122° where
there is a substituent, and 118° otherwise. The intermolecular O-H--O bonds,
indicated in Fig. 48, range in length from 2.73 to 2.76 Å.

Details of analysis

   The intensity data were recorded on integrated Weissenberg photographs (0*kl*
to 2*kl*; *h0l* to *h4l*) using CuK_α radiation. 484 reflexions were observed and
estimated visually. Refinement was by least squares to a final *R* index of 0.075.

TETRAHYDROXY-*p*-BENZOQUINONE DIHYDRATE

$C_6O_6H_4 \cdot 2H_2O$ F.W. = 208.1

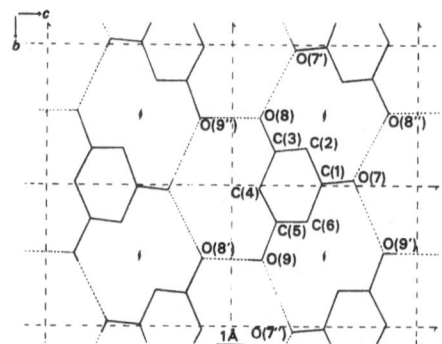

Fig. 48.   The structure of phloroglucinal projected along **a**. Hydrogen bonds
                are given as dotted lines. O(7') belongs to a molecule one repeat
                distance above the molecule of O(8"), and O(7") belongs to a molecule
                one repeat distance above the molecule of O(9').

    I. The crystal structure of tetrahydroxy–p–benzoquinone.   H.P. KLUG, 1965. *Acta
       Cryst.*, _19_, 983–992.

Monoclinic, a = 5.226±2, *b* = 5.118±2, *c* = 15.502±8 Å, β = 103.89±2°, [U = 402.5 Å³],
Z = 2, D_x = 1.717, (Calibration with quartz crystal with *a* = 4.2555±1 Å, *c/a* =
1.10000±5).

Space group $P2_1/c$ (C^5_{2h}) Molecular symmetry, centre.

Atomic positions

	x	y	z
C(1)	0.6587±6	0.2205±7	0.0310±2
C(2)	0.4690±6	0.1243±7	0.0809±2
C(3)	0.3204±6	−0.0858±7	0.0512±2
O(1)	0.7978±5	0.4118±5	0.0572±2
O(2)	0.4669±5	0.2661±5	0.1536±2
O(3)	0.1428±5	−0.1762±5	0.0939±2
O(4)	0.8523±5	0.6547±6	0.2418±2
H(2)	0.3100	0.2250	0.1700
H(3)	0.0700	0.6800	0.0750
H(4)	−0.0700	−0.1900	0.2400
H(4')	0.7500	0.5400	0.1850

    Anisotropic temperature factors are given for the non–hydrogen atoms, and the
thermal motion is analysed. The effect on the atomic positions is shown to be small.

Structure

    The tetrahydroxy–*p*–benzoquinone molecule is planar, except for the hydrogen
atoms, which lie 0.10 and 0.26 Å from the mean plane. Bond lengths and angles are
given in full; some mean bond lengths are: C=C, 1.342 Å;   C–C, 1.478 Å;   C=O,
1.229 Å;   C–OH, 1.344 Å;   (all ± 0.004 Å). The structure is held together by a
system of O–H··O bonds, involving the water molecules, with O...O distances ranging
from 2.654 to 2.968 Å. Some short intermolecular C...O distances (3.088 and 3.179 Å)
are believed to result from some charge–transfer interaction.

Details of analysis

    The intensity data were recorded on Weissenberg photographs using CuK$_\alpha$ radia-
tion. 703 of 841 accessible reflexions were observed and measured visually. Refinement was by full–matrix least squares to a final R index of 0.079.

1-BROMO-2,3,5,6-TETRAMETHYLBENZENE (α FORM)

$C_{10}H_{13}Br$ $Br(C_6H)(CH_3)_4$ F.W. = 213.1

I. Polymorphisme du bromo 1 tetraméthyl 2,3,5,6 benzène. G. CHARBONNEAU,
   J. BAUDOUR, J.-Cl. MESSAGER and J. MEINNEL, 1965. *Bull. Soc. Fr. Minéral.*
   Cristallogr., 88, 147-148. (The crystal structures of α and β forms are
   presented. The latter is said to be still undergoing refinement, and is not
   reported here.)

Orthorhombic, a = 14.62±5, b = 5.43±2, c = 12.05±4 Å, [U = 957 Å³], z = 4,

D_x = 1.48.

Space group $P2_12_12_1$ (D_2^4)

Atomic positions

	x	y	z
Br	0.2325	0.0698	0.0618
C(1)	0.334	0.054	0.150
C(2)	0.409	−0.125	0.137
C(3)	0.487	−0.130	0.207
C(4)	0.490	0.035	0.290
C(5)	0.420	0.188	0.310
C(6)	0.342	0.216	0.246
C(7)	0.390	−0.275	0.033
C(8)	0.565	−0.305	0.186
C(9)	0.425	0.381	0.407
C(10)	0.262	0.395	0.267

Structure

   The molecular geometry is consistent with expectation. There is no indication
of accuracy.

Details of analysis

   The structure was solved from b- and c-axis projections. The final R indices
were 0.138 and 0.183, respectively.

HEXAIODOBENZENE

C_6I_6 F.W. = 833.5

I. [Crystal structure of hexabromo- and hexaiodobenzene and the NQR spectra of
   Br79 and I127 in these compounds.] T.A. BABUSKINA, T.L. KHOCJANOVA and
   G.K. SEMIN, 1965. Ž. *Strukt. Khim. SSSR*, 6, 307-308 [J. *Struct. Chem.*,
   6, 285-286].

Monoclinic, a = 8.85±3, b = 4.28±2, c = 18.08±5 Å, β = 116°30±15', [U = 613 Å³],
z = 2, [D_x = 4.51].

Space group $P2_1/c$ (C_{2h}^5)

Atomic positions

	x	y	z
I(1)	0.415	0.233	0.069
I(2)	0.286	0	0.206
I(3)	−0.132	−0.299	0.139
C(1)	0.168	0.096	0.028
C(2)	0.116	0	0.084
C(3)	−0.054	−0.121	0.057

Structure

   The structure is isotypic with C_6Br_6 and C_6Cl_6 (1, 2).

Details of analysis

    253 reflexions in the *h0l* zone were recorded. The *x* and *z* coordinates were
refined by Fourier methods, and the *y* coordinate was deduced by assuming a planar
model of known dimensions.

1.   *Structure Reports*, <u>22</u>, 694.
2.   *Ibid.*, <u>24</u>, 648.

HEXA(BROMOMETHYL)BENZENE

$C_{12}H_{12}Br_6$ F.W. = 635.7

    I. Nouvelle détermination de la structure cristalline de l'hexa(bromométhyl)
       benzène. M.P. MARSAU, 1965. *Acta Cryst.*, <u>18</u>, 851–854. (For earlier work,
       and criticism thereof, see <u>1</u>, <u>2</u>.)

Rhombohedral, (the structure is reported in terms of the hexagonal cell, with
$a = 16.41\pm2$, $c = 5.38\pm2$ Å, $U = 1254.6$ Å^3, $D_m = 2.50$, $Z = 3$, $D_x = 2.53$.)

Space group $R\bar{3}$ (C_{3i}^2) Molecular symmetry, $\bar{3}$.

Atomic positions

	x	y	z
C (aromatic)	0.0920	0.0170	−0.0040
C (bromomethyl)	0.1935	0.0360	−0.0180
Br	0.2441	0.0402	0.3179

    Isotropic temperature factors are given also.

Structure

    The configuration of the molecule is shown in Fig. 49. The benzene ring is
slightly puckered; the ring atoms lie 0.02 Å from the mean plane z = 0. The bromomethyl carbon atoms lie 0.10 Å from this plane, on the same side as the corresponding carbon atoms. [Standard deviations are not given.] The distance Br–C is 1.97
Å, and the angle Br–C–C is 111°.

Details of analysis

    Intensity data were recorded on De Jong retigrams (*hk*0 to *hk*4) using CuKα
radiation. 330 independent reflexions were observed and measured visually. Refinement was by Fourier methods, using back-shift corrections, to a final *R* index of
0.087.

1.   *Strukturbericht*, <u>3</u>, 679.
2.   A.I. KITAJGORODSKIJ, 1957. *Organic Chemical Crystallography*, New York:
     Consultants Bureau.

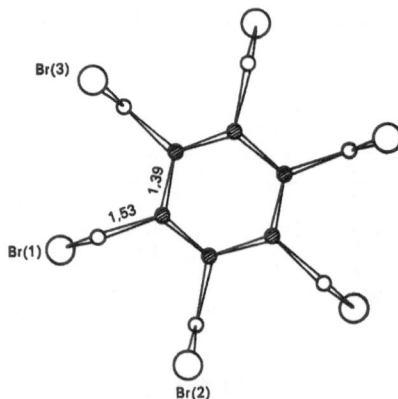

Fig. 49. Configuration of the hexa(bromomethyl)benzene molecule.

BENZIL

$C_{14}H_{10}O_2$ F.W. = 210.2

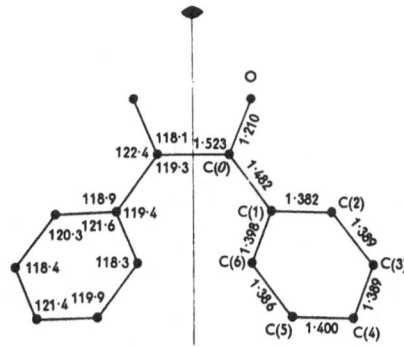

I. The crystal structure of benzil. C.J. BROWN and R. SADANAGA, 1965. *Acta Cryst.*,
   18, 158-164.

Trigonal, a = 8.376±9, c = 13.700±8 Å, [U = 832.4 Å³], D_m = 1.225, Z = 3, D_x = 1.256,
at 20°C. (At -92°C, a = 8.267±6, c = 13.407±26 Å).

Space group $P3_12_1$ (D_3^4)   or $P3_22_1$ (D_3^6) Molecular symmetry, two-fold axis.

Atomic positions

	x	y	z
C(0)	0.2270	0.1927	0.0525
C(1)	0.2313	0.0224	0.0766
C(2)	0.2841	0.0025	0.1693
C(3)	0.2862	−0.1570	0.1947
C(4)	0.2339	−0.2946	0.1249
C(5)	0.1798	−0.2752	0.0311
C(6)	0.1783	−0.1156	0.0061
O	0.2588	0.3111	0.1125

Positions and isotropic temperature factors are given for hydrogen atoms.
Anisotropic temperature factors for the non-hydrogen atoms are given and analysed,
and a table of atomic positions, corrected for rigid-body libration, is included.
This analysis of thermal motion is recognized to be of dubious validity, however.

Structure

The dimensions of the molecule are given in Fig. 50. (Corrections for
thermal motion would increase the bond lengths by 0.002 to 0.012 Å.) The
benzene rings are mutually inclined at 76.3°. The arrangement of molecules in
the unit cell is shown in Fig. 51. (The two-fold axes are in the plane of
projection.)

Details of analysis

Three-dimensional intensity data were recorded on Weissenberg and oscillation photographs, using CuK_α radiation. 539 of 758 accessible reflexions were
observed and measured photometrically. Refinement was by least squares to a
final R index of 0.078. The account indicates that a low-temperature analysis
is in progress.

Fig. 50. Bond lengths and angles of benzil, in Å and °. Standard deviations
         are 0.007 to 0.010 Å, and 0.4 to 0.6°.

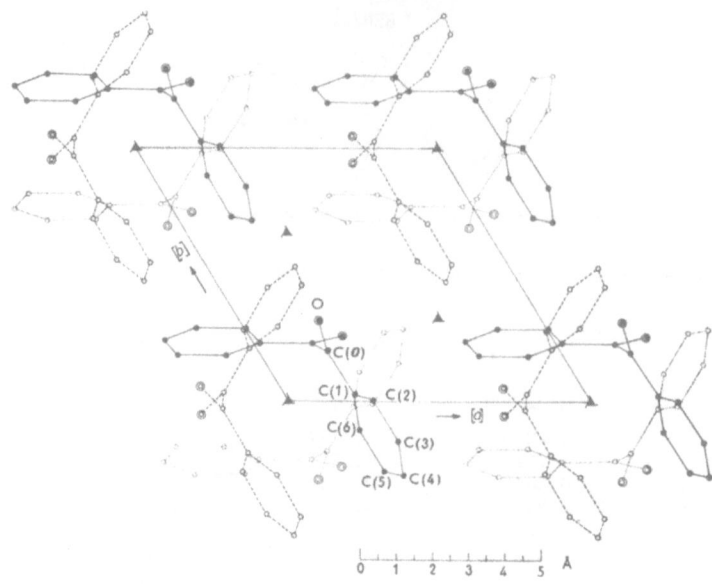

Fig. 51.   Projection of molecules of benzil on (0001). Molecules drawn with
                full lines have their mid-points at $z = 0$, those with broken lines
                at $z = 1/3$, and those with dotted lines at $z = 2/3$.

4,4'-DIHYDROXYTHIOBENZOPHENONE MONOHYDRATE

$C_{13}H_{10}O_2S \cdot H_2O$ F.W. = 248.3

H_2O

HO   OH

I. The crystal and molecular structure of 4,4'-dihydroxythiobenzophenone monohydrate. LJ. M. MANOJLOVIĆ and I.G. EDMUNDS, 1965. *Acta Cryst.*, 18, 543-548.

Monoclinic, $a = 5.62\pm1$, $b = 10.95\pm3$, $c = 20.24\pm6$ Å, $\beta = 103°30\pm15'$, $U = 1211$ Å³,
$D_m = 1.36$, $z = 4$, $D_x = 1.36$.
Space group $P2_1/c$ (C_{2h}^5)

Atomic positions

	x	y	z
S	0.1098	−0.1913	0.1941
O(1)	0.6395	0.2330	0.0443
O(2)	0.2640	0.1088	0.4874
O(H$_2$O)	−0.0352	0.3459	0.4746
C(1)	0.2408	−0.0610	0.2226
C(2)	0.3596	0.0166	0.1765
C(3)	0.5160	−0.0352	0.1393
C(4)	0.6110	0.0341	0.0954
C(5)	0.5489	0.1596	0.0881
C(6)	0.4050	0.2117	0.1257
C(7)	0.3060	0.1441	0.1700
C(8)	0.2446	−0.0123	0.2908

	x	y	z
C(9)	0.0576	−0.0498	0.3234
C(10)	0.0693	−0.0078	0.3906
C(11)	0.2648	0.0678	0.4233
C(12)	0.4500	0.1039	0.3907
C(13)	0.4325	0.0640	0.3240

Standard deviations range from 0.008 to 0.24 Å. Also given are isotropic temperature factors for the above atoms, and the positions of the hydrogen atoms.

Structure

Bond lengths and angles are given in full. The molecule of 4,4'-dihydroxythiobenzophenone can be considered as originating from a planar array, but with the benzene rings rotated by 30° and 47° about their adjacent C–CS and C–OH bonds. The sense of the rotations is such as to preserve approximate two-fold (rather than mirror) symmetry. The molecular packing is illustrated in Fig. 52. The distances indicated are:

O(2)	– O(1)	2.76±3 Å
O(1)	– O(H_2O)	2.70±3
O(H_2O)	– O(2)	3.07±3
O(H_2O)	– S	3.37±2

These are presumed to be hydrogen bonds.

Details of analysis

The intensity data for the zones normal to a and b were recorded on Weissenberg photographs, using CuK_α radiation. 388 of a possible 428 reflexions were observed and estimated visually. Refinement was by Fourier and least-squares methods to a final R index of 0.107 for ($h0l$) and 0.096 for ($hk0$).

Fig. 52.   The structure of 4,4'-dihydroxythiobenzophenone monohydrate projected along a. Dashed lines represent the hydrogen bonds.

METHYLPHENYLSULPHONE

$C_7H_8O_2S$ F.W. = 156.2

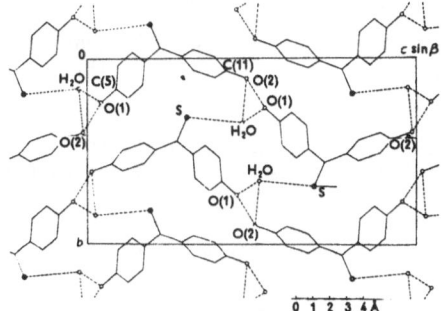

I. [The crystal structure of methylphenylsulfone.]   L.G. VORONCOVA,   1965.
   Kristallografija, 10, 187-193 [*Soviet Physics – Crystallography*, 10, 139-143].

Monoclinic, a = 8.35, b = 9.20, c = 10.98 Å, β = 112°, [U = 782 Å³], z = 4,
[D_x = 1.33].

Space group P2₁/c (C_{2h}^5)

Atomic positions

	x	y	z
S	0.369	0.137	0.169
O(1)	0.320	0.011	0.083
O(2)	0.456	0.109	0.310
C(1)	0.177	0.245	0.151
C(2)	0.022	0.228	0.039
C(3)	0.876	0.311	0.036
C(4)	0.896	0.423	0.130
C(5)	0.065	0.440	0.230
C(6)	0.210	0.365	0.237
C(CH₃)	0.515	0.231	0.116

Structure

   The molecule is not planar. C(1) (the atom at cc') lies 0.14 Å from
the plane of the remaining ring atoms, and the sulphur atom lies 0.22 Å from
this plane. Mean bond lengths and angles are:

a	=	1.41±2 Å	aa'	=	120°
b	=	1.40±2	ab	=	122
c	=	1.42±2	bc	=	116
d	=	1.82±2	cc'	=	123
e	=	1.455±15	cd	=	118
f	=	1.77±2	de	=	106
			ef	=	106
			ee'	=	120
			df	=	112

Details of analysis

   Intensity data were recorded photographically (hk0 to hk7; 0kl to 7kl; h0l)
using CuKα radiation.   764 reflexions were observed and estimated visually.   Refinement was by least squares to a final R index of 0.19.

OCTACHLOROCYCLOBUTANE

C_4Cl_8 F.W. = 331.7

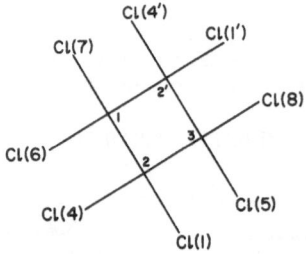

I. Refinement of cyclobutane structures. T.N. MARGULIS, 1965. *Acta Cryst.*, 19,
   857-859. This is a refinement of the structure described in 1, using the
   original crystal and intensity data.

Monoclinic, a = 8.00, b = 10.64, c = 6.28 Å, β = 107° 45', U = 509 Å³, z = 2,
D_x = 2.16.

Space group P2₁ (C_2^2) or P2₁/m (C_{2h}^2) P2₁/m confirmed by analysis. Molecular
symmetry, mirror plane.

Atomic positions

	x	y	z
Cl(1)	0.2407	0.0896	−0.0090
Cl(4)	0.2392	0.0173	0.4251
Cl(5)	0.4615	0.2500	0.6411
Cl(6)	0.0205	0.2500	0.4662
Cl(7)	−0.0837	0.2500	−0.0050
Cl(8)	0.5640	0.2500	0.2468
C(1)	0.096	0.2500	0.240
C(2)	0.239	0.147	0.251
C(3)	0.386	0.250	0.345

Temperature factors (anisotropic for the chlorine atoms) are given also.

Structure

The molecule is generally as described in $\underline{1}$, with a crystallographic mirror plane containing atoms C(1), C(3), Cl(5), Cl(6), Cl(7), Cl(8). Bond lengths and angles are given in full, but chemically equivalent quantities do not differ significantly. Mean values are: C–Cl, 1.75±1 Å, C–C, 1.58±3 Å; C–C–C, 88.4°; Cl–C–Cl, 108.6°. The cyclobutane ring is non-planar; the angle between the planes C(1) C(2) C(3) and C(2)C(1)(2') is 19°.

Details of analysis

Intensity data from $\underline{1}$ were refined by full-matrix least squares to a final R index of 0.12. The C–Cl distance was corrected (by 0.01 Å) for thermal motion.

1. *Structure Reports*, $\underline{15}$, 448.

1,2,3,4-TETRAPHENYLCYCLOBUTANE

$C_{28}H_{24}$ F.W. = 360.5

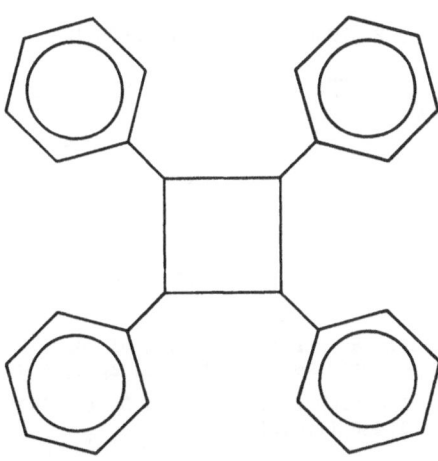

I. Refinement of cyclobutane structures. T.N. MARGULIS, 1965. *Acta Cryst.*, $\underline{19}$, 857–859. This is a refinement of the structure described in $\underline{1}$, using the original crystal and intensity data.

Monoclinic, a = 17.02±5, b = 5.775±20, c = 12.35±5 Å, β = 127±1°, U = 969.4 Å³, D_m = 1.23, Z = 2, D_x = 1.235.

Space group $P2_1/a$ (C_{2h}^5) Molecular symmetry, centre

Atomic positions

	x	y	z
C(1)	0.0503	0.1416	0.0153
C(2)	0.1602	0.1533	0.0963
C(3)	0.2219	−0.0254	0.1859
C(4)	0.3241	−0.0123	0.2559
C(5)	0.3641	0.1767	0.2376
C(6)	0.3028	0.3586	0.1484
C(7)	0.2029	0.3458	0.0802
C(8)	−0.0031	0.0751	0.0796
C(9)	0.0539	0.0391	0.2313
C(10)	0.0381	−0.1526	0.2846
C(11)	0.0810	−0.1654	0.4218
C(12)	0.1426	0.0104	0.5081
C(13)	0.1612	0.1994	0.4566
C(14)	0.1157	0.2143	0.3170

Anisotropic temperature factors are given also.

Structure

The structure is as described in 1. Bond lengths and angles are given in full, but chemically equivalent values do not differ significantly. Some average values are: C–C (aromatic) 1.394±4 Å (1.399 Å if corrected for thermal motion); C–C (cyclobutane ring) 1.570±15 Å; C–C (other) 1.510±15 Å. The cyclobutane ring is square within experimental error.

Details of analysis

Intensity data from 1 were refined by full-matrix least squares to a final R index of 0.14.

1.   *Structure Reports*, 12, 402.

2,2-DICHLORO-3-PHENYLCYCLOBUTENONE

$C_{10}H_6Cl_2O$ F.W. = 213.1

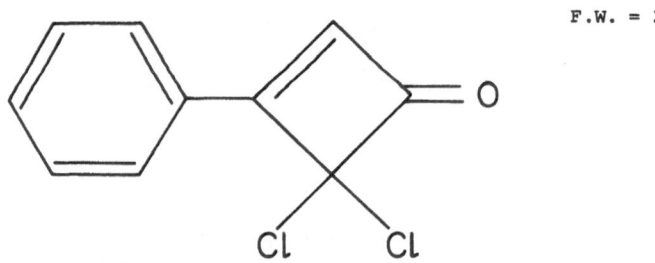

I. The crystal and molecular structure of 2,2-dichloro-3-phenylcyclobutenone.
I.L. KARLE and K. BRITTS, 1965. Z. *Kristallogr.*, 121, 190–203.

Monoclinic, a = 9.47±2, b = 9.68±2, c = 10.94±2 Å, β = 103°4±8', [U = 977 Å³], D_m = 1.393, Z = 4, D_x = 1.406 [1.448].

Space group $P2_1/c$ (C_{2h}^5)

Atomic positions

	x	y	z
Cl(1)	0.4409	0.9081	0.7097
Cl(2)	0.2535	0.8593	0.4647
C(3)	0.2640	0.9295	0.6172
C(4)	0.1376	0.8740	0.6777
O(5)	0.1110	0.7616	0.7057
C(6)	0.0934	1.0120	0.6822
C(7)	0.1973	1.0661	0.6240

	x	y	z
C(8)	0.2283	1.1956	0.5841
C(9)	0.3283	1.2095	0.5072
C(10)	0.3574	1.3376	0.4644
C(11)	0.2884	1.4520	0.4947
C(12)	0.1918	1.4436	0.5759
C(13)	0.1630	1.3162	0.6195

Isotropic temperature factors are given also.

Structure

Bond lengths and angles are given in full, with standard deviations ranging from 0.018 to 0.033 Å, and 0.7 and 1.8°. From the lengths of the bonds in the cyclo-butene ring, and of the inter-ring bridging bond, it is inferred that significant conjugation between single and double bonds occurs. Both rings are planar, but are rotated with respect to each other by about 12°. There is an intermolecular Cl...O contact of 3.08 Å.

Details of analysis

The intensity data were recorded on equi-inclination Weissenberg photographs about b. 963 of a possible 1199 reflexions were observed and measured visually. Refinement was by full-matrix least squares to a final R index of 0.17.

CYCLOPENTADIENE

C_5H_6 F.W. = 66.1

I. The crystal and molecular structure of cyclopentadiene. G. LIEBLING and R.E. MARSH, 1965. *Acta Cryst.*, <u>19</u>, 202–205.

Monoclinic, a = 7.89±2, b = 5.65±2, c = 10.45±3 Å, β = 114°10±20', [U = 425 Å³].
$D_m \sim 1.0$, z = 4, D_x = 1.031 (all at −150°C).

Space group $P2_1/n$ (C_{2h}^5)

Atomic positions

	x	y	z
C(1)	0.565±5	0.424±4	0.226±3
C(2)	0.483±5	0.604±4	0.247±4
C(3)	0.294±6	0.623±4	0.145±4
C(4)	0.259±4	0.428±3	0.059±4
C(5)	0.437±5	0.287±4	0.104±5

Isotropic temperature factors, and the positions of the hydrogen atoms, are given also.

Structure

The molecule is planar. Bond lengths and angles are given, and are consistent with the above formulation.

Details of analysis

The intensity data at −150°C were recorded on various zero-level precession photographs, using MoK$_\alpha$ radiation. 70 of 133 accessible reflexions were observed and measured visually. Refinement was by least squares to a final R index of 0.10.

1,2,3-TRIBROMO-6-(o-METHOXYPHENYL)FULVENE

$C_{13}H_9Br_3O$ F.W. = 420.9

I. The crystal structure of 1,2,3-tribromo-6-(o-methoxyphenyl)fulvene. Y. KATO, Y. SASADA and M. KAKUDO, 1965. *Bull. Chem. Soc. Japan.* **38**, 1761-1775.

Monoclinic, a = 11.46, b = 14.53, c = 7.85 Å, β = 96°35', [U = 1299 Å³], D_m = 2.17, Z = 4, D_X = 2.15.

Space group $P2_1/c$ (C^5_{2h})
(Crystal data for the isotypic chlorine derivative are: a = 11.65, b = 14.61, c = 7.64 Å, β = 98°58', [U = 1284 Å³], D_m = 1.50, Z = 4, D_X = 1.49.)

Space group $P2_1/c$ (C^5_{2h})

Atomic positions

	x	y	z
Br(1)	0.6048	-0.0195	0.2265
Br(2)	0.3089	0.0372	0.2778
Br(3)	0.3051	0.2582	0.4716
C(1)	0.5512	0.0874	0.3176
C(2)	0.4434	0.1080	0.3340
C(3)	0.4418	0.1975	0.4248
C(4)	0.5492	0.2303	0.4586
C(5)	0.6282	0.1630	0.3942
C(6)	0.7470	0.1593	0.4034
C(7)	0.8270	0.2338	0.4754
C(8)	0.7992	0.3288	0.4509
C(9)	0.8748	0.3942	0.5216
C(10)	0.9858	0.3683	0.6090
C(11)	1.0194	0.2752	0.6364
C(12)	0.9346	0.2099	0.5596
C(13)	1.0759	0.0862	0.6479
O	0.9584	0.1167	0.5790

Mean standard deviations are: Br, 0.003; O, 0.015; C, 0.022 Å. Temperature factors (anisotropic for the bromine atoms) are given also.

Structure

The conformation of the molecule is shown in Fig. 53. Bond lengths and angles are given in full. Selected values are:

    C(1) - C(2)   = 1.29±3 Å
    C(2) - C(3)   = 1.48±3
    C(3) - C(4)   = 1.32±3
    C(4) - C(5)   = 1.46±3
    C(5) - C(1)   = 1.49±3
    C(5) - C(6)   = 1.36±3
    C(6) - C(7)   = 1.49±3
    C(12) - O   = 1.39±3
    O   - C(13)   = 1.46±3
    C   - Br (mean) = 1.86±2

    C(5) - C(6) - C(7)   = 124.2°
    C(12) - O   - C(13) = 119.8

The molecule lies essentially in two planes, one containing C(1) to C(6) and Br(1) to Br(3), and the other containing C(6) to C(13) and O. The angle between them is 38.9°. Intermolecular distances are normal.

Details of analysis

The intensity data were recorded on equi–inclination Weissenberg photographs, (*hk*0 to *hk*6; *h*0*l*, *h*1*l*) using CuK_α radiation. 2060 of a possible 2900 reflexions were observed and measured visually. Refinement was by block–diagonal least squares to a final *R* index of 0.126.

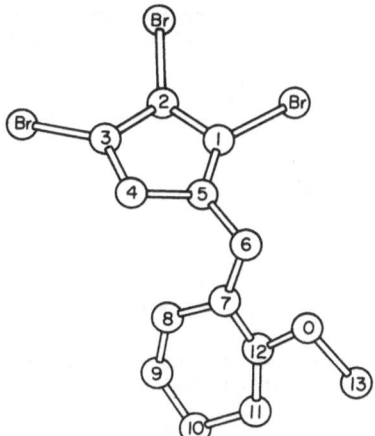

Fig. 53. View of the 1,2,3–tribromo–6–(*o*–methoxyphenyl)fulvene molecule.

1,3-*trans*-DIAMINOCYCLOHEXANE DIHYDROCHLORIDE

$C_6H_{14}N_2$,2HCl   F.W. = 187.1

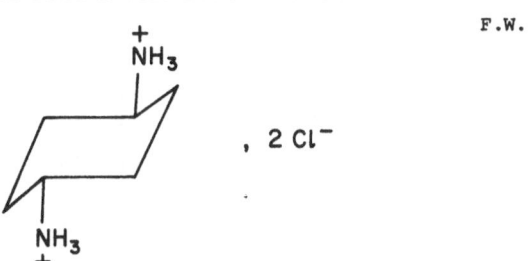

I. Die Kristallstruktur von 1,4–*trans*–Diaminocyclohexan–dihydrochloride. J.D. DUNITZ and P. STRICKLER, 1965. *Helv. Chim. Acta*, <u>48</u>, 1450–1456.

Monoclinic, *a* = 6.36±1, *b* = 5.27±1, *c* = 14.910±25 Å, γ = 100°0±20', *U* = 492.2 Å³, D_m = 1.248, *Z* = 2, D_x = 1.262.

Space group $P2_1/n$ (C_{2h}^5) (1st setting) Molecular symmetry, centre.

Atomic positions

	x	*y*	*z*
Cl	0.9632±3	0.9681±4	0.1541±1
N	0.2469±14	0.5248±14	0.1581±4
C(1)	0.4077±12	0.5717±14	0.0846±5
C(2)	0.5351±15	0.3521±17	0.0816±5
C(3)	0.7007±11	0.3975±16	0.0055±4

Anisotropic temperature factors are given, and have been analyzed in terms of rigid–body motion. The unrefined positions of the hydrogen atoms are given also.

Structure

The cyclohexane ring has the expected chair conformation, with a mean C–C–C angle of 110.7°, and mean C–C distance of 1.53±1 Å. The amino groups are in the equatorial position, with C–N = 1.49±1 Å. The analysis demonstrates the protonation of the amino groups as illustrated above. Each nitrogen atom is coordinated to three chlorine atoms at distances of 3.16 to 3.19 Å, and the C-N-Cl angles range from 109° to 114°.

Details of analysis

The intensity data were measured with a linear diffractometer and scintillation counter, using MoK$_\alpha$ radiation. 836 reflexions were observed above background in the sphere sinθ/λ<0.73. Refinement was by full-matrix least squares to a final R index of 0.093.

cis-1,4-AMINOMETHYLCYCLOHEXANE-CARBOXYLIC ACID HYDROBROMIDE

C$_8$H$_{16}$BrNO$_2$ F.W. = 238.1

I. *Cis-trans* relationship between the two amino-acids obtained by hydrogenation of *p*-aminomethyl benzoic acid. P. GROTH and O. HASSEL, 1965. *Acta Chem. Scand.*, **19**, 1709-1714.

Monoclinic, a = 5.48, b = 33.29, c = 7.96 Å, β = 131°, [U = 1096 Å³], Z = 4, D$_x$ = 1.44.

Space group P2$_1$/c (C$_{2h}^5$)

Atomic positions

	x	y	z
Br	0.3117	0.1996	0.2125
O(1)	0.6459	0.4923	0.2872
O(2)	0.2377	0.4691	-0.0482
N	0.5218	0.2930	0.4042
C(1)	0.2753	0.3615	0.2637
C(2)	0.0275	0.3860	0.0426
C(3)	-0.0324	0.4279	0.1054
C(4)	0.2733	0.4508	0.2655
C(5)	0.5310	0.4256	0.4644
C(6)	0.5800	0.3849	0.4085
C(7)	0.3105	0.3196	0.2023
C(8)	0.3807	0.4709	0.1539

Structure

The analysis established the *cis* configuration represented above. This is the inactive form of the amino acid. The NH$_3^+$ group is tetrahedrally coordinated to three bromine atoms at distances of 3.30, 3.32, and 3.33 Å. The COOH groups of neighbouring centrosymmetrically related amino-acid cations are held together by pairs of O-H...O bonds of length 2.69 Å. Bond lengths and angles are given in full, but there is nowhere any indication of accuracy.

Details of analysis

Intensity data for the zones [101] and [100] were recorded photographically, using MoK$_\alpha$ and CuK$_\alpha$ radiation. About 350 reflexions were observed and measured photometrically. Refinement was by least squares to a final R index of 0.065.

BICYCLOHEXYLIDENE

$C_{12}H_{10}$ F.W. = 164.3

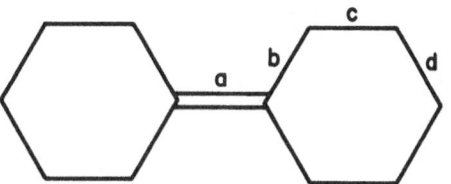

I. The crystal structure of bicyclohexylidene, $C_{12}H_{20}$. K. SASVÁRI and M. LÖW, 1965. *Acta Cryst.*, 19, 840–848.

Triclinic, $a = 5.32\pm1$, $b = 6.25\pm1$, $c = 8.36\pm1$ Å, $\alpha = 107.2\pm1$, $\beta = 79.1\pm1$, $\gamma = 105.6 \pm1°$, $[U = 254.4$ Å$^3]$, $D_m = 1.066$, $Z = 1$, $D_x = 1.073$.

Space group $P1$ (C_1^1) or $P\bar{1}$ (C_i^1) $P\bar{1}$ is confirmed by the analysis. Molecular symmetry, centre.

Atomic positions

	x	y	z
C(1)	0.0309±6	0.0498±5	0.0779±4
C(2)	0.1438±7	0.3019±5	0.1463±5
C(3)	0.4002±8	0.3451±6	0.2194±5
C(4)	0.3643±8	0.2183±7	0.3539±4
C(5)	0.2563±8	−0.0365±6	0.2845±4
C(6)	−0.0013±7	−0.0779±6	0.2129±4

Hydrogen positions are given also. Anisotropic temperature factors are given for the non–hydrogen atoms, and have been analysed in terms of rigid–body motion. Atomic positions which have been corrected for such motion are given also.

Structure

The conformation of the molecule is illustrated in Fig. 54. The rings are in the chair form, and the six innermost carbon atoms are coplanar. The molecule has non–crystallographic mirror symmetry; mean bond lengths are:

$$a = 1.332\pm6 \text{ Å}$$
$$b = 1.518\pm5$$
$$c = 1.531\pm5$$
$$d = 1.519\pm5$$

[The standard deviations are underestimated in the original.] The ring bond angles range from 110.4 to 111.9°; the angle ab is 124.7±3°. The intermolecular distances are consistent with van der Waals interaction.

Details of analysis

The intensity data were recorded on integrated Weissenberg photographs, using CuK_α radiation. 729 of 858 accessible reflexions were observed and measured photometrically and visually. Refinement was by full–matrix least squares to a final R index of 0.116. The corrections for rigid–body thermal motion have the effect of increasing bond lengths by 0.002 to 0.006 Å.

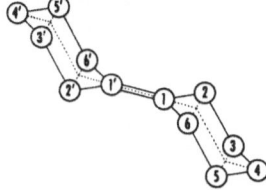

Fig. 54. The carbon skeleton of bicyclohexylidene showing the numbering scheme. The dotted line shows the non–crystallographic mirror plane of the molecule.

1,3,5,7-CYCLOOCTATETRAENECARBOXYLIC ACID

$C_9H_8O_2$ F.W. = 148.2

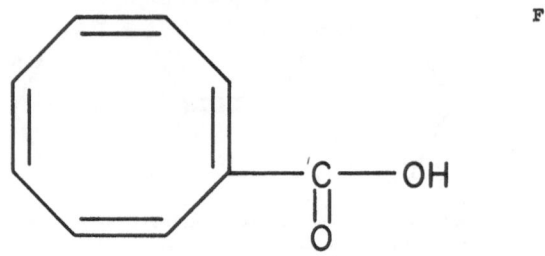

I. The crystal structure of cyclooctatetraenecarboxylic acid. D.P. SHOEMAKER,
   H. KINDLER, W.G. SLY and R.C. SRIVASTAVA, 1965. *J. Amer. Chem. Soc.*, <u>87</u>,
   482-487.

Monoclinic, a = 6.76±1, b = 14.64±2, c = 8.37±1 Å, β = 100°9.5±5', [U = 815.4 Å^3],
D_m = 1.256, Z = 4, D_x = 1.255 [1.21].

Space group $P2_1/c$ (C_{2h}^5)

Atomic positions

	x	y	z
C(1)	0.23518±58	0.19092±26	0.42007±48
C(2)	0.44308±68	0.19468±27	0.46905±57
C(3)	0.56964±68	0.26831±34	0.43536±60
C(4)	0.56371±74	0.35525±32	0.47711±69
C(5)	0.43044±82	0.39593±30	0.56145±68
C(6)	0.22139±87	0.39340±27	0.51120±69
C(7)	0.08106±69	0.34704±29	0.35493±64
C(8)	0.08733±62	0.26045±29	0.31732±53
C(9)	0.13403±62	0.10704±24	0.45367±48
O(1)	−0.06329±48	0.10006±20	0.40268±47
O(2)	0.25589±48	0.04298±20	0.53736±50

Also given are anisotopic temperature factors, and the assumed positions of
the hydrogen atoms.

Structure

The molecules are present as dimers, being hydrogen bonded across a centre of
symmetry *via* their carboxyl groups. The O–H..O distance is 2.604±10 Å. The eight-
membered ring has the "tub" conformation, with alternating single and double bonds,
and conforms closely to D_{2d} (42m) symmetry. The mean values of the ring bond
lengths and angles are 1.322±5 Å, 1.470±5 Å and 126.4±4°.

Details of analysis

The intensity data were recorded on equi-inclination Weissenberg photographs,
using CuK$_α$ radiation. About 1650 reflexions were observed and estimated visually.
Refinement was by full-matrix least squares to a final R index of 0.101.

OCTAPHENYLCYCLOOCTATETRAENE

$C_{56}H_{40}$

F.W. = 712.9
M.P. = 425-427°C

I. The crystal and molecular structure of octaphenylcyclooctatetraene. P.J. WHEATLEY, 1965. *J. Chem. Soc.*, 3136-3146. (For preliminary reports, see 1, 2.)

Tetragonal, a = 19.388±32, c = 10.606±14 Å, U = 3987 Å^3, D_m = 1.20, Z = 4, D_x = 1.188.

Space group $\overline{I}4_1/a$ (C_{6h}^4) Molecular symmetry, $\overline{4}$.

Atomic positions

	x	y	z
C(1)	0.0475	0.1865	0.1779
C(2)	0.0946	0.1660	0.2840
C(3)	0.0731	0.1799	0.4086
C(4)	0.1195	0.1648	0.5093
C(5)	0.1842	0.1364	0.4856
C(6)	0.2048	0.1220	0.3627
C(7)	0.1602	0.1370	0.2600
C(8)	−0.0198	0.1712	0.1729
C(9)	−0.0549	0.1233	0.2616
C(10)	−0.0256	0.0582	0.2880
C(11)	−0.0607	0.0135	0.3696
C(12)	−0.1250	0.0320	0.4212
C(13)	−0.1546	0.0957	0.3940
C(14)	−0.1187	0.1423	0.3123

Coordinates in the original are given in Å. Standard deviations range from 0.006 to 0.010 Å. Anisotropic temperature factors are given also.

Structure

The analysis confirms the formulation given above. The molecular conformation is illustrated in Fig. 55. The eight-membered ring consists of alternating single and double bonds of length 1.493 and 1.342 Å. The benzene rings are attached to the central ring by bonds of mean length 1.494 Å. The strain resulting from the crowding of the benzene rings causes significant angular distortion at the atoms of the central ring. The intermolecular distances are consistent with van der Waals interaction.

Details of analysis

The intensities were measured with a linear diffractometer, using MoK_α radiation, with the crystal mounted parallel to c. 1229 of a possible 3072 reflexions were used in the refinement, which was by block-diagonal least squares. The final R index was 0.117.

1. H.P. THRONDSEN, P.J. WHEATLEY and H. ZEISS, 1964. *Proc. Chem. Soc.*, 357.
2. G.S. PAWLEY, W.N. LIPSCOMB and H.H. FREEDMAN, 1964. *J. Amer. Chem. Soc.*, 86, 4725.

Fig. 55. The octaphenylcyclooctatetraene molecule, viewed along c.

[18] ANNULENE

$C_{18}H_{18}$ F.W. = 234.3

I. The crystal structure of [18] annulene. I. X-ray study. J. BREGMAN, F.S. HIRSHFELD, D. RABINOVICH and G.M.J. SCHMIDT, 1965. *Acta Cryst.*, 19, 227-234.

II. The crystal structure of [18] annulene. II. Results. F.S. HIRSHFELD and D. RABINOVICH, 1965. *Ibid.*, 19, 235-241.

Monoclinic, a = 14.889±4, b = 4.800±2, c = 10.235±3 Å, β = 111.60±14°, [U = 680.1 Å³], Z = 2, D_X = 1.144. (Cell and intensity data at 80°K. Cell constants at room temperature, a = 15.33, b = 4.88, c = 10.27 Å, β = 111.8°.)

Space group $P2_1/a$ (C_{2h}^5) Molecular symmetry, centre.

Atomic positions

	x	y	z
C(1)	−0.1147	−0.4574	0.0430
C(2)	−0.1068	−0.5089	0.1807
C(3)	−0.0441	−0.3627	0.2988
C(4)	0.0151	−0.1483	0.2903
C(5)	0.0836	−0.0082	0.3995
C(6)	0.1430	0.2077	0.3781
C(7)	0.1416	0.2877	0.2474
C(8)	0.1885	0.5084	0.2154
C(9)	0.1771	0.5884	0.0770
H(1)	−0.072	−0.278	0.031
H(2)	−0.143	−0.668	0.197
H(3)	−0.043	−0.427	0.390

	x	*y*	*z*
H(4)	0.006	−0.078	0.195
H(5)	0.095	−0.054	0.499
H(6)	0.187	0.320	0.464
H(7)	0.107	0.168	0.174
H(8)	0.227	0.626	0.297
H(9)	0.211	0.749	0.060

Standard deviations range from 0.0023 to 0.004 Å for carbon, and from 0.027 to 0.038 Å for hydrogen. Thermal displacement parameters (isotropic for hydrogen, anisotropic for carbon) are given.

Structure

The molecular structure is summarized in Fig. 56. The molecule is nearly but not precisely planar. Two types of C–C bond are present: the six independent 'inner' bonds (C(1)–C(2), etc.) have a mean length of 1.382±3 Å, whereas the three independent 'outer' bonds (C(2)–C(3) etc.) have a mean length of 1.419±4 Å. Inter-molecular distances are consistent with van der Waals interaction.

Details of analysis

Intensity data were recorded on equi–inclination Weissenberg photographs (h0l to h3l; 0kl) at 80°K, using CuK_{α} radiation. 883 reflexions were observed, and estimated visually. Refinement was by diagonal least squares, with the structure referred to nearly–orthogonal axes. The final R index was 0.076. An analysis of thermal motion indicates rigid-body vibrations, with resultant errors in bond lengths of not more than 0.001 Å.

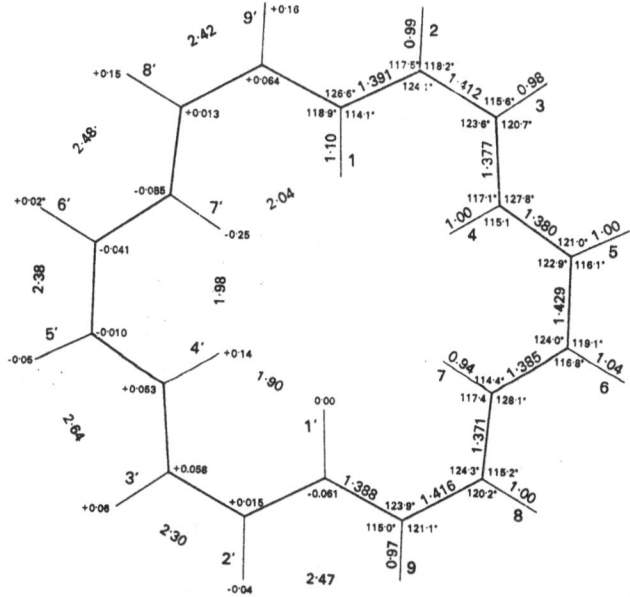

Fig. 56. Bond lengths and angles of [18] annulene. Standard deviations 0.005 Å for C–C, 0.035 Å for C–H. Figures near atomic positions give displacements (in Å) from the mean molecular plane.

1,5-DIMETHYLNAPHTHALENE

$C_{12}H_{12}$ F.W. = 156.2

CH₃... (structure diagram)

CH_3

g

b   c

a   d

f   e

CH_3

I. The crystal structure of 1,5,-dimethylnaphthalene. J. BEINTEMA, 1965.
   Acta Cryst., <u>18</u>, 647–654.

Monoclinic, a = 6.18±1, b = 8.91‡1, c = 16.77±2 Å, β = 101.40±5°, $[U$ = 905.2 Å³],
Z = 4, D_x = 1.145.

Space group $P2_1/c$ (C_{2h}^5)

Atomic positions

	x	y	z
C(1)	0.3758	0.4054	0.3264
C(2)	0.2398	0.3254	0.3665
C(3)	0.0523	0.2449	0.3239
C(4)	0.0024	0.2432	0.2399
C(5)	0.0898	0.3275	0.1090
C(6)	0.2216	0.4070	0.0688
C(7)	0.4082	0.4878	0.1106
C(8)	0.4593	0.4884	0.1951
C(9)	0.3209	0.4056	0.2385
C(10)	0.1434	0.3256	0.1963
C(1')	0.5676	0.4926	0.3728
C(5')	−0.1049	0.2381	0.0614

The mean standard deviation is estimated to be about 0.007 Å. The assumed
hydrogen positions are also given.

Structure

The molecule is very nearly planar, with some evidence for displacement of
the methyl groups to opposite sides of the mean plane of the aromatic nucleus.
Mean bond lengths and angles are:

a	=	1.425 Å		ab	=	121.9°
b	=	1.361		bc	=	117.6
c	=	1.454		cd	=	121.4
d	=	1.380		de	=	120.0
e	=	1.438		ef	=	118.4
f	=	1.384		fa	=	120.5
g	=	1.514		gc	=	121.9

(Distances ±0.008 Å, angles ±0.5°.) [These standard deviations may be optimistic.]
The packing of the molecules is illustrated. Intermolecular distances are consistent with van der Waals interaction.

Details of analysis

The structure was determined from projections. Intensity data were recorded
on Weissenberg photographs (0kl and h0l) using CuK_α radiation. Intensities were
estimated visually. Refinement was by least squares to a final R value of 0.09
for 0kl, and 0.14 for h0l.

3-BROMO-1,8-DIMETHYLNAPHTHALENE

$C_{12}H_{11}Br$ F.W. = 235.0

I. The crystal structure of an overcrowded aromatic compound: 3-bromo-1,8-dimethylnaphthalene. M.B. JAMESON and B.R. PENFOLD, 1965. *J. Chem. Soc.*, 528-536.

Triclinic, $a = 8.56\pm2$, $b = 8.59\pm2$, $c = 7.71\pm2$ Å, $\alpha = 118.8\pm1$, $\beta = 97.9\pm1$, $\gamma = 91.2\pm1°$, $U = 489.3$ Å³, $D_m = 1.57$, $Z = 2$, $D_x = 1.59$ (CuK_α, $\lambda = 1.5418$ Å).

Space group $P1$ (C_1^1) or $P\bar{1}$ (C_i^1) $P\bar{1}$ indicated by statistical test and confirmed by analysis.

Atomic positions

	x	y	z
Br	0.9106	0.9005	0.7312
C(1)	0.6226	0.4623	0.2615
C(2)	0.7509	0.5750	0.4005
C(3)	0.7279	0.7479	0.5456
C(4)	0.5868	0.8103	0.5539
C(5)	0.3012	0.7635	0.4315
C(6)	0.1710	0.6589	0.2988
C(7)	0.1811	0.4845	0.1529
C(8)	0.3253	0.4149	0.1263
C(9)	0.4639	0.5175	0.2607
C(10)	0.4533	0.6946	0.4148
C(11)	0.6611	0.2807	0.1112
C(12)	0.3222	0.2272	-0.0495

Standard deviations are given in full, in Å. Mean values are: Br, 0.0013; C, 0.01 Å. Ansiotropic temperature factors are given also.

Structure

Bond lengths and angles are given in full. The molecule is approximately planar, but small deviations from strict planarity presumably result from overcrowding. The most obvious indication of strain is the in-plane displacement of the methyl groups, which are separated by 2.92 Å. Some strain is communicated to the naphthalene nucleus: the separation of C(1) and C(8) is 2.56 Å, and that of C(4) and C(5) is 2.44 Å. The corresponding separation in naphthalene is 2.50 Å. Intermolecular distances are consistent with van der Waals interaction.

Details of analysis

Intensity data were recorded on equi-inclination Weissenberg photographs (*hk*0 to *hk*3; 0*kl* to 2*kl*) using CuK_α radiation. 1284 of a possible 1338 reflexions accessible on the films were observed and estimated visually. Crystal deterioration occurred, and three different specimens were used. Refinement was by full-matrix least squares to a final R index of 0.106.

1,8-DINITRONAPHTHALENE

$C_{10}H_6N_2O_4$ F.W. = 218.1

I. [Spatial difficulties and conformations of molecules communication 12. Crystal
and molecular structure of 1,8-dinitronaphthalene.] Z.A. AKOPJAN, A.I.
KITAJGORODSKIJ and J.T. STRUČKOV, 1965. Ž. Strukt. Khim. SSSR, 6, 729–744
[J. Struct. Chem., 6, 690–704]. For preliminary results, see 1 and 2.

Orthorhombic, a = 11.352±2, b = 14.934±2, c = 5.376±1 Å, U = 911 Å³, Z = 4, D_x = 1.59,
(Cell dimensions from 2.)

Space group $P2_12_12_1$ (D_2^4)

Atomic positions

	x	y	z
C(1)	0.6002	0.6476	0.1187
C(2)	0.6596	0.6093	0.3074
C(3)	0.6388	0.5171	0.3601
C(4)	0.5647	0.4678	0.2229
C(5)	0.4249	0.4525	−0.1256
C(6)	0.3648	0.4843	−0.3191
C(7)	0.3702	0.5784	−0.3731
C(8)	0.4389	0.6323	−0.2275
C(9)	0.5181	0.5972	−0.0367
C(10)	0.5042	0.5051	0.0166
N(1)	0.6381	0.7366	0.0332
N(2)	0.4236	0.7296	−0.2634
O(1)	0.6561	0.7503	−0.1895
O(2)	0.6583	0.7942	0.1930
O(3)	0.4083	0.7543	−0.4805
O(4)	0.4197	0.7772	−0.0741

Standard deviations are about 0.01 Å for all atoms. Also given are isotropic
temperature factors for the above atoms, and assumed positions for the hydrogen
atoms.

Structure

The geometry of the molecule is illustrated in Figs. 57 and 58. The deviations
from coplanarity of the atoms of the naphthalene nucleus are small but significant
(0.05 Å for C(1), 0.09 Å for C(8)). Corresponding values for N(1) and N(2) are 0.38
and 0.41 Å. The rotations of the nitro groups about their respective bonds are about
43°. Intermolecular distances are given in full, and appear to be normal van der
Waals contacts.

Details of analysis

Intensity data were recorded on equi-inclination Weissenberg photographs (hk0
to kh4; 0kl to 6kl) using CuK$_\alpha$ radiation. 915 independent reflexions were observed
and measured visually. Refinement was by least squares to a final R index of 0.165.

1.   Z.A. AKOPJAN and JU.T. STRUČKOV, 1964. Ž. Strukt. Khim. SSSR., 5, 496.
2.   V.M. KOKSIN, 1961. Ž. Strukt. Khim. SSSR, 2, 46.

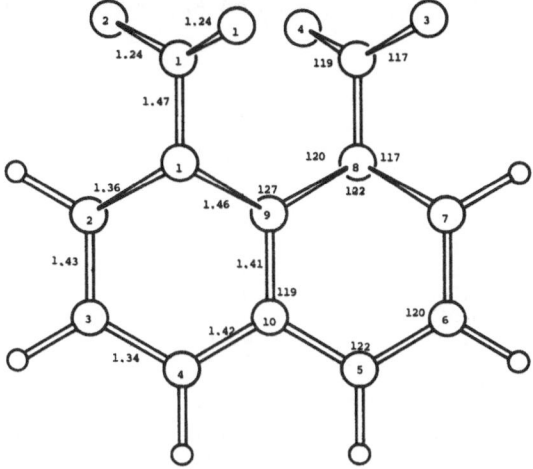

Fig. 57.   Bond lengths and angles of 1,8-dinitronaphthalene, averaged assuming
diad symmetry.   Standard deviations are 0.02 Å and 2°.

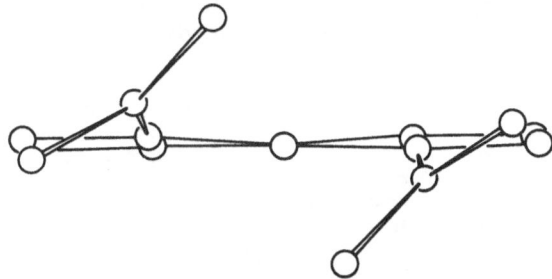

Fig. 58.   The molecule of 1,8-dinitronaphthalene viewed along C(9)–C(10).

1,5-DIBROMO-4,8-DICHLORONAPHTHALENE

$C_{10}H_4Br_2Cl_2$ F.W. = 354.9

I.   [Steric hindrance and conformation of molecules communication no. 11. Crystal
     and molecular structure of 1,5-dibromo-4,8-dichloronaphthalene.] M.A.
     DAVYDOVA and J.T. STRUČKOV, 1965. Ž. Strukt. Khim. SSSR, 6, 113–122 [J.
     Struct. Chem., 6, 98–106].

Monoclinic, $a = 9.57\pm2$, $b = 15.42\pm1$, $c = 7.17\pm1$ Å, $\beta = 90.0\pm5°$, $[U = 1058$ Å$^3]$, $D_m = 2.09$, $Z = 4$, $D_x = 2.24$.

Space group $P2_1/c$ (C_{2h}^5)

Atomic positions

	x	y	z
Br(1)	1.095	0.346	0.839
Br(2)	0.421	0.139	0.778
Cl(1)	0.680	0.020	0.873
Cl(2)	0.835	0.466	0.876
C(1)	0.950	0.262	0.838
C(2)	1.005	0.178	0.850
C(3)	0.912	0.103	0.848
C(4)	0.769	0.118	0.847
C(5)	0.557	0.223	0.807
C(6)	0.517	0.306	0.808
C(7)	0.600	0.378	0.837
C(8)	0.740	0.363	0.843
C(9)	0.802	0.281	0.836
C(10)	0.707	0.203	0.822
H(2)	1.106	0.165	0.850
H(3)	0.967	0.042	0.848
H(6)	0.407	0.319	0.808
H(7)	0.559	0.442	0.837

Standard deviations are: Br, 0.003; Cl, 0.004; C, 0.018 Å.

Structure

The structure is disordered; each molecule can adopt two alternative orientations, related to each other by two-fold rotation about the central bond. For the atomic positions given, chlorine appears to replace bromine (and *vice-versa*) in about 30% of the molecules. For this reason the bond lengths and angles (which are given in full) are not very accurate. The Br-Cl distances are 3.10 and 3.16 Å, imposing severe strain on the naphthalene nucleus. The molecule is not planar; carbon atoms lie as far as 0.25 Å, and halogen atoms as far as 0.48 Å, from the molecular plane.

Details of analysis

Intensity data were recorded on equi-inclination Weissenberg photographs (*hk*0 to *hk*5) using CuK$_\alpha$ radiation. 759 reflexions were observed and measured visually. Refinement was by least squares to a final R index of 0.184.

1,4-NAPHTHOQUINONE AND DERIVATIVES

I. Structure de l'α-naphtoquinone. J. GAULTIER and C. HAUW, 1965. *Acta Cryst.*,
   18, 179-183.

II. Structure de la bromo-2-naphtoquinone-1,4. *Idem*, 1965. *Ibid.*, 18, 604-608.

III. Structure des dérivés 2 et 2,3 de la naphtoquinone-1,4. IV. Le phtiocal-
   antagonisme par analogie structurale. *Idem*. 1965. *Ibid.*, 19, 919-926.

IV. Structures crystallines des dérivés 2 et 2,3 de la naphtoquinone-1,4. III.
   Chloro-2-amino-3-naphtoquinone-1,4. *Idem*, 1965. *Ibid.*, 19, 585-590.

V. Structures cristallines des dérivés 2 et 2,3 de la naphtoquinone-1,4. II.
   Chloro-2-hydroxy-3-naphtoquinone-1,4. *Idem*, 1965. *Ibid.*, 19, 580-584.

I. $R(2)$ = H; $R(3)$ = H

Monoclinic, a = 8.27±2, b = 7.76±2, c = 11.71±2 Å, β = 99°30±20', U = 741 Å³,
D_m = 1.42, z = 4, D_x = 1.417.

Space group $P2_1/c$ (C_{2h}^5)

Atomic positions

	x	y	z
C(1)	0.3436	-0.0934	0.1536
C(2)	0.2800	-0.2715	0.1554
C(3)	0.1759	-0.3325	0.0680
C(4)	0.1316	-0.2230	-0.0327
C(5)	0.1478	0.0555	-0.1337
C(6)	0.2165	0.2229	-0.1417
C(7)	0.3253	0.2845	-0.0496
C(8)	0.3591	0.1842	0.0520
C(9)	0.2984	0.0173	0.0561
C(10)	0.1935	-0.0471	-0.0395
O(1)	0.4312	-0.0387	0.2387
O(4)	0.0381	-0.2808	-0.1150

Temperature factors are given also.

II. $R(2)$ = Br; $R(3)$ = H

Monoclinic, a = 13.88±2, b = 3.98±2, c = 15.74±2 Å, β = 104°0±30', [U = 843.7
Å³], Z = 4, D_x = 1.87.

Space group $P2_1/c$ (C_{2h}^5)

Atomic positions

	x	y	z
C(1)	0.7207	0.0005	0.5301
C(2)	0.8265	0.1020	0.5491
C(3)	0.8879	0.0953	0.6280
C(4)	0.8486	-0.0351	0.7040
C(5)	0.7140	-0.3085	0.7542
C(6)	0.6201	-0.4309	0.7414
C(7)	0.5543	-0.4225	0.6566
C(8)	0.5872	-0.2708	0.5868
C(9)	0.6859	-0.1570	0.6037
C(10)	0.7482	-0.1555	0.6880
O(1)	0.6680	0.0209	0.4587
O(4)	0.9073	-0.0161	0.7780
Br	0.8690	0.2881	0.4553

Temperature factors (anisotropic for the bromine atom) are given also.

III. $R(2) = CH_3$; $R(3) = OH$

Monoclinic, $a = 11.85\pm2$, $b = 4.85\pm2$, $c = 7.71\pm2$ Å, $\beta = 90°30'$, $U = 443$ Å^3, $Z = 2$, $D_x = 1.41$

Space group $P2_1$ (C_2^2) or $P2_1/m$ (C_{2h}^2) $P2_1$ is confirmed by the analysis.

Atomic positions

	x	y	z
C(1)	0.2807	0.1873	0.4564
C(2)	0.1846	0.3781	0.4438
C(3)	0.1231	0.3920	0.2982
C(4)	0.1461	0.2155	0.1434
C(5)	0.2678	−0.1368	0.0127
C(6)	0.3605	−0.3109	0.0238
C(7)	0.4277	−0.3231	0.1678
C(8)	0.4006	−0.1614	0.3091
C(9)	0.3082	0.0131	0.3022
C(10)	0.2426	0.0278	0.1527
O(1)	0.3375	0.1720	0.5896
O(4)	0.0834	0.2333	0.0170
O(H)	0.0368	0.5732	0.2794
C(H3)	0.1599	0.5546	0.5993

Temperature factors are given also.

IV. $R(2) = Cl$; $R(3) = NH_2$

Monoclinic, $a = 8.11\pm2$, $b = 3.93\pm2$, $c = 14.84\pm3$ Å, $\beta = 113°$, $U = 435$ Å^3, $Z = 2$, $D_x = 1.59$.

Space group Pc (C_s^2) or $P2/c$ (C_{2h}^4) Pc is confirmed by structure analysis.

Atomic positions

	x	y	z
C(1)	0.3047	0.1689	0.1249
C(2)	0.2127	0.0231	0.0291
C(3)	0.2885	0.0003	−0.0372
C(4)	0.4744	0.1321	−0.0121
C(5)	0.7449	0.4183	0.1082
C(6)	0.8368	0.5732	0.1997
C(7)	0.7499	0.6046	0.2650
C(8)	0.5765	0.4571	0.2400
C(9)	0.4837	0.3091	0.1495
C(10)	0.5697	0.2854	0.0839
O(1)	0.2330	0.1815	0.1854
O(4)	0.5408	0.1028	−0.0737
N(H$_2$)	0.2083	−0.1424	−0.1286
Cl	0.0000	−0.1264	0.0000

Isotropic temperature factors are given also.

V. $R(2) = Cl$; $R(3) = OH$

Monoclinic, $a = 8.25\pm2$, $b = 3.92\pm2$, $c = 14.39\pm3$ Å, $\beta = 113°20'$, $U = 427$ Å^3, $Z = 2$, $D_x = 1.62$.

Space group Pc (C_s^2) or $P2/c$ (C_{2h}^4) Pc is confirmed by structure analysis.

Atomic positions

	x	y	z
C(1)	0.3050	0.1483	0.1219
C(2)	0.2090	0.0110	0.0254
C(3)	0.2703	-0.0100	-0.0478
C(4)	0.4528	0.1108	-0.0213
C(5)	0.7242	0.3980	0.0938
C(6)	0.8200	0.5515	0.1877
C(7)	0.7461	0.5900	0.2589
C(8)	0.5759	0.4394	0.2381
C(9)	0.4774	0.2918	0.1459
C(10)	0.5561	0.2721	0.0725
O(1)	0.2394	0.1672	0.1885
O(4)	0.5063	0.0814	-0.0920
O(H)	0.1858	-0.1609	-0.1347
Cl	0.0000	-0.1428	0.0000

Isotopic temperature factors are given also.

Structure

All the molecules investigated are essentially planar. Bond lengths and angles are given in full; some distances are:

	I	II	III	IV	V
C(1) - C(2)	1.48	1.48	1.47	1.44	1.40 Å
C(2) - C(3)	1.31	1.32	1.33	1.35	1.34
C(3) - C(4)	1.45	1.51	1.49	1.49	1.48
C(4) - C(10)	1.46	1.44	1.46	1.46	1.43
C(10)- C(9)	1.39	1.40	1.39	1.40	1.44
C(9) - C(1)	1.43	1.49	1.50	1.46	1.44
C(1) - O(1)	1.21	1.19	1.22	1.24	1.27
C(4) - O(4)	1.22	1.25	1.22	1.23	1.26
C(2) - R(2)		1.87	1.50	1.71	1.72
		(Br)	(CH₃)	(Cl)	(Cl)
C(3) - R(3)			1.36	1.37	1.31
			(OH)	(NH₂)	(OH)
C - C (mean aromatic)	1.39	1.40	1.38	1.40	1.41
R(3) - O(4) (Intramolecular)			2.67	2.67	2.67
R(3) - O(4') (Intermolecular)			2.79	2.85	2.75

From this distribution of distances it is inferred that, in the hydroxy- and amino-substituted derivatives, (III, IV, V) there is some contribution from tautomeric forms in which O(1) has acquired a hydrogen atom. [However, it is remarkable that in this entire series of papers, there is no indication of accuracy.] For these derivatives also, it is reasonably inferred that bifurcated hydrogen bonding occurs, with a hydroxyl or amino hydrogen atom coordinated to two ketonic oxygen atoms, one in the same, and one in a neighbouring molecule. [However, the hydrogen atoms have not been located.]

Details of analysis

The intensity data were recorded on De Jong retigrams, using CuK_{α} radiation. Some further details are: I, $0kl$ to $7kl$, 478 reflexions; II, $h0l$ to $h3l$, 560 reflexions; III, $h0l$ to $h4l$, 607 reflexions; IV, 533 reflexions; V, 386 reflexions. The intensities were estimated visually. Refinement was by least squares to final R indices as follows: I, 0.13; II, 0.09; III, 0.11 (including 170 unobserved); IV, 0.11; V, 0.11.

CORDEAUXIA-QUINONE
(7-ACETYL-2,5,8-TRIHYDROXY-3-METHOXY-6-METHYL-1,4-NAPHTHOQUINONE)

$C_{14}H_{12}O_7$ F.W. = 292.3

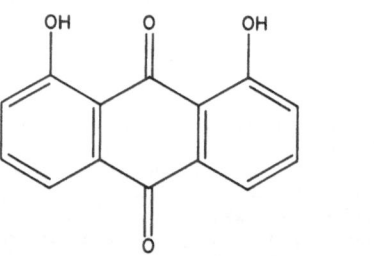

I. Die Struktur des Blattfarbstoffes Cordeauxia-Chinon. M. FEHLMANN and A.
   NIGGLI, 1965. *Helv. Chim. Acta*, <u>48</u>, 305–308.

Triclinic, a = 7.58, b = 10.84, c = 3.90 Å, α = 97.2, β = 93.2, γ = 101.2°,
U = 310 Å^3, z = 1, D_x = 1.57.

Space group P1 (C_1^1)

Atomic positions are not given.

Structure

The analysis establishes the positions of the substituents given above, and
favours the tautomeric form shown. The molecule is essentially planar except
for the methoxy and acetyl groups, which are rotated considerably out of the plane.
Intra- and intermolecular O–H..O bonds are inferred as indicated above. The latter
link molecules together to form endless chains.

Details of analysis

The intensity data were recorded with a linear diffractometer. Refinement
was by least squares to a final R index of 0.14.

CHRYSAZIN

$C_{14}H_8O_4$ F.W. = 240.2

I. The crystal and molecular structure of chrysazin. A. PRAKASH, 1965. Z.
   Kristallogr., <u>122</u>, 272–282.

Tetragonal, a = 5.760±15, c = 31.45±6 Å, [U = 1043 Å^3], D_m = 1.537, z = 4, D_x = 1.529.
(Crystal data are reported also in <u>1</u>.)

Space group P4$_1$ (C_4^2)   (or the enantiomorphic P4$_3$ (C_4^4))

Atomic positions

	x	y	z
C(1)	0.446	0.539	0.0951
C(2)	0.256	0.522	0.0661
C(3)	0.280	0.341	0.0327
C(4)	0.089	0.344	0.0008
C(5)	0.113	0.157	−0.0321

	x	y	z
C(6)	−0.075	0.153	−0.0623
C(7)	−0.054	−0.017	−0.0944
C(8)	0.109	−0.184	−0.0932
C(9)	0.300	−0.177	−0.0667
C(10)	0.290	−0.013	−0.0333
C(11)	0.477	−0.002	−0.0006
C(12)	0.452	0.176	0.0340
C(13)	0.631	0.194	0.0630
C(14)	0.610	0.375	0.0944
O(1)	−0.061	0.501	0.0000
O(2)	0.631	−0.153	−0.0038
O(3)	0.116	0.692	0.0651
O(4)	−0.242	0.317	−0.0615

The standard deviations are estimated to be about 0.03 Å.

Structure

The molecule is essentially planar. Bond lengths and angles are given in full, and are consistent with expectation. The hydrogen atoms have not been located, but intramolecular hydrogen bonding is inferred for both hydroxyl groups, with OH...O distances of 2.44 and 2.54 Å. There seems to be no intermolecular hydrogen bonding. However, there is a short contact (2.68 Å) between the ketonic oxygen atoms of molecules related by the translation −1, 1, 0.

Details of analysis

The intensities of the $0kl$ and $hk0$ reflexions were recorded on Weissenberg photographs, using CuK_α radiation, and were measured photometrically. Refinement was by Fourier and least-squares methods to a final R index of 0.16.

1.   *Structure Reports*, <u>21</u>, 608.

TRIKETOINDANE

$C_9H_4O_3$ F.W. = 160.1

I. The crystal structure of triketoindane. W. BOLTON, 1965. *Acta Cryst.*, <u>18</u>, 5-10.

Tetragonal, $a = 7.058\pm5$, $c = 28.77\pm1$ Å, $[U = 1433$ Å$^3]$, $D_m = 1.48$, $Z = 8$, $D_x = 1.49$.

Space group $I4_1cd$ (C_{4v}^{12}) Molecular symmetry, two-fold axis.
Crystal data are given for an orthorhombic form also. For details see <u>1</u>.

Atomic positions

	x	y	z
C(1)	0	0	0.0785
C(2)	0.1492	0.0852	0.0462
C(3)	0.0858	0.0485	−0.0019
C(4)	0.1754	0.0959	−0.0433
C(5)	0.0874	0.0489	−0.0841
O(1)	0	0	0.1197
O(2)	0.2936	0.1651	0.0595

Standard deviations range from 0.006 to 0.009 Å. Also given are anisotropic temperature factors for the above atoms, and assumed positions for the hydrogen atoms.

Structure

The molecule is planar, and its dimensions are as shown in Fig. 59. The structural arrangement is indicated in Fig. 60. The intermolecular distance C(1)–O(2) has the remarkably small value of 2.83±1 Å. Reasons for this are discussed.

Details of analysis

Intensity data were recorded on equi-inclination Weissenberg photographs (*h0l* to *h5l*) using CuK$_\alpha$ radiation. 380 of a possible 400 reflexions were observed and estimated visually. Refinement was by least squares to a final R index of 0.079.

1. *This volume*, p. 422.

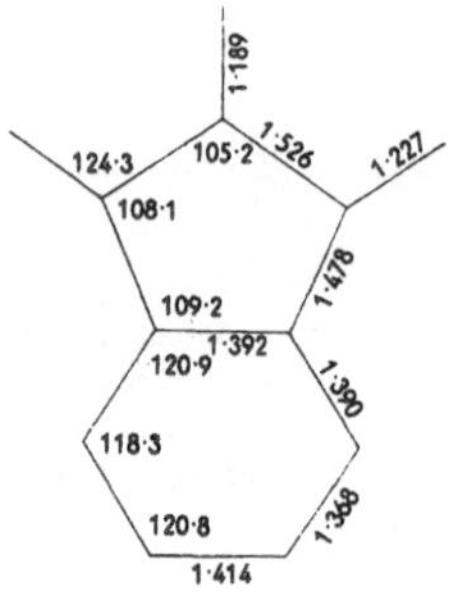

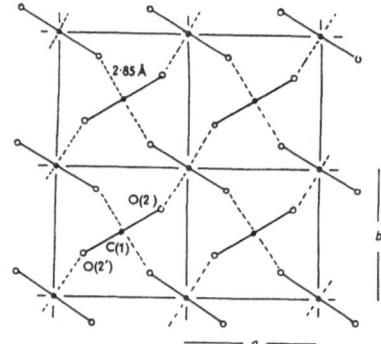

Fig. 59. Bond lengths and angles in triketoindane. Standard deviations range from 0.007 to 0.014 Å.

Fig. 60. A bounded projection of the triketoindane structure, of thickness *c*/4. Only C(1) and O(2) are shown. The broken lines indicate short inter-molecular contacts.

6,6-DIBROMO-2,3:4,5-DIMETHANO-2,4-DINITROCYCLOHEXANONE

$C_8H_6Br_2N_2O_5$ F.W. = 370.0

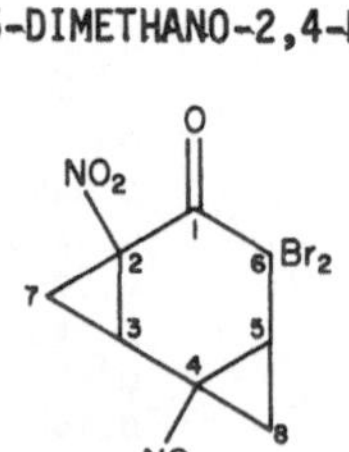

I. The crystal structure of 6,6-dibromo-2,3:4,5-dimethano-2,4-dinitrocyclohexanone. C.H. STAM and H. EVERS, 1965. *Rec. Trav. Chim. Pays-Bas*, **84**, 1496–1502.

Monoclinic, $a = 13.656\pm2$, $b = 6.255\pm6$, $c = 13.658\pm2$ Å, $\beta = 110.41$ 2°, $[U = 1093$ Å$^3]$, $Z = 4$, $[D_x = 2.25]$.

Space group $P2_1/n$ (C_{2h}^5)

Atomic positions

	x	y	z
Br (1)	0.40493	0.56249	0.15724
Br (2)	0.26110	0.18998	0.18114
O(1)	0.1870	0.6607	0.0324
O(2)	-0.0104	0.6083	-0.1150
O(3)	0.0392	0.5213	-0.2412
O(4)	0.2291	-0.1915	-0.1743
O(5)	0.3907	-0.1565	-0.0750
N(1)	0.0482	0.5064	-0.1506
N(2)	0.3007	-0.0828	-0.1130
C(1)	0.2022	0.4780	0.0117
C(2)	0.1270	0.3640	-0.0810
C(3)	0.1678	0.1876	-0.1328
C(4)	0.2784	0.1312	-0.0853
C(5)	0.3480	0.2163	0.0209
C(6)	0.2986	0.3584	0.0755
C(7)	0.0910	0.1349	-0.0805
C(8)	0.3601	0.2954	-0.0790

Standard deviations are: Br, 0.0025; other atoms, 0.02 Å. Temperature factors (anisotropic for bromine) are given also.

Structure

The analysis confirms the formulation given above. The conformation is illustrated in Fig. 61. Bond lengths and angles are given in full, and are consistent with expectation. The angles between the planes of the three-membered rings and the mean plane of the cyclohexane ring are 105 and 99°. The intermolecular distances appear to be normal.

Details of analysis

The intensity data were recorded on equi-inclination Weissenberg photographs (6 layers about [010], 2 about [101]) using CuK_α radiation. 1366 reflexions were observed; some were measured photometrically, but most visually. Refinement was by least squares to a final R index of 0.111.

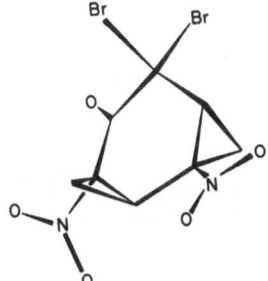

Fig. 61. The molecular conformation of $C_8H_6Br_2N_2O_5$.

TRICYCLO(5.3.0.0(2,6))DECAN-3,8-DIONE

$C_{10}H_{12}O_2$ F.W. = 164.2

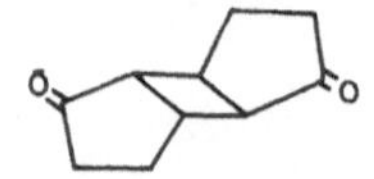

I. The crystal structure of the centrosymmetric photodimer of cyclopentanone.
   T.N. MARGULIS, 1965. *Acta Cryst.*, 18, 742-745.

Monoclinic, a = 6.78±3, b = 7.23±3, c = 8.67±3 Å, β = 98.9±2°, [U = 419.9 Å^3],
Z = 2, D_x = 1.30.

Space group $P2_1/n$ (C_{2h}^5) Molecular symmetry, centre.

Atomic positions

	x	y	z
C(1)	0.0285±13	-0.0986±18	0.0996±10
C(2)	-0.1582±12	0.0021±19	0.0151±10
C(3)	-0.2303±13	0.1388±18	0.1327±10
C(4)	-0.0563±14	0.1560±20	0.2676±11
C(5)	0.0723±12	-0.0087±18	0.2584±10
O	0.1957±10	-0.0638±13	0.3644± 8

Isotropic temperature factors for the above atoms, and positions of the hydrogen atoms are given also.

Structure

The conformation of the molecule is illustrated in Fig. 62. The cyclobutane ring is (of necessity) planar, but the five-membered ring is not; C(2) and C(3) lie 0.21 and 0.45 Å respectively from (and on the same side of) the plane containing C(5) and its adjacent atoms. Bond lengths and angles are given in Fig. 63. The intermolecular distances are consistent with van der Waals interaction.

Details of analysis

The intensity data were recorded on equi-inclination Weissenberg photographs (h0l to h4l) using CuK$_\alpha$ radiation. 354 reflexions were observed and measured visually. Refinement was by full-matrix least squares to a final R index of 0.157.

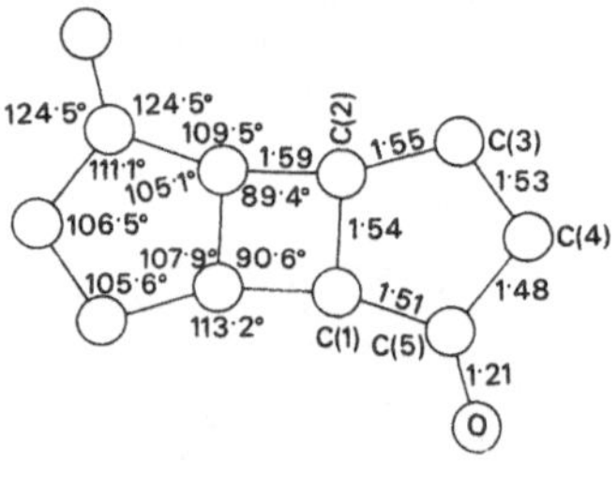

Fig. 62. Bond lengths and angles of tricyclo(5.3.0.0(2,6))decan-3,8-dione. The
          standard deviation of distances is 0.015 Å. It is stated that, because
          of thermal motion, the C-O distance may be underestimated by 0.01 Å.

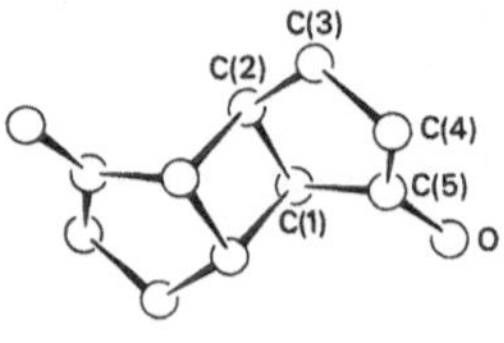

Fig. 63. The tricyclo(5.3.0.0(2,6))decan-3,8-dione molecule viewed along *b*.

PYRENE

$C_{16}H_{10}$

F.W. = 202.2

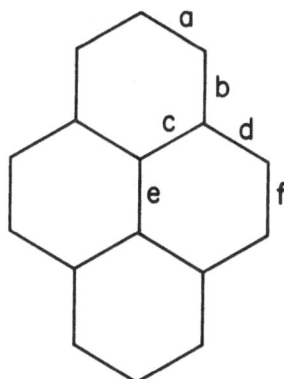

I. The crystal and molecular structure of pyrene. A. CAMERMAN and J. TROTTER, 1965. *Acta Cryst.*, 18, 636–643. (For an earlier determination from projections see 1.)

Monoclinic, $a = 13.649\pm10$, $b = 9.256\pm10$, $c = 8.470\pm10$ Å, $\beta = 100.28\pm4$, $U = 1052.9$ Å^3, $D_m = 1.27$, $Z = 4$, $D_x = 1.275$, (Cu$K_{\alpha1}$, $\lambda = 1.54051$ Å; Cu$K_{\alpha2}$, $\lambda = 1.54433$ Å).

Space group $P2_1/a$ (C_{2h}^5)

Atomic positions

	x	y	z
C(1)	0.2817	-0.0402	0.4119
C(2)	0.2947	0.0246	0.2717
C(3)	0.2296	-0.0077	0.1274
C(4)	0.2389	0.0567	-0.0238
C(5)	0.1783	0.0237	-0.1578
C(6)	0.0990	-0.0738	-0.1606
C(7)	0.0316	-0.1103	-0.3020
C(8)	-0.0449	-0.2090	-0.2966
C(9)	-0.0566	-0.2746	-0.1594
C(10)	0.0071	-0.2396	-0.0131
C(11)	-0.0030	-0.3070	0.1356
C(12)	0.0575	-0.2772	0.2706
C(13)	0.1389	-0.1735	0.2723
C(14)	0.2066	-0.1412	0.4161
C(15)	0.1514	-0.1091	0.1303
C(16)	0.0854	-0.1409	-0.0136

Standard deviations are given in full, and range from 0.004 to 0.010 Å. Anisotropic temperature factors for the carbon atoms are given and analysed in terms of rigid-body motion. The positions of the hydrogen atoms (with standard deviation 0.07 Å) are given also.

Structure

The molecule is planar, with non-crystallographic *mmm* symmetry. Mean bond lengths (corrected for thermal libration) and bond angles are:

a	=	1.380±11 Å	aa'	= 121.8°
b	=	1.420±9	ab	= 120.0
c	=	1.417±7	bc	= 118.9
d	=	1.437±9	cd	= 118.8
e	=	1.417±14	ec	= 119.8
f	=	1.320±14	df	= 121.5

(angles ±1.0 to 1.4°). [The standard deviations for bond lengths may be optimistic.]

Details of analysis

Intensity data were measured with a four-circle diffractometer and scintillation counter, using MoK_α radiation. 550 reflexions were observed in the sphere
$2\theta \leqslant 50.2°$; this is about 59% of the total accessible. Refinement was by differential syntheses to a final R index of 0.12.

1. *Structure Reports*, <u>11</u>, 700.

2,7-DIACETOXY-*trans*-15,16-DIMETHYL-15,16-DIHYDROPYRENE

$C_{22}H_{20}O_4$ F.W. = 348.4

I. The crystal structure of 2,7-diacetoxy-*trans*-15,16-dimethyl-15,16-dihydro-
   pyrene. A.W. HANSON, 1965. *Acta Cryst.*, <u>18</u>, 599-604.

Monoclinic, $a = 20.07\pm3$, $b = 7.56\pm1$, $c = 5.88\pm1$ Å, $\beta = 96.52\pm4°$, $U = 886.4$ Å^3,
$D_m = 1.31$, $Z = 2$, $D_x = 1.305$. (values above at room temperature. Cell dimensions
were also measured at $-130°C$; they are: $a = 19.93\pm6$, $b = 7.52\pm2$, $c = 5.81\pm2$ Å,
$\beta = 96.5\pm1°$).

Space group $P2_1/a$ (C_{2h}^5)

Atomic positions

	x	y	z
C(1)	0.1314	0.1480	-0.1461
C(2)	0.1035	0.3082	-0.2221
C(3)	0.0366	0.3516	-0.2180
C(4)	-0.0064	0.2356	-0.1234
C(5)	-0.0752	0.2643	-0.1247
C(6)	-0.1163	0.1428	-0.0300
C(7)	0.0239	0.0789	0.0128
C(8)	0.0930	0.0246	-0.0456
C(9)	0.0355	0.1446	0.2664
C(10)	0.1960	0.5046	-0.2212
C(11)	0.2308	0.6223	-0.3717
O(12)	0.1417	0.4269	-0.3389
O(13)	0.2115	0.4804	-0.0221

The numbering scheme is arbitrary. Standard deviations range from 0.0014
to 0.0025 Å. Also given are anisotropic temperature factors for the above atoms,
and the positions and isotropic temperature factors of the hydrogen atoms. These
data are for the structure at room temperature. Corresponding data for the
structure at $-130°C$ are given also.

Structure

The configuration of the molecule is shown in Fig. 64. Except for the acetoxy
group, the molecule has an approximate plane of symmetry passing through C(2),
C(7) and C(9). The sixteen peripheral atoms of the 15,16-dihydropyrene nucleus
are approximately coplanar, and the C-C distances in the ring range from 1.384 to
1.403 Å, with a mean value of 1.391±6 Å. The compound can therefore be considered
as 16-annulene with external and internal substituents. Bond lengths and angles
are given in full. Selected bond lengths are:

$$C(4) - C(7) = 1.518\pm2 \text{ Å}$$
$$C(8) - C(7) = 1.522\pm2$$
$$C(7) - C(7') = 1.527\pm3$$
$$C(7) - C(9) = 1.564\pm2$$

Corresponding data are given for the structure at $-130°C$, and the two sets are compared.

Details of analysis

For the structure at room temperature the intensity data were measured with a four-circle diffractometer, using CuK_α radiation, in the range $2\theta\leqslant164°$. 1509 of 1964 accessible reflexions were observed above background. Refinement was by block-diagonal least squares to a final R index of 0.044. For the low-temperature analysis, intensity data were recorded on equi-inclination Weissenberg photographs about b and c, covering essentially the entire copper sphere. 1495 of 1842 accessible reflexions were observed and measured visually. Refinement was by block-diagonal least squares to a final R index of 0.099.

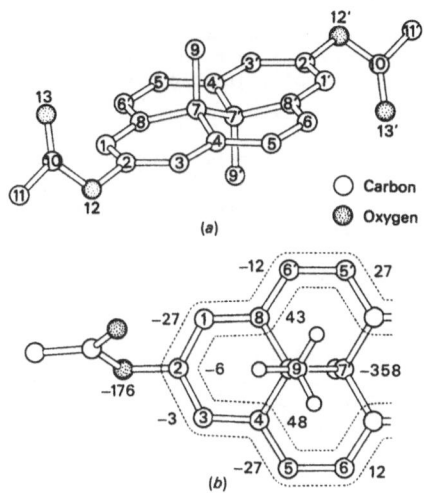

Fig. 64.   (a)   The $C_{22}H_{20}O_4$ molecule, viewed along b.
(b)   Part of the molecule viewed along the normal to the mean plane of the 16-ring. The distances of some atoms from this plane (in Å × 1000) are indicated.

CORONENE

$C_{24}H_{12}$ F.W. = 300.3

I. The crystal and molecular structure of coronene.   J.K. FAWCETT and J. TROTTER, 1965. *Proc. Roy. Soc.*, <u>A</u>, 289, 366–376. (For earlier work, see <u>1</u>.)

Monoclinic, $a = 16.119\pm6$, $b = 4.702\pm4$, $c = 10.102\pm6$ Å, $\beta = 110.9\pm1°$, $U = 715.3$ Å^3, $D_m = 1.38$, $Z = 2$, $D_x = 1.39$, (CuK$_{\alpha1}$, $\lambda = 1.54051$ Å; CuK$_{\alpha2}$, $\lambda = 1.54433$ Å).

Space group $P2_1/a$ (C_{2h}^5)

Atomic positions

	x	y	z
C(1)	-0.1201	-0.4079	0.0381
C(2)	-0.1122	-0.4788	0.1782
C(3)	-0.0497	-0.3600	0.2913
C(4)	0.0121	-0.1607	0.2786
C(5)	0.0799	-0.0339	0.3941
C(6)	0.1364	0.1555	0.3761
C(7)	0.1339	0.2444	0.2410
C(8)	0.1909	0.4490	0.2167
C(9)	0.1843	0.5286	0.0847
C(10)	-0.0606	-0.2029	0.0183
C(11)	0.0057	-0.0823	0.1380
C(12)	0.0666	0.1210	0.1206

Also given are the positions and isotropic temperature factors of the hydrogen atoms, and anisotropic temperature factors for the atoms listed above. The thermal motion has been analysed in terms of rigid-body modes.

Structure

The molecular structure is essentially as described in 1. Mean lengths for the four types of bond present are: $a = 1.346\pm4$ Å; $b = 1.415\pm3$ Å; $c = 1.430\pm3$ Å; $d = 1.422\pm4$ Å. (These values are corrected for thermal libration; corrections are 0.002 to 0.003 Å). The molecule is slightly but significantly non-planar, pairs of outer atoms lying alternately above and below the mean molecular plane. The maximum deviation is 0.029 Å.

Details of analysis

The intensity data were recorded with a four-circle diffractometer and scintillation counter, using CuK$_\alpha$ radiation. The θ-2θ scan method was used to measure reflexions in the range $2\theta\leqslant125$; of these 936, or 82% of those accessible, were observed. Refinement was by block-diagonal least squares to a final R index of 0.157.

1.   *Structure Reports*, 11, 709.

DODECAFLUORO-TRICYCLO(3,3,0,0(2,6))OCTANE

C_8F_{12} F.W. = 324.1

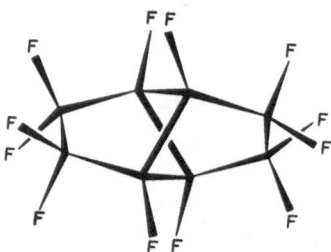

I. The structure of C_8F_{12}, a saturated dimer of hexafluoro-butadiene. I.L. KARLE, J. KARLE, T.B. OWEN and J.L. HOARD, 1965. *Acta Cryst.*, 18, 345-351.

Triclinic, $a = 6.02 \pm 1$, $b = 6.29 \pm 1$, $c = 7.27 \pm 1$ Å, $\alpha = 103.6 \pm 4$, $\beta = 107.9 \pm 4$, $\gamma = 106.4 \pm 4°$, [$U = 235.1$ Å^3], $Z = 1$, $D_X = 2.29$.

Space group P1 (C_1^1) or P$\bar{1}$ (C_i^1)   P$\bar{1}$ is suggested by intensity statistics and confirmed by structure analysis for a disordered structure in which two superposed molecules are related by a centre of symmetry.

Atomic positions

	x	y	z
C(1)	0.6084	0.6415	0.4799
C(2)	0.7401	0.8076	0.7048
C(3)	0.5998	0.6326	0.7911
C(4)	0.4445	0.4361	0.6001
C(5)	0.6128	0.4080	0.4862
C(6)	0.4382	0.2301	0.2666
C(7)	0.2212	0.3327	0.2338
C(8)	0.3408	0.5191	0.4380
F(1)	0.7046	0.7549	0.3672
F(2)	0.6984	1.0012	0.7447
F(3)	0.9926	0.8929	0.7892
F(4)	0.4648	0.7062	0.8838
F(5)	0.7533	0.5864	0.9323
F(6)	0.2953	0.2451	0.6328
F(7)	0.8125	0.3540	0.5544
F(8)	0.3615	−0.0022	0.2679·
F(9)	0.5442	0.2290	0.1337
F(10)	0.0054	0.1774	0.2000
F(11)	0.1837	0.4100	0.0726
F(12)	0.1875	0.6461	0.4456

Mean standard deviations are 0.020 Å for carbon, and 0.012 Å for fluorine. Anisotropic temperature factors are given also.

Structure

The analysis demonstrates the molecular structure indicated above. The average unit cell contents consists of two superposed molecules related by a centre of symmetry. Bond lengths and angles are given in full, but are of limited accuracy. Mean values of bond lengths are: C–C, 1.50; C–F, 1.34 Å. Angles are as nearly tetrahedral as the constraints of the system will allow, and range from 81.5° (for the internal angles of the cyclobutane ring) to 123.9°. The shortest intermolecular distance for the disordered structure is 2.79 Å. This corresponds to van der Waals separation, and it is therefore not necessary to postulate domains of order in the crystal.

Details of analysis

Intensity data were recorded on equi-inclination Weissenberg photographs about the [100], [010], and [110] directions, with CuK_α radiation. 753 of a possible 1048 reflexions were observed and estimated visually. Refinement was by full-matrix least squares, with precautions to prevent the coalescence of nearly-overlapping atoms. The final R index was 0.177.

anti-8-TRICYCLO[3,2,1,0^{2,4}]OCTYL *p*-BROMOBENZENESULPHONATE

$C_{14}H_{15}O_3SBr$ F.W. = 343.2

I. The crystal and molecular structure of *anti*-8-tricyclo[3,2,1,0^{2,4}]-octyl *p*-bromobenzenesulphonate. A.C. MacDONALD and J. TROTTER, 1965. *Acta Cryst.*, **18**, 243-249.

Monoclinic, a = 23.76±2, b = 7.06±1, c = 19.49±2 Å, β = 120.26±4°, U = 2825 Å^3, D_m = 1.62, Z = 8, D_x = 1.614, (CuK_α, λ = 1.5418 Å).

Space group   Cc   (C_s^4)   or   C2/c   (C_{2h}^6)   C2/c is confirmed by the analysis

Atomic positions

	x	y	z
Br	0.1131	0.0085	0.3837
S	0.2367	0.6539	0.6484
O(1)	0.1731	0.7434	0.6382
O(2)	0.2736	0.5709	0.7260
O(3)	0.2651	0.8014	0.6243
C(1')	0.1520	0.2010	0.4625
C(2')	0.1620	0.3794	0.4408
C(3')	0.1890	0.5176	0.4974
C(4')	0.2066	0.4703	0.5772
C(5')	0.1983	0.2911	0.5969
C(6')	0.1717	0.1543	0.5421
C(1)	0.0687	0.5776	0.5930
C(2)	0.0309	0.5082	0.6319
C(3)	0.0673	0.4213	0.7138
C(4)	0.0508	0.6318	0.7071
C(5)	0.1069	0.7625	0.7023
C(6)	0.0688	0.9054	0.6361
C(7)	0.0407	0.7838	0.5599
C(8)	0.1337	0.6328	0.6637

Temperature factors (some anisotropic) are given also.

Structure

The molecular configuration is illustrated in Fig. 65. Bond lengths and angles are given in full, and are consistent with the formulation given above. The angle at the bridgehead (C–CO–C) is 97°, and the remaining C–C–C angles in the norbornane nucleus range from 102 to 107°; considerable strain is inferred for the nucleus. The angles in the cyclopropane ring are all 60°. Intermolecular distances are consistent with van der Waals interaction.

Details of analysis

The intensity data were recorded with a four-circle diffractometer and scintillation counter, using the θ-2θ scan mode, and CuK_α radiation. 1034 of a possible 1062 reflexions in the range 2θ≤90 were observed. Refinement was by block-diagonal least squares to a final R index of 0.093.

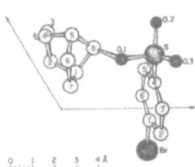

Fig. 65.   The C$_{14}$H$_{15}$O$_3$SBr molecule viewed along *b*.

ADAMANTANE
(TRICYCLO 3,3,1,1^{3,7} DECANE)

$C_{10}H_{16}$ F.W. = 136.2

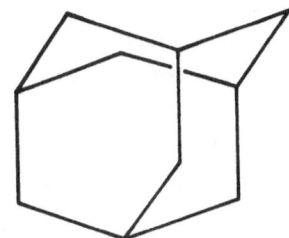

I. Phase transition and crystal structures of adamantane. C.E. NORDMAN and
   D.L. SCHMITKONS, 1965. *Acta Cryst.*, <u>18</u>, 764-767.

II. The crystal structure of adamantane: an example of a false minimum in least
    squares. J. DONOHUE and S.H. GOODMAN, 1965. *Acta Cryst.*, <u>22</u>, 352-354.

(I reports the determination of the crystal structure both at room temperature
and at -110°C. In II a more plausible interpretation of the data is presented for
the low-temperature structure. For earlier work on the room temperature form, see
<u>1</u>.)

AT ROOM TEMPERATURE

Cubic, a = 9.45 Å, U = 844 Å^3, Z = 4, D$_x$ = 1.08.

Space group F432 (O^3), F43m (T_d^2) or Fm3m (O_h^5) *Fm3m is consistent with the
structure analysis. Molecular symmetry, m3m.*

Atomic positions (from I)

		x	y	z
C(Methylene)	24(e)	0.182	0	0
1/2 C(Methine)	32(f)	0.095	(0.095)	(0.095)
1/2 H(Methylene)	96(k)	0.066	(0.066)	-0.252
1/2 H(Methine)	32(f)	0.159	(0.159)	(0.159)

Structure

The structure proposed is similar to that described in <u>1</u>, but is disordered in
that two orientations are present, each related to the other by a 90° rotation about
an axis through two opposite methylene groups (*e.g.*, about any cell edge).

Details of analysis

The intensity data were recorded on Weissenberg (CuK_α radiation) and precession (MoK_α radiation) photographs. 39 independent reflexions were observed
and measured visually. Refinement was by least squares to a final R index of
0.069. (The corresponding figure for an ordered structure is 0.140.)

AT -110°C

Tetragonal, a = 6.60, c = 8.81 Å, U = 384 Å^3, Z = 2, D$_x$ = 1.18.

Space group P$\bar{4}2_1c$ (D_{2d}^4) *Molecular symmetry, $\bar{4}$.*

Atomic positions (from II)

	x	y	z
C(1)	-0.0296±14	0.1870±13	0.0999±9
C(2)	0.1580±10	0.2187±12	-0.0009±19
C(3)	0	0	0.2006±15
H(1)	-0.054 ±12	0.317 ±12	0.164 ±11
H(2)	0.141 ±8	0.017 ±14	0.267 ±8

H(3)	0.290 ±13	0.237 ±11	0.070 ±10
H(4)	0.129 ±11	0.354 ±12	-0.064 ±10

Structure

The structure at -110°C is related to that at room temperature in that $a_{tetr.}$ lies along the a-b diagonal of the cubic cell ($a_{tetr.} \sim a_{cubic}/\sqrt{2}$) and $c_{tetr.}$ corresponds to c_{cubic}. In addition, the molecule is rotated 9° about c, so that the methylene groups in the plane $z = 0$ do not, (as they otherwise would), lie on the a-b diagonals. (In I it is reported that the molecule is significantly squashed in the c direction. However, in II it is shown that the same intensity data are in better accord with an undistorted molecule.) Bond lengths and angles are given in full: C–C distances are equal, within experimental error, with a mean value of 1.536±11 Å, and the C–C–C angles are all within experimental error of the tetrahedral value.

Details of analysis

The intensity data were recorded at -110°C on precession photographs, using CuK_α radiation. 53 reflexions were observed and measured visually. Refinement was by full-matrix least squares to a final R index of 0.039. (Attempted refinement of the "squashed" model converged to a false minimum, with a final R index of 0.080.)

1.   *Structure Reports*, <u>10</u>, 247.

1-p-BROMOBENZENESULPHONYLOXYMETHYL-5-METHYLBICYCLO-[3,3,1]NONAN-9-OL

$C_{17}H_{23}BrO_4S$

F.W. = 403.3

M.P. = 116–118 C

I. Molecular conformations. Part I. The bicyclo[3,3,1]nonane system: X-ray analysis of 1-p-bromobenzenesulphonyloxy-methyl-5-methylbicyclo[3,3,1]nonan-9-ol. W.A.C. BROWN, J. MARTIN and G.A. SIM, 1965. *J. Chem. Soc.*, 1844-1857.

Triclinic, $a = 7.40$, $b = 12.06$, $c = 11.34$ Å, $\alpha = 112°30$, $\beta = 109°31$, $\gamma = 72°41'$, $U = 862$ Å^3, $D_m = 1.53$, $Z = 2$, $D_x = 1.55$.

Space group $P1$ (C_1^1) or $P\bar{1}$ (C_i^1) $P\bar{1}$ confirmed by analysis.

Atomic positions

	x	y	z
C(1)	0.5149	0.4307	0.2186
C(2)	0.7011	0.3530	0.1666
C(3)	0.6584	0.2407	0.0483
C(4)	0.5351	0.1692	0.0572
C(5)	0.3523	0.2486	0.1132
C(6)	0.1974	0.3103	0.0150
C(7)	0.2590	0.4079	-0.0068
C(8)	0.3631	0.4984	0.1238
C(9)	0.4206	0.3481	0.2418
C(10)	0.2597	0.1645	0.1394
C(11)	0.5763	0.5247	0.3525
C(12)	0.8198	0.7893	0.3964
C(13)	1.0083	0.7481	0.3809
C(14)	1.0827	0.8101	0.3434
C(15)	0.9715	0.9188	0.3141
C(16)	0.7836	0.9651	0.3352
C(17)	0.7077	0.9024	0.3764

	x	y	z
O(1)	0.2564	0.4259	0.2984
O(1')	0.5677	0.2904	0.3390
O(2)	0.6549	0.6126	0.3308
O(3)	0.8779	0.6643	0.5518
O(4)	0.5613	0.7954	0.5111
S	0.7268	0.7162	0.4600
Br	1.0781	1.0024	0.2532

Standard deviations (in Å) are given in full. Mean values are: Br, 0.003; S, 0.005; O, 0.013, C, 0.020 Å. (As a result of packing disorder, O(1) appears to occupy two chemically reasonable positions. The standard deviations of these positions are about 0.026 Å). Anisotropic temperature factors are given also.

Structure

Each cyclohexane ring of the bicyclo[3,3,1]nonane group adopts the chair conformation, but is somewhat distorted to allow the reasonable contact distance of 3.06 Å between C(3) and C(7). Bond lengths and angles are given in full, and are consistent with expectation. The hydroxyl oxygen atom randomly occupies two stereochemically acceptable positions on opposite sides of the plane containing C(1), C(5) and C(9). For a preliminary account see 1.

Details of analysis

The intensity data were recorded on equi-inclination Weissenberg photographs ($hk0$ to $hk7$; $0kl$ to $1kl$; CuK_α radiation) and precession photographs ($h0l$; MoK_α radiation). 1633 reflexions were observed and measured visually. Refinement was by least squares to a final R index of 0.129.

1. W.A.C. BROWN, J. MARTIN and G.A. SIM, 1964. *Proc. Chem. Soc.*, 57.

1,6-METHANO-CYCLODECAPENTAENE-2-CARBOXYLIC ACID

$C_{12}H_{10}O_2$ F.W. = 186.1

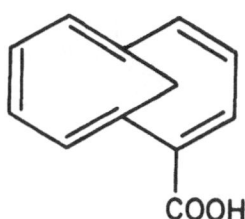

COOH

I. Die Kristallstruktur der 1,6-Methano-cyclodecapentaen-2-carbonsaure. M. DOBLER and J.D. DUNITZ, 1965. *Helv. Chim. Acta*, <u>48</u>, 1429-1440.

Monoclinic, a = 7.59, b = 8.75, c = 15.02 Å, (All ±0.15 Å), γ = 108°5', U = 949 Å³, Z = 4, D_x = 1.28.

Space group $P2_1/c$ (C_{2h}^5) (1st setting)

Atomic positions

	x	y	z
C(1)	0.7608±8	0.0190±6	0.2599±4
C(2)	0.6989±7	-0.0066±6	0.1699±3
C(3)	0.5134±8	-0.0478±7	0.1482±4
C(4)	0.3553±9	-0.1112±8	0.2025±4
C(5)	0.3556±9	-0.1442±8	0.2929±5
C(6)	0.5100±9	-0.0746±7	0.3483±4
C(7)	0.5602±10	-0.1465±8	0.4230±4
C(8)	0.7424±11	-0.1210±8	0.4466±4
C(9)	0.8992±9	-0.0647±8	0.3906±4
C(10)	0.9052±9	-0.0220±8	0.3016±4

C(11)	0.6413±9	0.0797±7	0.3185±4
C(12)	0.8324±8	−0.0030±6	0.0965±4
O(1)	0.7711±5	−0.0218±6	0.0174±2
O(2)	1.0012±6	0.0226±6	0.1134±3

Anisotropic temperature factors are given, and have been analysed in terms of rigid-body motion. The unrefined positions of the hydrogen atoms are given also.

Structure

The molecule has the conformation and dimensions given in Fig. 66. Adjacent centrosymmetrically-related molecules are joined by pairs of O–H..O bonds of length 2.62 Å. The C–O bonds are of equal length, and the bonding hydrogen atoms appear to be midway between the oxygen atoms. This situation is attributed to disorder of the C=O and C–OH groups, rather than to the presence of a symmetrical hydrogen bond.

Details of analysis

The intensity data were measured with a linear diffractometer and scintillation counter (0kl to 8kl), using MoK_α radiation. 878 of a possible 2300 reflexions were observed above background in the region $\sin\theta/\lambda < 0.65$. Refinement was by full-matrix least squares to a final R index of 0.068.

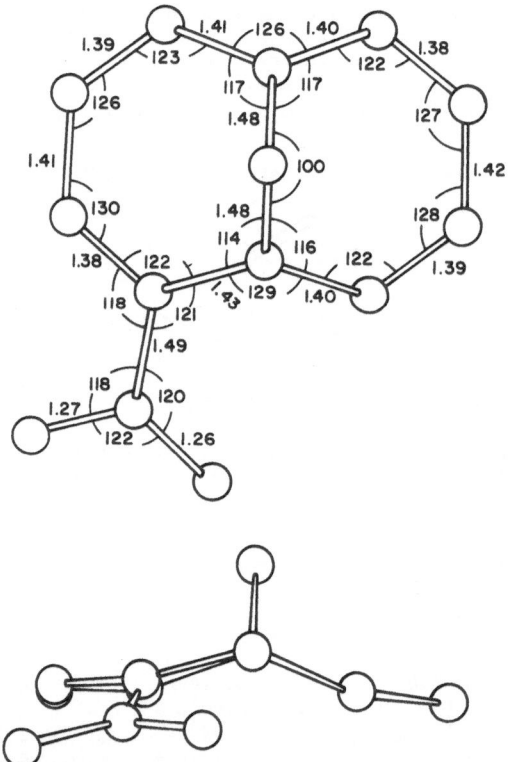

Fig. 66. The molecular geometry of $C_{12}H_{10}O_2$. The standard deviations of distances are 0.01 Å.

CONGRESSANE

$C_{14}H_{20}$ F.W. = 188.3

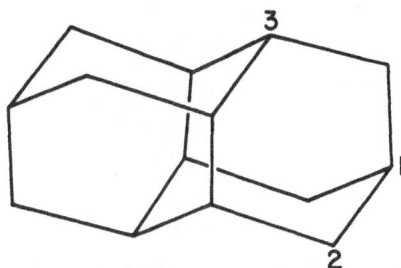

I. The crystal and molecular structure of congressane, $C_{14}H_{20}$, by X-ray diffraction. I.L. KARLE and J. KARLE, 1965. *J. Amer. Chem. Soc.*, **87**, 918-920.

Cubic, a = 10.109±20 Å, [U = 1033 Å³], z = 4, D_x = 1.210.

Space group Pa3 (T_h^6) Molecular symmetry, 3.

Atomic positions

	x	y	z
C(1)	0.1333	0.1333	0.1333
C(2)	0.2193	0.0456	0.0434
C(3).	0.1312	−0.0421	−0.0438
H(1)	0.1907	0.1907	0.1907
H(2,1)	0.2587	0.0973	−0.0320
H(2,2)	0.2523	−0.0120	0.0911
H(3)	0.1824	−0.1085	−0.0920

Structure

The molecule has the configuration depicted above, with C–C distances ranging from 1.532 to 1.538 Å (± 0.005 Å), and C–C–C angles from 108°43' to 110°10' (± 25'). It can thus be considered to consist of a portion of the diamond structure. The smallest intermolecular distance is 4.14 Å.

Details of analysis

The intensity data were recorded on equi-inclination Weissenberg photographs (0kl to 7kl), using CuK_α radiation. 325 independent reflexions were observed and measured visually. Refinement was by least squares to a final R index of 0.11.

trans-(1,4),(5,8)-DIMETHYLENE-*cis*, anti, *cis*-PERHYDROANTHRAQUINONE

$C_{16}H_{20}O_2$ F.W. = 244.3

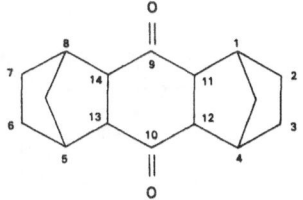

I. The crystal and molecular structures of *trans*-(1,4),(5,8)-dimethylene-*cis,anti, cis*-perhydroanthraquinone. H.G. NORMENT, 1965. *Acta Cryst.*, **18**, 627-635.

Orthorhombic, a = 10.526±6, b = 6.464±3, c = 18.30±12 Å, U = 1245.7 Å³, D_m = 1.299, z = 4, D_x = 1.302.

Space group Pcab (D_{2h}^{15}) Molecular symmetry, centre.

Atomic positions

	x	y	z
C(1)	−0.1441±8	−0.0225±16	0.3571±4
C(2)	−0.0294±8	−0.1753±14	0.3607±4
C(3)	0.0726±8	−0.0519±18	0.3177±5
C(4)	0.0650±7	0.1461±15	0.3639±4
C(5)	−0.0821±8	0.1970±18	0.3609±4
C(6)	0.0291±8	−0.1571±14	0.4394±4
C(7)	−0.0714±8	−0.2072±18	0.4981±4
C(8)	0.0959±6	0.0546±14	0.4409±4
O(1)	−0.1300±7	−0.3681±13	0.4951±3

Also given are anisotropic temperature factors for the above atoms, and the positions and isotropic temperature factors of the hydrogen atoms.

Structure

The configuration of the molecule is shown in Fig. 67. Bond lengths and angles are given in full. Mean values, assuming mirror symmetry, are:

a	=	1.53 Å	ga	=	120.2°
b	=	1.54	aa'	=	119.6
c	=	1.57	ab	=	120.1
d	=	1.57	ac	=	111.6
e	=	1.56	bc	=	104.1
f	=	1.54	[cd	=	107.7]
g	=	1.21	ed	=	103.6
			cf	=	99.4
			df	=	101.3
			ff'	=	96.5

(Standard deviations 0.01 Å and 0.7°).

There are four sets of nearly coplanar atoms. The angles between them are:

ab to bc,	125.7°
bc to ff',	122.3
bc to de,	112.5
de to ff',	125.1

Intermolecular distances are consistent with van der Waals interaction.

Details of analysis

Intensity data were recorded on equi-inclination Weissenberg photographs (h0l to h3l, radiation not stated). 787 independent reflexions were observed and estimated visually. Refinement was by full-matrix least squares to a final R index of 0.103.

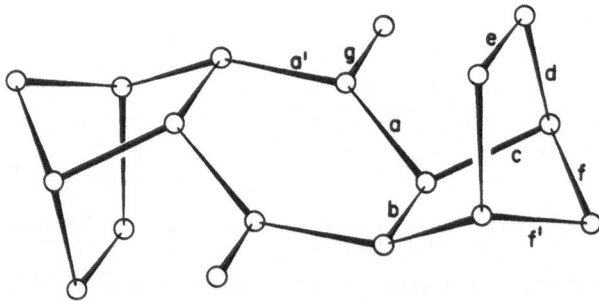

Fig. 67. Configuration of the $C_{16}H_{20}O_2$ molecule.

[3.3]PARACYCLOPHANE

C$_{18}$H$_{20}$ F.W. = 236.4

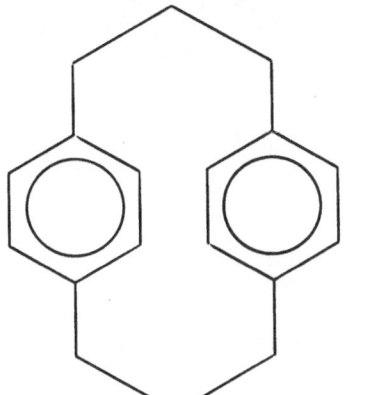

I. The crystal and molecular structure of [3.3] paracyclophane. P.K. GANTZEL
   and K.N. TRUEBLOOD, 1965. *Acta Cryst.*, 18, 958–968.

Monoclinic, $a = 9.715\pm10$, $b = 8.138\pm10$, $c = 8.524\pm10$ Å, $\beta = 90.69\pm3°$, $[U = 673.9$ Å$^3]$,
$D_m = 1.156$, $Z = 2$, $D_x = 1.165$.
(Calibration with sodium chloride, $a = 5.639$ Å).

Space group $P2_1/n$ (C_{2h}^5) Molecular symmetry, centre.

Atomic positions

	x	y	z
C(1)	0.0890	0.0273	0.2167
C(2)	-0.0339	-0.1111	0.2043
C(3)	-0.0578	-0.2253	0.0859
C(4)	0.0537	-0.2664	-0.0068
C(5)	0.1775	-0.1842	0.0071
C(6)	0.1949	-0.0556	0.1124
C(7)	0.3165	0.0582	0.1054
C(8)	0.3159	0.1692	-0.0401
C(9)	0.2000	0.2945	-0.0535

Standard deviations are in the range 0.0025 to 0.003 Å. Anisotropic tempera-.
ture factors are given, and have been analysed in terms of rigid–body motion.
Coordinates which have been corrected for such motion are given also, as are the
positions of the hydrogen atoms.

Structure

The shape of the molecule and the bond lengths and angles are given in Fig.
68. The aromatic rings are distorted: C(6) and C(3) lie about 0.08 Å from the
mean plane of the remaining four atoms. Intermolecular distances are normal
van der Waals values.

Details of analysis

Intensities were recorded on equi–inclination Weissenberg photographs about
[010] and [101], using CuK$_\alpha$ radiation. Of 1480 accessible reflexions 1095 were
observed. Refinement was by full–matrix least squares to a final *R* index of 0.11.

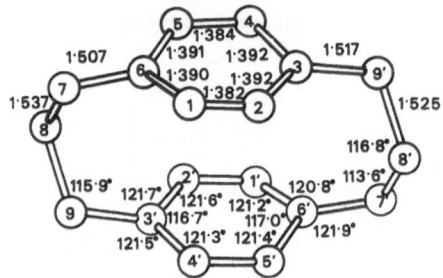

Fig. 68.   Bond distances and angles in [3.3] paracyclophane after correction
for thermal motion.   Standard deviations 0.004 Å and 0.3°.

α-TRICHLOROMETHYL-N-METHYLOLETHYLENIMINE

$C_4H_6Cl_3NO$ F.W. = 190.5

$$H_2C \diagdown \atop H_2C \diagup \!\!\! C \!\!-\!\! N \!\!-\!\! \overset{H}{\underset{CCl_3}{C}} \!\!-\!\! OH$$

I. [Crystal structure of α-trichloromethyl-N-methylolethylenimine $C_4H_6ONCl_3$.]
R.P. ŠIBAEVA and L.O. ATOVMJAN, 1965. *Dokl. Akad. Nauk SSSR*, 160, 334-336
[*Proceedings of the Academy of Sciences of the U.S.S.R.*, 160, 334-336.]

Orthorhombic, a = 10.50±5, b = 9.25±3, c = 7.75±3 Å, [U = 753 Å³], z = 4, D_x = 1.69.

Space group $P2_12_12_1$ (D_2^4)

Atomic positions

	x	y	z
Cl(1)	0.3090	0.8572	0.4245
Cl(2)	0.4620	0.5998	0.4223
Cl(3)	0.2512	0.6245	0.6577
C(1)	0.3747	0.7194	0.5538
C(2)	0.4639	0.7866	0.6900
C(3)	0.6030	0.7101	0.9472
C(4)	0.4611	0.6596	0.9861
N	0.5149	0.6741	0.8027
O	0.4043	0.9006	0.7847

Isotropic temperature factors are given also.   Standard deviations range from
0.004 Å for chlorine to 0.016 Å for carbon.

Structure

Bond lengths and angles are given in full.   Distances in the ring are all
1.49 Å;   other distances have their expected values.   The extra-ring C-N bond
makes an angle of about 126° with the ring plane.   The configuration of bonds to
the carbon atoms of the chain is approximately tetrahedral.   The structure is
held together by O-H...N bonds of length 2.75 Å.

Details of analysis

The intensity data were recorded photographically (*hk*0; 0*kl*; *h*0*l* to *h*4*l*)
using CuK$_\alpha$ radiation.   345 reflexions were observed.   Refinement was by least
squares to a final R index of 0.157.

3-METHOXYCARBONYL-*trans*-3,5-DIMETHYL-Δ'-PYRAZOLINE HYDROBROMIDE

$C_7H_{13}N_2O_2Br$ F.W. ≈ 237.1

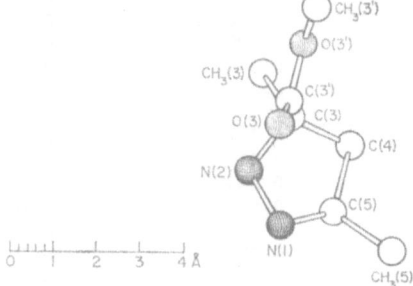

I. The crystal and molecular structure of the hydrobromide of 3-methoxycarbonyl-
   trans-3,5-dimethyl- Δ'-pyrazoline. H. LUTH and J. TROTTER, 1965. *Acta Cryst.*,
   19, 614-619.

Monoclinic, $a = 8.28 \pm 3$, $b = 10.31 \pm 3$, $c = 13.92 \pm 3$ Å, $\beta = 122.4 \pm 3°$, $U = 1003$ Å^3,
$D_m = 1.56$, $Z = 4$, $D_x = 1.57$ (CuK_α, $\lambda = 1.5418$ Å; MoK_α, $\lambda = 0.7107$ Å).

Space group $P2_1/c$ (C_{2h}^5)

Atomic positions

	x	y	z
Br$^-$	0.1222	0.2002	0.0180
N(1)	-0.1812	0.0642	0.1647
N(2)	-0.0067	0.0271	0.1662
C(3)	0.1834	0.0136	0.2867
C(4)	0.0907	-0.0073	0.3627
C(5)	-0.1059	0.0170	0.2821
CH3(3)	0.2971	-0.1053	0.2878
CH3(5)	-0.2740	0.0339	0.3061
C(3')	0.2456	0.1510	0.3130
O(3)	0.1625	0.2575	0.2735
O(3')	0.4395	0.1482	0.3997
CH3(3')	0.5212	0.2712	0.4393

Standard deviations in Å are given in full. Mean values are: bromine,
0.006 Å; other atoms, 0.032 to 0.054 Å. Anisotropic temperature factors are
given.

Structure

Details of the organic cation are given in Fig. 69 and Fig. 70. The ring is
significantly non-planar; all the ring distances correspond to single bonds, and
C(5) is very nearly trigonal. This atom is involved also in a very short intermolecular contact of 2.76 Å with O(3) in the adjacent molecule related by a two-
fold screw axis. It is suggested that C(5) bears a positive charge, which attracts
the relatively negative oxygen atom. There are two short Br-N(2) distances of 3.22
and 3.31 Å. The shorter of these is believed to be a hydrogen bond.

Details of analysis

Intensity data were recorded on equi-inclination Weissenberg photographs
(h0l to h5l), using CuK_α radiation. 409 reflexions were observed, and estimated
visually. Refinement was by block-diagonal least squares to a final R index of
0.12.

Fig. 69. A perspective drawing of the $C_7H_{13}N_2O_2Br$ molecule.

124   ORGANIC COMPOUNDS

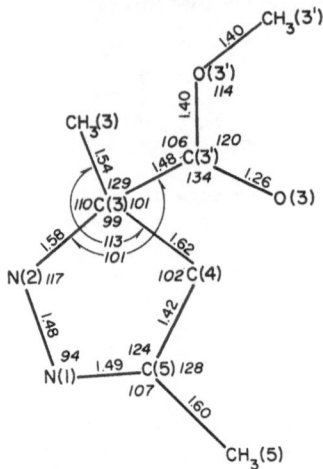

Fig. 70. Bond lengths and angles in the $C_7H_{13}N_2O_2Br$ molecule. Standard deviations
are 0.06 Å and 3°.

1-(4-CHLOROBENZYL)-1-NITROSO-1(4,5-DIHYDRO-2-IMIDAZOLYL)
HYDRAZINE MONOHYDRATE

$C_{10}H_{12}ClN_5O.H_2O$ F.W. = 271.7

I. The structure of 1-(4-chlorobenzyl)-1-nitroso-2-(4,5-dihydro-2-imidazolyl)
hydrazine monohydrate. G.J. PALENIK, 1965. *Acta Cryst.*, 19, 47-56.

Monoclinic, a = 10.434±4, b = 11.352±4, c = 11.245±4 Å, β = 108.34°, [U = 1264.4 Å³],
D_m = 1.41, Z = 4, D_x = 1.428, (CuK$_{\alpha1}$, λ = 1.54050 Å; CuK$_{\alpha2}$, λ = 1.54434 Å).

Space group $P2_1/c$ (C_{2h}^5)

Atomic positions

	x	y	z
Cl	0.1161	0.1292	0.1559
C(1)	−0.0242	0.1745	0.1945
C(2)	−0.1475	0.1230	0.1370
C(3)	−0.2583	0.1559	0.1670
C(4)	−0.2474	0.2386	0.2598
C(5)	−0.1236	0.2904	0.3182
C(6)	−0.0110	0.2600	0.2847
C(7)	−0.3683	0.2646	0.3048
C(8)	−0.6041	0.4318	0.1140
C(9)	−0.8075	0.4971	−0.0152
C(10)	−0.8282	0.4388	0.0960
N(1)	−0.4174	0.3859	0.2754
N(2)	−0.4737	0.4168	0.1498
N(3)	−0.3836	0.4600	0.3683
N(4)	−0.6724	0.4586	−0.0073
N(5)	−0.6900	0.4207	0.1786
O(1)	−0.4206	0.5638	0.3397
O(2)	−0.6593	0.3189	0.4214

Standard deviations are: Cl, 0.0022 Å; C, 0.0066 Å; N and O, 0.0055 Å. Anisotropic temperature factors for the above atoms, and the positions of the hydrogen atoms are given also.

Structure

Bond lengths and angles are given in Fig. 71, and the molecular packing is illustrated in Fig. 72. The structure is held together by a system of hydrogen bonds. One CH...O contact appears to be a hydrogen bond, with distances C..O = 3.25 Å, and H..O = 2.27 Å.

Details of analysis

The intensity data were recorded with a four-circle diffractometer and proportional counter, using CuK$_\alpha$ radiation. Within the range 2θ≤100°, 1593 of a possible 1885 reflexions were observed above background. Refinement was by block-diagonal least squares to a final R index of 0.096.

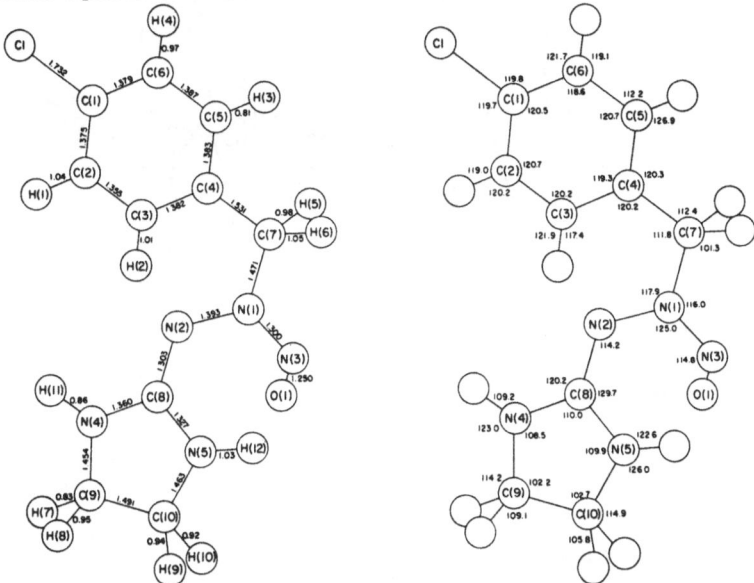

Fig. 71.   Bond lengths and angles in C$_{10}$H$_{12}$ClN$_5$O.H$_2$O.   (Standard deviations of distances are about 0.009 Å for bonds not involving Cl or H.)

DL-ALLANTOIN

C$_4$H$_6$O$_3$N$_4$ F.W. = 158.1

```
        O(4)  H(5)   H(6)   H(9)
          \    |     /     /
           C(4)   N(6)   N(8)
            \     /     \
            C(5)   C(7)   H(8)
             |      ‖
H(3)—N(3)   N(1)   O(7)
             \    /
             C(2)   H(1)
              ‖
             O(2)
```

I. The crystal structure of DL-allantoin. D. MOOTZ, 1965. *Acta Cryst.*, 19, 726–734.

Monoclinic, a = 8.024±8, b = 5.153±4, c = 14.797±8 Å, β = 93.01±4°, U = 611.0 Å³, D$_m$ = 1.722, Z = 4, [D$_x$ = 1.72]. (CuK$_{\alpha 1}$, λ = 1.5405 Å; CuK$_{\alpha 2}$, λ = 1.5443 Å).

Space group   P2$_1$/c   (C$_{2h}^5$)

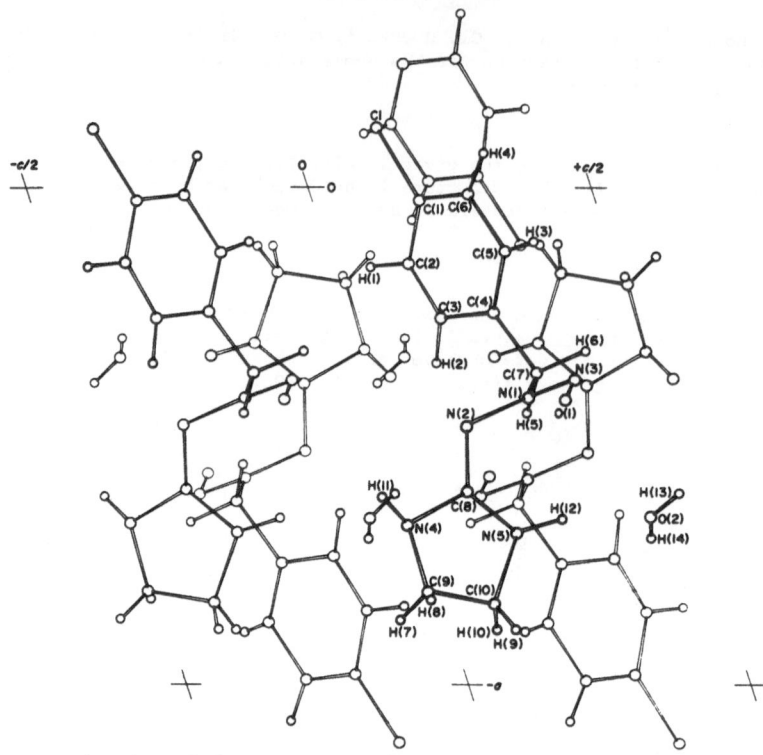

Fig. 72. A projection of the structure of $C_{10}H_{12}ClN_5O \cdot H_2O$ on the (010) plane
          illustrating the molecular packing.

Atomic positions

	x	y	z
C(2)	0.0007±4)	0.2996± 6)	0.0876±2)
C(4)	0.2834±4)	0.3529± 6)	0.0932±2)
C(5)	0.2116±4)	0.5506± 6)	0.1583±2)
C(7)	0.3149±4)	0.3118± 5)	0.2902±2)
N(1)	0.0343±3)	0.4843± 6)	0.1497±2)
N(3)	0.1514±3)	0.2267± 5)	0.0517±2)
N(6)	0.2881±3)	0.5444± 5)	0.2474±2)
N(8)	0.3923±4)	0.3192± 5)	0.3723±2)
O(2)	-0.1354±3)	0.2125± 5)	0.0629±2)
O(4)	0.4296±3)	0.3210± 4)	0.0792±2)
O(7)	0.2640±3)	0.1062± 4)	0.2538±1)
H(1)	-0.045 ±8)	0.561 ±15)	0.181 ±5)
H(3)	0.156 ±8)	0.076 ±14)	0.014 ±5)
H(5)	0.239 ±8)	0.736 ±13)	0.131 ±5)
H(6)	0.303 ±9)	0.722 ±13)	0.275 ±5)
H(8)	0.429 ±9)	0.189 ±13)	0.387 ±5)
H(9)	0.431 ±8)	0.474 ±14)	0.392 ±4)

Anisotropic temperature factors are given for the non-hydrogen atoms.

Structure

      The bond lengths and angles are given in Fig. 73. The structure is held
together by an intricate system of inter-molecular N-H..O bonds, of which full
details are given. Part of this bonding system, and the shape of the molecule,
are illustrated in Fig. 74. The five ring atoms and the two adjacent oxygen
atoms are approximately coplanar, as are the four remaining non-hydrogen atoms.
The ring atom C(5) is approximately tetrahedral, and rotation about the bond
C(5)-N(6) is such that the angle between the ring plane and the chain plane is
80°.

Details of analysis

Intensities were recorded on integrated equi—inclination Weissenberg photographs (0*kl* to 5*kl*; *h*0*l* to *h*4*l*), using Cu*K*$_\alpha$ radiation. 1130 of 1292 accessible reflexions were observed and measured visually. Refinement was by full-matrix least squares to a final *R* index of 0.087.

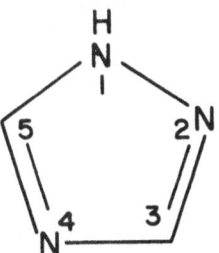

Fig. 73. Bond lengths and angles in DL-allantoin. Standard deviations are
0.004 Å and 0.3° where only heavy atoms are involved and 0.07 Å
and 4° (7° for H-N-H) where hydrogen atoms are involved.

1,2,4-TRIAZOLE

$C_2H_3N_3$ F.W. = 69.1

I. Die Roentgenstrukturanalyse von 1,2,4-Triazol. H. DEUSCHL, 1965. *Ber.*
Bunsengesellsch. phys. Chem., *Dtsch.*, <u>69</u>, 550-557.

Orthorhombic, *a* = 9.69±4, *b* = 9.38±4, *c* = 7.14±3 Å, [*U* = 649 Å³], *D*$_m$ = 1.392,
Z = 8, *D*$_x$ = 1.401.

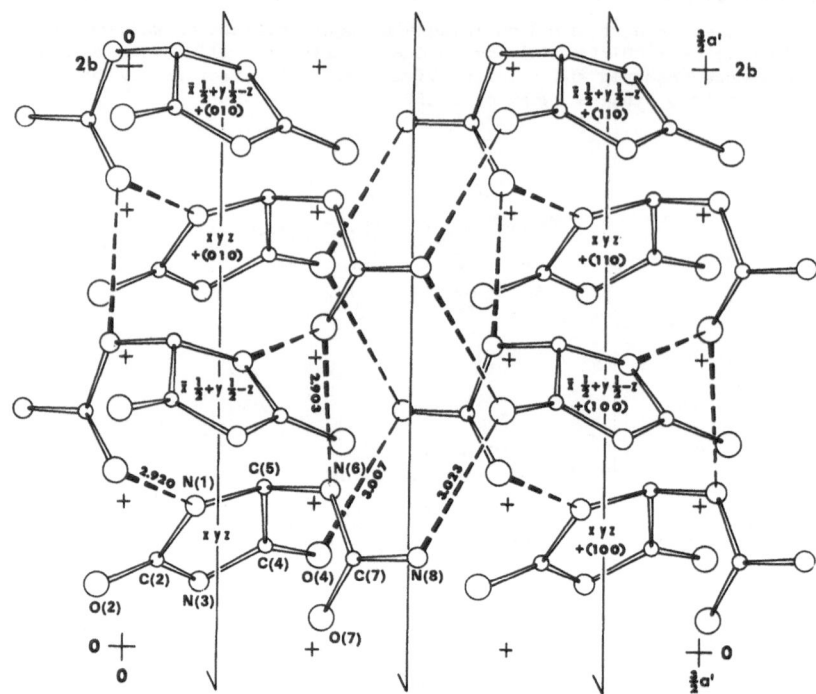

Fig. 74. A part of the structure of DL-allantoin, viewed along [101]. For
clarity all molecules generated from the original by centres of symmetry
or glide planes have been omitted. Thus the figure does not show the
N(3)-H..O(2) bonds which link enantiomorphous molecules into pairs across
centres of symmetry.

Space group Pbca (D_{2h}^{15})

Atomic positions

	x	y	z
N(1)	-0.0058	-0.0864	0.2091
N(2)	0.0261	-0.2044	0.1087
C(3)	0.1622	-0.1925	0.0923
N(4)	0.2164	-0.0757	0.1758
C(5)	0.1067	-0.0103	0.2550

[This numbering scheme differs from that of the original.] Assumed positions
for H(3) and H(5) are given, as are anisotropic temperature factors for the above
atoms.

Structure

The molecule is planar. The C-C and C-N bond lengths all lie in the range
1.330 to 1.354 Å (±0.014 Å), and no distinction can be made between single and
double bonds. Adjacent molecules are linked by N-H..N bonds, of length 2.82 Å,
to form infinite chains parallel to *a*. N(1) and N(4) are involved in this hydrogen
bonding; it is suggested that tautomerism may occur, with the hydrogen atom shared
between N(1) and N(4).

Details of analysis

The intensity data were recorded on equi-inclination Weissenberg photographs
(0*kl* to 8*kl*; *hk*0 to *hk*6) using CuK$_\alpha$ radiation. 645 reflexions were observed and

measured with an integrating photometer. Refinement was by least squares and Fourier methods to a final R index of 0.195.

PYRIDINE HYDROGEN NITRATE

$C_5H_6N_2O_3$ F.W. = 142.1

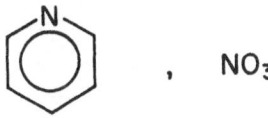

, NO_3

I. The crystal structure of pyridine hydrogen nitrate. A.J. SEREWICZ, B.K. ROBERTSON and E.A. MEYERS, 1965. *J. Phys. Chem.*, 69, 1915-1921.

Monoclinic, a = 3.905, b = 12.286, c = 13.470 Å, β = 90°30', [U = 646.2 Å^3], D_m = 1.432. Z = 4, D_x = 1.454 [1.461].

Space group $P2_1/c$ (C_{2h}^5)

Atomic positions

	x	y	z
O(1)	-0.4946±22	0.4847± 6	0.1157± 7
O(2)	-0.6149±25	0.5572± 8	0.2545± 7
O(3)	-0.3671±24	0.4026± 9	0.2497± 7
N(1)	-0.1080±24	0.3131± 9	0.0465± 9
N(2)	-0.4896±25	0.4825± 9	0.2093± 8
C(1)	-0.0287±34	0.2230±14	0.0990± 8
C(2)	0.1346±30	0.1393±10	0.0554±10
C(3)	0.2092±29	0.1462±10	-0.0418±11
C(4)	0.1248±30	0.2352±13	-0.0934± 8
C(5)	-0.0305±32	0.3199±10	-0.0475±12

Anisotropic temperature factors are given also, and they have been analysed to determine errors in the bond lengths. [In view of the restrictions on the measured data, the thermal parameters, and the deductions therefrom, are of questionable validity.]

Structure

The NO_3 group is planar, with average O-N-O angle 120°. The pyridine ring appears to be a regular hexagon. The two moieties are joined by a hydrogen bond (N...O) of length 2.76±1 Å. (The hydrogen atom has not been located.) Bond lengths and angles are given in full; the former have been corrected for thermal motion. [The corrections are large, and probably unreliable.]

Details of analysis

The intensity data were recorded by various photographic techniques (principally integrated Weissenberg, $0kl$ to $2kl$) using CuK_α radiation. The intensities were measured visually and photometrically. Refinement was by least squares to a final R index of 0.085.

3-PICOLYLAMINE DIHYDROCHLORIDE

$C_6H_8N_2 \cdot 2HCl$ F.W. = 181.1

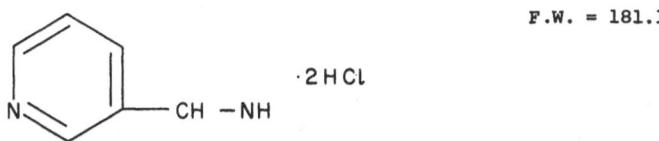

I. Contribution à l'étude de la liaison hydrogène dans quelques chlorhydrates d'amines. F. GENET, 1965. *Bull. Soc. Fr. Minér. Crist.*, 88, 463-482.

Monoclinic, a = 14.80±5, b = 12.65±5, c = 4.59±2 Å, β = 94±1°, U = 856 Å^3, D_m = 1.418, Z = 4, D_x = 1.402.

Space group $P2_1/a$ (C_{2h}^5)

Atomic positions

	x	y	z
Cl(1)	0.1233	0.0118	0.1603
Cl(2)	0.6023	0.1475	0.9071
N(1)	0.3572	0.0392	0.4842
N(8)	0.5190	0.3509	0.5878
C(2)	0.4117	0.1158	0.3957
C(3)	0.3985	0.2193	0.4786
C(4)	0.3286	0.2381	0.6694
C(5)	0.2588	0.1550	0.7668
C(6)	0.2930	0.0549	0.6709
C(7)	0.4555	0.3041	0.3670

Isotropic temperature factors are given also.

Structure

The molecule of 3-picolylamine is planar except for the amino nitrogen atom, which lies about 1.2 Å from the plane of the remaining non-hydrogen atoms. Bond lengths and angles are given in full, and are consistent with expectation, but there is no indication of accuracy. The hetero nitrogen atom is hydrogen bonded to a chlorine atom (N...Cl = 3.05 Å) which lies 0.65 Å from the ring plane. The amino nitrogen atom is hydrogen bonded to three chlorine atoms at distances of 3.11, 3.17, and 3.20 Å; the arrangement of bonds to this nitrogen atom is roughly tetrahedral. One chlorine atom is bonded to a hetero nitrogen atom of one molecule and to an amino nitrogen of another. The other chlorine atom is bonded to two amino nitrogen atoms of different molecules. The structure is thereby held together in chains extended along b.

Details of analysis

The intensity data were recorded on Weissenberg photographs, using CuK_α radiation, and measured visually. (A few reflexions were measured on a diffractometer.) Refinement was by least squares to a final R index of 0.16.

PICOLINIC ACID HYDROCHLORIDE

$C_6H_6ClNO_2$ F.W. = 159.5

I. Structure crystalline du chlorhydrate de l'acide picolique. A. LAURENT, 1965. *Acta Cryst.*, **18**, 799–806.

Orthorhombic, a = 7.76±2, b = 13.85±2, c = 6.65±2 Å, U = 715 Å³, D_m = 1.49. Z = 4, D_x = 1.47, ($CuK_{\alpha1}$, λ = 1.5405 Å).

Space group $Pbn2_1$ (C_{2v}^9) or $Pbnm$ (D_{2h}^{16}) *Pbnm* assumed.   Molecular symmetry, mirror plane.

Atomic positions

	x	y	z
Cl	0.1285±4	0.1228±2	1/4
O(1)	0.0270±11	0.7214±6	1/4
O(2)	0.1558±14	0.8620±9	1/4
N	0.8610±13	0.9562±7	1/4
C(1)	0.8557±16	0.8607±9	1/4
C(2)	0.7039±16	0.8115±9	1/4
C(3)	0.5489±18	0.8635±10	1/4
C(4)	0.5550±17	0.9630±10	1/4
C(5)	0.7118±17	0.0112±9	1/4
C(6)	0.0312±16	0.8149±9	1/4

Anisotropic temperature factors are given also.

Structure

Symmetry requires that the picolinic acid molecule be planar. Bond lengths and angles are given in full, and it is noted that the double bond $C(6)=O(2)$ is exceptionally short (1.17 Å). [This distance has not been corrected for the anomalously intense out-of-plane thermal motion of $O(2)$, for which $B_{33} = 15.0$ Å^2, and $B_{11} = 4.31$ Å^2. Another possibility, apparently not considered, is that the molecule is non-planar, and that either the structure is disordered, or the space group is $Pbn2_1$.] The molecules in a given plane are linked by hydrogen bonds to the chlorine atom to form zigzag chains as illustrated in Fig. 75. Distances between molecules in adjacent planes are consistent with van der Waals interaction.

Details of analysis

The intensity data were recorded on equi-inclination Weissenberg photographs ($hk0$ to $hk6$; $0kl$ to $7kl$) using CuK$_\alpha$ radiation. 841 of 1080 accessible reflexions were observed and measured visually. Refinement was by full-matrix least squares to a final R index of 0.156.

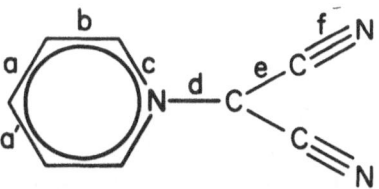

Fig. 75.   Distances between molecules of picolinic acid hydrochloride in a
           single plane. Broken lines are hydrogen bonds, dotted lines are
           other short intermolecular contacts.

PYRIDINIUM DICYANOMETHYLIDE

$C_8H_5N_3$ F.W. = 143.1

I. The crystal structure of pyridinium dicyanomethylide, $C_8H_5N_3$. C. BUGG and
   R.L. SASS, 1965. *Acta Cryst.*, <u>18</u>, 591–594.

Monoclinic, $a = 7.87\pm2$, $b = 12.512\pm4$, $c = 3.86\pm1$ Å, $\beta = 114.8\pm1°$, $[U = 345.0$ Å$^3]$, $D_m = 1.36$, $Z = 2$, $D_x = 1.37$, (CuK$_\alpha$, $\lambda = 1.5418$ Å).

Space group $P2_1$ (C_2^2) or $P2_1/m$ (C_{2h}^2). $P2_1/m$ is confirmed by the structure analysis. Molecular symmetry, mirror plane.

Atomic positions

	x	y	z
C(1)	0.9574±23	0.2500	1.0984±48
C(2)	0.8703±15	0.3458±8	0.9364±34
C(3)	0.6980±14	0.3456±8	0.6206±32
C(4)	0.4441±19	0.2500	0.1413±42
C(5)	0.3594±14	0.3475±8	0.0340±33
N(1)	0.6163±15	0.2500	0.4640±33
N(2)	0.2893±13	0.4243±7	0.1782±27

Isotropic temperature factors are given also.

Structure

Bond lengths and angles are as follows:

a	=	1.39 Å	aa' =	119.0°
b	=	1.39	ab =	120.5
c	=	1.37	bc =	119.5
d	=	1.42	cd =	119.4
e	=	1.41	de =	120.0
f	=	1.13	ef =	180.0

(Standard deviations 0.01 Å and 1.0°.)

The molecule deviates from planarity by displacement of the cyano groups from the plane of the rest of the molecule. The nitrogen atoms lie 0.13 Å from this plane, and the carbon atom adjacent to the cyano groups is thus slightly pyramidal. Intermolecular distances are consistent with van der Waals interaction.

Details of analysis

The intensity data were recorded on equi-inclination Weissenberg photographs ($hk0$ to $hk2$) using CuK_α radiation. 320 reflexions were observed and measured visually. Refinement was by full-matrix least squares to a final R index of 0.129.

α-PYRIDOIN

(1,2-DI-2-PYRIDYLETHENDIOL-1,2)

$C_{12}H_{10}N_2O_2$ F.W. = 314.2

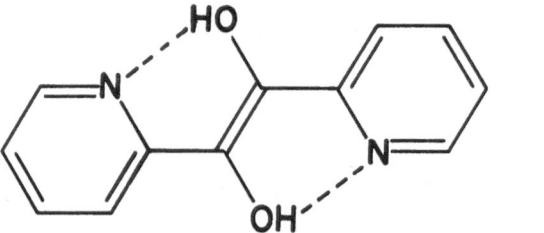

I. The crystal structure of α-pyridoin, 1,2-di-2-pyridyl-ethendiol-1,2.
   T. ASHIDA, S. HIROKAWA, and Y. OKAYA, 1965. *Acta Cryst.*, 18, 122–127.

Monoclinic, $a = 16.62\pm2$, $b = 4.72\pm1$, $c = 13.18\pm2$ Å, $\beta = 100.1\pm3°$, $[U = 1017.9$ Å$^3]$, $D_m = 1.39$, $Z = 4$, $D_x = 1.40$.

Space group Cc (C_s^4) or $C2/c$ (C_{2h}^6) $C2/c$ suggested by statistical test and confirmed by analysis. Molecular symmetry, centre.

Atomic positions

	x	y	z
C(1)	0.08192	0.27042	0.05965
C(2)	0.15301	0.39503	0.03572
C(3)	0.19575	0.58759	0.10386
C(4)	0.16681	0.65534	0.19301
C(5)	0.09679	0.52340	0.21183
C(6)	0.03462	0.06524	−0.00991

	x	y	z
N	0.05509	0.33529	0.14760
O	0.06637	0.01314	−0.09699
H(2)	0.170	0.350	−0.026
H(3)	0.248	0.683	0.089
H(4)	0.195	0.786	0.241
H(5)	0.073	0.575	0.271
H(0)	0.029	−0.107	−0.130

Standard deviations (in Å), are given in full, and range from 0.021 to 0.036 Å for hydrogen and 0.0014 to 0.0031 Å for other atoms. Temperature factors (anisotropic for the non—hydrogen atoms), are given also.

Structure

The bond lengths and angles are given in Fig. 76. The molecule as a whole is very nearly planar; the oxygen atoms lie about 0.015 Å from the plane of the remaining non—hydrogen atoms. The structure is stabilized by intramolecular O—H..N bonds as shown, with the O..N distance 2.599 Å. Intermolecular distances correspond to van der Waals contacts. The arrangement of the molecules in the unit cell is shown in Fig. 77.

Details of analysis

Intensity data were recorded on equi-inclination Weissenberg photographs ($h0l$ to $h3l$; $hk0$ to $hk8$), using CuK_α radiation. More than 90% of accessible reflexions were observed and estimated visually. Refinement was by full—matrix least squares to a final R index of 0.083.

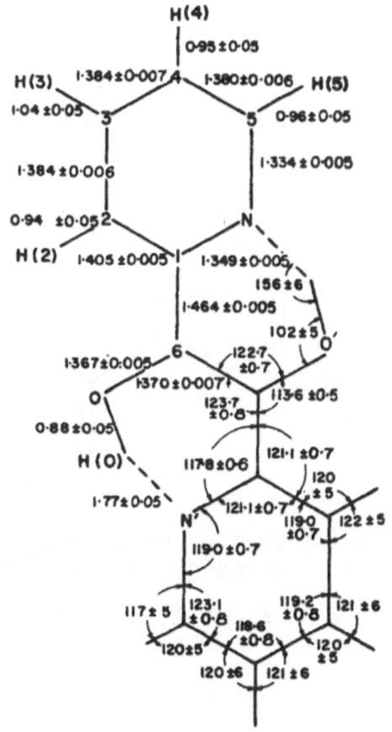

Fig. 76. Bond lengths and angles in α—pyridoin, with their standard deviations, in Å and degrees.

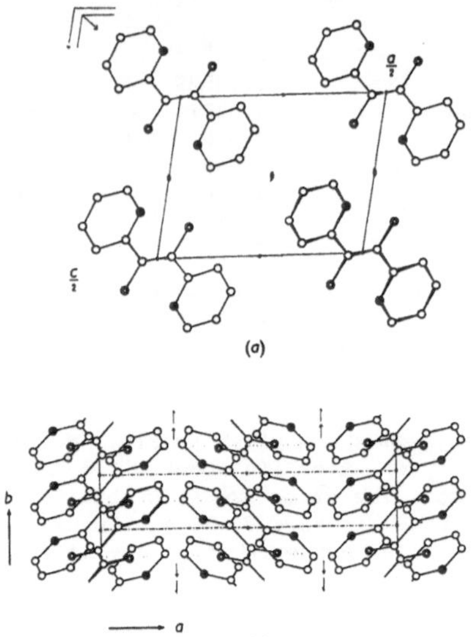

Fig. 77.　(a)　The structure of α-pyridoin projected along *b*.
　　　　　(b)　The structure of α-pyridoin projected along *c*.

s-TRIPHENYLTRIAZINE

$C_{21}H_{15}N_3$ F.W. = 309.4

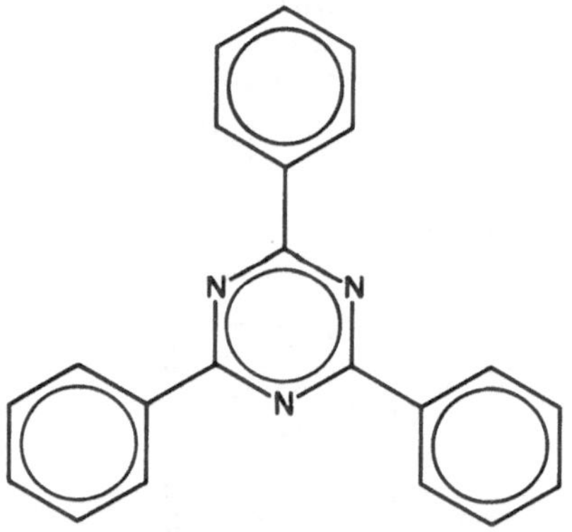

I. The crystal and molecular structure of *s*-triphenyltriazine. A. DAMIANI, E. GIGLIO and A. RIPAMONTI, 1965. *Acta Cryst.*, <u>19</u>, 161–168.

Monoclinic, $a = 10.94$, $b = 3.91$, $c = 35.84$ Å, $\beta = 90°38'$, $[U = 1533$ Å$^3]$, $z = 4$, $D_x = 1.326$.

Space group $P2_1/c$　(C_{2h}^5)

Atomic positions

	x	y	z
C(1)	−0.1725	0.6726	0.1911
C(2)	−0.1157	0.8267	0.2214
C(3)	−0.1764	0.8794	0.2544
C(4)	−0.2983	0.7825	0.2570
C(5)	−0.3555	0.6305	0.2272
C(6)	−0.2955	0.5713	0.1941
C(7)	−0.1704	0.2790	0.0637
C(8)	−0.2926	0.1638	0.0663
C(9)	−0.3520	0.0435	0.0349

	x	y	z
C(10)	−0.2953	0.0339	0.0010
C(11)	−0.1737	0.1412	−0.0017
C(12)	−0.1132	0.2611	0.0296
C(13)	0.2039	0.6374	0.1235
C(14)	0.2679	0.5201	0.0925
C(15)	0.3948	0.5612	0.0920
C(16)	0.4573	0.7120	0.1216
C(17)	0.3930	0.8337	0.1524
C(18)	0.2657	0.7919	0.1535
C(19)	0.0694	0.5875	0.1247
C(20)	−0.1065	0.4094	0.0972
C(21)	−0.1073	0.6001	0.1565
N(1)	0.0151	0.4634	0.0945
N(2)	−0.1698	0.4743	0.1276
N(3)	0.0149	0.6626	0.1564

Standard deviations are given in full. They range from 0.003 to 0.006 Å for nitrogen, and from 0.004 to 0.009 Å for carbon. Temperature factors are given also. [These are anisotropic, but seem to assume that the principal axes of the thermal ellipsoids invariably lie in or normal to (010), and are probably meaningless.]

Structure

Bond lengths and angles are given in full. The more important mean values are:

    C − N   = 1.335±6 Å
    C(triazine) − C(phenyl)   = 1.475±7 Å
    C − N − C   = 115.5±5°
    N − C − N   = 124.5±5°

(The estimates of accuracy may be optimistic.)

All rings are planar, but each phenyl group is twisted slightly out of the plane of the triazine group. Angles of twist are 7.6, 10.9, and 6.9°, the sense of the third being opposite to that of the others. Intermolecular distances are consistent with van der Waals interactions.

Details of analysis

Intensity data were recorded on equi-inclination Weissenberg photographs ($h0l$ to $h2l$; $0kl$) using CuK_α radiation and estimated visually. 1152 reflexions were observed. Refinement was by differential syntheses to a final R index of 0.114.

2,4,6-TRI(p-CHLOROPHENYL)-s-TRIAZINE

$C_{21}H_{12}Cl_3N_3$ F.W. = 412.7

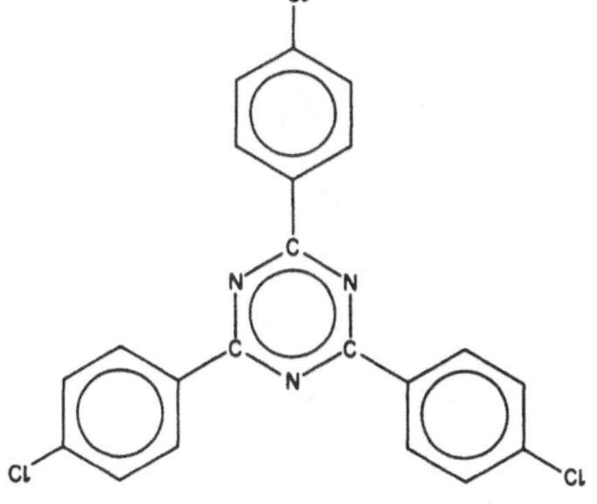

I. The molecular structure of 2,4,6-tri(*p*-chlorophenyl)-*s*-triazine. D. BELITSKUS
   and G.A. JEFFREY, 1965.· *Spectrochim. Acta.*, $\underline{21}$, 1563–1567.

Orthorhombic, a = 21.49, b = 21.75, c = 4.04 Å, [U = 1889 Å^3], D_x = 1.39, z = 4,
D_x = 1.41, [1.45].

Space group Pna2$_1$ (C_{2v}^9) or Pnma (D_{2h}^{16}) Pnma confirmed by analysis. Molecular
symmetry, mirror plane.

Atomic positions

	x	y
Cl(1)	-0.275	0.250
Cl(2)	0.169	-0.044
N(1)	-0.033	0.195
N(2)	0.078	0.250
C(1)	-0.203	0.250
C(2)	-0.176	0.193
C(3)	-0.119	0.194
C(4)	-0.088	0.250
C(5)	-0.027	0.250
C(6)	0.051	0.198
C(7)	0.078	0.138
C(8)	0.047	0.083
C(9)	0.076	0.029
C(10)	0.134	0.028
C(11)	0.164	0.082
C(12)	0.136	0.138

The mean standard deviations are: Cl, 0.01 Å, C and N, 0.03 Å. Isotropic
temperature factors are given also.

Structure

The molecule has the configuration indicated above, and is, within the limita-
tions of the two-dimensional structure determination, planar.

Details of analysis

The intensity data were recorded on a Weissenberg photograph, using CuK_α
radiation. 151 reflexions in the *h0l* zone were observed and measured visually.
Refinement was by least squares to a final R index of 0.14.

1,8-DIAZACYCLOTETRADECANE DIHYDROBROMIDE

$C_{12}H_{26}N_2$,2HBr   F.W. = 360.2

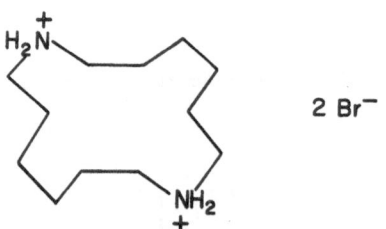

2 Br$^-$

I. Die Strukturen der mittleren Ringverbinderung. IX. 1,8–Diazacyclotetradecan-
   dihydrobromid. J.D. DUNITZ and E.F. MEYER, 1965. *Helv. Chim. Acta*, $\underline{48}$, 1441–
   1449.

Orthorhombic, a = 13.58, b = 14.81, c = 8.01 Å, (all ±0.15%), U = 1611 Å^3, D_m =
1.50, z = 4, D_x = 1.485.

Space group Pccn (D_{2h}^{10}) Molecular symmetry, centre.

Atomic positions

	x	y	z
Br	0.0817	0.1308	0.1516
C(1)	0.1873	0.6268	-0.0005
C(2)	0.1986	0.5293	0.0195
C(3)	0.1263	0.4924	0.1451
C(4)	0.1308	0.3867	0.1330
N(5)	0.0549	0.3493	0.2465
C(6)	-0.0532	0.3720	0.2134
C(7)	-0.0858	0.3319	0.0515

These coordinates were obtained by Fourier refinement. Parameters obtained by least-squares refinement are given also, but are not considered reliable. There is no estimate of error.

Structure

The analysis is consistent with the formulation given above, in which the nitrogen atoms are protonated. The conformation of the 14-membered ring corresponds closely to an idealized conformation based on the diamond structure. The structure consists of stacks of rings extended in the c direction. Adjacent rings in a stack are joined by N-H··Br··H-N bridges, with Br...N distances 3.28 and 3.34 Å.

Details of analysis

The intensity data were measured with a linear diffractometer and scintillation counter, ($hk0$ to $hk9$), using MoK$_\alpha$ radiation. 734 of 1500 accessible reflexions were observed above background. Refinement was by Fourier methods to a final R index of 0.107. Results of a least-squares refinement are given also, but are said to be unreliable because of systematic errors in the intensity data.

ISOQUINOLINE HYDROCHLORIDE
ISOQUINOLINE HYDROCHLORIDE MONOHYDRATE

$C_9H_7N.HCl$ F.W. = 165.6

$C_9H_7N.HCl.H_2O$ F.W. = 183.6

I. Contribution a l'étude de la liaison hydrogène dans quelques chlorhydrates d'amines. F. GENET, 1965. *Bull. Soc. Fr. Minér. Crist.*, <u>88</u>, 463-482.

ISOQUINOLINE HYDROCHLORIDE

Monoclinic, a = 9.10±5, b = 17.7±2, c = 5.12±2 Å, β = 100±2°, U = 812 Å^3, D_m = 1.31, Z = 4, D_x = 1.35.

Space group $P2_1/a$ (C_{2h}^5)

Atomic positions

	x	y	z
Cl	0.1569	0.1245	0.3760
N(2)	0.4250	0.3571	0.6706
C(1)	0.3813	0.4178	0.6934
C(3)	0.3739	0.2849	0.7118
C(4)	0.2759	0.2731	0.8771
C(5)	0.1170	0.3291	0.1735
C(6)	0.0685	0.3940	0.2851
C(7)	0.1280	0.4679	0.2526
C(8)	0.2290	0.4774	0.0822
C(9)	0.2783	0.4124	0.9603
C(10)	0.2217	0.3378	0.0051

Isotropic temperature factors are given also.

Structure

The isoquinoline molecule is planar. Bond lengths and angles are given in full, and have reasonable values, but there is no indication of accuracy. The chlorine atom lies close to the plane of the molecule, and appears to be hydrogen-bonded to the nitrogen atom (N...Cl = 2.92 Å). The arrangement of bonds to the nitrogen atom is thereby approximately trigonal.

Details of analysis

The intensity data were recorded on Weissenberg photographs using CuK_α radiation. 916 reflexions were observed and measured with a densitometer. Refinement was by full-matrix least squares to a final R index of 0.15.

ISOQUINOLINE HYDROCHLORIDE MONOHYDRATE

Triclinic, $a = 7.32\pm1$, $b = 7.36\pm1$, $c = 9.00\pm5$ Å, $\alpha = 95.0\pm5$, $\beta = 107.0\pm5$, $\gamma = 90\pm1°$, $U = 461$ Å^3, $D_m = 131$, $Z = 2$, $D_x = 1.33$.

Space group $P\,1$ (C_1^1) or $P\bar{1}$ (C_i^1) $P\bar{1}$ is consistent with analysis.

Atomic Positions

	x	y	z
Cl	0.8318	0.8438	0.1021
O	0.3991	0.8359	0.1193
N(2)	0.8457	0.6892	0.7859
C(1)	0.7826	0.5160	0.7456
C(3)	0.8707	0.9900	0.6773
C(4)	0.8295	0.7371	0.5233
C(5)	0.7216	0.4861	0.3203
C(6)	0.6519	0.3107	0.2827
C(7)	0.6394	0.2032	0.3988
C(8)	0.6824	0.2657	0.5502
C(9)	0.7485	0.4482	0.5920
C(10)	0.7602	0.5565	0.4747

Isotropic temperature factors are given also.

Structure

The isoquinoline molecule is essentially as reported above. The N...Cl distance is 3.00 Å, and the chlorine atom lies about 0.75 Å from the molecular plane. The distance of the chlorine atom from the water molecule is 3.21 Å.

Details of analysis

The intensity data were recorded on Weissenberg photographs, using CuK_α radiation. 950 reflexions were observed and measured visually. Refinement was by least squares to a final R index of 0.18.

8-AZAGUANINE MONOHYDRATE

$C_4H_4N_6O.H_2O$ F.W. = 170.1

I. The crystal and molecular structure of 8-azaguanine monohydrate. W.M. MacINTYRE, P. SINGH and M.S. WERKEMA, 1965. *Biophys. J., U.S.A.*, **5**, 697-713. (For a determination of this structure, see <u>1</u>.)

Monoclinic, a = 3.57±1, *b* = 11.41±2, *c* = 16.53±2 Å, β = 95.3±1°, [*U* = 670 Å³], D_m = 1.70, Z = 4, D_x = 1.69.

Space group P2₁/c (C_{2h}^5)

Atomic positions

	x	y	z
C(2)	0.7155±22	0.3190±8	0.1362±4
C(4)	0.7959±19	0.5086±7	0.1113±4
C(5)	0.6797±22	0.5458±8	0.1849±5
C(6)	0.5733±22	0.4596±7	0.2424±4
N(1)	0.5977±20	0.3486±7	0.2121±4
N(2)	0.7237±23	0.2059±7	0.1187±4
N(3)	0.8220±18	0.3976±6	0.0838±3
N(7)	0.6890±21	0.6629±6	0.1885±4
N(8)	0.8136±22	0.7024±6	0.1214±4
N(9)	0.8746±19	0.6081±6	0.0729±4
O(6)	0.4583±19	0.4787±6	0.3092±3
O(w)	0.8691±32	0.3944±8	0.4678±5

Also given are the positions of the hydrogen atoms, and anisotropic temperature factors for the atoms given above.

Structure

The azaguanine molecule is accurately planar. Bond lengths and angles are given in full, with standard deviations of 0.02 Å and 1.2°. (Greater precision is achieved in <u>1</u>.) The hydrogen atoms have been located, and demonstrate the formulation given above. The structure is held together by the network of hydrogen bonds shown in Fig. 78. Molecules related by the a translation overlap each other with interplanar spacing of 3.25 Å. The presumed interaction is discussed in terms of the cell-poisoning action of 8-azaguanine.

Details of analysis

The intensity data were recorded with a four-circle diffractometer, using MoK$_\alpha$ radiation. In the range 2θ<40°, 1143 of a possible 1500 reflexions were observed. Refinement was by block-diagonal least squares to a final R index of 0.12.

1.   J. SLETTEN, E. SLETTEN and L.H. JENSEN, 1968. *Acta Cryst.*, <u>B24</u>, 1692.

2,3,7,8-TETRACHLOROPHENAZINE

$C_{12}H_4N_2Cl_4$ F.W. = 318.0

I. Die Kristall und Molekülstruktur des 2,3,7,8-Tetrachlorphenazins. V. RIGANTI, S. LOCCHI, R. CURTI and B. BOVIO, 1965. *J. Heterocyc. Chem. U.S.A.*, <u>2</u>, 87-90.

Monoclinic, a = 3.85±2, *b* = 6.04±2, *c* = 25.33±4 Å, β = 91°46±15', [*U* = 589 Å³], Z = 2, [D_x = 1.80].

Space group P2₁/c (C_{2h}^5) Molecular symmetry, centre.

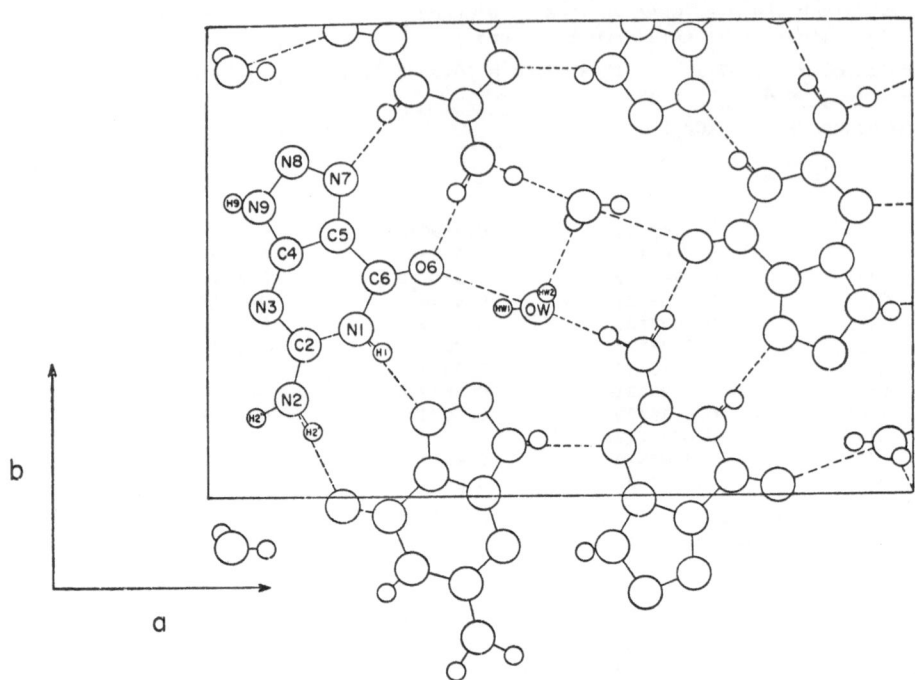

Fig. 78. The structure of 8-azaguanine monohydrate viewed along *a*, showing hydrogen
          bonds (standard deviations, 0.016 to 0.024 Å).

Atomic positions

	x	*y*	*z*
C(1)	0.0127	−0.0532	0.0511
C(2)	0.0117	−0.1080	0.1066
C(3)	−0.1183	0.0452	0.1414
C(4)	−0.2636	0.2496	0.1235
C(5)	−0.2781	0.3046	0.0696
C(6)	−0.1380	0.1518	0.0335
N	−0.1496	0.2038	−0.0192
Cl(1)	−0.1178	−0.0183	0.2078
Cl(2)	−0.4324	0.4324	0.1687

     Standard deviations (in Å), are given in full; mean values are: Cl, 0.004;
N, 0.011; C, 0.013 Å. Anisotropic temperature factors are given also.

Structure

     The molecule is planar within experimental error. Mean bond lengths and angles
are:

a = 1.353±15 Å	aa' = 115.6±1.2°
b = 1.430±17	ab = 117.5±1.2
c = 1.433±18	ac = 122.2±1.2
d = 1.395±18	bc = 120.3±1.2
e = 1.424±19	bd = 118.3±1.2
f = 1.728±13	de = 121.4±1.2
	df = 119.1±1.2
	ef = 119.5±1.2

Intermolecular distances are consistent with van der Waals interaction.

Details of analysis

The intensity data were recorded on integrated Weissenberg photographs, using CuK_α radiation, and measured visually and photometrically.   624 reflexions were recorded.   Refinement was by least squares to a final R index of 0.11.

1,4,6,9-TETRACHLOROPHENAZINE

$C_{12}H_4N_2Cl_4$ F.W. = 318.0

I. Die Kristall- und Molekülstruktur des 1,4,6,9-Tetrachlorphenazins. V. RIGANTI, S. LOCCHI, R. CURTI and B. BOVIO, 1965. *J. Heterocyc. Chem. U.S.A.*, **2**, 176-180.

Monoclinic, $a = 3.95\pm2$, $b = 14.06\pm3$, $c = 10.80\pm3$ Å, $\beta = 94°37\pm15'$, $[U = 598$ Å$^3]$, $Z = 2$, $[D_X = 1.76]$.

Space group $P2_1/n$ (C^5_{2h}) Molecular symmetry, centre.

Atomic positions

	x	*y*	*z*
C(1)	0.1047	0.0082	0.1218
C(2)	0.2277	0.0202	0.2505
C(3)	0.4065	0.0986	0.2851
C(4)	0.4844	0.1704	0.1981
C(5)	0.3849	0.1588	0.0750
C(6)	0.1944	0.0779	0.0313
N	0.0786	0.0689	-0.0877
Cl(1)	0.1390	-0.0661	0.3544
Cl(2)	0.4970	0.2417	-0.0326

Anisotropic temperature factors are given also.

Structure

The molecule is essentially planar. Mean bond lengths and angles are:

a	=	1.337±22 Å		aa'	=	118.5±1.5°
b	=	1.434±25		ac	=	120.7±1.5
c	=	1.447±25		bc	=	118.6±1.5
d	=	1.356±25		bd	=	120.4±1.5
e	=	1.429±25		de	=	120.9±1.5
f	=	1.718±17		bf	=	118.4±1.5

Intermolecular distances are consistent with van der Waals interaction.

Details of analysis

The intensity data were recorded on integrated Weissenberg photographs ($0kl$ to $2kl$) using CuK_α radiation, and precession photographs ($hk0$; $h0l$) using MoK_α radiation. Refinement was by least squares to a final R index of 0.106.

DODECAHYDRO-1,4,7,9b-TETRAAZAPHENALENE
TRIHYDROCHLORIDE HEMIHYDRATE

$C_9H_{18}N_4.3HCl.\frac{1}{2}H_2O$ F.W. = 300.7

3 HCl
$\frac{1}{2}H_2O$

I. The structure of dodecahydro-1,4,7,9b-**tetraazaphenalene** trihydrochloride hemihydrate, $C_9H_{18}N_4.3HCl.\frac{1}{2}H_2O$. A.E. SMITH, 1965. *Acta Cryst.*, <u>19</u>, 248-255.

Orthorhombic, $a = 19.21$, $b = 11.20$, $c = 6.917$ Å, [$U = 1488$ Å^3], $D_m = 1.3\pm1$, $Z = 4$, $D_m = 1.3$, (CuK_α, $\lambda = 1.5418$ Å).

Space group $Pna2_1$ (C_{2v}^9) or $Pnam$ (D_{2h}^{16}) $Pna2_1$ confirmed by analysis.

Atomic positions

	x	*y*	*z*
C(1)	0.4461±6	0.4229±10	0.0202±28
C(1')	0.3828±6	-0.0087±10	-0.0177±34
C(1'')	9.1980±6	0.3012±10	0.0042±34
C(2)	0.4660±6	0.2979±11	-0.0568±29
C(2')	0.3089±6	0.0233±11	-0.0749±25
C(2'')	0.2507±6	0.3890±11	-0.0638±24
C(3)	0.4124±5	0.2081±10	0.0195±24
C(3')	0.2920±5	0.1488± 9	0.0024±33
C(3'')	0.3231±5	0.3578±10	0.0250±26
N(1)	0.3731±5	0.4460± 9	-0.0605±21
N(1')	0.4317±5	0.0918±10	-0.0685±23
N(1'')	0.2233±5	0.1837± 8	-0.0631±20
N(2)	0.3426±4	0.2387± 7	-0.0519±18
Cl(1)	0.1220±2	-0.0239± 3	0.0024±13
Cl(2)	0.5710±1	-0.0386± 3	-0.0071±13
Cl(3)	0.3084±2	0.6991± 3	0.0000 0
O(1)	0.4783±2	0.7124±24	-0.0870±56

Also given are isotropic temperature factors for the above atoms, and the positions of the hydrogen atoms.

Structure

A view of the structure is given in Fig. 79. Bond lengths and angles are given in full, and appear to be normal. The organic molecule is puckered, as shown, and the central nitrogen atom is pyramidal, with C-N-C = 110°. Other angles are close to tetrahedral. Mean values of bond lengths are: C-N (pyramidal), 1.468±14; C-N (other), 1.493±16; C-C, 1.530±15 Å. Each peripheral ring nitrogen is hydrogen bonded to two chlorine atoms, one in the molecular plane, and another above or below the molecule. All Cl-N distances are 3.10±2 Å. The water molecule (which has an occupancy of only about one half) is coordinated to three chlorine atoms, and appears to be hydrogen bonded to two of them. The Cl-O distances are 3.36, 3.32 and 3.55 Å.

Details of analysis

Intensity data were recorded on equi-inclination Weissenberg photographs using CuK_α (*h0l* to *h5l*) and MoK_α (*h6l* to *h9l*) radiation. 1144 independent reflexions were observed, and estimated visually. Absorption corrections were applied. Refinement was by full-matrix least squares to a final R index of 0.093 Å.

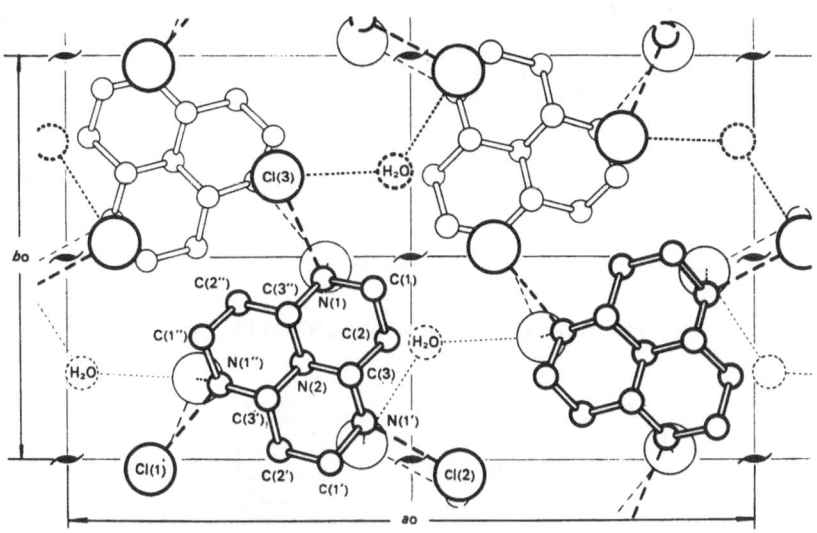

Fig. 79. The structure of $C_9H_{18}N_4 \cdot 3HCl \cdot \frac{1}{2}H_2O$ viewed along c. Broken and dotted lines indicate hydrogen bonds.

$asym$-$\alpha\beta$-NAPHTHAZINE

$C_{20}H_{12}N_2$ F.W. = 280.3

I. Refinement of the structure of $asym$-$\alpha\beta$-naphthazine. B. BOVIO and S. LOCCHI, 1965. Z. Kristallogr., 121, 306-311. (This is a refinement of the structure reported in 1, 2.)

Monoclinic, $a = 10.97$, $b = 4.71$, $c = 14.10$ Å, $\beta = 113.9°$, $[U = 666$ Å$^3]$, $z = 2$, $D_x = 1.40$.

Space group $P2_1/c$ (C_{2h}^5) Molecular symmetry, centre

Atomic positions

	x	y	z
C(1)	0.04003	0.12490	0.09249
C(2)	0.08671	0.26536	0.19147
C(3)	0.18634	0.46211	0.22018
C(4)	0.24657	0.54028	0.14870
C(5)	0.35090	0.74595	0.17866
C(6)	0.40998	0.80862	0.11040
C(7)	0.36850	0.67942	0.01208
C(8)	0.26507	0.47799	-0.01712
C(9)	0.20486	0.39337	0.05052
C(10)	0.09824	0.19233	0.02229
N	0.05779	0.06987	-0.07087

The nominal standard deviations range from 0.008 to 0.018 Å, but are reported to be somewhat optimistic. Anisotropic temperature factors are given also.

Structure

The molecule is planar. Bond lengths and angles are given in full, and are
consistent with expectation. The angle C–N–C is 116±1°. Intermolecular distances
are consistent with van der Waals interaction.

Details of analysis

The three-dimensional intensity data used in 1, 2 have been refined by least
squares analysis to a final R index of 0.098.

1.   *Structure Reports*, 27, 989.
2.   V. RIGANTI, S. LOCCHI, R. CURTI and B. BOVIO, 1963. *Rend. Accad. Naz.
     Lincei.*, 34, 261.

7-(*p*-IODOBENZENESULPHONYL)-7-AZABICYCLO[4.1.0]HEPTANE

$C_{12}H_{14}INO_2S$ F.W. = 363.2

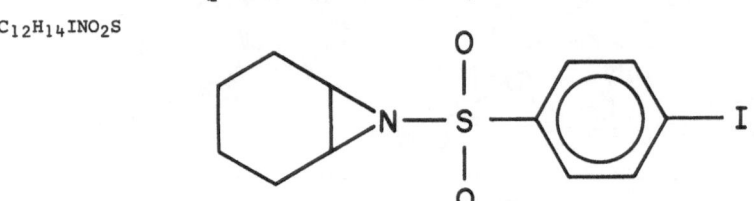

I. Crystal and molecular structure of 7-(*p*-iodobenzenesulfonyl)-7-azabicyclo
    [4.1.0]heptane. L.M. TREFONAS and R. MAJESTE, 1965. *J. Heterocyc. Chem.,
    U.S.A.*, 2, 80-86.

Monoclinic, a = 10.35, b = 16.33, c = 8.22 Å [all presumably ±0.005 Å],
β = 99°40±5', [U = 1370 Å³], D_m = 1.75, Z = 4, D_x = 1.76.

Space group $P2_1/n$ (C_{2h}^5)

Atomic positions

	x	y	z
I	0.8554	0.2908	0.6161
S	0.2293	0.3851	0.5616
O(1)	0.1541	0.3108	0.5845
O(2)	0.1871	0.4282	0.4195
N	0.2457	0.4509	0.7180
C(1)	0.1224	0.4740	0.7794
C(2)	0.2126	0.4155	0.8777
C(3)	0.3067	0.4410	0.0264
C(4)	0.3047	0.5310	0.0668
C(5)	0.2263	0.5904	0.9532
C(6)	0.1205	0.5624	0.8230
C(7)	0.4716	0.3994	0.4774
C(8)	0.6025	0.3749	0.4736
C(9)	0.6581	0.3141	0.5889
C(10)	0.5763	0.2703	0.6981
C(11)	0.4443	0.2900	0.6744
C(12)	0.3907	0.3562	0.5781

The reported mean standard deviations are: I, 0.02 Å; S, 0.03 Å; other atoms,
less than 0.045 Å. [The ratio of magnitudes seems anomalous.] Anisotropic temperature factors are given also.

Structure

The conformation of the molecule is illustrated in Fig. 80. Bond lengths and
angles are given in full. The intermolecular distances are consistent with van der
Waals interaction.

Details of analysis

The intensity data were recorded on precession photographs (*hk*0 to *hk*4; *h*0*l* to *h*5*l*) using MoK$_\alpha$ radiation. 1272 reflexions were observed and measured visually. Refinement was by full–matrix least squares to a final *R* index of 0.115.

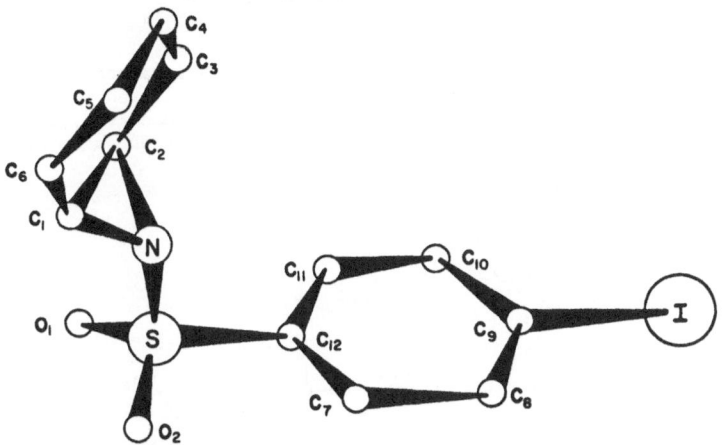

Fig. 80. Molecular conformation of $C_{12}H_{14}INO_2S$.

2,6-DIMETHYL-γ-PYRONE HYDROBROMIDE MONOHYDRATE

$C_7H_9BrO_2 \cdot H_2O$ F.W. = 223.1

, HBr , H_2O

I. The crystal structure of 2,6-dimethyl-γ-pyrone (2,6-dimethyl-4H-pyran-4-one) hydrobromide monohydrate. H. HOPE, 1965. *Acta Chem. Scand.*, <u>19</u>, 217-222.

Triclinic, *a* = 7.00±3, *b* = 8.33±3, *c* = 9.47±3 Å, α = 109.9±5, β = 92.9±5, γ 106.0±5°, [*U* = 493 Å³], D_m = 1.50, *Z* = 2, D_x = 1.50.

Space group P1 (C_1^1) or P$\bar{1}$ (C_i^1). P$\bar{1}$ confirmed by analysis.

Atomic positions

	x	*y*	*z*
O(1)	0.8362	0.7504	0.3074
C(2)	0.8077	0.6584	0.1567
C(3)	0.7398	0.4769	0.0907
C(4)	0.6997	0.3811	0.1894
C(5)	0.7350	0.4812	0.3450
C(6)	0.8020	0.6571	0.4051
C(7)	0.8503	0.7824	0.5651
C(8)	0.8563	0.7846	0.0670
O(9)	0.6318	0.2036	0.1250
O(10)	0.5921	0.0548	0.3185
Br(11)	0.3059	0.6816	0.3257

The standard deviations are 0.003 Å for bromine and 0.02 to 0.03 Å for the other atoms. Temperature factors (anisotropic for bromine) are given also.

Structure

The organic molecule is planar. Bond lengths and angles are given in full. The C-O distance of 1.32 Å suggests that the keto oxygen has acquired a proton, and the hydrogen-bonding scheme of Fig. 81 is proposed. The hydrogen-bonded distances are: O...O, 2.53 Å; Br...O, 3.23 and 3.25 Å.

Details of analysis

The intensity data were recorded on integrated Weissenberg photographs (0*kl* and *h*0*l*) using CuK_α radiation. The intensities were measured photometrically. In the 0*kl* zone, 174 of a possible 186 reflexions were observed, and in the *h*0*l* zone, 133 of 150. Refinement was by full-matrix least squares to final *R* indices of 0.11 for 0*kl* and 0.076 for *h*0*l*.

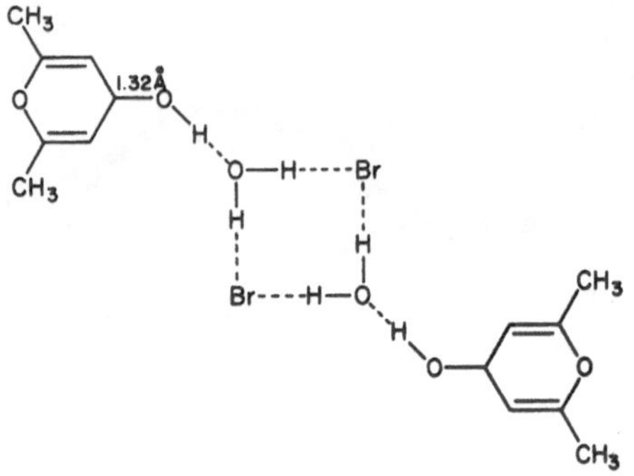

Fig. 81.   The proposed hydrogen bonding scheme in $C_7H_9BrO_2.H_2O$.

3-BROMO-4-HYDROXYCOUMARIN MONOHYDRATE

$C_9H_5BrO_3.H_2O$ F.W. = 259.1

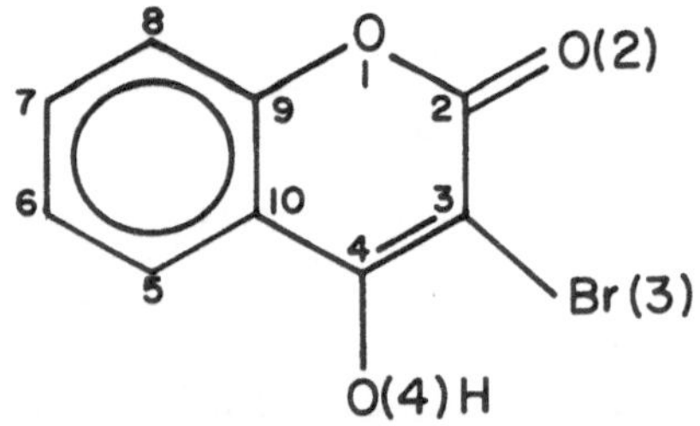

I. Structure crystalline et moléculaire de la bromo-3-hydroxy-4-coumarine mono-hydratée.   J. GAULTIER and C. HAUW, 1965. *Acta Cryst.*, <u>19</u>, 927-933.
   (For a preliminary report, see <u>1</u>.)

Monoclinic, *a* = 7.11±2, *b* = 13.73±2, *c* = 10.14±2 Å, β = 92°30', *U* = 989 Å^3, *z* = 4, D_x = 1.73.

Space group $P2_1/n$ (C_{2h}^5)

Atomic positions

	x	*y*	*z*
O(1)	0.1946	−0.0507	0.3712
C(2)	0.2016	0.0283	0.2991
C(3)	0.2613	0.1195	0.3550
C(4)	0.3055	0.1254	0.4917
C(5)	0.3274	0.0402	0.7079
C(6)	0.3091	−0.0477	0.7760
C(7)	0.2518	−0.1324	0.7111
C(8)	0.2145	−0.1343	0.5781
C(9)	0.2340	−0.0467	0.5105

	x	y	z
C(10)	0.2891	0.0401	0.5714
O(2)	0.1655	0.0187	0.1763
O(4)	0.3666	0.2082	0.5498
Br(3)	0.2634	0.2307	0.2473
W	0.0167	0.1378	−0.0381

Anisotropic temperature factors are given also.

Structure

The molecule is essentially planar, although individual atoms deviate by as much as 0.085 Å from the mean plane. Bond lengths and angles are given in full; some values of interest are:

$$C(2) - C(3) = 1.43 \text{ Å}$$
$$C(3) - C(4) = 1.41$$
$$C(2) - O(2) = 1.26$$
$$C(4) - O(4) = 1.34$$

This distribution of distances is attributed to the contribution of a tautomer in which the hydroxyl group lies at 2 rather than 4. However, there is no indication of the accuracy achieved.

Details of analysis

The intensity data were recorded on De Jong retigrams ($hk0$ to $hk6$). 720 reflexions were observed and measured visually. Refinement was by least squares to a final R index of 0.09.

1. J. GAULTIER and C. HAUW, 1965. *C.R. Acad. Sci. Paris*, <u>260</u>, 3666.

3-PHENYL-1,2-DITHIOLIUM IODIDE

$C_9H_7IS_2$ F.W. = 306.2

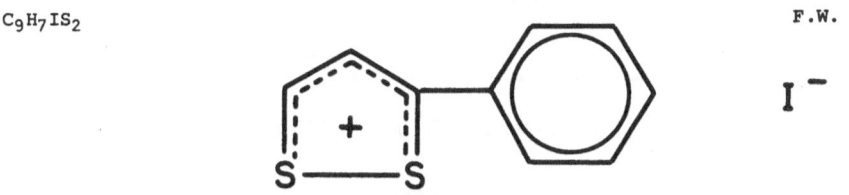

I. The crystal and molecular structure of 3-phenyl-1,2-dithiolium iodide. A. HORDVIK and H.M. KJOGE, 1965. *Acta Chem. Scand.*, <u>19</u>, 935-946.

Monoclinic, $a = 8.28$, $b = 5.51$, $c = 22.52$ Å, (all ±0.5%), $\beta = 98.4°$, $[U = 1016 \text{ Å}^3]$, $D_m = 1.99$, $Z = 4$, $D_x = 2.00$. (The crystal data have been reported in <u>1</u>; the values given here have been redetermined.

Space group $P2_1/c$ (C_{2h}^5)

Atomic positions

	x	y	z
I	0.7474	0.5490	−0.0763
S(1)	0.6376	0.2307	0.0394
S(2)	0.5712	0.0342	0.1064
C(1)	0.8189	0.0955	0.0422
C(2)	0.8612	−0.0869	0.0843
C(3)	0.7417	−0.1428	0.1194
C(4)	0.7668	−0.2953	0.1697
C(5)	0.8652	−0.5097	0.1704
C(6)	0.8814	−0.6773	0.2159
C(7)	0.8031	−0.6377	0.2626
C(8)	0.6989	−0.4344	0.2651
C(9)	0.6769	−0.2616	0.2192

Assumed hydrogen positions are given also. Anisotropic temperature factors are given for the non-hydrogen atoms.

Structure

Bond lengths and angles are given in full, with standard deviations ranging from 0.01 to 0.05 Å. The S–S bond distance is 2.00±1 Å. The five- and six-membered rings of the 3-phenyl-1,2,-dithiolium ion are mutually inclined at 27°. The arrangement of ions in the unit cell is shown in Fig. 82. The sulphur and iodine atoms are roughly coplanar; the sulphur-iodine distances shown are all less than the sum of van der Waals radii (4.00 Å) and are therefore indicative of weak covalent bonding.

Details of analysis

The intensity data were recorded on Weissenberg photographs (*h0l* to *h2l*; *0kl*) using CuK$_\alpha$ radiation. 913 of a possible 1221 reflexions were observed and measured visually. Refinement was by block-diagonal least squares to a final *R* index of 0.10.

1. A. HORDVIK, 1963. *Acta Chem. Scand.*, **17**, 1809.

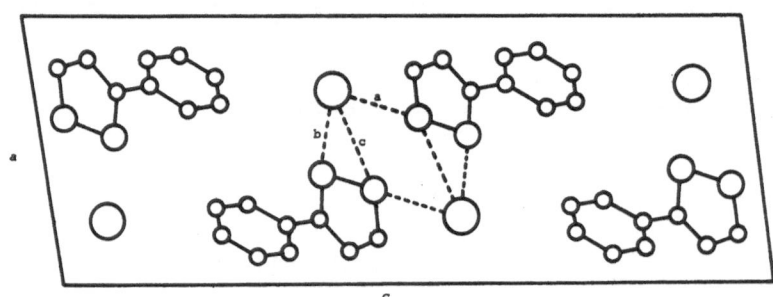

Fig. 82. The structure of 3-phenyl-1,2-dithiolium iodide viewed along *b*. The S–I distances shown are: a, 3.37; b, 3.49; c, 3.62 Å; all ±0.01 Å.

3,5-DIAMINO-1,2-DITHIOLIUM IODIDE

$C_3H_5IN_2S_2$ F.W. = 260.1

I. The sulphur-sulphur bond in 3,5-diamino-1,2-dithiolium iodide. A. HORDVIK, 1965. *Acta Chem. Scand.*, **19**, 1039–1044.

Orthorhombic, *a* = 5.46, *b* = 9.23, *c* = 14.17 Å, (all ±0.5%), [*U* = 714 Å³], *D*$_m$ = 2.25±5, *Z* = 4, *D*$_X$ = 2.33.

Space group Pn2$_1$a (C_{2v}^9)   or   Pnma (D_{2h}^{16}) Pnma is consistent with deduced structure. Molecular symmetry, mirror plane. (The crystal data were reported in 1.)

Atomic positions

	y	*z*
I	0.2500	−0.1050
S	0.1373	0.0790
N	−0.011	0.159
C(1)	0.250	0.154
C(2)	0.094	0.144

Temperature factors are given also.

Structure

The length of the S-S bond is 2.08±2 Å. (This bond is required by symmetry to be parallel to *b*, and thus in the plane of the projection studied.) [This conclusion is of course contingent on the correct choice of space group.]

Details of analysis

The intensity data for the [100] zone were recorded on a Weissenberg photograph, using CuK_α radiation. 69 of a possible 90 reflexions were observed and measured visually. Refinement was by block–diagonal least squares to a final *R* index of 0.11.

1. A. HORDVIK, 1963. *Acta Chem. Scand.*, <u>17</u>, 1809.

5-PHENYL-3-(5-PHENYL-1,2-DITHIOL-3-YLIDENMETHYL)-1,2-DITHIOLIUM IODIDE

$C_{19}H_{13}IS_4 \cdot CH_2OH$ F.W. = 527.5

I. The structure of 5-phenyl-3-(5-phenyl-1,2-dithiol-3-ylidenmethyl)-1,2-dithiolium iodide. A. HORDVIK, 1965. *Acta Chem. Scand.*, <u>19</u>, 1253-1254. (This is a preliminary communication.)

Monoclinic, $a = 5.25$, $b = 17.43$, $d_{001} = 23.26$ Å, (only the 0*kl* zone has been studied, thus β is not known, and the *b* and *c* axes have not been unambiguously identified.) $[U = 2128$ Å$^3]$, $D_m = 1.645$, $Z = 4$, $D = 1.64$.

Plane group *Pgg* The space group is assumed to be $P2_1/c$ (C_{2h}^5).

Atomic positions

	y	z		y	z
I	0.1601	0.1886	C(9)	0.395	0.222
S(1)	0.5167	0.1952	C(10)	0.364	0.248
S(2)	0.4932	0.2520	C(11)	0.361	0.287
S(3)	0.4499	0.3342	C(12)	0.318	0.309
S(4)	0.4180	0.3869	C(13)	0.331	0.348
C(1)	0.475	0.081	C(14)	0.275	0.377
C(2)	0.468	0.045	C(15)	0.207	0.351
C(3)	0.400	0.022	C(16)	0.171	0.384
C(4)	0.338	0.594	C(17)	0.192	0.426
C(5)	0.343	0.102	C(18)	0.256	0.462
C(6)	0.424	0.113	C(19)	0.300	0.431
C(7)	0.425	0.145	C(20)	0.121	0.076
C(8)	0.388	0.176	C(21)	0.520	0.478

Both non-hydrogen atoms of the assumed methanol of solvation have been treated as carbon atoms.

Structure

The analysis confirms the formulation given above. The disulphide rings appear to be coplanar. Comparison of projected S-S distances suggests that there is partial bonding between the innermost sulphur atoms.

Details of analysis

There are few details. Refinement of the 0*kl* zone was by least squares to an *R* index of 0.10.

trans-2,3-DICHLORO-1,4-DITHIANE

$C_4H_6Cl_2S_2$ F.W. - 189.1

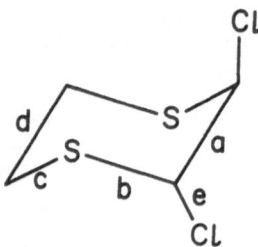

I. The conformation of non-aromatic ring compounds. XIV. The crystal structure
of *trans-2,3-dichloro-1,4-dithiane* at –180°C. H.T. KALFF and C. ROMERS, 1965.
Acta Cryst., <u>18</u>, 164–168.

Monoclinic, $a = 7.174\pm4$, $b = 7.511\pm3$, $c = 6.726\pm4$ Å, $\beta = 93.93\pm7°$, $[U = 361.6$ Å$^3]$,
$Z = 2$, $D_x = 1.75$, (all at –180°C; calibration with aluminium powder, with (at 20°C),
$a = 4.9089$ Å).

Corresponding data at room temperature are: $a = 7.30$, $b = 7.60$, $c = 6.88$ Å,
$\beta = 93.67°$, $D_m = 1.68$, $D_x = 1.66$, and for the isomorphous 2-bromo-3-chloro derivative,
$a = 7.40$, $b = 7.60$, $c = 6.96$ Å, $\beta = 93.50°$, $D_m = 2.03$, $D_x = 1.99$.   Crystal data for
the corresponding 2,3-dibromo derivative are given also, and are reported in <u>1</u>.

Space group Pn (C_s^2) or $P2/n$ (C_{2h}^4) Pn is confirmed by the analysis.

Atomic positions

	x	y	z
S(1)	0.3749	0.2037	0.2124
S(2)	0.3302	0.3241	-0.2880
Cl(1)	0.0000	0.1575	0.0000
Cl(2)	0.3563	0.6200	0.0090
C(1)	0.1778	0.3159	0.0886
C(2)	0.2238	0.4327	-0.0894
C(3)	0.4709	0.0933	-0.0020
C(4)	0.5320	0.2172	-0.1537

Standard deviations are given in full. Mean values are: S and Cl, 0.006 Å;
C, 0.022 Å.   Temperature factors for the three projections are given also.

Structure

The molecule has the chair form indicated above.   Bond lengths and angles
are given in full.   Mean values are:

a	=	1.54±3 Å		ab	=	115.8±14°
b	=	1.79±2		bc	=	100.7±9°
c	=	1.84±2		cd	=	112.4±14
d	=	1.47±3		ae	=	107.4±14
e	=	1.80±2		be	=	112.0±12

(The standard deviations given are those corresponding to individual values.)
Intermolecular distances are consistent with van der Waals interaction.

Details of analysis

Intensity data were recorded on Weissenberg photographs about [100], [010], and
[101] at –180°C using MoK_α radiation.   Refinement was by least squares to final R
indices of 0.081, 0.069, and 0.061 for the three zones.

1.   This volume, p. 414.

TETRACYANO-1,4-DITHIIN

$C_8N_4S_2$ F.W. = 216.2

I. The crystal structure of tetracyano-1,4-dithiin. W.A. DOLLASE, 1965. *J. Amer. Chem. Soc.*, 87, 979–982.

Monoclinic, $a = 6.953 \pm 2$, $b = 7.024 \pm 2$, $c = 18.498 \pm 5$ Å, $\gamma = 90.52 \pm 2°$, $[U = 903.4$ Å$^3]$, $D_m = 1.59$, $Z = 4$, $[D_x = 1.59]$. (The unique axis has been designated "*c*".)

Space group $P2_1/n$ (C_{2h}^5)

Atomic positions

	x	y	z
S(1)	1.12828	0.86637	0.38321
S(2)	0.69162	0.87199	0.43700
C(1)	0.81627	0.65762	0.42583
C(2)	1.00138	0.65682	0.40402
C(3)	0.76401	0.98015	0.35580
C(4)	0.94632	0.97578	0.33248
C(5)	0.71673	0.48242	0.44073
C(6)	1.10018	0.47957	0.39289
C(7)	0.61445	1.05608	0.31141
C(8)	1.00005	1.05644	0.26499
N(1)	0.64328	0.34112	0.45504
N(2)	1.17862	0.33685	0.38939
N(3)	0.49649	1.12137	0.27535
N(4)	1.04361	1.11883	0.20957

The mean standard deviations are: S, 0.0015 to 0.0027 Å; C and N, 0.006 to 0.010 Å. Isotropic temperature factors are given also.

Structure

The molecule has approximate *mm*2 symmetry, and has a folded, or dihedral conformation, with the sulphur atoms lying in the fold. The dihedral angle is 124°, and each half of the molecule is effectively planar. Mean bond lengths and angles are:

a	=	1.755±3 Å		aa'	=	97.3±2°
b	=	1.344±5		ab	=	121.8±7
c	=	1.432±12		ac	=	117.2±11
d	=	1.150±7		bc	=	120.8±5
				cd	=	177.2±22

[The length of the C–N bond ("*d*") may be underestimated, because of thermal motion.] Some rather short intermolecular C–N distances (3.13 to 3.35 Å) are attributed to the closeness of the packing.

Details of analysis

The intensity data were measured with an equi-inclination diffractometer and counter, using CuK$_\alpha$ radiation. 850 reflexions were observed above background. Refinement was by full-matrix least squares and difference syntheses to a final R index of 0.117.

ORGANIC COMPOUNDS

1,3,5-TRITHIANE

$C_3H_6S_3$ F.W. = 138.3

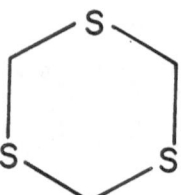

I. The crystal structure of 1,3,5-trithiane. G. VALLE, G. CARAZZOLO and M. MAMMI, 1965. *Ric. Sci.*, 2, A, *Ital.*, 8, 1469–1483. (For earlier work, see 1, 2.)

Orthorhombic, a = 7.668±5, b = 7.003±5, c = 5.285±5 Å, U = 284 Å^3, D_m = 1.59, z = 2, D_x = 1.618.

Space group Pmn2$_1$ (C_{2v}^7) or Pmmn (D_{2h}^{13}). Pmn2$_1$ is confirmed by the analysis. Molecular symmetry, mirror plane.

Atomic positions

	x	y	z
S(1)	0.1996	0.3230	0.1457
S(2)	0	0.0065	−0.1293
C(1)	0.1810	0.1740	−0.1323
C(2)	0	0.4610	0.1293

Coordinates from 1 are given also; there are significant differences. Temperature factors (anisotropic for sulphur) are given.

Structure

The molecule has the chair form, with crystallographic mirror symmetry, and (within experimental error), non-crystallographic trigonal symmetry. The sulphur and carbon planes are parallel, and separated by 0.64 Å. The mean bond lengths and angles are: S–C, 1.812±5 Å; C–S–C, 100.6±8°; S–C–S 115.8±8°. The molecular packing is discussed; intermolecular distances are consistent with van der Waals interaction.

Details of analysis

The intensity data for the three principal zones were recorded photographically (hk0, Weissenberg, CuK_α radiation; 0kl, h0l, precession camera, MoK_α radiation.) 156 of a possible 172 reflexions were observed and measured with a densitometer. Refinement was by Fourier methods to final R indices of 0.048 to 0.065 for the three zones.

1. *Strukturbericht*, 5, 157.
2. *Structure Reports*, 10, 284.

2,2'-p-PHENYLENEBIS-(5-PHENYLOXAZOLE)

$C_{24}H_{16}N_2O_2$ F.W. = 364.4

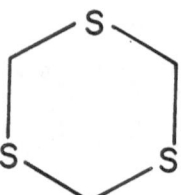

I. The crystal structure of 2,2'-p-phenylenebis-(5-phenyl oxazole) – POPOP. I. AMBATS and R.E. MARSH, 1965. *Acta Cryst.*, 19, 942–948.

Monoclinic, a = 9.2300±3, b = 5.2850±8, c = 19.3220±7 Å, β = 92.088±2°, U = 941.9 Å^3, D_m = 1.306, z = 2, D_x = 1.285 (CuK_α, λ = 1.5418 Å). [The estimate of accuracy seems optimistic; the temperature is not stated.]

Space group P2$_1$/c (C_{2h}^5) Molecular symmetry, centre.

Atomic positions

	x	y	z
C(1)	$0.1126^{\pm 2}$	-0.1681 ± 5	0.0146 ± 1
C(2)	$0.0791^{\pm 2}$	0.0145 ± 5	0.0623 ± 1
C(3)	-0.0335 ± 2	0.1859 ± 4	0.0482 ± 1
C(4)	-0.0669 ± 2	0.3769 ± 4	0.0998 ± 1
C(5)	-0.0771 ± 3	0.6166 ± 5	0.1873 ± 1
C(6)	-0.1881 ± 2	0.6851 ± 4	0.1441 ± 1
C(7)	-0.3026 ± 2	0.8734 ± 5	0.1464 ± 1
C(8)	-0.4134 ± 3	0.8827 ± 5	0.0957 ± 1
C(9)	-0.5200 ± 3	1.0657 ± 6	0.0995 ± 2
C(10)	-0.5187 ± 3	1.2383 ± 6	0.1531 ± 2
C(11)	-0.4099 ± 3	1.2291 ± 5	0.2035 ± 1
C(12)	-0.3016 ± 3	1.0489 ± 5	0.2004 ± 1
O	-0.1830 ± 2	0.5305 ± 3	0.0868 ± 1
N	-0.0001 ± 2	0.4205 ± 4	0.1586 ± 1
H(1)	0.1985 ± 29	-0.2793 ± 53	0.0252 ± 14
C(2)	0.1386 ± 29	0.0282 ± 53	0.1073 ± 14
H(3)	-0.0524 ± 29	0.6850 ± 55	0.2345 ± 14
H(4)	-0.4142 ± 28	0.7490 ± 55	0.0572 ± 14
H(5)	-0.5947 ± 30	1.0704 ± 58	0.0652 ± 15
H(6)	-0.5985 ± 31	1.3575 ± 61	0.1574 ± 15
H(7)	-0.3990 ± 31	1.3422 ± 62	0.2438 ± 15
H(8)	-0.2265 ± 29	1.0323 ± 56	0.2332 ± 15

Temperature factors (anisotropic for the non-hydrogen atoms) are given also.

Structure

    Bond lengths and angles are given in Fig. 83. The distances have not been
corrected for thermal motion, the effects of which may be appreciable. Each ring
in the molecule is planar, but the end phenyl ring makes an angle of 6.4° with
the oxazole ring, which in turn makes an angle of 3.8° (in the same sense) with the
central phenyl ring. The molecule is compared to a screw, or a propeller. [However,
the comparison is inappropriate, as the molecule is centrosymmetrical.] A view of
the structure is given in Fig. 84. Intermolecular distances are consistent with van
der Waals interaction.

Details of analysis

    The intensity data were recorded on Weissenberg photographs ($h0l$ to $h5l$) using
CuK_α radiation. 1370 of 1984 accessible reflexions were observed and measured visually. Refinement was by block-diagonal least squares to a final R index of 0.058.

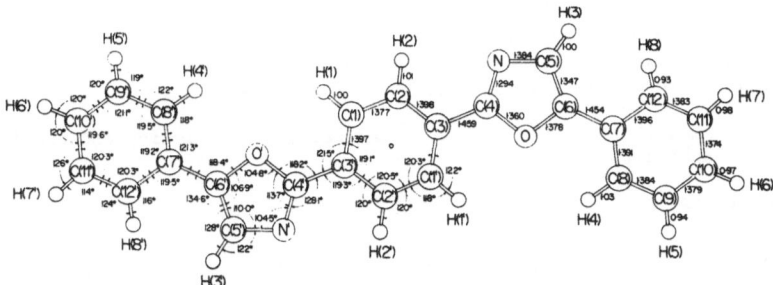

Fig. 83. Bond lengths and angles in $C_{24}H_{16}N_2O_2$. For quantities not involving
          hydrogen the standard deviations are 0.002 to 0.004 Å, and 0.3 to 0.6°.
          For the rest they are 0.3 Å and 3 to 4°.

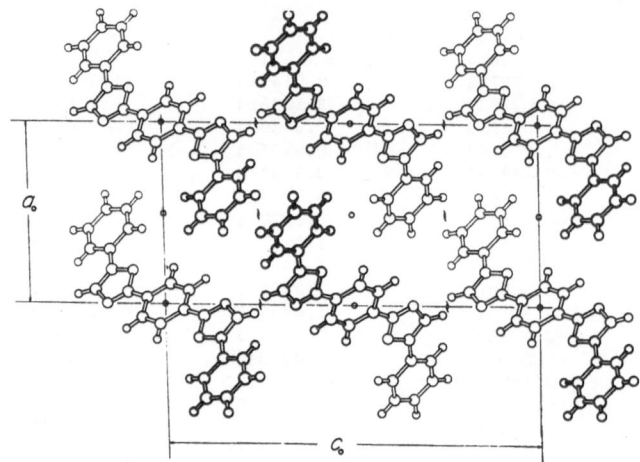

Fig. 84.   The structure of $C_{24}H_{16}N_2O_2$, viewed along b.

2-p-METHOXYPHENYL-3,4-DIBENZYL-1,3,4-THIADIAZOLIDINE-5-THIONE

$C_{23}H_{22}N_2OS_2$ F.W. = 406.6

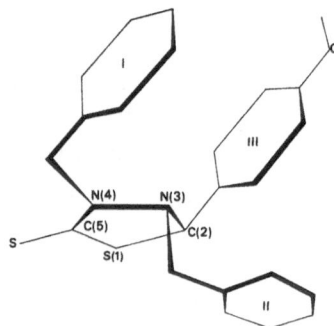

I. The structure of 2-p-methoxyphenyl-3,4-dibenzyl-1,3,4-thiadiazolidine-5-thione.
$C_{23}H_{22}N_2OS_2$.   I.L. KARLE and J. KARLE, 1965. *Acta Cryst.*, 19, 92-99.
(For a preliminary report, see 1.)

Triclinic, $a = 7.74\pm2$, $b = 12.07\pm3$, $c = 12.32\pm3$ Å, $\alpha = 111°20\pm20$, $\beta = 88°10\pm20$,
$\gamma = 100°43\pm20'$, $[U = 1053$ Å$^3]$, $D_m = 1.264$, $Z = 2$, $D_x = 1.282$.

Space group $P1$ (C_1^1) or $P\bar{1}$ (C_i^1). $P\bar{1}$ indicated by statistics and confirmed
by structure analysis.

Atomic positions

	x	y	z
S(1)	0.0540	0.5000	0.2168
C(2)	0.1503	0.6627	0.2754
N(3)	0.3078	0.6740	0.2015
N(4)	0.3795	0.5678	0.1769
C(5)	0.2656	0.4673	0.1755
S(6)	0.3099	0.3304	0.1407
C(7)	0.5676	0.5821	0.1709
C(8)	0.6439	0.6485	0.2902
C(9)	0.6291	0.5847	0.3670
C(10)	0.6961	0.6453	0.4816
C(11)	0.7783	0.7639	0.5125

	x	y	z
C(12)	0.7941	0.8284	0.4365
C(13)	0.7236	0.7691	0.3220
C(14)	0.2745	0.6824	0.0883
C(15)	0.2120	0.7990	0.1118
C(16)	0.3262	0.9094	0.1602
C(17)	0.2654	1.0168	0.1808
C(18)	0.0979	1.0153	0.1532
C(19)	−0.0180	0.9055	0.1047
C(20)	0.0438	0.7967	0.0834
C(21)	0.1948	0.7095	0.4034
C(22)	0.1590	0.6420	0.4736
C(23)	0.1979	0.6936	0.5932
C(24)	0.2701	0.8132	0.6371
C(25)	0.3131	0.8831	0.5667
C(26)	0.2729	0.8297	0.4497
O(27)	0.3132	0.8747	0.7515
C(28)	0.2729	0.8074	0.8302

Mean standard deviations are: S, 0.0016; C, 0.008; N and O, 0.005 Å. Aniso-tropic temperature factors are given also.

Structure

The configuration of the molecule is as represented above, and bond lengths and angles are given in Fig. 85. Rings I and III are approximately parallel to each other, and approximately perpendicular to the five-membered ring. Four atoms of this ring are approximately coplanar, and the fifth (C(2)) lies 0.58 Å from their mean plane. The bonds to N(3) are in an approximately tetrahedral arrangement, while those to N(4) lie in a plane. Intermolecular distances are consistent with van der Waals interaction.

Details of analysis

Intensity data were recorded on equi-inclination Weissenberg photographs ($0kl$ to $6kl$) with CuK_α radiation. 3780 reflexions were observed and estimated visually. Refinement was by full-matrix least squares to a final R index of 0.107.

1. I.L. KARLE, J. KARLE and R.M. MORIARTY, 1964. *Tet. Lett.* 3579.

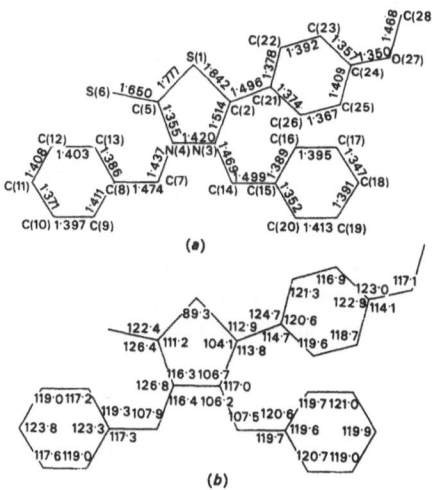

Fig. 85. Bond lengths and angles in $C_{23}H_{22}N_2OS_2$. Standard deviations range from 0.007 to 0.010 Å, and from 0.4 to 0.9°.

2,5-DIPHENYLTHIODIAZOLE

$C_{14}H_{10}N_2S$ F.W. = 238.3

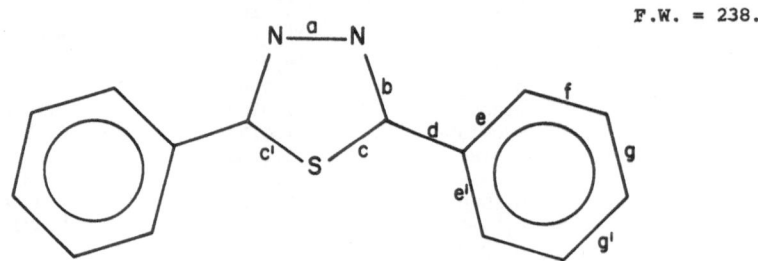

I. [Atomic structure of 2,5-diphenylthiodiazole.] Z.V. ZVONKOVA and A.N. KHVATKINA,
    1965. *Kristallografija*, <u>10</u>, 734-737. [*Soviet Physics, Crystallography*, <u>10</u>,
    614-615.]

Monoclinic, a = 27.84, b = 5.77, c = 7.37 Å, β = 107°05', [U = 1131.7 Å^3], D_m = 1.36,
z = 4, D_x = 1.39.

Space group   *Cc* (C_s^4)   or *C2/c* (C_{2h}^6).   *C2/c* is consistent with deduced structure.
Molecular symmetry, two-fold axis.

Atomic positions

	x	y	z
S	0	0.1315	0.250
N	0.0216	0.5594	0.2531
C(5)	0.0418	0.3632	0.2487
C(6)	0.0959	0.3050	0.2640
C(7)	0.1320	0.4732	0.3342
C(8)	0.1818	0.4139	0.3452
C(9)	0.1962	0.2143	0.2699
C(10)	0.1590	0.0522	0.2040
C(11)	0.1080	0.1009	0.1901

Structure

    The two-fold axis contains the sulphur atom and the mid-point of the N-N bond.
The phenyl rings are inclined to the thiodiazole ring by about 25°, giving the mole-
cule a propeller shape. Mean bond lengths and angles (assuming *mm*2 symmetry for the
C - C$_6$H$_5$ group) are:

a	=	1.19 Å	ab	=	110.5°
b	=	1.27	bc	=	112.8
c	=	1.77	cc'	=	89.5
d	=	1.51	bd	=	129.8
e	=	1.38	cd	=	118.4
f	=	1.42	de	=	119.8
g	=	1.38	ee'	=	120.6
			ef	=	117.9
			fg	=	123.0
			gg'	=	115.4

    There is no reliable indication of error.

    Intermolecular distances are consistent with van der Waals interaction.

Details of analysis

    The intensity data were recorded on equi-inclination Weissenberg photographs
(*h*0*l* to *h*4*l*; *hk*0 to *hk*7) using CuK$_\alpha$ radiation. 742 reflexions were observed.
Refinement was by Fourier methods. There is no indication of accuracy.

THIURET HYDROBROMIDE

$C_2H_4BrN_3S_2$ F.W. = 214.1

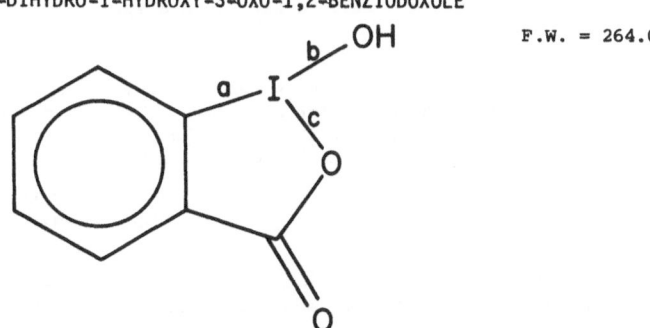

I. The crystal and molecular structure of thiuret hydrobromide. A. HORDVIK and S. JOYS, 1965. *Acta Chem. Scand.*, <u>19</u>, 1539-1548.

Monoclinic, a = 5.11, b = 12.76, c = 10.47 Å, β = 110°, (all ±0.5%), U = 642 Å^3, D_m = 2.22, Z = 4, D_x = 2.22.

Space group $P2_1/c$ (C^5_{2h})

Atomic positions

	x	y	z
Br	0.7724	0.1954	-0.0628
S(1)	0.8607	-0.0109	0.1614
S(2)	0.9702	-0.1268	0.3103
N(1)	0.5011	0.1338	0.1687
N(2)	0.6000	0.0129	0.3426
N(3)	0.7406	-0.1211	0.5041
C(1)	0.6379	0.0499	0.2297
C(2)	0.7492	-0.0736	0.3883

Assumed positions for the hydrogen atoms, and anisotropic temperature factors for the non-hydrogen atoms, are given also.

Structure

The thiuret ion is accurately planar, with bond lengths and angles as follows:

a	=	1.34±3 Å	aa'	=	112±2°
b	=	1.730±25	ab	=	112±2
c	=	1.35 ±3	ac	=	120±2
d	=	2.081±8	bc	=	118±2
			bd	=	92±1

The structural arrangement is illustrated in Fig. 86. Each bromine atom forms four weak covalent bonds with sulphur atoms, of length 3.38 to 3.45 Å. (The corresponding sum of van der Waals radii is 3.80 Å.) The structure is additionally stabilized by N-H...N and N-H...Br bonds.

Details of analysis

The intensity data were recorded on Weissenberg photographs ($h0l$, $0kl$, $1kl$) using CuK$_\alpha$ radiation. 451 reflexions were observed and measured visually. Refinement was by block-diagonal least squares to a final R index of 0.09.

1,3-DIHYDRO-1-HYDROXY-3-OXO-1,2-BENZIODOXOLE

$C_7H_5IO_3$ F.W. = 264.0

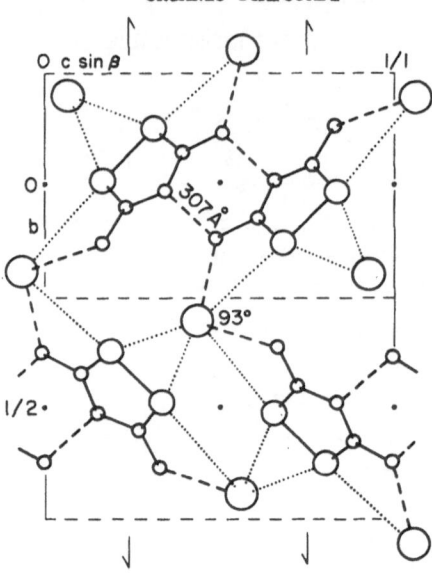

Fig. 86. The structure of thiuret hydrobromide, viewed along *a*. Broken lines
indicate N–H...N and N–H...Br bonds, and dotted lines indicate the
partial covalent bonding between bromine and sulphur.

I. Crystal and molecular structure of 1,3–dihydro–1–hydroxy–3–oxo–1,2–benziodoxole.
   E. SHEFTER and W. WOLF, 1965. *J. Pharm. Sci.*, 54, 104–107.

Monoclinic, *a* = 12.89±1, *b* = 4.10±1, *c* = 14.05±1 Å, β = 96.73±5°, [*U* = 737 Å³],
D_m = 2.3, *Z* = 4, D_x = 2.38.

Space group $P2_1/c$ (C_{2h}^5)

Atomic positions.

	x	*y*	*z*
I	0.0890	1.1305	0.1492
C(1)	0.236	0.894	0.135
C(2)	0.267	1.010	0.051
C(3)	0.368	0.838	0.032
C(4)	0.429	0.750	0.106
C(5)	0.391	0.634	0.195
C(6)	0.294	0.760	0.212
C(7)	0.201	1.145	-0.024
O(1)	0.235	1.196	-0.108
O(2)	0.119	1.265	-0.004
O(3)	0.091	0.926	0.279

Isotropic temperature factors are given also.

Structure

    The analysis confirms the above formulation. Bond lengths and angles are
given in full, but with rather low accuracy. [The e.s.d. of distances between
light atoms is given as 0.01 Å; it seems likely that 0.1 Å is meant.] The bond
lengths and angles involving the iodine atom are:

              a = 2.16±5 Å   ac =   77±5°
              b = 2.00±5   bc =   89±5
              c = 2.30±5   ac = 165±5

    As illustrated in Fig. 87, the structure is stabilized by intermolecular
O–H...O bonds of length 2.77 Å, and I...O contacts of 2.90 Å.

Details of analysis

The intensity data were recorded on Weissenberg photographs (*h0l* to *h2l*) using CuKα radiation. 520 reflexions (about 35% of those accessible to the radiation used) were observed and measured photometrically and visually. Refinement was by least squares to a final *R* index of 0.137.

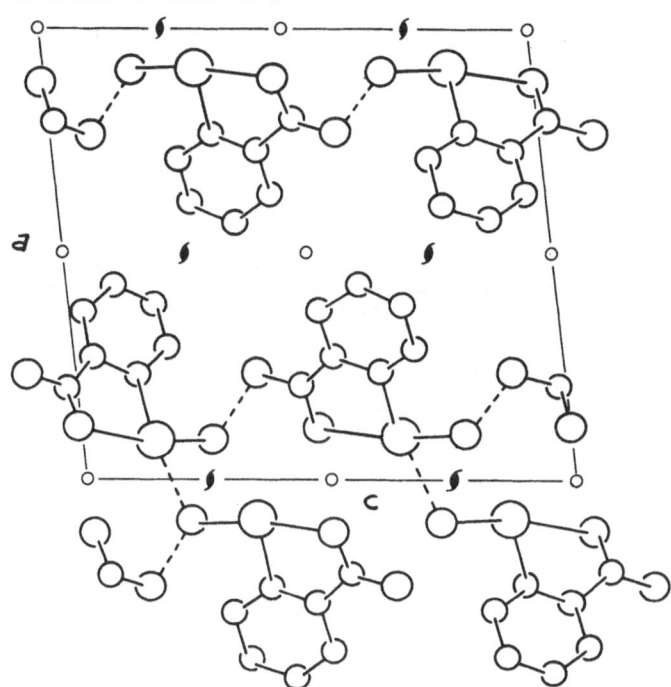

Fig. 87. The structure of $C_7H_5IO_3$ viewed along *b*.

5,5-DIHYDROXY BARBITURIC ACID
(ALLOXAN)

$C_4H_4N_2O_5$ F.W. = 160.1

I. The structure of the pyrimidines and purines. VIII. The crystal structure of alloxan, $C_4H_4N_2O_5$. C. SINGH, 1965. *Acta Cryst.*, **19**, 759–767.

Triclinic, a = 5.770±6, b = 6.812±7, c = 8.206±8 Å, α = 74°24±4, β = 81°52±5, γ = 66°29±4', U = 281.5 Å^3, D_m = 1.777, Z = 2, D_x = 1.880.

Space group P1 (C_1^1) or P$\bar{1}$ (C_i^1). P$\bar{1}$ is suggested by intensity statistics, and confirmed by analysis.

Atomic positions

	x	y	z
C(2)	−0.2369	0.4447	0.1783
C(4)	0.1393	0.2660	0.3472
C(5)	0.1932	0.0469	0.3016
C(6)	−0.0283	0.0436	0.2196
N(1)	−0.2120	0.2452	0.1555
N(3)	−0.0559	0.4468	0.2689
O(2)	−0.4128	0.6129	0.1199
O(4)	0.2700	0.2802	0.4425
O(5)	0.2528	−0.1271	0.4434
O(6)	−0.0296	−0.1247	0.1985
O(7)	0.3923	0.0317	0.1818

Mean standard deviations are: C, 0.0034; N, 0.0030; O, 0.0027 Å. The positions of the hydrogen atoms are given also. Anisotropic temperature factors are given for the non-hydrogen atoms; they have been used in an analysis of rigid-body motion, and the positions reported have been appropriately corrected. (The positions given here were obtained from a three-dimentional analysis of photographic data. Also given in the original are positions obtained from a two-dimensional analysis of counter data.)

Structure

The molecule has approximate mirror symmetry. Mean bond lengths and angles are:

a	=	1.221 Å	ab	=	121.4°
b	=	1.373	bb'	=	117.2
c	=	1.374	bc	=	126.7
d	=	1.539	cd	=	116.3
e	=	1.208	dd'	=	113.9
f	=	1.392	ff'	=	113.2
(All ±0.005 Å)			cd	=	121.4
			ed	=	121.5
			df	=	109.8

The molecule is puckered, with the tetrahedral carbon atom (dd') lying on one side of the mean plane of the ring, and the adjacent ketonic oxygen atoms lying on the other side. The structure is held together by a system of hydrogen bonds, for details of which the original should be consulted.

Details of analysis

The intensity data for the three-dimensional analysis were recorded on equi-inclination Weissenberg photographs (0kl to 4kl; h0l to h5l) using CuKα radiation. 1173 reflexions were observed and presumably measured visually. Refinement was by full-matrix least squares to a final R index of 0.087. A two-dimensional analysis of counter data is described also.

5,5-DIHYDROXYBARBITURIC ACID TRIHYDRATE

$C_4H_4N_2O_5 \cdot 3H_2O$ F.W. = 214.1

I. The crystal structure of 5,5-dihydroxybarbituric acid trihydrate. D. MOOTZ
and G.A. JEFFREY, 1965. *Acta Cryst.*, 19, 717–725.

Monoclinic, a = 9.579±5, b = 12.267±7, c = 7.400±10 Å, β = 90.75±25°, U = 869.5 Å^3,
D_m =1.639, Z = 4, D_x = 1.635, (Cu$K_{\alpha1}$, λ = 1.5405 Å, Cu$K_{\alpha2}$, λ = 1.5443 Å).

Space group $C2$ (C_2^3), Cm (C_s^3), or $C2/m$ (C_{2h}^3). $C2/m$ suggested by intensity
statistics and confirmed by structure analysis. Molecular symmetry, mirror plane
for 5,5-dihydroxybarbituric acid, two-fold axis for one of three water molecules.

Atomic positions

	x	y	z
C(2)	0.3371±5	0	−0.0889±6
C(4)	0.1821±3	0.1042±2	0.1042±4
C(5)	0.1374±4	0	0.2002±5
N(3)	0.2862±3	0.0961±2	−0.0205±3
O(2)	0.4228±4	0	−0.2074±5
O(4)	0.1295±3	0.1916±2	0.1437±4
O(5)	−0.0064±3	0	0.2162±4
O(7)	0.2118±4	0	0.3654±4
W(1)	0.1534±3	0.1611±2	−0.3883±4
W(2)	0.5	0.1534±3	−0.5

Also given are anisotropic temperature factors for the atoms above, and the
positions of some of the hydrogen atoms.

Structure

The 5,5-dihydroxybarbituric acid molecule has a mirror plane normal to the
pyrimidine ring, and passing through C(2), O(2), C(5) and the OH groups. C(5) and
the midpoint of the hydroxyl oxygen atoms are displaced from the mean plane of the
rest of the ring by 0.21 and 0.47 Å. The keto oxygen atoms lie on the opposite side
of the plane, at distances 0.14 Å for O(4) and 0.09 Å for O(2). Mean bond lengths
for the molecule are:

    C-N   = 1.373±4 Å
    C=O   = 1.215±4
    C-C   = 1.526±5
    C-OH = 1.395±5

The structure is held together by a hydrogen bonding system (illustrated in
Fig. 88) which involves all the oxygen, nitrogen and hydrogen atoms.

Details of analysis

The intensity data were recorded on integrated equi-inclination Weissenberg
photographs (0kl to 7kl; h0l to h4l; hk0 to hk3) using CuK_α radiation. 874 of
a possible 982 reflexions were observed and estimated visually. Refinement was
by full-matrix least squares to a final R index of 0.097.

5-(6'-BROMO-3'-ETHYL-2'-METHYLBENZIMIDAZOLIUM) BARBITURATE MONOHYDRATE

$C_{14}H_{13}N_4O_3Br.H_2O$ (Structure formula on p. 162)   F.W. = 383.2

I.   The crystal structure of the condensation product of alloxan hydrate and the
bromo derivative of *o*-amino-diethylaniline. B.W. MATHEWS, 1965. *Acta Cryst.*,
18, 151–157.

Monoclinic, a = 8.687±4, b = 23.730±4, c = 14.995±3 Å, β = 95°5.3±15', [U = 3079 Å^3],
D_m = 1.652, Z = 8, D_m = 1.652.

Space group $P2_1/n$ (C_{2h}^5)

Atomic positions

Atomic positions and isotropic temperature factors are given for the 46 non-
hydrogen atoms. The standard deviations are 0.0013 Å for bromine, and 0.006 to
0.016 Å for the other atoms.

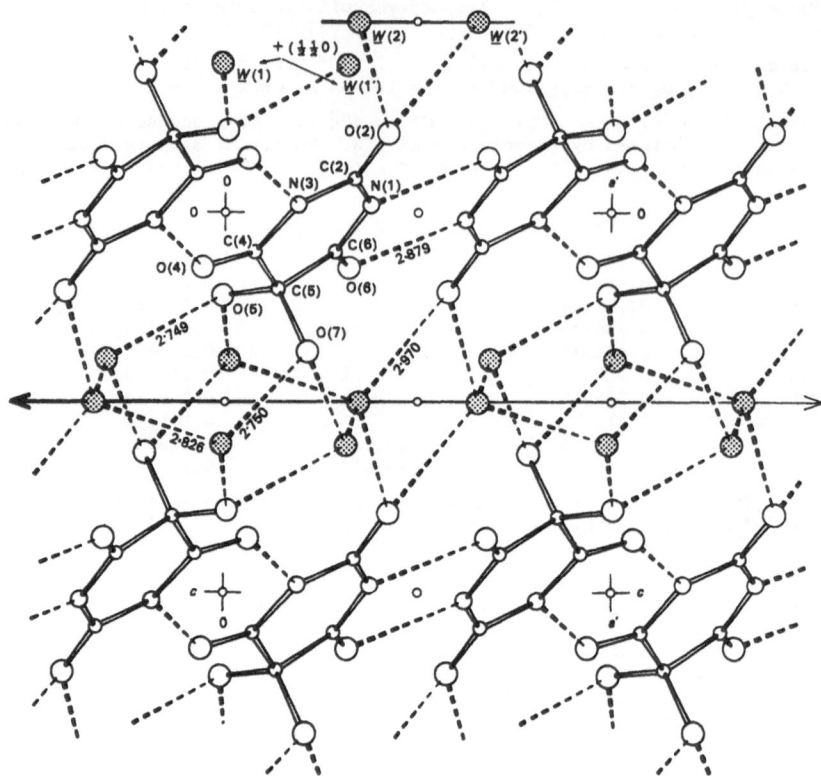

Fig. 88. Part of the structure of 5,5-dihydroxybarbituric acid trihydrate viewed
along [110], showing the hydrogen-bonding network.

Structure formula

Structure

     Bond lengths and angles are given in full for the two independent molecules,
and are consistent with the formulation given above. There are no significant
differences between the two molecules. Both the benzimidazole nucleus and the

barbiturate ring are essentially planar, and the dihedral angle between them is 68.2°. The structure is stabilized by a complex system of hydrogen bonds (five bonds for one molecule, six for the other) in which the water molecules are involved.

Details of analysis

Intensity data were recorded on equi-inclination Weissenberg photographs (0kl to 7kl; h0l to h4l) using CuK_α radiation. 4804 of a possible 6765 reflexions were observed and measured visually. Refinement was by differential syntheses to a final R index of 0.173.

POTASSIUM 5,5'-DIETHYLBARBITURATE

$C_8H_{11}KN_2O_3$ F.W. = 222.3

I. Structure crystalline du sel de potassium de l'acide 5,5'-diéthylbarbiturique (veronal). J. BERTHOU, B. RERAT and C. RÉRAT, 1965. *Acta Cryst.*, 18, 768-777.

Orthorhombic, a = 12.35±5, b = 12.97±5, c = 6.83±2 Å, U = 1094 Å^3, D_m = 1.36, Z = 4, D_x = 1.35.

Space group $P2_12_12_1$ (D_2^4)

Atomic positions

	x	y	z
K	−0.1547±2	−0.0115±3	0.4024±6
O(1)	0.0093±15	0.1097±16	0.4652±40
O(2)	0.1742±9	0.4180±9	0.2649±23
O(3)	0.1030±13	0.2846±13	0.0418±29
N(1)	0.0883±12	0.2624±13	0.4755±31
N(2)	0.1453±12	0.3494±13	0.7490±31
C(1)	0.0429±14	0.1784±14	0.5512±31
C(2)	0.1324±16	0.3534±16	0.5650±38
C(3)	0.0977±16	0.2726±17	0.8674±39
C(4)	0.0404±24	0.1812±27	0.7825±59
C(5)	0.1092±26	0.0838±29	0.8501±64
C(6)	0.2329±26	0.0881±30	0.7844±80
C(7)	−0.0697±22	0.1789±21	0.8585±57
C(8)	−0.1462±31	0.2732±38	0.7877±89

Isotropic temperature factors are given for all atoms.

Structure

The diethylbarbiturate ion has the form indicated above. The ethyl groups and the adjacent ring carbon are coplanar; the six ring atoms and the three oxygen atoms are nearly so. The angle between the planes is 88°. Bond lengths and angles are given in full, but are rather inaccurate. Adjacent diethylbarbiturate ions are joined by a hydrogen bond N-H..O of length 2.98 Å. The potassium ion is co-ordinated to three oxygen atoms in different molecules, and each oxygen atom in a given molecule is coordinated to a different potassium ion. K-O distances are 2.60, 2.60 and 2.75 Å. Full details of the ionic and hydrogen bonding scheme are given.

Details of analysis

Intensity data were recorded with the de Jong retigraph (*hk*0 to *hk*6; *h*0*l* to *h*11*l*) using Cu*K*$_\alpha$ radiation. 1189 reflexions or about 80% of those accessible in the copper sphere, were observed and measured visually. Refinement was by diagonal least squares to a final *R* index of 0.246.

RUBIDIUM VIOLURATE
POTASSIUM VIOLURATE DIHYDRATE

$C_4H_2N_3O_4Rb$ F.W. = 241.6

$C_4H_2N_3O_4K.2H_2O$ F.W. = 213.2

Rb^+

K^+ $.2H_2O$

I. Contribution au phénomène de chromoisomérie. I. Détermination de la structure cristalline du violurate de rubidium et du violurate dihydrate de potassium. H. GILLIER, 1965. *Bull. Soc. Chim. Fr.*, 2373-2384.

II. Contribution au phénomène de chromoisomérie. II. Description de la structure cristalline du violurate de rubidium et du violurate dihydrate de potassium. *Ibid.*, 1965. *Ibid.*, 2385-2395.

RUBIDIUM VIOLURATE

Triclinic, a = 4.79±2, b = 7.86±3, c = 9.62±3 Å, α = 103°50±30, β = 85°40±30, γ = 111°25±30', U = 327.4 Å^3, D_m = 2.38, Z = 2, D_x = 2.43.

Space group P1 (C_1^1) or P$\bar{1}$ (C_i^1). P$\bar{1}$ is confirmed by the analysis.

Atomic positions

	x	y	z
Rb	0.3621±6	0.2417±4	0.1288±3
O(1)	−0.176 ±3	0.616 ±2	0.087 ±1
N(1)	0.095 ±3	0.736 ±2	0.108 ±2
C(1)	0.240 ±4	0.743 ±2	0.227 ±2
C(2)	0.536 ±4	0.884 ±2	0.246 ±2
O(2)	0.656 ±3	0.992 ±2	0.167 ±1
N(2)	0.694 ±3	0.904 ±2	0.366 ±2
C(3)	0.596 ±4	0.796 ±2	0.464 ±2
O(3)	0.738 ±3	0.819 ±2	0.571 ±1
N(3)	0.313 ±3	0.660 ±2	0.441 ±2
C(4)	0.129 ±4	0.622 ±2	0.325 ±2
O(4)	−0.119 ±3	0.495 ±2	0.319 ±1

Isotropic temperature factors are given also.

Structure

The violurate ion is planar except for the isonitroso oxygen atom, which lies 0.1 Å from the ring plane. Bond lengths and angles are given in full, with standard deviations of 0.04 Å and 0.5°. All bond lengths are intermediate between accepted double- and single-bond values. The rubidium ion is coordinated to six oxygen atoms of five different violurate ions, with Rb – O distances ranging from 2.90 to 3.06 Å. A seventh oxygen atom lies at a distance of 3.28 Å from the rubidium atom, the co-ordination can thus be considered either as six- or seven-fold. Adjacent violurate ions are linked by pairs of N-H..O bonds (of length 2.79 to 2.81 Å) to form endless chains as shown in Fig. 89.

Details of analysis

The intensity data were recorded on integrated Weissenberg photographs $(0kl$ to $4kl)$, using CuK_α radiation, and were measured photometrically. Refinement was by least squares to a final R index of 0.13.

POTASSIUM VIOLURATE DIHYDRATE

Monoclinic, $a = 6.47\pm2$, $b = 11.93\pm3$, $c = 11.39\pm3$ Å, $\beta = 107°30\pm30'$, $U = 838.5$ Å^3, $D_m = 1.81$, $Z = 4$, $D_x = 1.82$.

Space group $P2_1/a$ (C_{2h}^5)

Atomic positions

	x	y	z
K+	0.2670±5	0.9956±3	0.1542±3
O(1)	0.1462±11	0.2236±6	0.1431±6
N(1)	0.1575±12	0.3116±7	0.2057±8
C(1)	0.1988±14	0.3026±8	0.3277±8
C(2)	0.2099±14	0.4092±8	0.3913±8
O(2)	0.1825±11	0.5017±6	0.3399±6
N(2)	0.2538±12	0.4053±7	0.5167±7
C(3)	0.2728±14	0.3099±8	0.5870±8
O(3)	0.2986±11	0.3149±6	0.6957±6
N(3)	0.2607±12	0.2107±7	0.5235±7
C(4)	0.2237±14	0.1994±8	0.3991±8
O(4)	0.2215±11	0.1053±6	0.3561±6
(H$_2$O)(1)	0.4973±11	0.8197±6	0.0940±6
(H$_2$O)(2)	0.1784±11	0.0193±6	0.9059±6

Isotropic temperature factors are given also.

Structure

The violurate ion is planar except for the oxygen atoms, some of which lie as much as 0.07 Å from the ring plane. Bond lengths and angles are given in full, with standard deviations of 0.02 Å and 1.4°. Mean bond lengths are: C-C, 1.46 Å; C-O, 1.22 Å; C-N(ring), 1.38 Å; C-NO, 1.34 Å; N-O, 1.26 Å. The potassium ion is coordinated to seven oxygen atoms, (three belonging to water molecules), at distances of 2.73 to 2.88 Å. As illustrated in Fig. 90, adjacent violurate ions are linked by pairs of N-H..O bonds, (of length 2.77 and 2.90 Å), to form endless chains which are cross-linked by O-H..O bonds involving the water molecules.

Details of analysis

The intensity data were recorded on integrated Weissenberg photographs, using CuK_α radiation, and were measured photometrically and visually. Refinement was by least squares to a final R index of 0.18.

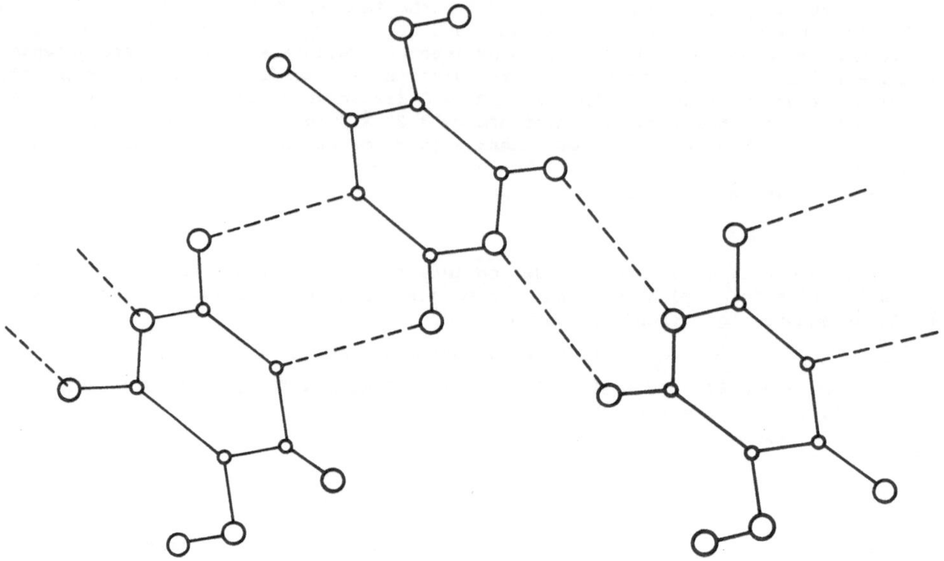

Fig. 89.   Hydrogen bonding in rubidium violurate.

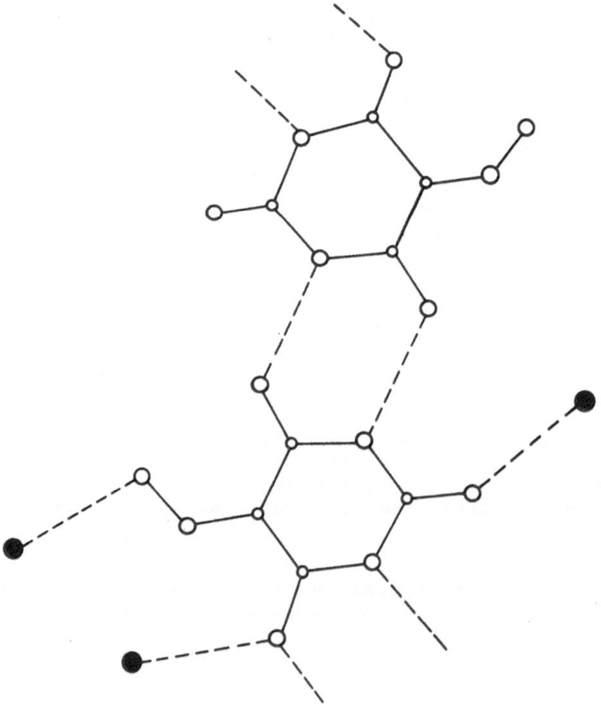

Fig. 90.   Hydrogen bonding in hydrated potassium violurate.   The solid circles
           represent water molecules.

DIALURIC ACID MONOHYDRATE

$C_4H_4N_2O_4 \cdot H_2O$ F.W. = 162.1

I. A reinvestigation of the crystal structure of dialuric acid monohydrate.
W. BOLTON, 1965. *Acta Cryst.*, 19, 1051-1055. (For earlier work, see 1.
For a recent neutron-diffraction analysis, see 2.)

Monoclinic, $a = 12.714\pm3$, $b = 3.676\pm3$, $c = 12.949\pm4$ Å, $\beta = 95°24\pm5'$, $[U = 602.5 \text{ Å}^3]$, $D_m = 1.812$, $Z = 4$, $D_x = 1.786$. (D_m from 1.)

Space group $P2_1/n$ (C_{2h}^5)

Atomic positions

	x	y	z
C(2)	0.1389	0.2960	0.3341
C(4)	−0.0460	0.2594	0.3635
C(5)	−0.0719	0.4248	0.2656
C(6)	0.0072	0.5215	0.2077
N(1)	0.1092	0.4581	0.2424
N(3)	0.0596	0.2041	0.3910
C(2)	0.2311	0.2415	0.3645
O(4)	−0.1119	0.1663	0.4224
O(5)	−0.1738	0.5069	0.2336
O(6)	−0.0025	0.6853	0.1176
O(7) (H₂O)	−0.3159	[0.4593]*	0.4461
H(1)	0.162	0.433	0.207
H(3)	0.078	0.069	0.449
H(5)	−0.217	0.475	0.288
H(6)	−0.077	0.723	0.093
H(7)	−0.262	0.300	0.428
H(8)	−0.300	0.580	0.500

*[This value is given incorrectly in original.]
Anisotropic temperature factors are given for the non-hydrogen atoms. Standard
deviations for these atoms average 0.003 Å. No estimate of error is given for the
positions of the hydrogen atoms, and some of these are regarded as tentative.

Structure

The molecule is essentially planar. The bond lengths are:

N(1) − C(2) = 1.350 Å	C(4) − C(5) = 1.417
C(2) − O(2) = 1.218	C(5) − O(5) = 1.357
C(2) − N(3) = 1.347	C(5) − C(6) = 1.357
N(3) − C(4) = 1.371	C(6) − O(6) = 1.308
C(4) − O(4) = 1.233	C(6) − N(1) = 1.352 All ±0.005 Å

The hydrogen bonding network is shown in Fig. 91.

Details of analysis

Intensity data were recorded on integrated equi-inclination Weissenberg photographs, using CuK_α radiation. 950 of a possible 1250 reflexions were observed and
measured photometrically. Refinement was by full-matrix least squares to a final R
index of 0.08.

1. *Structure Reports, 20*, 625.
2. B.M. CRAVEN and T.M. SABINE, 1969. *Acta Cryst.,* B25, 1970.

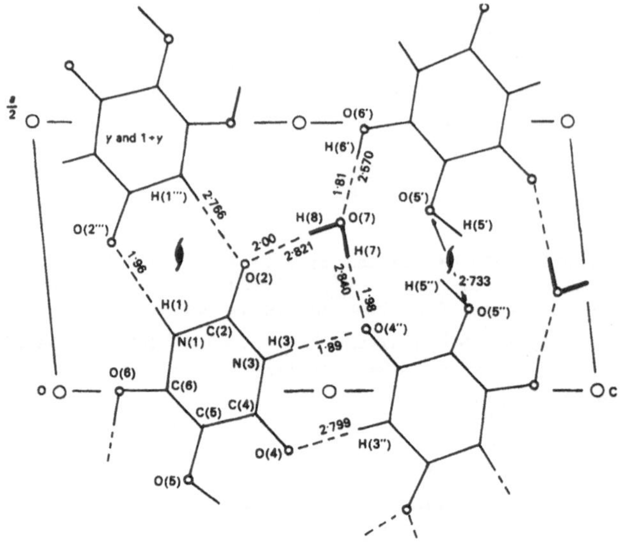

Fig. 91. The structure of dialuric acid monohydrate viewed along *b*, showing the
hydrogen bonds. (H(5) is poorly defined in the difference map.)

ALLOXANTIN DIHYDRATE

$C_8H_6N_4O_8 . 2H_2O$ F.W. = 322.2

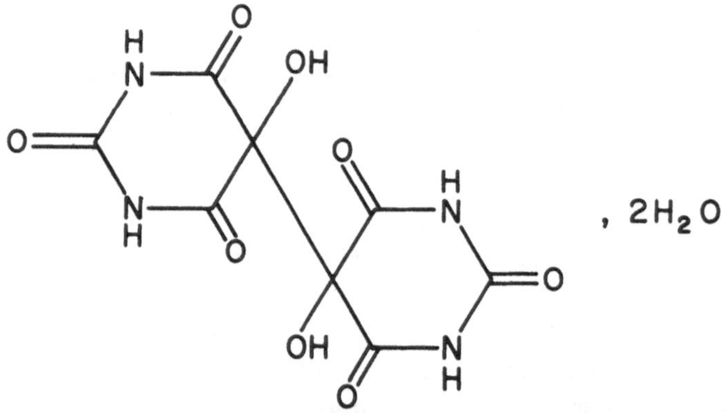

, $2H_2O$

I. The structure of the pyrimidines and purines. IX. The crystal structure of
alloxantin dihydrate. C. SINGH, 1965. *Acta Cryst.,* 19, 767–774.

Triclinic, a = 6.703±3, *b* = 6.879±7, *c* = 7.311±7 Å, α = 67°5±4', β = 83°51±5',
γ = 71°0±4', *U* = 293 Å³, D_m = 1.87, *z* = 1, D_x = 1.83.

Space group P1 (C_1^1) or P$\bar{1}$ (C_i^1). P$\bar{1}$ is confirmed by analysis. Molecular
symmetry, centre.

Atomic positions

	x	y	z
C(2)	0.2254	-0.2338	0.4243
C(4)	-0.0544	-0.2375	0.2365
C(5)	0.0518	-0.1274	0.0410
C(6)	0.2870	-0.1852	0.0769
N(1)	0.3504	-0.2060	0.2586
N(3)	0.0246	-0.2346	0.4036
O(2)	0.2908	-0.2439	0.5794
O(4)	-0.2096	-0.2850	0.2374
O(5)	0.0207	-0.2129	-0.0951
O(6)	0.4106	-0.2005	-0.0540
O(7)	0.2493	0.3469	0.2845

The mean standard deviations are: Carbon, 0.011; Nitrogen, 0.010; oxygen, 0.008 Å. The standard deviations for y are considerably greater than for x and z. The hydrogen positions are given also.

Structure

The analysis confirms the formulation given above. Bond lengths and angles are given in full, and appear to be normal. The pyrimidine ring is puckered; the tetra-hedral carbon atom lies on one side of the ring plane, while the adjacent ketonic oxygen atoms lie on the other side. The structure is held together by a network of hydrogen bonds.

Details of analysis

The intensity data were recorded on zero-level Weissenberg photographs about a, b, c, and [011], using CuK_α radiation. About 175 reflexions were observed and measured visually. Refinement was by Fourier methods to a final R index of 0.046.

CYCLOBUTANE-1,5-SPIRO-2,4,6-TRIKETOHEXAHYDROPYRIMIDINE

$C_7H_8N_2O_3$ F.W. = 168.2

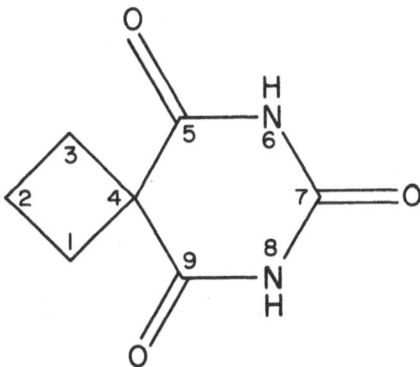

I. Analisi strutturale della 1-1' ciclobutandicarbonil-urea. G. GIACOMELLO, P. CORRADINI and C. PEDONE, 1965. *Gazz. Chim. Ital.*, 95, 1100-1114.

Monoclinic, $a = 5.52\pm2$, $b = 15.25\pm5$, $c = 9.10\pm4$ Å, $\beta = 106°6\pm30'$, $[U = 736$ Å$^3]$, $z = 4$, $D_x = 1.58$ [1.52].

Space group $P2_1/n$ (C_{2h}^5)

Atomic positions

	x	y	z
N(1)	0.287	-0.015	0.173
N(3)	0.669	-0.015	0.364
O(2)	0.390	-0.118	0.363
O(4)	0.972	0.085	0.367
O(6)	0.220	0.079	-0.025
C(2)	0.440	-0.050	0.304
C(4)	0.746	0.062	0.312
C(5)	0.560	0.112	0.191
C(6)	0.342	0.059	0.101
C(7)	0.694	0.172	0.096
C(8)	0.481	0.195	0.262
C(9)	0.641	0.251	0.187

Assumed hydrogen positions are given also.

Structure

Bond lengths and angles are given in full, and are consistent with the above formulation. The 4- and 6-membered rings are mutually perpendicular. Neither is planar; C(5) lies 0.20 Å from the plane of the remainder of the pyrimidine ring, and the cyclobutane ring is folded, with a dihedral angle of about 160°. As indicated in Fig. 92, the structure consists of chains of molecules, extended along c, linked together by pairs of N-H..O bonds of length 2.90 Å.

Details of analysis

The intensity data for the three principal zones were recorded on Weissenberg photographs, using CuK_α radiation. Refinement was by Fourier methods to R indices of 0.164 to 0.184 for the three zones.

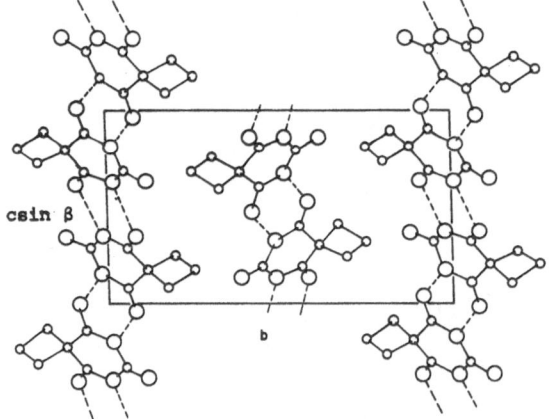

Fig. 92. The structure of $C_7H_8N_2O_3$ viewed along a.

2-(4'-AMINO-5'-AMINOPYRIMIDYL)-2-PENTENE-4-ONE

$C_9H_{12}N_4O$ F.W. = 192.2

I. The crystal structure of 2-(4'-amino-5'-aminopyridyl)-2-pentene-4-one.
   J. SILVERMAN and N.F. YANNONI, 1965. *Acta Cryst.*, <u>18</u>, 756-764.
   (For a preliminary report, see <u>1</u>.)

Monoclinic, a = 7.02±9, b = 12.42±4, c = 12.28±2 Å, β = 108.4±1°, U = 1017 Å³,
D_m = 1.26, Z = 4, D_x = 1.252.

Space group $P2_1/c$ (C_{2h}^5)

Atomic positions

	x	y	z
O	[0.1773±9]*	0.5770±5	0.4856±5
N(1)	0.0460±11	0.2328±5	0.7913±7
N(2)	0.0359±10	0.3756±5	0.9173±6
N(3)	0.1009±10	0.5524±5	0.8798±6
N(4)	0.2164±9	0.4802±5	0.6859±6
C(1)	0.4391±16	0.6821±8	0.4572±10
C(2)	0.3524±13	0.6117±6	0.5308±8
C(3)	0.4672±14	0.5879±7	0.6453±9
C(4)	0.4046±13	0.5244±7	0.7188±8
C(5)	0.5427±14	0.5006±10	0.8403±10
C(6)	0.1486±10	0.4098±6	0.7568±7
C(7)	0.0949±11	0.4455±6	0.8510±7
C(8)	0.0157±13	0.2730±7	0.8834±8
C(9)	0.1192±13	0.3027±6	0.7290±9

*An incorrect value of 0.1173 is given in the original.
Anisotropic temperature factors are given also.

Structure

The molecule lies approximately in two planes, one containing the six-
membered ring and the adjacent nitrogen atoms, and the other containing the carbon
chain and the adjacent nitrogen and oxygen atoms. The angle between these planes
is about 70°. Bond lengths and angles are given in full, and are consistent with
the assumed formulation. The molecular conformation and packing, and the proposed
system of hydrogen bonding are illustrated in Fig. 93. The distances indicated
are:

Intermolecular:	N(1)...N(3)	=	3.02 Å
	N(2)...N(3)	=	3.07
	N(4)...O	=	2.67
Intramolecular:	N(4)...O	=	2.98

Details of analysis

The intensity data were recorded on precession photographs (0kl to 1kl, MoK$_\alpha$
radiation) and equi-inclination Weissenberg photographs (hk0 to hk,10, CuK$_\alpha$
radiation). 1005 reflexions, about 51% of the accessible total, were observed
and estimated visually. Refinement was by full-matrix least squares to a final R
index of 0.128.

1. N.F. YANNONI and J. SILVERMAN, 1964. *Nature, Lond.*, <u>203</u>, 484.

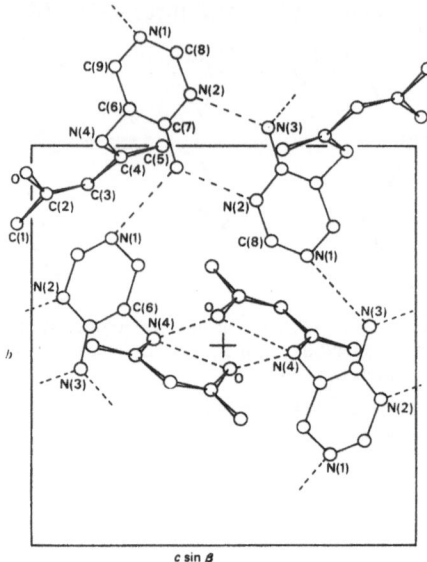

$c \sin \beta$

Fig. 93. The $C_9H_{12}N_4O$ structure viewed along a. The dotted lines are proposed hydrogen bonds.

ISOCYTOSINE

$C_4H_5N_3O$ F.W. = 111.1

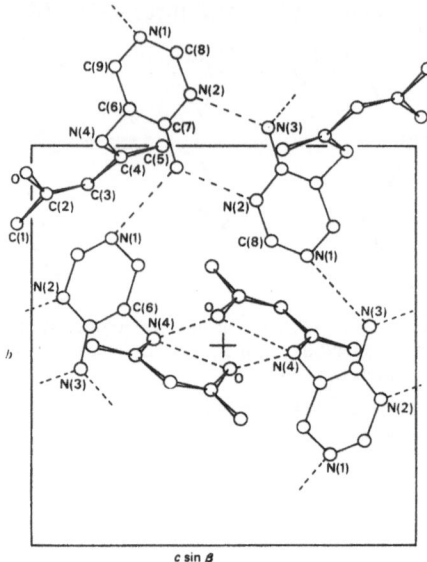

I. The crystal and molecular structure of isocytosine. B.D. SHARMA and J.F. McCONNELL, 1965. *Acta Cryst.*, 19, 797–806.
(For a preliminary report, see 1).

Monoclinic, $a = 8.745\pm1$, $b = 11.412\pm2$, $c = 10.441\pm1$ Å, $\beta = 99.79\pm1°$, $U = 1038$ Å^3, $D_m = 1.403$, $Z = 8$, $D_X = 1.421$, (CuKα, $\lambda = 1.5418$ Å).

Space group $P2_1/n$ (C_{2h}^5)

Atomic positions

Molecule A

	x	y	z
N(1A)	0.7724±2	0.8326±2	0.3948±2
C(2A)	0.6861±2	0.7350±2	0.3709±2
N(3A)	0.5667±2	0.7102±1	0.4376±1
C(4A)	0.5274±2	0.7852±2	0.5309±2
C(5A)	0.6204±2	0.8875±2	0.5582±2
C(6A)	0.7379±3	0.9086±2	0.4883±2
N(7A)	0.7217±2	0.6630±2	0.2787±2
O(8A)	0.4120±2	0.7636±1	0.5895±1
H(9A)	0.853 ±3	0.859 ±2	0.344 ±2
H(10A)	0.595 ±3	0.942 ±2	0.619 ±2
H(11A)	0.811 ±2	0.977 ±2	0.500 ±2
H(12A)	0.660 ±3	0.595 ±2	0.262 ±3
H(13A)	0.797 ±3	0.679 ±2	0.230 ±2

Molecule B

	x	y	z
N(1B)	0.1166±2	0.4282±2	0.3863±2
C(2B)	0.2168±2	0.5132±2	0.4181±2
N(3B)	0.3525±2	0.5228±1	0.3623±2
C(4B)	0.3948±2	0.4478±2	0.2684±2
C(5B)	0.2881±2	0.3566±2	0.2349±2
C(6B)	0.1578±2	0.3519±2	0.2966±2
N(7B)	0.1867±2	0.5924±2	0.5050±2
O(8B)	0.5185±2	0.4654±1	0.2205±2
H(9B)	0.420 ±3	0.586 ±2	0.387 ±2
H(10B)	0.308 ±3	0.303 ±2	0.173 ±2
H(11B)	0.091 ±2	0.285 ±2	0.279 ±2
H(12B)	0.260 ±3	0.644 ±2	0.531 ±2
H(13B)	0.086 ±3	0.582 ±2	0.547 ±3

Temperature factors (anisotropic for non-hydrogen atoms) are included.

Structure

   The two independent molecules in the asymmetric unit are identifiable as the
two tautomeric forms shown above. (It is emphasized that this situation is not the
result of deliberate co-crystallization of the two components.) Bond lengths and
angles for the two molecules are given in Fig. 94. In each case the rings are
effectively planar, but the non-ring atoms are displaced from the ring planes by as
much as 0.06 Å. The tautomers A and B are joined by three hydrogen bonds to form a
base pair analogous to that proposed for guanine-cytosine in deoxyribonucleic acid.
The base pair is not quite planar; A and B are mutually inclined by 9°. Each base
pair is hydrogen bonded to others, as indicated in Fig. 95. For full details of
the hydrogen bonding network the original should be consulted. A further feature
of the structure is that each base pair is largely overlapped by a parallel pair
distant 3.36 Å, as shown in Fig. 96.

Details of analysis

   Intensity data were recorded on equi-inclination Weissenberg photographs ($0kl$
to $7kl$; $h0l$ to $h6l$; $hk0$ to $hk8$) using CuK_α radiation. Of 2218 reflexions in the
copper sphere, 1848 were observed and measured visually. Refinement was by full-
matrix least squares to a final R index of 0.061.

1.   J.F. McCONNELL, B.D. SHARMA and R.E. MARSH, 1964. *Nature, Lond.*, 203, 399.

PURINE

C$_5$H$_4$N$_4$ F.W. = 120.1

I. The crystal and molecular structure of purine. D.G. WATSON, R.M. SWEET and
   R.E. MARSH, *Acta Cryst.*, 19, 573-580.

   The results of two independent investigations are given. Only the arithmetic
mean values of numerical results are reported here.

Orthorhombic, $a = 15.552\pm2$, $b = 9.37\pm1$, $c = 3.66\pm1$ Å, $[U = 534.2$ Å$^3]$, $D_m = 1.491$,
$z = 4$, $D_x = 1.493$, $(CuK_\alpha, \lambda = 1.5418$ Å$)$.

Space group $Pna2_1$ (C_{2v}^9) or $Pnam$ (D_{2h}^{16}). $Pna2_1$ selected on basis of packing
considerations and confirmed by analysis.

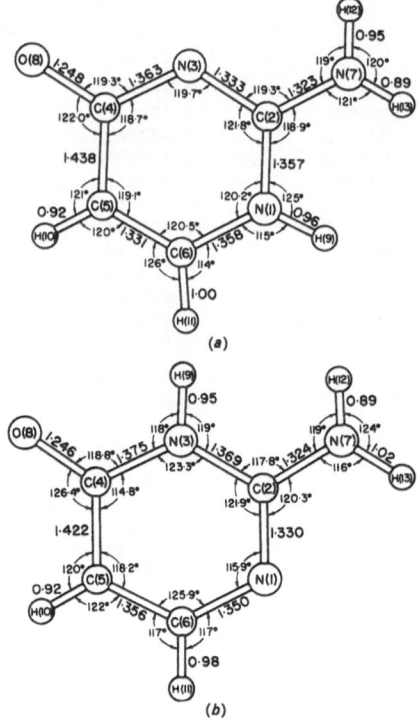

Fig. 94. Bond lengths and angles for the tautomeric molecules of isocytosine.
          (a) Molecule A; (b) Molecule B. Standard deviations are 0.003 Å
          and 0.1° for non-hydrogen values, 0.05 Å and 1° for values involving
          hydrogen.

Fig. 95. Details of the hydrogen bonds within the isocytosine base pair, and those
          linking the base pairs to one another.

Atomic positions

	x	y	z
N(1)	0.0034±3	0.3044±5	0.1266±30
C(2)	0.0157±4	0.1691±6	0.2396±31
N(3)	0.0888±3	0.0954±4	0.2434±25
C(4)	0.1561±3	0.1700±5	0.1214±34
C(5)	0.1503±3	0.3110±5	-0.0044±29
C(6)	0.0702±3	0.3766±5	0.0006
N(7)	0.2320±3	0.3466±5	-0.1128±27
C(8)	0.2810±3	0.2331±6	-0.0456±37
N(9)	0.2398±3	0.1236±5	0.0928±28

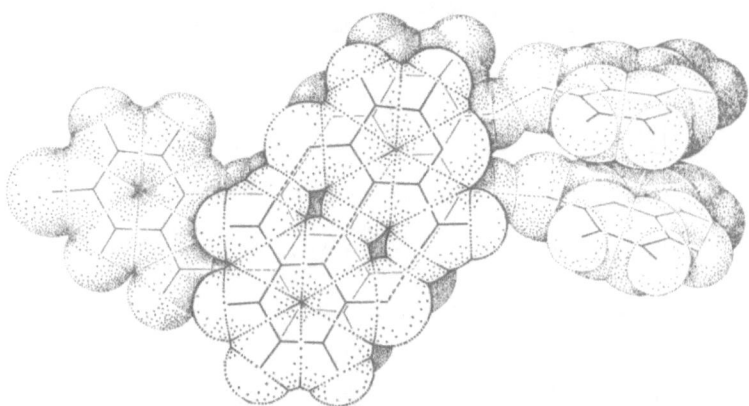

Fig. 96. View of part of structure of isocytosine showing overlap of base pairs.

Also given are anisotropic temperature factors for the non-hydrogen atoms, and the positions of the hydrogen atoms. The z coordinate of C(6) was held constant during refinement.

Structure

The molecule is planar; bond lengths and angles are given in full. Mean values of the bond lengths are:

$$
\begin{aligned}
N(1) &- C(2) = 1.349 \text{ Å}\\
C(2) &- N(3) = 1.332\\
N(3) &- C(4) = 1.337\\
C(4) &- C(5) = 1.403\\
C(5) &- C(6) = 1.389\\
C(6) &- N(1) = 1.330\\
C(5) &- N(7) = 1.374\\
N(7) &- C(8) = 1.330\\
C(8) &- N(9) = 1.312\\
N(9) &- C(4) = 1.374
\end{aligned}
$$

(All ±0.008 Å)

The positions of the hydrogen atoms confirm the formulation given above.

The molecular packing is shown in Fig. 97. The hydrogen bonds depicted there (length N-H..N = 2.85 Å) hold molecules together to form chains along [011]. The molecules stack along c with an interplanar spacing of 3.39 Å.

Details of analysis

Intensity data were recorded on equi-inclination Weissenberg photographs (for one determination, $hk0$ to $hk3$, 327 reflexions, for the other, $hk0$ to $hk3$, $h0l$ to $h5l$, 570 reflexions) using CuK_α radiation. Intensity measurement was by visual estimation. In both cases refinement was by full-matrix least squares to final R indices of 0.048 and 0.070, respectively.

GUANINE HYDROCHLORIDE DIHYDRATE

$C_5H_5N_5O.HCl.2H_2O$ F.W. = 223.6

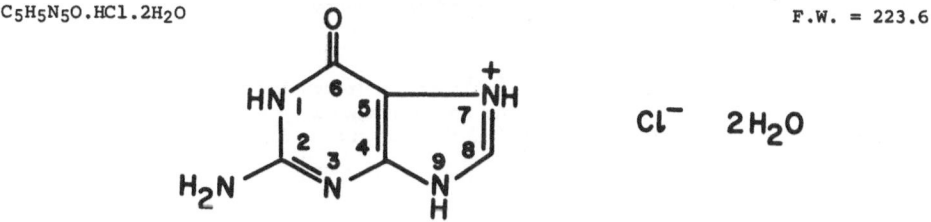

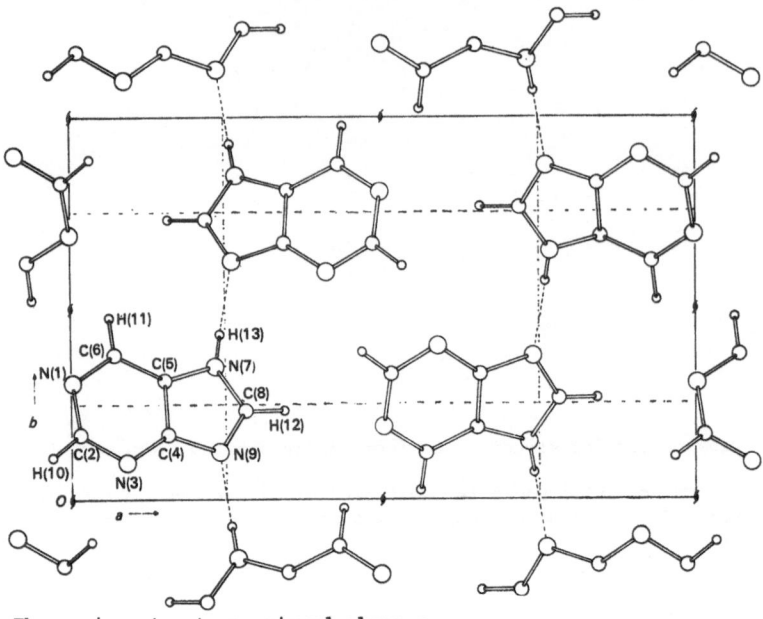

Fig. 97. The purine structure, viewed along c.

I. The crystal and molecular structure of guanine hydrochloride dihydrate.
   J. IBALL and H.R. WILSON, 1965. *Proc. Roy. Soc.*, A. <u>288</u>, 418-439.
   (For a preliminary report, see <u>1</u>. For the corresponding monohydrate see <u>2</u>.)

Monoclinic, $a = 14.69 \pm 1$, $b = 13.40 \pm 1$, $c = 4.840 \pm 5$ Å, $\beta = 93.8 \pm 1°$, $[U = 950.6$ Å$^3]$, $D_m = 1.562$, $Z = 4$, $D_x = 1.569$. (D_m is taken from <u>2</u>.)

Space group $P2_1/a$ (C_{2h}^5)

Atomic positions

	x	y	z
Cl	0.3339	0.0148	0.6179
O(H$_2$O(1))	0.4814	0.0921	0.2122
O(H$_2$O(2))	0.2376	0.1668	0.1962
O	0.3094	0.2922	0.8131
N(1)	0.3493	0.3946	0.4647
N(3)	0.4940	0.4003	0.2763
N(7)	0.5094	0.2140	0.7933
N(9)	0.5295	0.2794	0.4993
N(10)	0.3780	0.5066	0.1210
C(2)	0.4095	0.4323	0.2861
C(4)	0.5147	0.3268	0.4590
C(5)	0.4589	0.2850	0.6452
C(6)	0.3683	0.3194	0.6579
C(8)	0.5923	0.2127	0.7024

   (The positions in the original are given in Å.) Mean standard deviations are:
Cl, 0.001 Å; other, 0.003 to 0.004 Å. Also given are anisotropic temperature
factors for these atoms, and the positions of the hydrogen atoms.

Structure

   The analysis shows the guanine molecule to be protonated at N(7), as shown
above. It is essentially planar, with no non-hydrogen atom deviating from the
mean plane by more than 0.024 Å. Bond lengths and angles are:

N(1) - C(2) = 1.374±5 Å		C(2) - N(1) - C(6) = 125.6°
N(1) - C(6) = 1.390±5		N(1) - C(2) - N(10) = 116.0
C(2) - N(3) = 1.318±5		N(10) - C(2) - N(3) = 120.6
C(2) - N(10)= 1.339±5		N(1) - C(2) - N(3) = 123.4
N(3) - C(4) = 1.345±5		C(2) - N(3) - C(4) = 112.8
C(4) - C(5) = 1.377±5		N(3) - C(4) - N(9) = 126.3
C(4) - N(9) = 1.375±5		N(3) - C(4) - C(5) = 127.6
C(5) - C(6) = 1.414±5		C(5) - C(4) - N(9) = 106.2
C(5) - N(7) = 1.378±5		C(4) - N(9) - C(8) = 108.6
C(6) - O = 1.237±5		N(9) - C(8) - N(7) = 109.6
N(9) - C(8) = 1.335±6		C(8) - N(7) - C(5) = 108.2
C(8) - N(7) = 1.322±5		N(7) - C(5) - C(4) = 107.4
		N(7) - C(5) - C(6) = 132.7
		C(4) - C(5) - C(6) = 119.9
		C(5) - C(6) - N(1) = 110.8
		C(5) - C(7) - O = 128.9
		O - C(6) - N(1) = 120.3

Adjacent centrosymmetrically related molecules are coplanar, and are linked by pairs of N-H--N bonds. These guanine dimers are then linked to water molecules and chlorine atoms, as shown in Fig. 98, to form nearly planar ribbons which extend throughout the crystal. Adjacent symmetry-related ribbons are cross-linked by further hydrogen bonds. The lengths of the hydrogen bonds are:

$$
\begin{aligned}
\text{N-H...N} &= 3.045\pm5 \text{ Å} \\
\text{N-H...O} &= 2.656\pm5 \\
&\ \ \ 2.702\pm5 \\
\text{O-H...O} &= 2.763\pm4 \\
\text{O-H...Cl} &= 3.129\pm4 \\
&\ \ \ 3.152\pm4 \\
&\ \ \ 3.191\pm4 \\
\text{N-H...Cl} &= 3.140\pm3
\end{aligned}
$$

(It is to be noted that all the hydrogen atoms have been unequivocally located.)

Details of analysis

The intensity data were recorded on uni-dimensionally integrated Weissenberg photographs about several different axes, using CuK_α radiation. The weaker reflexions were measured visually, and the rest photometrically. 1600 reflexions were observed. Refinement was by least squares to a final R index of 0.073.

1. J. IBALL and H.R. WILSON, 1963. *Nature, Lond.*, **198**, 1193.
2. *Structure Reports*, **15**, 494.

α-METHYL-D-GALACTOSIDE 6-BROMOHYDRIN

$C_7H_{13}BrO_5$ F.W. = 257.1

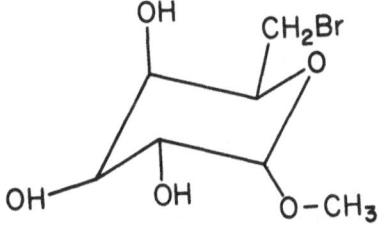

I. The crystal structure of α-methyl-D-galactoside 6-bromohydrin. J.H. ROBERTSON, and B. SHELDRICK, 1965. *Acta Cryst.*, **19**, 820-826.

Orthorhombic, a = 11.142±5, b = 7.815±3, c = 10.612±10 Å, $[U$ = 924.0 Å³$]$, D_m = 1.86, z = 4 $[D_x$ = 1.85$]$, (cell and intensity data at 125°K).

Space group $P2_12_12$ (D_2^3)

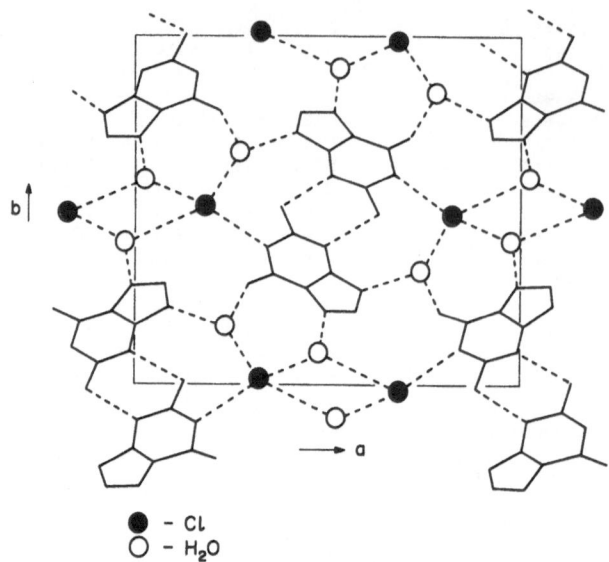

Fig. 98. The structure of guanine hydrochloride dihydrate viewed along *c*.

Atomic positions

	x	*y*	*z*
Br	0.0806	0.2279	-0.0116
O(1)	0.2562	0.0685	0.7078
O(2)	0.8673	0.0106	0.4910
O(3)	0.5109	0.2209	0.4111
O(4)	0.3202	0.0079	0.3208
O(5)	0.6650	0.2082	0.7229
C(1)	0.7071	0.0798	0.6390
C(2)	0.8148	0.1448	0.5673
C(3)	0.9140	0.2082	0.6580
C(4)	0.3620	0.1565	0.2565
C(5)	0.2559	0.2339	0.1870
C(6)	0.1972	0.1107	0.0986
C(7)	0.3536	0.1536	0.7706

These are given in Å in the original. Standard deviations are given in full;
average values are: Br, 0.0015; O, 0.008; C, 0.013 Å. Hydrogen positions are given
also, but have not been refined. Anisotropic temperature factors are given for the
non-hydrogen atoms.

Structure

The configuration of the molecule is illustrated in Fig. 99. The bond lengths
are:

Br –C(6)	1.974±14 Å
C(1)–C(2)	1.509±16
C(2)–C(3)	1.551±17
C(3)–C(4)	1.502±17
C(4)–C(5)	1.524±15
C(5)–C(6)	1.492±18
O(1)–C(1)	1.430±14
O(1)–C(7)	1.436±16
O(2)–C(2)	1.448±15
O(3)–C(3)	1.414±14
O(4)–C(4)	1.425±14
O(5)–C(5)	1.461±13
O(5)–C(1)	1.421±14

Bond angles are given in full, and range from 107 to 114°, with standard deviation 1°. All hydroxyl groups are involved in a network of intermolecular hydrogen bonds with lengths ranging from 2.733 to 2.824 Å.

Details of analysis

Intensity data were recorded, at 125°K, on equi-inclination Weissenberg photographs using CuK_α radiation. 1203 of 1265 possible reflexions were observed and measured. Refinement was by least squares to a final *R* index of 0.096.

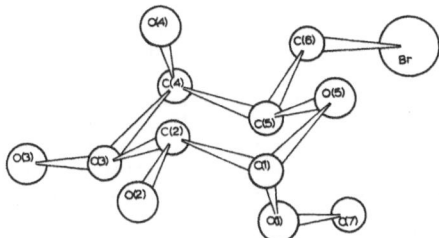

Fig. 99. The configuration of the $C_7H_{13}BrO_5$ molecule.

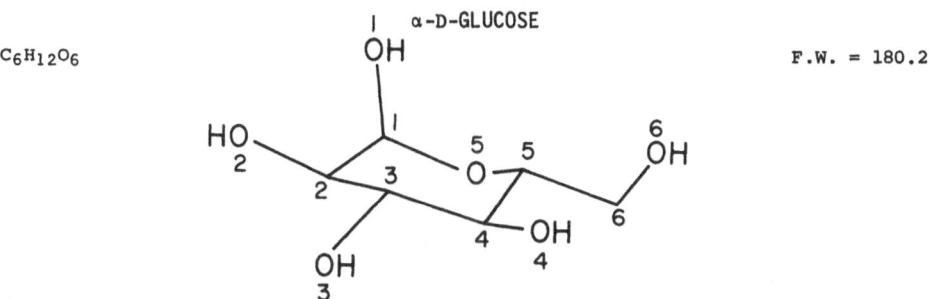

$C_6H_{12}O_6$ F.W. = 180.2

I. α-D-Glucose: precise determination of crystal and molecular structure by neutron diffraction analysis. G.M. BROWN and H.A. LEVY, 1965. *Science*, 147, 1038-1039. (This appears to be a preliminary communication. It is a redetermination of the structure reported in 1.)

Orthorhombic, a = 10.36±2, b = 14.84±2, c = 4.97±2 Å, U = 764 Å³, D_m = 1.56, Z = 4, D_x = 1.56.

Space group $P2_12_12_1$ (D_2^4) (Unit cell and space group data from 1.)

Atomic positions

	x	y	z
C(1)	0.3329±1	0.6007±1	-0.0762±3
C(2)	0.4184±1	0.6746±1	0.0455±3
C(3)	0.5611±1	0.6514±1	0.0154±3
C(4)	0.5860±1	0.5569±1	0.1200±3
C(5)	0.4956±1	0.4886±1	-0.0123±3
C(6)	0.5141±1	0.3953±1	0.1026±4
O(1)	0.3470±2	0.6007±1	-0.3541±4
O(2)	0.3880±1	0.7598±1	-0.0657±5
O(3)	0.6334±1	0.7154±1	0.1634±5
O(4)	0.7176±1	0.5348±1	0.0688±4
O(5)	0.3654±1	0.5154±1	0.0369±4
O(6)	0.4458±2	0.3311±1	-0.0518±5
H(1)	0.2327±2	0.6124±2	-0.0154±9
H(2)	0.3958±3	0.6776±2	0.2599±7
H(3)	0.5888±3	0.6543±2	-0.1995±7
H(4)	0.5677±3	0.5565±2	0.3385±6
H(5)	0.5136±3	0.4865±2	-0.2312±7

	x	y	z
H(6)	0.6172±3	0.3792±2	0.1032±12
H(7)	0.4804±5	0.3947±3	0.3116±9
H(8)	0.2753±4	0.5674±3	−0.4299±9
H(9)	0.4485±3	0.7776±2	−0.2031±9
H(10)	0.7247±2	0.7137±2	0.1178±8
H(11)	0.7517±3	0.5082±3	0.2326±8
H(12)	0.4127±3	0.2861±2	0.0699±10

Structure

     The structure is essentially as described in 1, but has been determined more
accurately, and is completed by the precise location of the hydrogen atoms. Bond
lengths and angles are given in detail, with standard deviations of 0.003 to 0.006
Å and 0.1 to 0.4°. The C-C, C-H, and O-H distances deviate only slightly from
their mean values of 1.523, 1.098, and 0.968 Å. The distance C(1)-O(1) is 1.389
±6 Å, significantly less than the mean value for the other six C-O distances of
1.420±3 Å. The valence angle of the ring oxygen atom is 113.8°. The five O-H..O
bonds range in length from 2.707 to 2.847 Å. The conformation is 1a2e3e4e5e.

Details of analysis

     Neutron intensity data were recorded with a four-circle automatic diffracto-
meter. 1619 reflexions were observed above background. Refinement was by full-
matrix least squares to a final R index of 0.060.

1.   *Structure Reports*, 16, 479.

α-D-GLUCOSAMINE HYDROCHLORIDE
α-D-GLUCOSAMINE HYDROBROMIDE

$C_6H_{14}ClNO_5$ F.W. = 215.6

$C_6H_{14}BrNO_5$ F.W. = 260.1

I. The crystal structure of α-D-glucosamine hydrochloride and hydrobromide.
   S.S.C. CHU and G.A. JEFFREY, 1965. *Proc. Roy. Soc.*, A, 285, 470-479.
   (This is a refinement, using original data, of the structure reported in 1.)

HYDROCHLORIDE

Monoclinic, a = 7.680±5, b = 9.180±5, c = 7.110±5 Å, β = 112°29 15', [U = 463.2 Å³],
D_m = 1.55, Z = 2, D_x = 1.546.

HYDROBROMIDE

Monoclinic, a = 7.960±5, b = 9.290±5, c = 7.180±5 Å, β = 112°35±20', [U = 490.2 Å³],
D_m = 1.75, Z = 2, D_x = 1.748.

Space group $P2_1$ (C_2^2)

Atomic positions

Hydrochloride:

	x	y	z
Cl	0.1494	0.0000	0.0250
O(1)	0.6505	0.4788	0.4985
O(3)	0.2995	0.2964	−0.0762
O(4)	0.6697	0.1921	−0.0209
O(6)	0.8394	−0.0334	0.4185
O(ring)	0.6962	0.2282	0.5056

	x	y	z
N	0.2756	0.4582	0.2596
C(1)	0.5718	0.3399	0.4904
C(2)	0.4025	0.3279	0.2850
C(3)	0.4627	0.3190	0.1055
C(4)	0.6035	0.1969	0.1417
C(5)	0.7670	0.2291	0.3458
C(6)	0.9160	0.1091	0.4019

Standard deviations are: Cl, 0.004 Å; O and N, 0.013 Å; C, 0.017 to 0.021 Å.

Hydrobromide:

	x	y	z
Br	0.1395	0.0000	0.0187
O(1)	0.6557	0.4783	0.4927
O(3)	0.3150	0.3020	-0.0693
O(4)	0.6709	0.1827	-0.0157
O(6)	0.8235	-0.0398	0.4197
O(ring)	0.6931	0.2232	0.5087
N	0.2938	0.4591	0.2620
C(1)	0.5748	0.3403	0.4837
C(2)	0.4139	0.3299	0.2896
C(3)	0.4698	0.3193	0.1088
C(4)	0.6036	0.1954	0.1457
C(5)	0.7654	0.2191	0.3462
C(6)	0.9047	0.0956	0.3998

Standard deviations are: Br, 0.002 Å; O and N, 0.018 Å; C, 0.024 Å.

Temperature factors (anisotropic for the halogen atoms) are given also.

Structure

Bond lengths and angles are given in full, and are consistent with expectation. The configuration was confirmed to be as shown above, and as originally reported in **1**. Although the hydrogen positions were not determined, a complex system of hydrogen bonds, dominated by the coordination about the NH_3^+ group, has been inferred.

Details of analysis

(The original intensity data were recorded on oscillation photographs taken about the principal axes.) Refinement was by full-matrix least squares to final R indices of 0.12 and 0.11.

1.   *Strukturbericht, 7, 272.*

D-β-GLUCOSE-*p*-BROMOPHENYLHYDRAZONE

$C_{12}H_{17}BrN_2O_5$ F.W. = 349.2

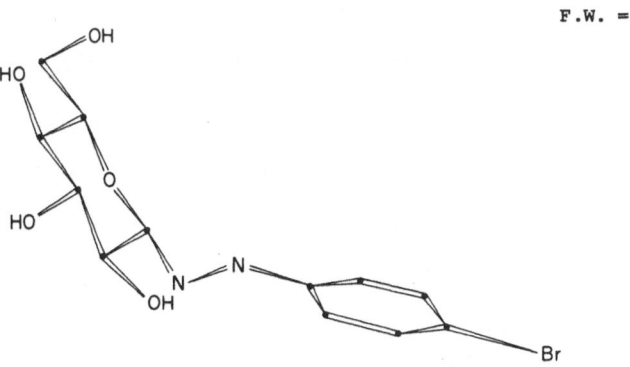

I. Crystal and molecular structure of D-β-glucose-p-bromo-phenylhydrazone.
   T. DUKEFOS and A. MOSTAD, 1965. *Acta Chem. Scand.*, <u>19</u>, 685–696.

Orthorhombic, $a = 6.82$, $b = 32.52$, $c = 6.19$ Å, (all ±0.5%), [$U = 1373$ Å^3],
$D_m = 1.678$, $Z = 4$, $D_x = 1.69$.

Space group $P2_12_12_1$ (D_2^4)

Atomic positions

	x	y	z
Br	0.153	0.1641	0.351
N(1)	0.361	0.3413	0.137
N(2)	0.298	0.3732	0.280
O(8)	0.250	0.3868	0.721
O(9)	0.463	0.4585	0.890
O(10)	0.847	0.4663	0.702
O(11)	0.570	0.4164	0.275
O(12)	0.974	0.4280	0.080
C(1)	0.221	0.2188	0.283
C(2)	0.283	0.2288	0.074
C(3)	0.316	0.2700	0.029
C(4)	0.305	0.3000	0.187
C(5)	0.228	0.2911	0.394
C(6)	0.175	0.2500	0.437
C(7)	0.449	0.3900	0.398
C(8)	0.368	0.4151	0.587
C(9)	0.537	0.4317	0.725
C(10)	0.678	0.4555	0.584
C(11)	0.742	0.4304	0.383
C(12)	0.872	0.4536	0.222

Temperature factors (anisotropic for bromine) are given. The positions of
the hydrogen atoms are given also, but there is no indication of how they were
determined.

Structure

The analysis established the conformation given above. Bond lengths and angles
are given in full, with standard deviations of 0.03 Å and 2 - 3°. The molecular
packing is described, and details of some probable hydrogen bonds are given.

Details of analysis

The intensity data for the zones [001] and [100] were recorded
on integrated Weissenberg photographs, using CuK_α radiation. For the first zone,
258 of a possible 293 reflexions were observed, and for the second, 168 of 261.
Intensities were measured photometrically. Refinement was by difference syntheses
and least squares to a final R index of 0.11.

CALCIUM 5-KETO-D-GLUCONATE DIHYDRATE

$C_{12}H_{18}CaO_{14}.2H_2O$ F.W. = 462.4

I. The crystal structure of calcium 5-keto-D-gluconate (calcium F-*xylo*-5-
hexulosonate). A.A. BALCHIN and C.H. CARLISLE, 1965. *Acta Cryst.*, <u>19</u>,
103-111.

Monoclinic, $a = 9.39\pm4$, $b = 8.03\pm2$, $c = 12.37\pm5$ Å, $\beta = 107.9\pm1.3°$, $[U = 888 \text{ Å}^3]$,
$D_m = 1.75$, $Z = 2$, $D_x = 1.735$.

Space group $A2$ (C_2^3), Am (C_s^3), or $A2/m$ (C_{2h}^3). A2 is the only space group
possible for an ordered structure, and is consistent with the analysis. Molecular
symmetry, two-fold axis. (Crystal data for calcium 2-keto-D-gluconate trihydrate,
$Ca(C_6H_9O_7)_2 \cdot 3H_2O$, are given in <u>1</u>.)

Atomic positions

	x	y	z
Ca	0.0000	0.0000±6	0.0000
O(1a)	0.1538±10	-0.2525±12	0.0271±7
O(1b)	0.3136±10	-0.4328±13	0.1360±7
O(2)	0.1279±9	-0.0760±11	0.1996±7
O(3)	0.4371±10	-0.0513±13	0.2168±8
O(4)	0.3101±11	-0.1052±13	0.4603±8
O(5)	0.0598±10	0.0453±12	0.3396±8
O(6)	0.0548±10	0.2341±13	0.1345±8
C(1)	0.2308±14	-0.3035±17	0.1204±11
C(2)	0.2305±14	-0.2160±17	0.2269±11
C(3)	0.3778±14	-0.1365±18	0.2942±11
C(4)	0.3260±13	-0.0145±19	0.3672±10
C(5)	0.1722±14	0.0477±17	0.2882±10
C(6)	0.1810±14	0.2151±18	0.2348±10
H2O	0.2407±11	-0.4235±14	0.4884±9
(W)	0.500	-0.135	0.000

W is a small, anomalous residual peak in the electron density distribution.
It may indicate the presence of a non-stoichiometric water molecule. Anisotropic
temperature factors are given also.

Structure

[The description given for the conformation of the 5-keto-D-gluconate ion is
incorrect and misleading. The five-membered ring is said to be 'almost planar',
with C(1) and O(5) near the ring plane. In fact C(3) lies 0.61 Å from the mean
plane of the remaining four (nearly coplanar) ring atoms, and the conformation is
as indicated above. The bonding environment of all the ring carbon atoms is essentially tetrahedral.] Bond lengths and angles are given in full, and appear to
be normal. The calcium ion lies between two interlocking 5-keto-D-gluconate ions,
and is coordinated to four oxygen atoms, two hydroxyl groups, and two water molecules, lying at the vertices of a regular triakistetrahedron. The Ca-O distances
range from 2.39 to 2.47 Å. A view of the structure, showing the extensive hydrogen
bonding system, is given in Fig. 100.

Details of analysis

The intensity data were recorded on equi-inclination Weissenberg photographs,
using CuK_α radiation. 1022 independent reflexions were observed, and measured
visually. Refinement was by least squares to a final R index of 0.11.

1. This volume, p. 427.

1-*o*-(*p*-BROMOBENZENESULPHONYL)-4,5,7-TRI-*o*-
ACETYL-2,6-ANHYDRO-3-DEOXY-D-GLUCOHEPTITOL

$C_{19}H_{23}O_{10}SBr$ F.W. = 523.4

I. Configuration of heptitols. The crystal and molecular structure of 1-*o*-
(*p*-bromobenzenesulphonyl)-4,5,6-tri-*o*-acetyl-2,6-anhydro-3-deoxy-D-*gluco*-
heptitol.   A. CAMERMAN and J. TROTTER, 1965. *Acta Cryst.*, <u>18</u>, 197–203.

Orthorhombic, $a = 13.71 \pm 3$, $b = 29.37 \pm 8$, $c = 5.79 \pm 1$ Å, $U = 2331$ Å^3, $D_m = 1.5$,
$Z = 4$, $D_x = 1.49$, (CuK$_\alpha$, $\lambda = 1.5418$ Å).

Space group $P2_12_12_1$ (D_2^4)

Atomic positions

Atomic positions and temperature factors (some anisotropic) are given for the
31 non-hydrogen atoms.

Structure

Bond lengths and angles are given in full. The accuracy is not high (standard
deviations of bond length range from 0.015 to 0.044 Å), but all values are consistent with expectations. A view of the molecule is given in Fig. 101. As the compound was derived from D-glucose this figure depicts the correct absolute configuration. All intermolecular distances are consistent with van der Waals interaction.

Details of analysis

Three-dimensional intensity data were measured with a four-circle diffracto-
meter and scintillation counter, using CuK$_\alpha$ radiation. The range $2\theta \leqslant 90°$ was scanned
and 851 reflexions (about 74% of those accessible) were observed. Refinement was
by block-diagonal least squares to a final R index of 0.090.

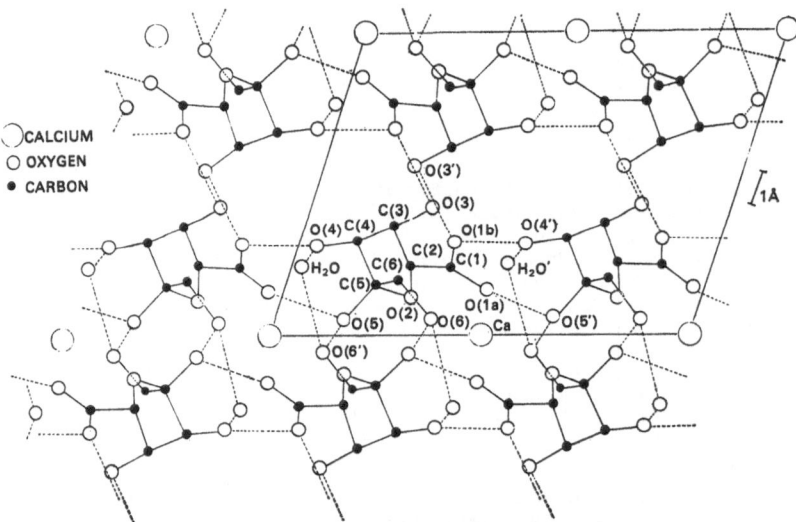

Fig. 100. Projection of the $C_{12}H_{18}CaO_{14} \cdot 2H_2O$ structure along *b*.

2-*o*-(*p*-BROMOBENZENESULPHONYL)-1,4:3,6-DIANHYDRO-D-GLUCITOL 5-NITRATE

$C_{12}H_{12}O_8N_5Br$ F.W. = 410.2

I. The crystal and molecular structure of 2-o-(p-bromobenzenesulphonyl)-1,4:3,
6-dianhydro-D-glucitol 5-nitrate. A. CAMERMAN, N. CAMERMAN and J. TROTTER,
1965. *Acta Cryst.*, 19, 449–456.

Monoclinic, a = 26.16±4, b = 11.11±2, c = 5.34±1 Å, β = 90°16±5', U = 1552 Å^3,
D_m = 1.7, Z = 4, D_x = 1.755, (CuK_α, λ = 1.5418 Å; MoK_α, λ = 0.7107 Å).

Space group $P2_1$ (C_2^2) or $P2_1/m$ (C_{2h}^2). $P2_1$ established by observation of optical
activity.

Atomic positions

Atomic positions and isotropic temperature factors of the 46 non-hydrogen atoms
are given.

Structure

The conformations of the two independent molecules are shown in Fig. 102.
Average and extreme values for bond lengths and angles of various types are given.
These are in reasonable accord with normal values, considering that standard deviations range from 0.03 to 0.06 Å, and 2 to 4°. In each molecule there is a relatively
short intramolecular contact (of 2.90 Å) between an oxygen atom of the nitro group
and the oxygen atom of the non-adjacent 5-membered ring.

Details of analysis

Intensity data were measured with a four-circle diffractometer and scintillation counter, using CuK_α radiation, in the range 2θ⩽119°. Of 2400 accessible
reflexions, 1565 were observed above background. Refinement was by block–diagonal
least squares to a final R index of 0.157.

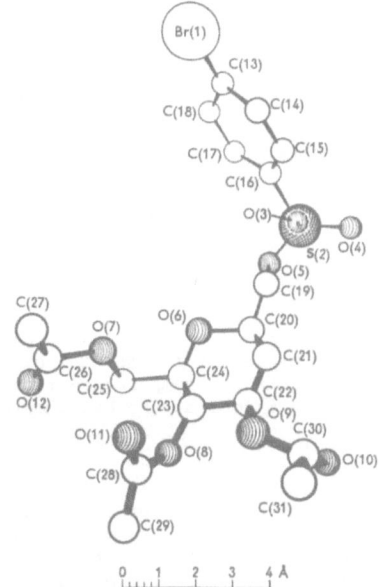

Fig. 101. A perspective drawing of the C$_{19}$H$_{23}$O$_{10}$SBr molecule.

186   ORGANIC COMPOUNDS

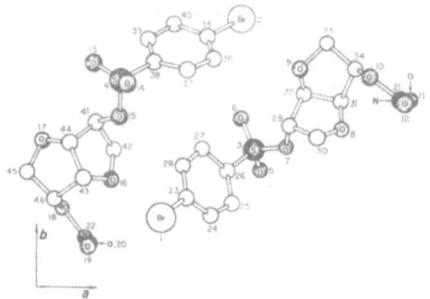

Fig. 102. A view of the two independent molecules of $C_{12}H_{12}O_8N_5Br$.

CYCLOHEXAAMYLOSE, POTASSIUM ACETATE COMPLEX

$C_{36}H_{60}O_{30},1.54(C_2H_3O_2K),9.7H_2O$ F.W. = 1298

$$\cdot 1.54\,(CH_3COOK)\cdot 9.7\,H_2O$$

I. The crystal and molecular structure of the cyclohexaamylose-potassium acetate
   complex. A. HYBL, R.E. RUNDLE and D.E. WILLIAMS, 1965. *J. Amer. Chem. Soc.*,
   87, 2779-2788.

Orthorhombic, $a = 21.89$, $b = 16.54$, $c = 8.03$ Å, $[U = 3005$ Å$^3]$, $D_m = 1.434$, $Z = 2$,
$[D_x = 1.434]$.

Space group $P2_12_12$ (D_2^3) Molecular symmetry, two-fold axis.

Atomic positions

   Atomic positions and their standard deviations are given for some 44 non-
hydrogen atoms. Some of these are present in non-stoichiometric ratios, and
occupancy factors are given. Anisotropic temperature factors are given for the
above atoms, as are the assumed positions of the hydrogen atoms of the cyclohexaamylose molecule.

Structure

   The molecule of cyclohexaamylose has approximate six-fold symmetry. The
individual α-D-glucose units are all in the pyranose staggered chair form with
the conformation C1 (1a2e3e4e5e). In addition to the α-(1,4)-glucosidic linkages,
contiguous glucose residues are joined by O-H..O bonds (of length 2.852±10 Å)
between atoms O(1) and O(2). Bond lengths and angles are given in full; mean C-C
and C-O distances are 1.528 and 1.426 Å respectively.

   The structure consists of a rigid framework of cyclohexaamylose molecules
permeated by continuous channels and cavities which are occupied (in some case non-
stoichiometrically) by potassium and acetate ions and water molecules. A view of
the structure is given in Fig. 103. An analogous detailed structure for amylose
is proposed.

Details of analysis

The intensity data were recorded with a four-circle diffractometer, using CuK_α radiation. Crystal deterioration necessitated the use of three different specimens to record the intensities of 2559 independent reflexions. Refinement was by block-diagonal least squares to a final R index of 0.10.

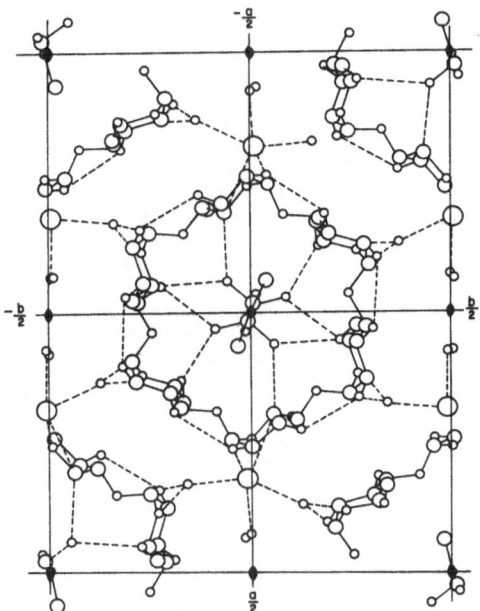

Fig. 103.   The cyclohexaamylose, potassium acetate complex, viewed along c.

SPERMINE PHOSPHATE HEXAHYDRATE

$C_{10}H_{32}N_4O_8P_2.6H_2O$ F.W. = 506.4

$$NH_2^+(CH_2)_3NH_2^+(CH_2)_3NH_2^+(CH_2)_3NH_3^+ .2(HPO_4)^{2-}.6H_2O$$

I. The crystal structure of spermine phosphate hexahydrate. Y. IITAKA and
   Y. HUSE, 1965. *Acta Cryst.*, <u>18</u>, 110-121.

Monoclinic, $a = 7.95(5)$, $b = 23.21(6)$, $c = 6.87(0)$ Å, $\beta = 113°39'$, $[U = 1161 \text{ Å}^3]$, $D_m = 1.443$, $Z = 2$, $D_x = 1.445$.

Space group $P2_1/a$ (C_{2h}^5) Molecular symmetry, centre

Atomic positions

	x	y	z
P	−0.0010	0.1967	0.3665
O(1)	−0.0691	0.2110	0.1318
O(2)	−0.0205	0.2463	0.4995
O(3)	−0.0941	0.1420	0.3983
O(4)	0.2100	0.1804	0.4485
W(1) (H_2O)	0.1170	0.0465	0.5452
W(2)	0.4185	0.0536	0.4424
W(3)	0.6706	0.1470	0.5961

N(1)	0.7135	0.0942	−0.0075
N(2)	0.1676	0.2115	−0.0657
C(1)	0.9650	0.0313	−0.0110
C(2)	0.7806	0.0340	0.0015
C(3)	0.5231	0.0950	−0.0097
C(4)	0.4482	0.1560	−0.0293
C(5)	0.2585	0.1535	−0.0291

Standard deviations, in Å, are given in full. Mean values are: P, 0.0015; O (phosphate), 0.005; other atoms, 0.005 to 0.008 Å. Isotropic temperature factors are given also.

Structure

Bond lengths and angles in the spermine molecule are given in Fig. 104. The molecule is essentially planar except for N(2), which lies 0.19 Å from the plane of the remaining atoms. The structure consists of parallel sheets of spermine molecules separated by sheets of phosphate ions and water molecules (Fig. 105). The configuration of bonds about N(1) and N(2) is tetrahedral, suggesting that these atoms are protonated and exist as NH_2^+ and NH_3^+. The geometry of the phosphate-water sheet is discussed in detail. Bond lengths in the phosphate ion are: P–O(1), 1.518; P–O(2), 1.517, P–O(3), 1.529; P–O(4), 1.589 Å; (all ±0.005 Å). The long P–O(4) distance suggests that O(4) is bonded to a hydrogen atom. It is concluded that the formulation of the compound is that given above.

Details of analysis

Intensity data were recorded on equi-inclination Weissenberg photographs (0*kl*; *h0l* to *h2l*; *hk0* to *hk5*) using CuK$_\alpha$ radiation. 1747 of a possible 2163 reflexions were observed and estimated visually. Refinement was by diagonal least squares to a final *R* index of 0.127.

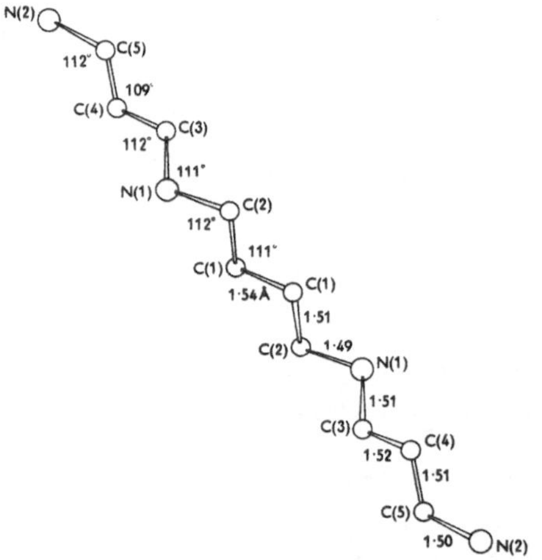

Fig. 104. Bond lengths and angles in spermine molecule. Standard deviations are 0.01 Å and 0.6°.

CALCIUM 1-NAPHTHYL PHOSPHATE TRIHYDRATE

$C_{20}H_{16}CaO_8P_2 \cdot 3H_2O$ F.W. = 540.4

$$Ca(C_{10}H_7HPO_4)_2 \cdot 3H_2O$$

I. The crystal and molecular structure of calcium 1-naphthyl phosphate trihydrate, $Ca(C_{10}H_7HPO_4)_2 \cdot 3H_2O$. C.-T. LI and C.N. CAUGHLAN, 1965. *Acta Cryst.*, 19, 637–645.

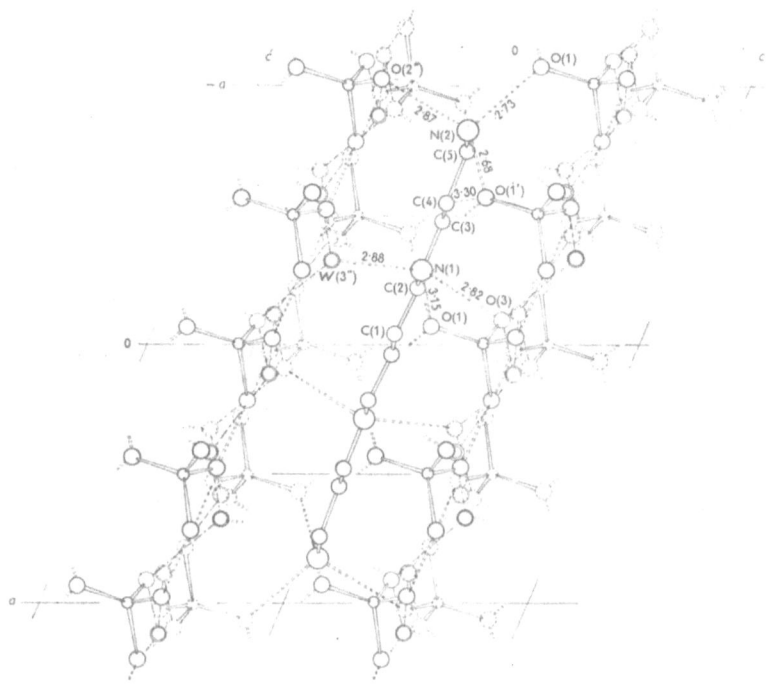

Fig. 105. Projection of the spermine phosphate hexahydrate structure along *b*.
Hydrogen bonds are indicated by double broken lines and single chain
lines. Some short intermolecular contacts are indicated by single
broken lines.

Triclinic, a = 7.244±2, b = 8.994±3, c = 18.725±4 Å, α = 95°41±4', β = 101°28±3',
γ = 88°52±5', [U = 1189.6 Å³], D_m = 1.530, z = 2, D_x = 1.508.

Space group P1 (C_1^1)   or   P$\bar{1}$ (C_i^1).   P$\bar{1}$ suggested by intensity statistics and
confirmed by analysis.

Atomic positions

Atomic positions and anisotropic temperature factors are given for the 34 non-
hydrogen atoms.

Structure

Bond lengths and angles are given in full; many of these are shown in Fig. 106,
107, and 108. The two phosphate groups are not identical, but the following re-
marks are true of both. The longest P-O bond (mean 1.592±8 Å) involves an oxygen
atom bonded to a hydrogen atom. The two shortest, and nearly equal, P-O bonds
(mean 1.487±8 Å) involve oxygen atoms coordinated to calcium. The calcium ion is
coordinated to seven oxygen atoms arranged in a distorted pentagonal bipyramid.
Four of the oxygen atoms, including the apical atoms, are contributed by phosphate
groups and three by water molecules. The basal oxygen atoms are approximately
coplanar, and are linked by hydrogen bonds as depicted in Fig. 108.

Details of analysis

Intensity data were recorded on precession and equi-inclination Weissenberg
photographs (h0l to h5l) using CuK_α radiation. 1750 of a possible 2774 reflexions
were observed and measured visually. Absorption corrections were applied. Refinement was by full-matrix least squares to a final R index of 0.097.

Fig. 106. The structure of calcium 1-naphthyl phosphate trihydrate viewed along
a. Water molecules are not shown. Standard deviations are 0.008 Å
for P-O and O-O, 0.013 Å for C-O.

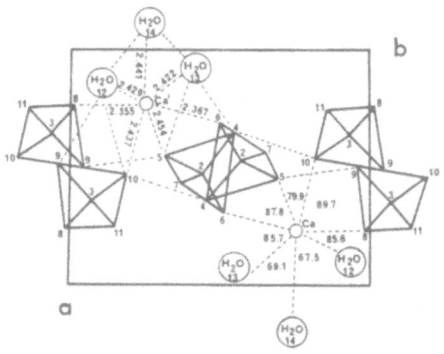

Fig. 107. The structure of calcium 1-naphthyl phosphate trihydrate viewed along
c. Standard deviations are 0.006 Å and 0.3°.

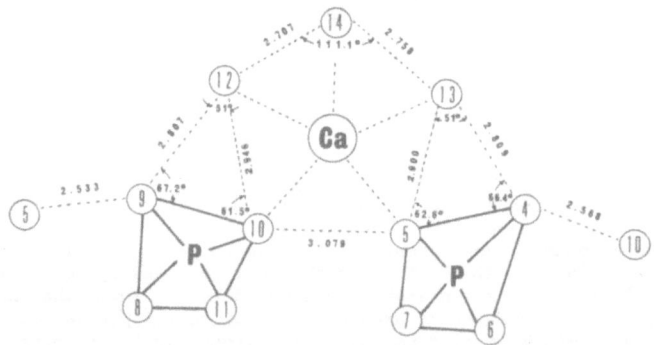

Fig. 108. The hydrogen-bonding network in the basal plane of the pentagonal bipyramid of calcium 1-naphthyl phosphate trihydrate.

DIPOTASSIUM GLUCOSE-1-PHOSPHATE DIHYDRATE

$C_6H_{11}K_2O_9P.2H_2O$ F.W. = 372.4

I. The crystal structure of dipotassium glucose-1-phosphate dihydrate. C.A.
BEEVERS and G.H. MACONACHIE, 1965. *Acta Cryst.*, <u>18</u>, 232-236.

Monoclinic, a = 10.440±5, b = 9.025±5, c = 7.518±5 Å, β = 110°24±6', U = 665 Å^3,
D_m = 1.85, Z = 2, D_X = 1.86.

Space group $P2_1$ (C_2^2)   ($P2_1/m$ rejected because of optical activity and piezo-
electricity.)

Atomic positions

	x	y	z
K(1)	0.948	0.454	0.439
K(2)	0.421	0.002	0.986
P	0.203	0.510	0.984
C(1)	0.266	0.271	0.193
C(2)	0.223	0.193	0.353
C(3)	0.323	0.252	0.544
C(4)	0.471	0.209	0.572
C(5)	0.502	0.281	0.405
C(6)	0.656	0.248	0.438
O(1)	0.238	0.419	0.177
OH(2)	0.087	0.225	0.337
OH(3)	0.284	0.195	0.694
OH(4)	0.551	0.298	0.735
O(5)	0.406	0.227	0.221
OH(6)	0.684	0.302	0.275
O(7)	0.084	0.609	0.980
O(8)	0.332	0.599	0.003
O(9)	0.161	0.405	0.823
W(1)	0.956	0.204	0.692
W(2)	0.852	0.430	0.023

The mean standard deviations are: potassium, 0.008 Å; phosphorus, 0.009 Å;
oxygen, 0.024 Å; carbon, 0.034 Å.

Structure

Bond lengths (with probable errors 0.03 to 0.05 Å) are given in full for the
glucose phosphate anion, and appear to be normal. The bond angles are tetrahedral,
except for P-O(1)-C(1), which is 124°.   The crystal structure is held together
by an elaborate system of bonds involving the two potassium ions, the water molecules, and the hydroxyl groups, and the oxygen atoms O(7), O(8), and O(9) of the
phosphate group, as shown in Fig. 109.

Details of analysis

Intensity data were recorded on normal-beam Weissenberg photographs about a, b,
c, and the shorter ac diagonal. The radiation is not stated. 970 reflexions, cor-

rected for absorption, were used in the refinement, which was by least squares to a final R index of 0.172.

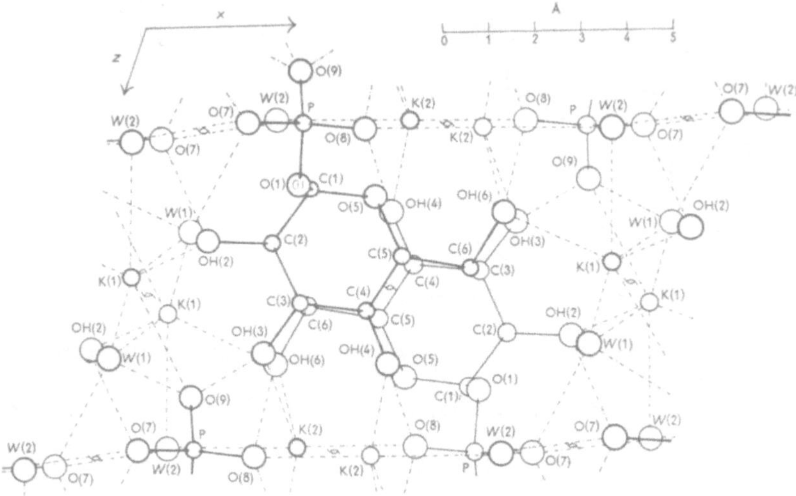

Fig. 109. Projection of the dipotassium glucose-1-phosphate dihydrate structure along b.

L-α-GLYCEROPHOSPHORYLCHOLINE CADMIUM CHLORIDE TRIHYDRATE

$C_8H_{20}NO_6P.CdCl_2.3H_2O$ F.W. = 494.6

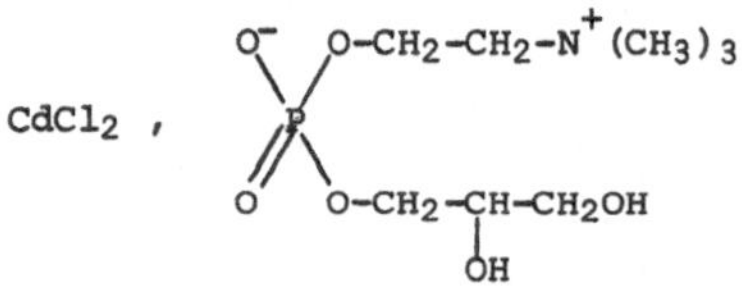

I. Crystal and molecular structure of a phospholipid component: L-α-glycerophos-phorylcholine cadmium chloride trihydrate. M. SUNDARALINGAM and L.H. JENSEN, 1965. *Science*, 150, 1035-1036.

Orthorhombic, $a = 9.452\pm2$, $b = 27.038\pm4$, $c = 7.391\pm2$ Å, $[U = 1889$ Å$^3]$, $D_m = 1.731$, $z = 4$, $D_x = 1.739$.

Space group $P2_12_12_1$ (D_2^4)

Atomic positions

Atomic positions are not given.

Structure

The cadmium chloride framework of the structure consists of a pleated ribbon extended along z. Each cadmium atom is coordinated to four chlorine atoms in a square-planar array with adjacent squares sharing Cl-Cl edges. Distorted octa-hedral coordination of the cadmium atoms is completed by the two available phos-phate oxygen atoms of the glycerophosphorylcholine molecule; each cadmium atom is coordinated to an oxygen atom of a different molecule, and the two oxygen atoms of a given molecule are coordinated to adjacent cadmium atoms. Mean Cd-Cl and Cd-O distances are 2.62 and 2.32 Å. The structure is further stabilized by O-H..O bonds involving the water molecules. The glycerophosphorylcholine molecule is roughly L-shaped, with the phosphate group at the corner of the L. The projected angle between the adjacent C-O bonds of the glycerol residue is about 60°.

Details of analysis

The intensity data were recorded on uni-dimensionally integrated equi-inclination Weissenberg photographs (*hk*0 to *hk*6), using Cu*K*$_\alpha$ radiation. 1200 reflexions were observed and measured with a densitometer. Refinement was by full-matrix least squares to a final *R* index of 0.115.

TRIPHENYL PHOSPHATE

$C_{18}H_{15}O_4P$ F.W. = 326.3

I. Refinement of the crystal structure of triphenyl phosphate. G. SVETICH and C.N. CAUGHLAN, 1965. *Acta Cryst.*, 19, 645–650. (For an account of an earlier, two-dimensional analysis, see 1.)

Monoclinic, *a* = 17.124±48, *b* = 5.833±36, *c* = 16.970±42 Å, β = 105°21±15', [*U* = 1634 Å³], *D*$_m$ = 1.302, *Z* = 4, *D*$_x$ = 1.331. (Calibration with sodium chloride for which *a* = 5.6402 Å).

Space group $P2_1/a$ (C_{2h}^5)

Atomic positions

	x	y	z
P(1)	0.3888±2	0.3454±7	0.2539±2
O(2)	0.4388±5	0.5459±16	0.2253±5
O(3)	0.3229±5	0.2812±18	0.1752±6
O(4)	0.3317±5	0.4742±18	0.2966±5
O(5)	0.4396±5	0.1687±17	0.2989±5
C(6)	0.5246±7	0.5460±30	0.2358±8
C(7)	0.5652±7	0.3702±24	0.2114±8
C(8)	0.6479±9	0.3800±33	0.2189±9
C(9)	0.6902±9	0.5835±34	0.2557±9
C(10)	0.6444±9	0.7644±25	0.2789±9
C(11)	0.5622±9	0.7391±31	0.2689±9
C(12)	0.3420±8	0.2102±27	0.1022±8
C(13)	0.3794±9	0.0086±25	0.1017±9
C(14)	0.4013±9	0.9551±34	0.0289±12
C(15)	0.3846±14	0.0917±55	0.9629±12
C(16)	0.3453±14	0.2890±48	0.9694±12
C(17)	0.3244±8	0.3680±28	0.0411±9
C(18)	0.3516±7	0.5172±28	0.3794±8
C(19)	0.3234±10	0.3638±34	0.4261±9
C(20)	0.3383±10	0.4037±47	0.5062±11
C(21)	0.3840±11	0.5970±42	0.5453±11
C(22)	0.4105±10	0.7457±30	0.4944±13
C(23)	0.3954±9	0.7099±31	0.4133±11

Anisotropic temperature factors are given for all atoms. Assumed hydrogen positions are given also.

Structure

Bond lengths and angles are given in full. Mean values of the more important bond lengths are:

$$P-OC_6H_5 \quad 1.57\pm2 \text{ Å}$$
$$P-O \quad\quad\; 1.43\pm1$$
$$C-O \quad\quad\; 1.41\pm2$$

These distances are not corrected for thermal motion; it is shown that corrections to individual values may be as much as 0.035 Å. The phosphate tetrahedron is distorted: $H_5C_6O-P-OC_6H_5$ angles range from 96.6 to 104.0°, and $O-P-OC_6H_5$ angles from 112.9 to 119.1°.

Details of analysis

Intensity data were recorded on precession photographs ($hk0$ and $0kl$, MoK_α radiation) and uni-dimensionally integrated Weissenberg photographs ($h0l$ to $h3l$, CuK_α radiation). Of 3642 reflexions in the copper sphere, 762 were observed and measured visually. Refinement was by full-matrix least squares to a final R index of 0.109.

1.   *Structure Reports*, 27, 916.

METHYL ETHYLENE PHOSPHATE

$C_3H_7O_4P$ F.W. = 138.1

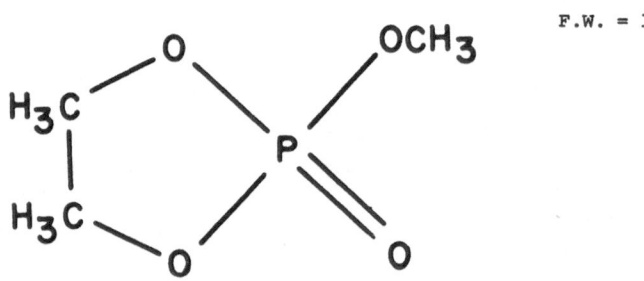

I. Molecular structure of methyl ethylene phosphate. T.A. STEITZ and W.N. LIPSCOMB, 1965. *J. Amer. Chem. Soc.*, 87, 2488-2489.

II. The molecular and crystal structure of $P_2O_4C_{12}H_{18}$, 1,8-diethoxy-3a,4,7,7a-tetrahydro-4,7-phosphinidenephosphindiole 1,8-dioxide. A Diels-Alder dimer of 1-ethoxyphosphole 1-oxide. Y.H. CHIU and W.N. LIPSCOMB, 1965. *J. Amer. Chem. Soc.*, 91, 4150-4155.

(An appendix to II gives the results of further refinement of the structure reported in I.)

Monoclinic, a = 11.29, b = 5.96, c = 9.09 Å, β = 113°, [U = 563 Å³], Z = 4, D_x = 1.47 [1.63] (all at -40°C).

Space group Cc (C_s^4) or $C2/c$ (C_{2h}^6). Cc is confirmed by analysis.

Atomic positions

	x	y	z
P(1)	0.2500	0.0241±4	0.2500
O(2)	0.2425±11	0.1787±16	0.3886±10
O(3)	0.1293±9	0.1219±15	0.1073±11
O(4)	0.3751±9	0.0899±14	0.2308±13
O(5)	0.2495±19	-0.2176±13	0.2713±21
C(6)	0.1261±16	0.3104±27	0.3367±18
C(7)	0.0682±14	0.3045±23	0.1501±16
C(8)	0.3944±18	0.3205±29	0.1936±22

Anisotropic temperature factors are given also.

Structure

The arrangement of bonds to the phosphorus atom is approximately tetrahedral;
the angle subtended by the ring oxygen atoms is 98°, and the other O-P-O angles
range from 106° to 118°. Bond lengths and angles are given in full; the phosphorus-
oxygen distances are: P-OCH$_3$, 1.54±1 Å; P-OCH$_2$, 1.58±1 Å and 1.59±1 Å; P=O,
1.45±1 Å. The five-membered ring is significantly non-planar.

Details of analysis

The intensity data were recorded at -40°C on Weissenberg photographs (6 self-
correlating levels about [110] using CuK$_\alpha$ radiation. 533 reflexions were observed.
Refinement was by full-matrix least squares to a final R index of 0.078.

DIPHENYLPHOSPHINIC ACID

C$_{12}$H$_{11}$O$_2$P   F.W. = 218.2

(C$_6$H$_5$)$_2$POOH

I. [The crystal and molecular structure of diphenylphosphinic acid.] T.-T. LIANG
and K.-C.CHI, 1965. *Acta Chim. Sinica*, 31, 155-164.

Monoclinic, a = 11.49, b = 6.09, c = 15.78 Å, β = 100°, U = 1088 Å^3, z = 4,
[D_x = 1.33].

Space group $P2_1/c$ (C_{2h}^5)

Atomic positions

	x	y	z
P	0.1602	0.1063	0.2508
O(1)	0.1406	0.3477	0.2461
O(2)	0.0977	-0.0594	0.2914
C(1)	0.3109	0.1238	0.3031
C(2)	0.3664	-0.0375	0.3531
C(3)	0.4852	-0.0156	0.3906
C(4)	0.5469	0.1797	0.3836
C(5)	0.4906	0.3516	0.3328
C(6)	0.3711	0.3289	0.2938
C(7)	0.1602	0.0320	0.1391
C(8)	0.2070	0.1813	0.0859
C(9)	0.2070	0.1234	0.0000
C(10)	0.1602	-0.0797	-0.0320
C(11)	0.1333	-0.2281	0.0211
C(12)	0.1333	-0.1719	0.1070

Assumed positions for the phenyl hydrogen atoms are given also.

Structure

Bond lengths and angles are given in full, and are consistent with expectation.
The configuration of bonds to the phosphorus atom is roughly tetrahedral, with P-O
distances 1.45 and 1.49 Å. O-H..O bonds of length 2.74 Å link the molecules into endless chains extended along b.

Details of analysis

The intensity data for the principal zones were recorded on Weissenberg photographs, using CuK$_\alpha$ radiation. Refinement was by Fourier methods to final R indices
ranging from 0.16 to 0.20.

ADENOSINE, 5'-BROMOURIDINE MONOHYDRATE
(1:1 HYDROGEN-BONDED COMPLEX)

$C_{19}H_6BrN_7O_{10}.H_2O$ F.W. = 591.0

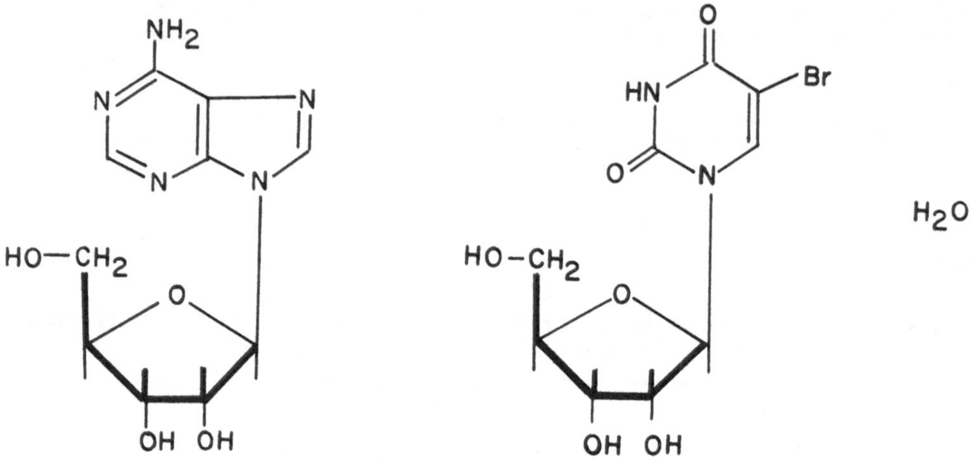

I. The crystal structure of a hydrogen bonded complex of adenosine and 5'-bromo-
   uridine. A.E.V. HASCHEMEYER and H.M. SOBELL, 1965. *Acta Cryst.*, 18, 525-532.
   (For a preliminary report, see 1.)

Orthorhombic, $a = 4.82\pm1$, $b = 15.19\pm1$, $c = 31.76\pm3$ Å, $[U = 2325$ Å$^3]$, $D_m = 1.71$,
$Z = 4$, $D_x = 1.686$, (CuK_α, $\lambda = 1.5418$ Å).

Space group $P22_12_1$ (D_2^3)

Atomic positions

Atomic positions and isotropic temperature factors of the non-hydrogen atoms
are given in full. The positions of five of the nucleosidic hydrogen atoms are
given also.

Structure

Bond lengths and angles are given in full, and are in reasonable agreement
with those found in other structures. The adenine and bromouracil residues are
nearly planar, and the sugar residues are puckered. The structure is held together
by the system of hydrogen bonds illustrated in Fig. 110. Two of these bonds join
the adenine and bromouracil residues to form a nearly-planar base pair. Another
view of the structure is given in Fig. 111. The position of the water molecule is
disordered.

Details of analysis

The intensity data were recorded on equi-inclination Weissenberg photographs
(0kl to 4kl) using CuK_α radiation. 1986 of a possible 2511 reflexions were observed, and measured visually. Refinement was by full-matrix least squares to a
final R index of 0.14.

1. A.E.V. HASCHEMEYER and H.M. SOBELL, 1963. *Proc. Nat. Acad. Sci., Wash.*, 50,
   872.

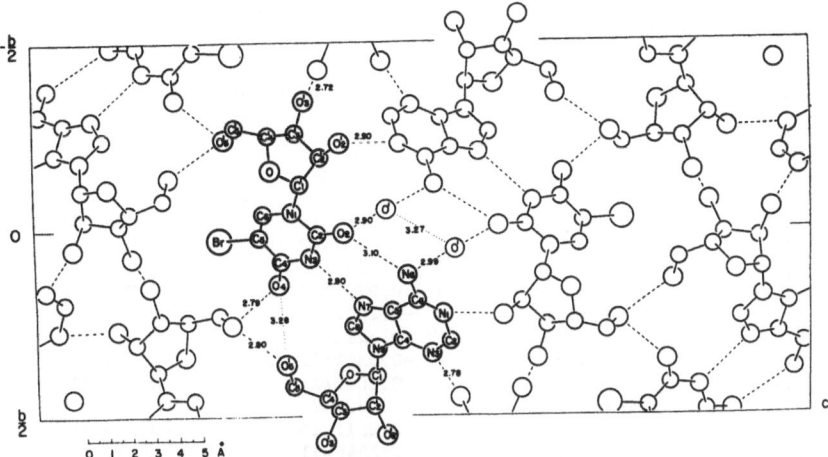

Fig. 110. The structure of $C_{19}H_6BrN_7O_{10} \cdot H_2O$ viewed along a. Hydrogen bonds are shown as broken lines, and other contacts of interest as dotted lines. The average water position is denoted by O'.

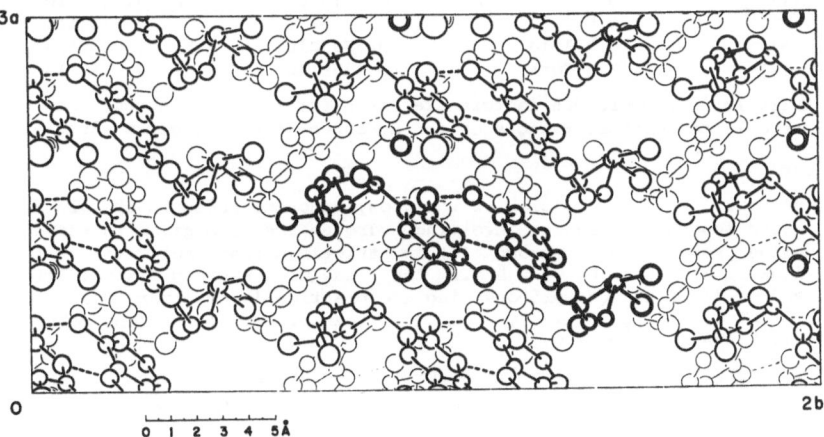

Fig. 111. The structure of $C_{19}H_6BrN_7O_{10} \cdot H_2O$ viewed along c. One base pair is drawn with heavy lines. The average water molecule is shown as a double circle.

DEOXYADENOSINE MONOHYDRATE

$C_{10}H_{13}N_5O_3 \cdot H_2O$ F.W. = 269.3

I. The crystal structure of deoxyadenosine monohydrate. D.G. WATSON, D.J. SUTOR, and P. TOLLIN, 1965. *Acta Cryst.*, <u>19</u>, 111–124.

Monoclinic, a = 16.060±7, b = 7.866±3, c = 4.700±2 Å, β = 96°4±1', [U = 590.0 Å³], D_m = 1.510, Z = 2, D_x = 1.514.

Space group $P2_1$ (C_2^2) or $P2_1/m$ (C_{2h}^2). $P2_1$ is required by lack of molecular symmetry.

Atomic positions

	x	y	z
N(1)	0.9042±4	−0.4782±10	−0.4005±15
C(2)	0.8404±5	−0.5202±12	−0.2608±19
N(3)	0.7994±4	−0.4253±9	−0.0884±14
C(4)	0.8316±4	−0.2763±10	−0.0655±14
C(5)	0.8982±4	−0.2040±10	−0.2005±16
C(6)	0.9359±4	−0.3212±11	−0.3754±16
N(7)	0.9126±4	−0.0361±10	−0.1308±14
C(8)	0.8569±4	−0.0000±11	0.0437±16
N(9)	0.8047±4	−0.1321±9	0.0827±12
N(10)	1.0010±5	−0.2811±10	−0.5153±18
C(1')	0.7271±4	−0.1349±11	0.2249±14
C(2')	0.6516±4	−0.1499±10	0.0061±16
C(3')	0.5878±4	−0.0370±11	0.1250±14
C(4')	0.6434±4	0.1041±9	0.2659±14
C(5')	0.6572±5	0.2496±12	0.0676±18
O(1')	0.7202±3	0.0210±8	0.3692±10
O(3')	0.5494±3	−0.1325±8	0.3324±13
O(5')	0.7012±4	0.3795±10	0.2373±13
O(11)	0.5798±4	−0.4589±10	0.5132±16

Anisotropic temperature factors are given for the above atoms. The positions and isotropic temperature factors of the hydrogen atoms are given also.

Structure

The bond lengths and angles of the deoxyadenosine molecule are given in Fig. 112. (Bond lengths and angles involving hydrogen are also given in the original.) The adenine group (atoms 1 through 10) is strictly planar, but C(1') of the glycosidic link is displaced by 0.22 Å from the plane of the group. Atoms C(1'), C(2'), C(4') and O(1') of the deoxyribose ring are practically coplanar; C(3') and O(3') lie 0.55 and 1.97 Å respectively from their mean plane. This plane makes an angle of 69° with the adenine plane.

The packing in the crystal is determined by hydrogen bonding in which all available groups (including the water molecule) participate. For details the original should be consulted.

Details of analysis

All reflexions accessible to CuK_α radiation were recorded on equi-inclination Weissenberg photographs about b and c, and on oscillation photographs about a and c. 91% of the accessible reflexions were observed and estimated visually. Refinement was by full-matrix least squares to a final R index of 0.078.

CYTIDINE

$C_9H_{13}O_5N_3$ F.W. = 243.2

I. A refinement of the crystal structure of cytidine. S. FURBERG, C.S. PETERSEN, and CHR. RØMMING, 1965. *Acta Cryst.*, <u>18</u>, 313-230. (For earlier, two-dimensional work, see <u>1</u>).

Orthorhombic, $a = 13.99\pm2$, $b = 14.786\pm2$, $c = 5.116\pm1$ Å, $[U = 1058.4$ Å$^3]$, $z = 4$.

Space group $P2_12_12_1$ (D_2^4)

Atomic positions

	x	y	z
C(6)	0.6795	0.2135	0.0400
C(2)	0.5289	0.2461	0.2096
C(4)	0.6180	0.1218	0.3820
C(5)	0.6895	0.1390	0.2133
C(1')	0.4550	0.1563	0.5593
C(4')	0.4405	0.0132	0.7512
C(3')	0.4026	0.0079	0.4704
C(2')	0.3755	0.1049	0.4186
C(5')	0.5101	-0.0605	0.8247
O(2)	0.4539	0.2916	0.2102
O(1')	0.4866	0.1005	0.7660
O(5')	0.5823	-0.0751	0.6389
O(3')	0.3252	-0.0522	0.4386
O(2')	0.2859	0.1298	0.5297
N(1)	0.6026	0.2669	0.0466
N(3)	0.5376	0.1745	0.3808
N(6)	0.7471	0.2343	-0.1330

    Standard **deviations** are given in full, and range from 0.003 to 0.005 Å. Anisotropic temperature factors and the positions of the hydrogen atoms (standard deviation 0.06 Å) are given also.

Structure

    The molecular geometry is illustrated in Fig. 113. The cytosine moiety is

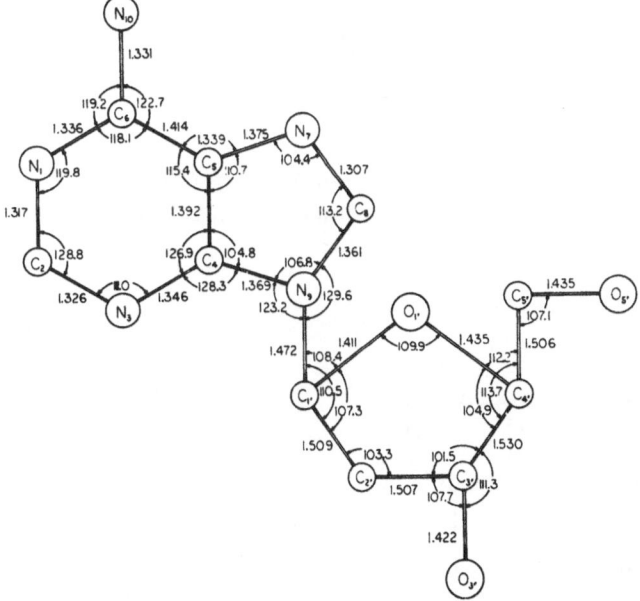

Fig. 112. Bond lengths and angles in deoxyadenosine monohydrate. Mean standard deviations are 0.010 Å and 0.8°.

essentially planar, and the ribose ring is puckered; the angle between their
mean planes is 80°. The structure is stabilized by the system of hydrogen
bonds illustrated in Fig. 114.

Details of analysis

   The intensity data were recorded on integrated equi-inclination Weissenberg
photographs (*hk*0 to *hk*2; *h0l* to *h8l*; *0kl*). 1030 of 1195 accessible reflexions
were observed and measured photometrically. Refinement was by block-diagonal
least squares to a final *R* index of 0.056.

1. *Structure Reports*, 14, 571.

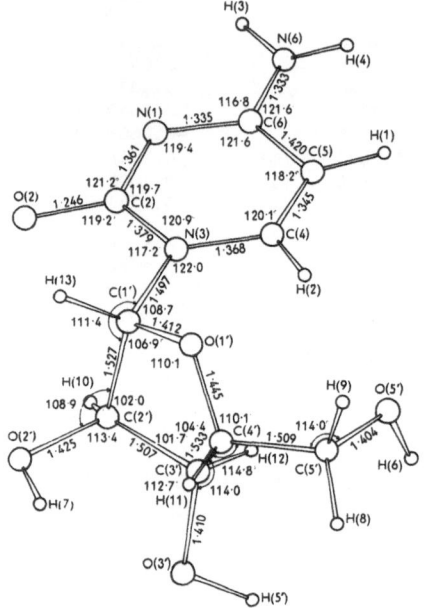

Fig. 113. The cytidine molecule.

DEOXYGUANOSINE, 5-BROMODEOXYCYTIDINE
(1:1 HYDROGEN-BONDED COMPLEX)

$C_{19}H_{25}N_8O_8Br$ F.W. = 573.4

I. The crystal structure of a hydrogen bonded complex of deoxyguanosine and 5-
   bromodeoxycytidine. A.E.V. HASCHEMEYER and H.M. SOBELL, 1965. *Acta Cryst.*,
   19, 125-130. (For a preliminary report, see 1.)

Orthorhombic, $a = 5.14\pm2$, $b = 19.11\pm2$, $c = 23.66\pm3$ Å, $[U = 2324$ Å$^3]$, $D_m = 1.63$,
$Z = 4$, $D_x = 1.638$.

Space group $P22_12_1$ (D_2^3)

Atomic positions

     Atomic positions are given for the 36 non-hydrogen atoms. O(5') of the 5-
bromodeoxycytidine molecule is disordered, and is given as two fractional atoms.
The standard deviations of the atoms other than bromine range from 0.02 to 0.07 Å.
Isotropic temperature factors are given also.

Structure

     Bond lengths and angles are given in full, and are consistent with expectation.
For each of the nucleosides the deoxyribose ring is puckered, with C(1'), O(1'),
C(3'), and C(4') coplanar, and C(2') lying 0.5 Å from their mean plane. The cytosine
ring of 5-bromodeoxycytidine is planar, and the substituent atoms lie in or near the
mean plane. The same is true of the guanine residue of deoxyguanosine, except that
C(1') lies 0.33 Å from the guanine plane. This deviation may be related to the unusual orientation of the sugar, which is *syn* with respect to the base. The dihedral
angle between the guanine and sugar planes is 69°. O(5') of the deoxyribose group
appears to be disordered. For 5-bromodeoxycytidine the orientation of the sugar is
anti with respect to the base, and the dihedral angle between cytosine and sugar
planes is 77°. Adjacent guanosine and cytosine residues are approximately coplanar,
and are linked by three hydrogen bonds to form a base pair. These and other hydrogen
bonds are illustrated in Fig. 115.

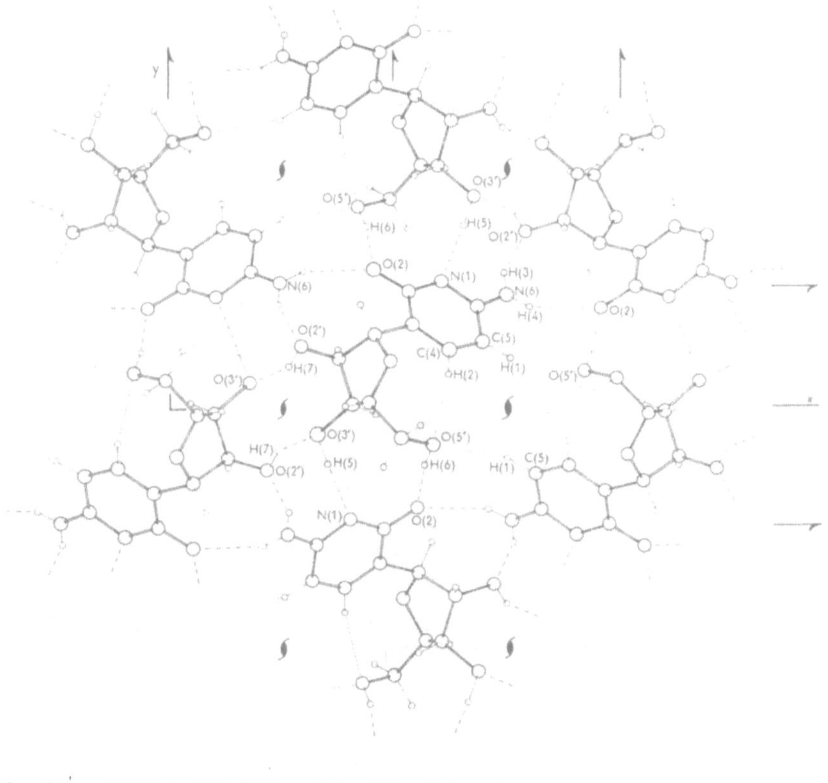

Fig. 114. The cytidine structure projected along *c*.

Details of analysis

The intensity data were recorded on oscillation and Weissenberg photographs, using CuK_α radiation. About half of the theoretically accessible reflexions were observed. Refinement was by full-matrix least squares to a final R index of 0.135.

1.   A.E.V. HASCHEMEYER and H.M. SOBELL, 1964. *Nature, Lond.,* <u>202</u>, 969.

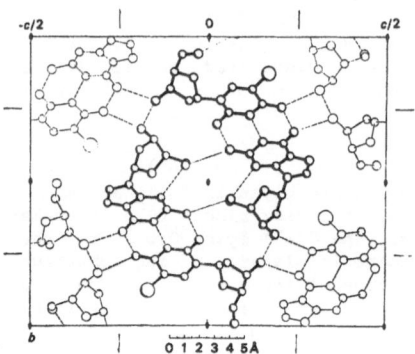

Fig. 115. The $C_{19}H_{25}N_8O_8Br$ structure viewed along a. Dashed lines indicate hydrogen bonds, and dotted lines indicate other short contacts.

CYTIDYLIC ACID B
(CYTIDINE-3'-PHOSPHATE)

$C_9H_{14}N_3O_8P$ F.W. = 323.2

I. Stereochemistry of nucleic acid constituents. I.   Refinement of the structure of cytidylic acid b. M. SUNDARALINGAM and L.H. JENSEN, 1965. *J. Molec. Biol.,* <u>13</u>, 914-920.   (This is a redetermination; for earlier work, see <u>1</u>.)

Orthorhombic, $a = 8.778\pm1$, $b = 21.649\pm3$, $c = 6.847\pm1$ Å, $[U = 1301.2$ Å$^3]$, $D_m = 1.66$ (from <u>1</u>), $Z = 4$, $D_x = 1.55$, $(CuK_\alpha, \lambda = 1.5418$ Å$)$.

Space group $P2_12_12$ (D_2^3)

Atomic positions

	x	y	z
P	0.06124	0.24612	0.11564
O(6)	0.19560	0.20607	0.06524
O(7)	-0.03692	0.25015	-0.07236
O(8)	-0.02640	0.22765	0.29130
O(3')	0.11926	0.31637	0.13753
O(2')	0.38205	0.39188	0.08896
O(1')	0.06387	0.42731	0.40293
O(5')	0.28315	0.37445	0.70685
O(2)	0.18955	0.53474	-0.02972
N(1)	0.22500	0.50558	0.29079
N(3)	0.28111	0.60506	0.18441
N(4)	0.38438	0.67854	0.38896
C(3')	0.19705	0.33674	0.30956
C(2')	0.29050	0.39419	0.25659
C(1')	0.16687	0.44288	0.25119
C(4')	0.08213	0.36267	0.45772
C(5')	0.12721	0.35787	0.67012
C(6)	0.26336	0.52248	0.47406
C(5)	0.31405	0.57971	0.51841
C(4)	0.32719	0.62262	0.36245
C(2)	0.22813	0.54711	0.13389

The positions of the hydrogen atoms (located with significant precision) are given also. Temperature factors (anisotropic for the non-hydrogen atoms) are given. The mean standard deviations are: P, 0.0011 Å; O, 0.0036 Å; N, 0.0042 Å; C, 0.005 Å; H, 0.06 Å.

Structure

The analysis establishes the formulation indicated above. (The observation that N(3) is protonated refutes the assumptions of 1.) Bond lengths and angles are given in full. The cytosine moiety is appreciably non-planar, with some atoms lying as much as 0.05 Å from the mean plane, and the pyrimidine ring is slightly boat-shaped. The ribose ring is puckered, with no four atoms coplanar. The angle between the mean planes of the rings is 60.5°. The structure is held together by a system of O-H..O and N-H..O bonds, with lengths ranging from 2.53 to 2.98 Å (Fig. 116).

Details of analysis

The intensity data were recorded on uni-dimensionally integrated Weissenberg photographs (*hk*0 to *hk*6) using CuK$_\alpha$ radiation. 1207 of a possible 1260 reflexions were observed and measured with a microdensitometer. The refinement was by full-matrix least squares to a final R index of 0.045.

1.   *Structure Reports*, <u>23</u>, 711.

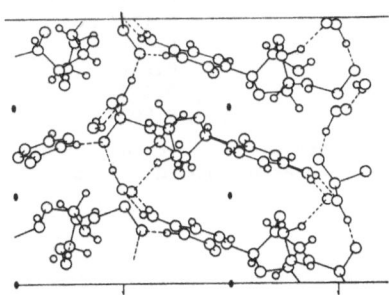

Fig. 116. Part of the structure of cytidylic acid b, viewed along *c*.

5-IODO-2'-DEOXYURIDINE

$C_9H_{11}N_2O_5I$ F.W. = 354.1

I. The crystal and molecular structure of 5-iodo-2'-deoxyuridine. N. CAMERMAN
and J. TROTTER, 1965. *Acta Cryst.*, 18, 203-211.

Triclinic, a = 4.98±1, b = 6.83±1, c = 9.60±2 Å, γ = 101°40±5', β = 109°18±5',
γ = 98°20±5', U = 293 Å^3, D_m = 2.014, z = 1, D_x = 2.008, (CuK_α, λ = 1.5418 Å,
MoK_α, λ = 0.7107 Å).

(Crystal data for the isomorphous bromine derivative are given also. They
are: a = 4.87±1, b = 6.72±1, c = 9.56±2 Å, α = 100°10±5', β = 107°24±5', γ = 98°31±5',
U = 285 Å^3, D_x = 1.789.)

Space group   P1 (C_1^1)

Atomic positions

	x	y	z
I	0.0000	1.0000	1.0000
N(1)	0.5809	0.6132	1.1563
C(2)	0.5719	0.4787	1.0257
O(2)	0.7339	0.3577	1.0370
N(3)	0.3761	0.4918	0.8902
C(4)	0.2192	0.6401	0.8804
C(5)	0.2357	0.7770	1.0065
C(6)	0.4290	0.7627	1.1581
O(6)	0.4554	0.8796	1.2769
C(1')	0.3567	0.3480	0.7458
O(1')	0.4699	0.4619	0.6613
C(2')	0.0386	0.2374	0.6368
C(3')	0.0782	0.1796	0.4804
O(3')	0.2026	0.0019	0.4789
C(4')	0.3312	0.3672	0.5011
C(5')	0.2177	0.5277	0.4163
O(5')	-0.0101	0.5904	0.4526

Anisotropic temperature factors are given for the above atoms. Standard
deviations are given in full, and range from 0.012 to 0.020 Å for the atoms other
than iodine.

Structure

Bond lengths and angles are given in full, and are sufficiently accurate to
confirm the formulation indicated above. A view of the molecule is given in Fig.
117. The pyrimidine base is nearly planar; the deoxyribose ring is puckered, with
C(2') lying 0.59 Å from the plane of the other four ring atoms. The plane of the

sugar ring makes an angle of 81° with that of the base. The intermolecular packing is shown in Fig. 118. There are N-H..O and O-H..O bonds as shown, and a very
short I-O contact (2.96 Å) suggesting charge-transfer interaction. The biological significance of this interaction is discussed.

Details of analysis

The intensity data were measured with a four-circle diffractometer and scintillation counter, using MoK$_\alpha$ radiation. 1434 of a possible 1445 reflexions were
observed in the sphere 2θ<57.3. Refinement was by block-diagonal least squares to
a final *R* index of 0.054.

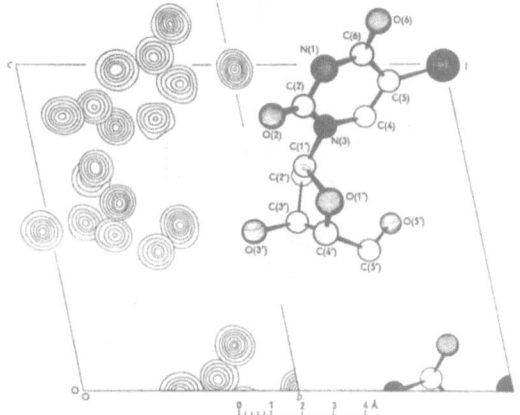

Fig. 117. A view of the 5-iodo-2'-deoxyuridine molecule.

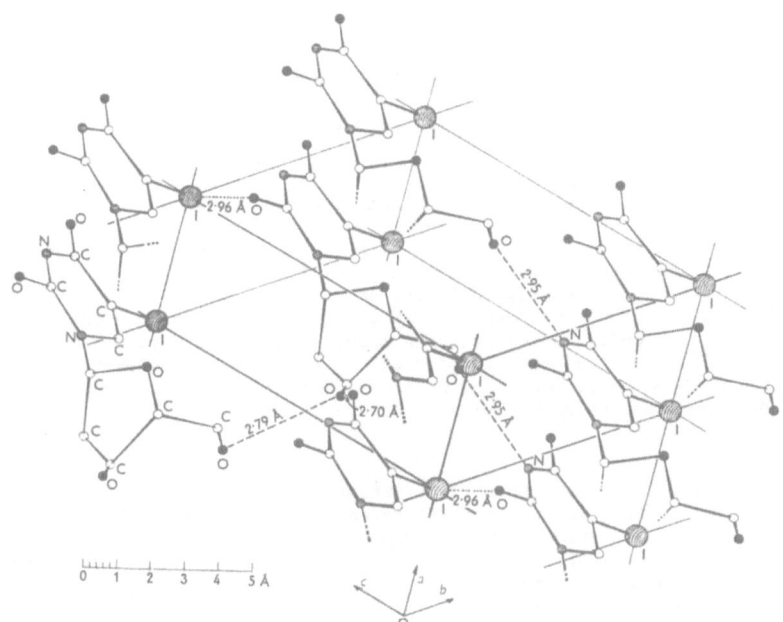

Fig. 118. The molecular packing of 5-iodo-2'-deoxyuridine. The dashed lines are
    the hydrogen bonds, and the dotted lines the short I-O contacts.

D(+)-BARIUM URIDINE-5'-PHOSPHATE HEPTAHYDRATE

$C_9H_{11}N_2O_9PBa \cdot 7H_2O$ F.W. = 585.6

I. The crystal and molecular structure of D(+)-barium uridine-5'-phosphate hepta-
   hydrate. E. SHEFTER and K.N. TRUEBLOOD, 1965. *Acta Cryst.*, 18, 1067-1077.
   (*Corrigendum, Idem*, 1967. *Ibid.*, 23, 673.)

Orthorhombic, a = 21.11±2, b = 9.06±3, c = 20.98±2 Å, [U = 4013 Å^3], D_m = 2.05,
Z = 8, [D_x = 1.94].

Space group $C222_1$ (D_2^5) Molecular symmetry, two barium atoms and two water molecules on two-fold axes.

Atomic positions

     Atomic positions are given for the 31 non-hydrogen atoms. Mean standard devia-
tions are: Ba, 0.002 Å; P, 0.005 Å; other, 0.015 to 0.025 Å. Temperature factors
(anisotropic for barium and phosphorus) are given also.

Structure

     Bond lengths and angles for the nucleotide are given in full, with standard
deviations of 0.03 Å and 2°, and are consistent with expectation. The uracil ring
is essentially planar, but O(2) and C(1') lie 0.14 and 0.12 Å, respectively, from
the ring plane, and on opposite sides of it. Atoms C(1'), O(1'), C(3'), and C(4')
of the ribose residue are nearly coplanar, and C(2') lies 0.52 Å from their mean
plane. The dihedral angle between the planes of the uracil and furanose rings is
67°. The absolute configuration of the nucleotide has been determined, and is that
indicated in Fig. 119. The coordination of oxygen atoms around each barium atom is
ten-fold, with Ba-O distances ranging from 2.7 to 3.3 Å. The structure is held
together by a complex system of hydrogen bonds.

Details of analysis

     The intensity data were recorded on integrated equi-inclination Weissenberg
photographs (h0l to h6l) using nickel-filtered CuK$_\alpha$ radiation. 1502 *hkl*, and 1000
hkl reflexions were observed and measured photometrically and visually; the observed reflexions constitute about 53% of those accessible to CuK$_\alpha$ radiation. The
refinement was by block-diagonal least squares to a final R index of 0.098. The
absolute configuration was determined by observing the effect of the anomalous
scattering of the barium atom.

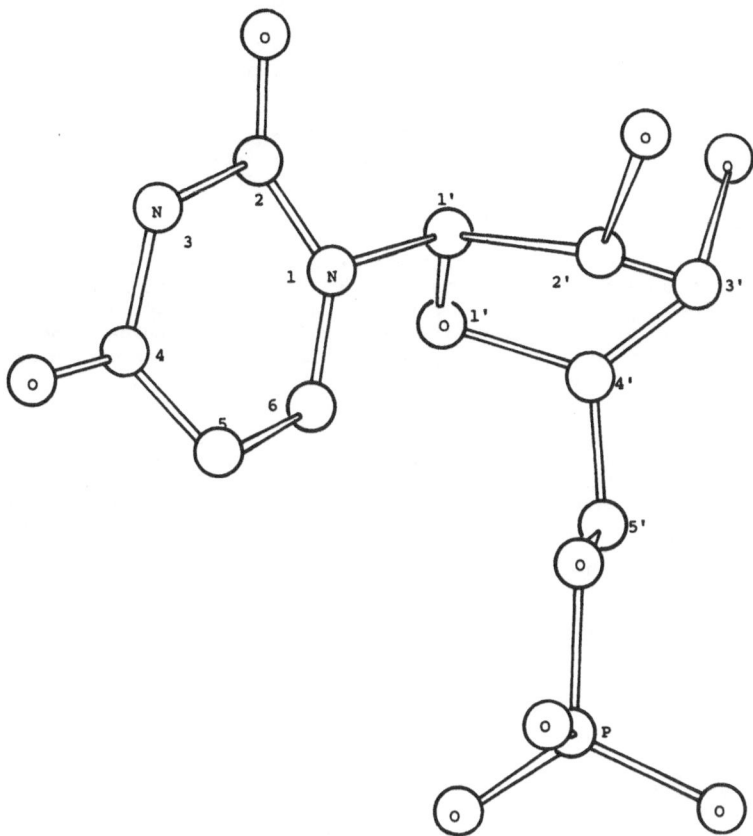

Fig. 119. The absolute configuration of the nucleotide.

β-ALANINE

$C_3H_7NO_2$ F.W. = 89.1

$$H_3\overset{+}{N}—CH_2—CH_2—\overset{\overset{\displaystyle O}{\|}}{C}—O^-$$

I. The crystal and molecular structure of β-alanine. P. JOSE and L.M. PANT, 1965.
   Acta Cryst., <u>18</u>, 806–810.

Orthorhombic, a = 9.865, *b* = 13.81, *c* = 6.07 Å, [*U* = 827 Å³], D_m = 1.412, *Z* = 8,
D_x = 1.423. (D_m from <u>1</u>. This reference also gives cell dimensions in good agreement with those found here.)

Space group Pbca (D_{2h}^{15})

Atomic positions

	x	y	z
N	0.2955±6	0.0663±4	−0.0450±10
C(1)	0.2318±7	0.1599±5	0.0120±12
C(2)	0.0850±7	0.1601±5	−0.0796±12
C(3)	−0.0214±7	0.1107±5	0.0717±12
O(1)	−0.1393±5	0.0969±4	−0.0156±9
O(2)	0.0174±5	0.0851±4	0.2655±9

Structure

To the accuracy of the analysis, C–C and C–O bonds, and C–C–O angles are equal. Mean values are

C – N = 1.478 Å	N – C – C = 108.4°
C – C = 1.552	C – C – C = 114.4
C – O = 1.290	C – C – O = 116.5
	O – C – O = 127.0

(Standard deviations, 0.010 Å and 0.6°.) The carboxyl group and the adjacent carbon atom are coplanar. The angle between their plane and the one containing the nitrogen atom and the two adjacent carbon atoms is 75.3°. The structure is extensively hydrogen bonded. The nitrogen atom appears to form three strong hydrogen bonds (lengths 2.78, 2.78 and 2.76 Å) with oxygen atoms of neighbouring molecules. These bonds, and the C–N bond form an approximately tetrahedral array. It is concluded that β-alanine is in the zwitterion form depicted above.

Details of analysis

Intensity data were recorded on equi-inclination Weissenberg photographs ($hk0$ to $hk4$; $0kl$ to $6kl$) using CuK_α radiation. 240 of a possible 917 reflexions were observed and measured visually. Refinement was by least squares to a final R index of 0.157.

1. *Structure Reports*, <u>13</u>, 482.

L–CYSTEINE ETHYL ESTER HYDROCHLORIDE–UREA (1:1 COMPLEX)

$C_6H_{16}ClN_3O_3S$ F.W. = 245.7

$$CH_3CH_2OCOCH(NH_3)^+ \ CH_2SHCl^-, \ CO(NH_2)_2$$

I. Intermolecular complexes. III. L–Cysteine ethyl ester hydrochloride–urea (1:1). D.J. HAAS, 1965. *Acta Cryst.*, <u>19</u>, 860–861.

Monoclinic, a = 9.974±5, b = 12.091±5, c = 5.127±5 Å, β = 97.00±5°, [U = 613.7 Å^3], D_m = 1.34, [Z = 2], D_x = 1.324.

Space group $P2_1$ (C_2^2) or $P2_1/m$ (C_{2h}^2). $P2_1$ required by molecular asymmetry.

Atomic positions

	x	y	z
Cl	0.1646	0.5011	0.9230
S	0.4885	0.5971	0.6921
O(1)	0.6511	0.3466	0.6494
O(2)	0.4654	0.2950	0.8386
O(3)	0.1876	0.1757	0.4292
N(1)	0.2972	0.3805	0.4459
N(2)	0.2065	0.1633	0.8741
N(3)	0.0251	0.0772	0.6109
C(1)	0.4812	0.5169	0.4010
C(2)	0.4451	0.3914	0.4249
C(3)	0.5210	0.3385	0.6630
C(4)	0.7424	0.2851	0.8501
C(5)	0.8609	0.3282	0.8852
C(6)	0.1366	0.1441	0.6423

Standard deviations are: Cl and S, 0.006 Å; light atoms except C(4) and C(5), 0.02 Å; C(4), 0.03 Å; C(5), 0.05 Å. Isotropic temperature factors are given also.

Structure

The structure is held together by the network of hydrogen bonds depicted in Fig. 120. The bond joining N(1) to the chlorine ion has electrostatic as well as hydrogen bonding character. The shortest hydrogen bond is N(1)–H..O(3), in keeping with the positive charge on N(1).

Details of analysis

Intensity data were measured with a four–circle diffractometer, using CuK_α radiation. Within the limit of resolution of 1 Å, 530 of a possible 744 reflexions were observed. Refinement was by block–diagonal least squares to a final R index of 0.092.

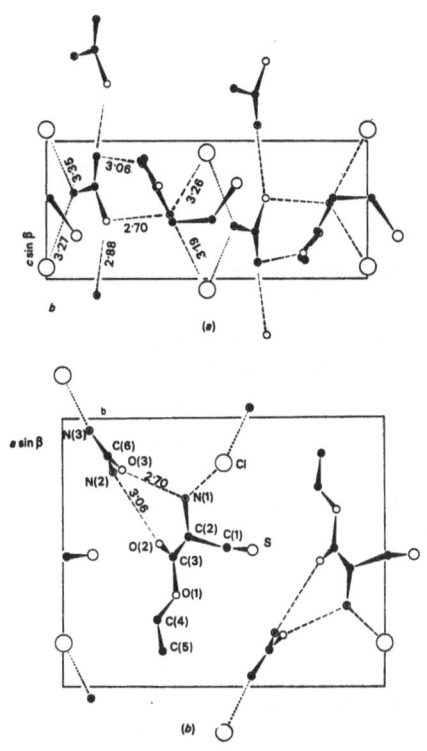

Fig. 120. (a) The $C_6H_{16}ClN_3O_3S$ structure viewed along a.
　　　　　(b) The $C_6H_{16}ClN_3O_3S$ structure viewed along c.
　　　　　Dotted lines indicate contacts which cross the cell faces parallel to the plane of projection. Standard deviations, 0.02 Å for distances involving chlorine, 0.03 Å otherwise.

L-PROLINE

$C_5H_9NO_2$ F.W. = 115.1

1. [X-ray determination of the structure of L-proline.] R.L. KAJUSINA and B.K. VAJNSTEJN, 1965. *Kristallografija*, <u>10</u>, 833-844 [*Soviet Physics - Crystallography*, <u>10</u>, 698-706]. (For preliminary determination of unit cell and space group, see <u>1</u>).

Orthorhombic, a = 11.55, b = 9.02, c = 5.20 Å, (all ±0.2%), [U = 542 Å^3], Z = 4, [D_x= 1.41].

Space group $P2_12_12_1$ (D_2^4)

Atomic positions

	x	*y*	*z*
C(1)	0.029	0.268	0.131
C(2)	0.055	0.295	0.415
C(3)	0.153	0.406	0.450
C(4)	0.259	0.305	0.413
C(5)	0.226	0.163	0.557
N	0.099	0.148	0.523
O(1)	0.057	0.144	0.031
O(2)	−0.023	0.364	−0.001

Standard deviations are: O, 0.008; N, 0.010; C, 0.012 Å. Also given are iso-tropic temperature factors for the above atoms and the positions of the hydrogen atoms.

Structure

The configuration and the bond lengths of the molecule are shown in Fig. 121. The carboxyl group is symmetrical indicating that the molecule is in the zwitterion form depicted. The group C(2) C(3) C(5) N is approximately planar. C(4) lies 0.60 Å from their plane, and is *cis* to the carboxyl group. The structure is held to-gether by intermolecular N-H..O bonds of length 2.69 and 2.71 Å.

Details of analysis

Intensity data were recorded on equi-inclination Weissenberg photographs (*hk*0 to *hk*4) using CuK_α radiation. 506 reflexions were observed and estimated visually. Refinement was by least squares to a final *R* index of 0.169.

1.   *Structure Reports,* 12, 417.

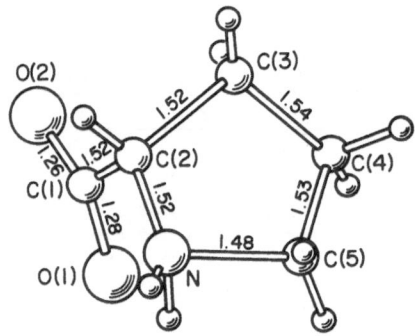

Fig. 121. Configuration and bond lengths in L-proline. Standard deviations 0.015 to 0.018 Å.

D,L-HOMOCYSTEINE THIOLACTONE HYDROCHLORIDE

C_4H_8ClNOS F.W. = 153.6

I. Crystal structures of D,L-homocysteine thiolactone hydrochloride: two poly-morphic forms and a hybrid. S.T. FREER and J. KRAUT, 1965. *Acta Cryst.,* 19, 992-1002. There are two polymorphic forms and a hybrid of the two.

1. *Orthorhombic*, a = 9.806±3, b = 9.321±2, c = 7.321±2 Å, [U = 669.2 Å³],
D_m = 1.519, Z = 4, D_x = 1.525, (CuK$_\alpha$, λ = 1.5418 Å; CuK$_{\alpha 1}$, λ = 1.54051 Å).

Space group   Pbc2₁ (C_{2v}^5) or   Pbcm (D_{2h}^{11}).   Pbc2₁ is confirmed by the analysis.

Atomic positions

	x	y	z
Cl	0.8192±2	0.0839±2	0.2615*
S	0.6160±2	0.6969±2	0.2053±5
O	0.8700±5	0.7020±5	0.0877±9
N	0.9100±6	0.4102±7	0.1743±12
C(1)	0.7842±7	0.6419±7	0.1763±12
C(2)	0.8063±7	0.4953±6	0.2750±14
C(3)	0.6671±9	0.4239±8	0.2778±17
C(4)	0.5633±9	0.5394±8	0.3395±15

    * No standard deviation is available for this coordinate which was used to
define the origin. Also given are anisotropic temperature factors for the atoms
above, and the positions of the hydrogen atoms.

2. *Orthorhombic*, a = 19.512±2, b = 9.296±3, c = 7.272±1 Å, [U = 1319 Å³],
D_m = 1.543, Z = 8, D_x = 1.547.

Space group   Pbca (D_{2h}^{15})

Atomic positions

	x	y	z
Cl	0.4066±1	0.0760±1	0.2615±1
S	0.3114±1	0.6973±1	0.2246±1
O	0.4380±1	0.6970±2	0.0963±3
N	0.4553±1	0.4008±2	0.1786±4
C(1)	0.3950±1	0.6347±3	0.1843±4
C(2)	0.4041±1	0.4889±3	0.2794±3
C(3)	0.3340±1	0.4193±3	0.2911±4
C(4)	0.2855±1	0.5382±3	0.3512±4
H(1)	0.417 ±1	0.511 ±3	0.399 ±3
H(2)	0.331 ±1	0.333 ±3	0.371 ±4
H(3)	0.318 ±1	0.377 ±3	0.172 ±4
H(4)	0.290 ±1	0.562 ±3	0.487 ±4
H(5)	0.237 ±1	0.517 ±3	0.307 ±4
H(6)	0.458 ±1	0.304 ±3	0.231 ±5
H(7)	0.504 ±1	0.431 ±3	0.201 ±4
H(8)	0.451 ±2	0.398 ±3	0.059 ±5

Anisotropic temperature factors are given for the non-hydrogen atoms.

3. *Orthorhombic*, a = 19.55±3, b = 9.29±2, c = 7.23±2 Å.
(This is the hybrid form.)

    For the cell dimensions given above, it has been assumed that for CuK$_{\alpha 1, 2}$
λ = 1.5418 Å, and for CuK$_{\alpha 1}$, λ = 1.54041 Å.

Structure

    Both structures are discussed in detail. The remarkable similarity in molec-
ular packing permits formation of the hybrid. The molecular geometry does not
differ significantly in the two forms. As the analysis of the second form (Pbca)
appears to be the more accurate, the results presented here are for that form only.

    A view of the D,L-homocysteine thiolactone molecule is given in Fig. 122.
The bond lengths and angles are:

S	-	C(1)	1.756±3 Å	C(1) - S - C(4)	94.3±1°
S	-	C(4)	1.814±3	O - C(1) - S	125.3±2
O	-	C(1)	1.204±3	O - C(1) - C(2)	125.8±2
N	-	C(2)	1.486±3	S - C(1) - C(2)	109.0±2
C(1)	-	C(2)	1.532±3	N - C(2) - C(1)	110.1±2
C(2)	-	C(3)	1.515±4	N - C(2) - C(3)	113.5±2
C(3)	-	C(4)	1.520±4	C(1) - C(2) - C(3)	107.4±2
				C(2) - C(3) - C(4)	105.5±2
				S - C(4) - C(3)	105.8±2

(values involving hydrogen atoms are given also). The five-membered ring is puckered, with four atoms lying nearly in a plane and a fifth (C(3)) displaced about
0.7 Å from this plane. The N-H..Cl bonding scheme is illustrated in Fig. 123.
Each ammonium group is coordinated to three chloride ions and *vice-versa*. N...Cl
distances are 3.222, 3.178 and 3.185 Å, all ±0.003 Å.

Details of analysis

The intensity data were measured with a four-circle diffractometer, using
CuK_α radiation. All unique reflexions within the range of the instrument ($2\theta \leqslant 160°$)
were examined. For 1, ($Pbc2_1$) 796 reflexions were measured, with 86% observed;
for 2, ($Pbca$) 1413 were measured, with 90% observed; for the hybrid form, 1452
were measured, and 66% observed. By ill chance the first specimen examined was of
the hybrid form, and much labour was expended before the true nature of the
problem was recognized. The refinement of 1 and 2 was by full-matrix least squares
to final R indices of 0.074 and 0.050.

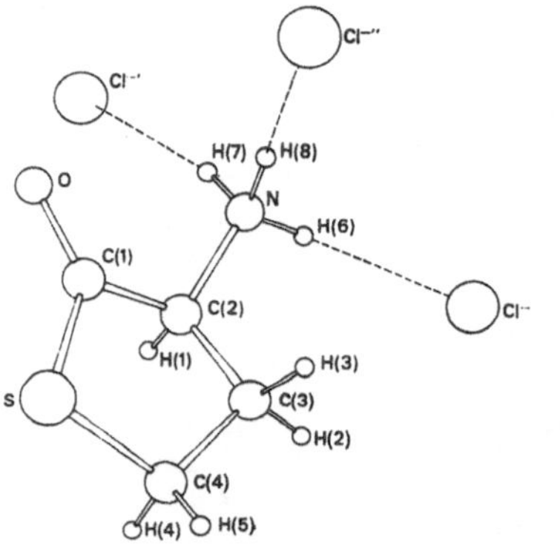

Fig. 122. The D,L-homocysteine thiolactone molecule, showing chloride ion coordination.

CUCURBITINE PERCHLORATE

$C_5H_{11}ClN_2O_6$ F.W. = 230.6

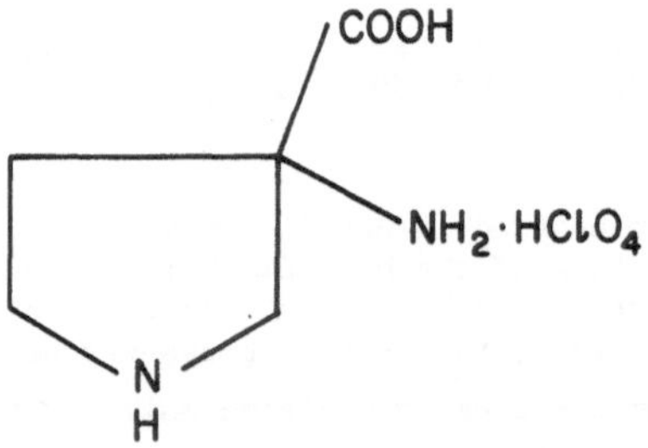

I. The crystal structure of cucurbitine perchlorate and the absolute configuration of the cucurbitine molecule. H.-F. FAN and C.-C. LIN, 1965. *Acta Phys.
Sinica*, 21, 253-262.

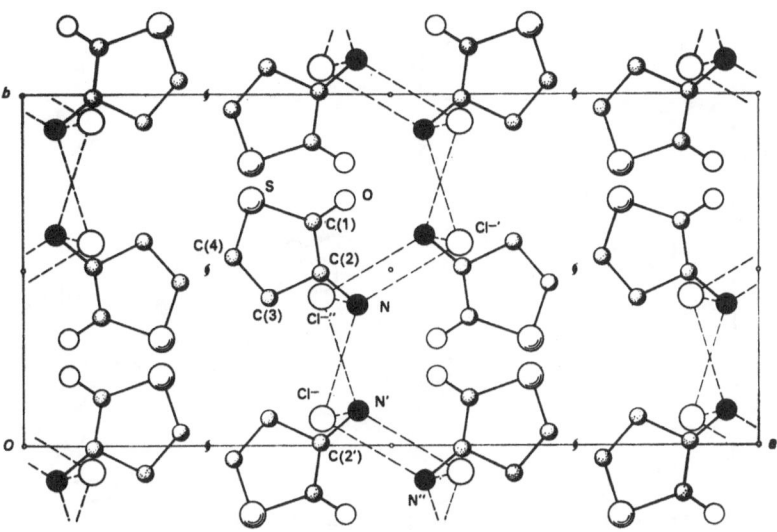

Fig. 123. The structure of D,L–homocysteine thiolactone hydrochloride, (2, *Pbca*)
viewed along *c*, showing N–H..Cl bonding.

Monoclinic, $a = 5.68\pm1$, $b = 12.78\pm1$, $c = 6.23\pm1$ Å, $\beta = 107°26\pm12'$, $[U = 431.5 \text{ Å}^3]$,
$z = 2$, $[D_x = 1.77]$.

Space group $P2_1$ (C_2^2). ($P2_1/m$ is ruled out by the asymmetry of the molecule.)

Atomic positions

	x	y	z
Cl	0.957	0.998	0.023
C(1)	0.520	0.806	0.437
C(2)	0.617	0.700	0.556
C(3)	0.635	0.687	0.168
C(4)	0.436	0.765	0.193
C(5)	0.306	0.843	0.532
N(1)	0.743	0.646	0.408
N(2)	0.723	0.888	0.459
O(1)	0.891	0.890	0.992
O(2)	0.883	0.055	0.815
O(3)	0.855	0.053	0.184
O(4)	0.222	0.005	0.133
O(5)	0.341	0.931	0.616
O(6)	0.133	0.789	0.513

Structure

The bond lengths and angles are given in full, and are consistent with the
formulation above. The structure is held together by a system of N–H..O bonds,
ranging in length from 2.57 to 3.02 Å, and illustrated in Fig. 124. The absolute
configuration of the cucurbitine molecule is as shown in the figure.

Details of analysis

The structure was solved by three–dimensional Patterson methods, and refined
with two–dimensional Fourier syntheses. The absolute configuration of the structure
was determined from the anomalous scattering of CuK_α radiation by chlorine. The
final R factors are 0.14, 0.19, and 0.20 for the three projections.

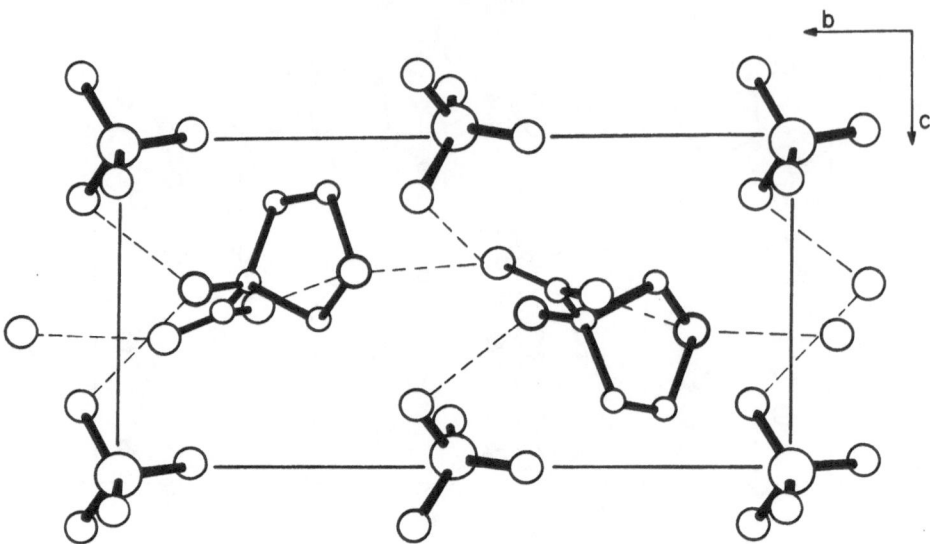

Fig. 124. The structure of cucurbitine perchlorate, viewed along *a*.

5-HYDROXY TRYPTAMINE CREATININE SULPHATE MONOHYDRATE
(SEROTONIN-CREATININE SULPHATE MONOHYDRATE (1:1 COMPLEX))

$C_{14}H_{21}N_5O_6S.H_2O$ F.W. = 405.4

I. The crystal and molecular structure of the serotonin-creatinine sulphate complex. I.L. KARLE, K.S. DRAGONETTE and S.A. BRENNER, 1965. *Acta Cryst.*, 19, 713-716. (This appears to be a preliminary communication.)

Monoclinic, a = 10.75±3, b = 9.70±2, c = 35.26±8 Å, β = 99°20±10', [U = 3628 Å³], D_m = 1.489, Z = 8, D_x = 1.485.

Space group Cc (C_s^4)   or   *C2/c* (C_{2h}^6).   C2/c is consistent with deduced structure, except that the position of the water molecule appears to be disordered.

Atomic positions

	x	y	z
S	0.3390	0.5497	0.0657
O(1)	0.2025	0.5644	0.0621
O(2)	0.3710	0.4257	0.0446
O(3)	0.3928	0.6765	0.0513
O(4)	0.3976	0.5324	0.1072
N(1)	0.3659	0.2449	0.1248
C(2)	0.3213	0.1488	0.0968
C(3)	0.2558	0.0427	0.1102
C(4)	0.2011	0.0127	0.1797
C(5)	0.2165	0.0793	0.2159
C(6)	0.2880	0.1971	0.2234
C(7)	0.3439	0.2631	0.1952
C(8)	0.3235	0.2009	0.1579
C(9)	0.2555	0.0781	0.1507

	x	y	z
C(10)	0.1907	−0.0778	0.0904
C(11)	0.1836	−0.0663	0.0469
N(12)	0.1016	−0.1819	0.0285
O(5)	0.1616	0.0144	0.2440
N(13)	0.0752	0.4605	0.1605
C(14)	0.0362	0.3668	0.1345
N(15)	−0.0326	0.2632	0.1484
C(16)	−0.0377	0.2930	0.1873
C(17)	0.0319	0.4243	0.1973
N(18)	0.0574	0.3654	0.0982
C(19)	0.1503	0.5841	0.1564
O(6)	−0.0926	0.2192	0.2073
O(7)	0.3542	0.1867	−0.0046
O(8)	0.5405	−0.0616	−0.0031

Isotropic temperature factors are given also. Refinement is not complete, and there is no estimate of accuracy.

Structure

Bond lengths and angles are given in full, and appear to be consistent with the above formulation. The creatinine molecule is nearly planar, as is the indole portion of the serotinin molecule. The water molecule appears to be shared between the positions O(7) and O(8). The structure is held together by the network of hydrogen bonds illustrated in Fig. 125. The proposed hydrogen bonding scheme suggests an =NH$_2^+$ group on the creatinine, and an -NH$_3^+$ group on the serotonin, molecule.

Details of analysis

The intensity data were recorded on equi-inclination Weissenberg photographs (rotation axis parallel to [110], layers 0 through 5) using CuK$_\alpha$ radiation. 2680 of a possible 3360 reflexions were observed and measured visually. Refinement was by full-matrix least squares, continued only until the general correctness of the structure seemed assured. The final R index is 0.185, and further refinement is planned.

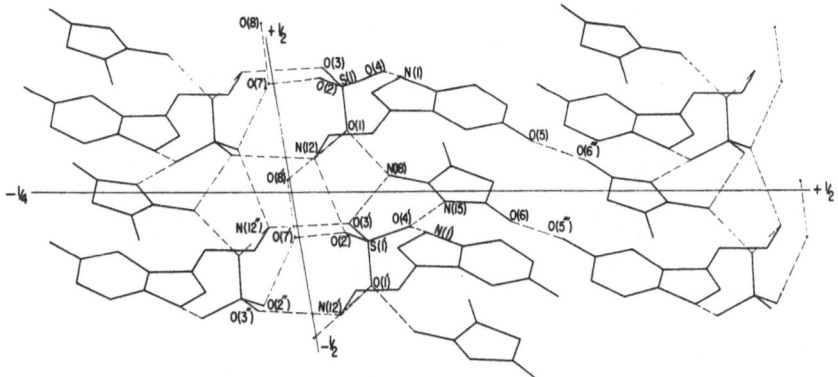

Fig. 125. The C$_{14}$H$_{21}$N$_5$O$_6$S.H$_2$O structure viewed along b, showing the hydrogen bonding network.

PORPHINE

$C_{20}H_{14}N_4$ F.W. = 310.3

I. Crystal structure of porphine. L.E. WEBB and E.B. FLEISCHER, 1965. *J. Chem. Phys.*, <u>43</u>, 3100-3111.
   (For a preliminary report, see <u>1</u>.)

Monoclinic, $a = 12.36\pm1$, $b = 12.12\pm2$, $c = 10.272\pm2$ Å, $\beta = 102.1\pm1°$, $[U = 1504.6$ Å$^3]$, $D_m = 1.336$, $Z = 4$, $D_x = 1.340$.

Space group $P2_1/a$ (C_{2h}^5)

Atomic positions

	x	y	z
'C(1)	0.1978±2	0.4489±2	0.3114±2
C(2)	0.2661±2	0.5349±2	0.3825±3
C(3)	0.2884±2	0.6059±2	0.2912±3
C(4)	0.2351±2	0.5645±2	0.1614±2
C(5)	0.2396±2	0.6123±3	0.0405±3
C(6)	0.1903±2	0.5735±2	-0.0848±2
C(7)	0.1966±2	0.6234±2	-0.2098±3
C(8)	0.1393±2	0.5596±2	-0.3069±3
C(9)	0.0949±2	0.4684±2	-0.2449±2
C(10)	0.0292±2	0.3836±2	-0.3099±2
C(11)	-0.0154±2	0.2962±2	-0.2520±2
C(12)	-0.0850±2	0.2097±2	-0.3221±3
C(13)	-0.1105±2	0.1407±2	-0.2309±3
C(14)	-0.0566±2	0.1829±2	-0.1016±2
C(15)	-0.0621±2	0.1362±2	0.0199±3
C(16)	-0.0126±2	0.1756±2	0.1444±3
C(17)	-0.0182±2	0.1253±2	0.2697±3
C(18)	0.0418±2	0.1870±2	0.3669±3
C(19)	0.0862±2	0.2780±2	0.3046±2
C(20)	0.1549±2	0.3604±2	0.3702±2
N(21)	0.1811±2	0.4682±2	0.1782±2
N(22)	0.1283±2	0.4794±2	-0.1108±2
N(23)	0.0007±2	0.2770±2	-0.1188±2
N(24)	0.0517±2	0.2686±2	0.1698±2

Also given are isotropic temperature factors for the above atoms, and the positions of the hydrogen atoms.

Structure

The molecule is essentially planar, with statistical symmetry 4/*mmm* (D_{4h}). Bond lengths and angles are given in full, with standard deviations 0.003 to 0.004 Å, and 0.2 to 0.3°. Mean values are:

a	=	1.442 Å	ab	=	107.6°	cc' = 107.8
b	=	1.342	ac	=	108.6	dd' = 126.8
c	=	1.366	ad	=	125.9	
d	=	1.386	cd	=	125.5	

The two inner hydrogen atoms appear as four half-atoms, as though shared equally by the four nitrogen atoms, and possible reasons for this are discussed. There are no unusually short intermolecular contacts; molecules related by the centre of symmetry partially overlap each other with an interplanar spacing of 3.42 Å.

Details of analysis

The intensity data were measured with a four-circle diffractometer and scintillation counter, using the stationary-crystal stationary-counter method, with CuK_α radiation. 2421 of a possible 2996 reflexions were observed above background. Refinement was by full-matrix least squares to a final R index of 0.049. A residual peak in the difference map, at the centre of the molecule, was attributed to the presence of 5 to 10% of some unknown metalloporphyrin.

1.   L.E. WEBB and E.B. FLEISCHER, 1965. *J. Amer. Chem. Soc.*, <u>87</u>, 667.

α-CHLOROHEMIN

$C_{34}H_{32}N_4O_4FeCl$ F.W. = 652.0

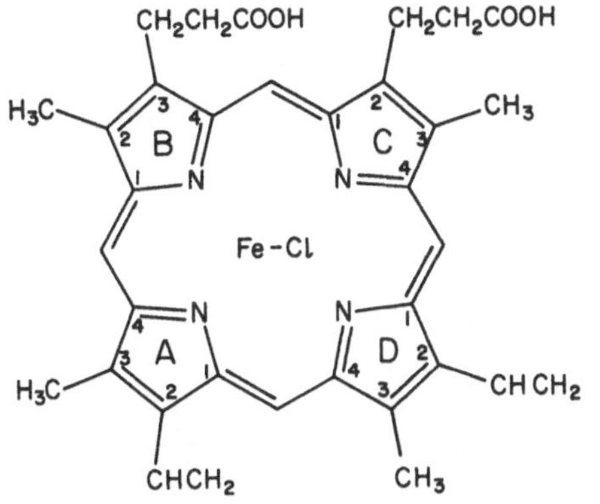

I. The structure of α-chlorohemin. D.F. KOENIG, 1965. *Acta Cryst.*, <u>18</u>, 663-673.

Triclinic, a = 11.494±20, b = 14.097±20, c = 10.854±20 Å, α = 98.56±4, β = 108.49±4, γ = 107.65±4°, U = 1530 Å^3, D_m = 1.42, Z = 2, D_x = 1.415.

Space group P1 (C_1^1) or   P$\bar{1}$ (C_i^1). Statistical analysis suggests P$\bar{1}$, which was assumed for the analysis. However, the correct space group may be P1.

Atomic positions

Atomic positions and temperature factors (some anisotropic) are given for the non-hydrogen atoms. The assumed positions of some hydrogen atoms are given also.

Structure

The molecule is as depicted above, except that because of disorder, or the possibly incorrect assumption that the structure is centrosymmetrical, the vinyl groups at A2 and D2 appear to be shared equally with positions A3 and D3. Bond lengths and angles are given in full, and appear to be normal. The square pyramidal coordination of the iron atom is illustrated in Fig. 126; the nitrogen atoms are coplanar. Each pyrrole residue is planar, but is tilted by about 7° to the plane of the nitrogen atoms, so that the porphyrin ring is convex to the iron atom. Pairs of adjacent molecules are hydrogen-bonded together *via* their carboxyl groups.

Details of analysis

The intensity data were measured with a four-circle diffractometer and scintillation counter, using MoK_α radiation. 1566 reflexions were observed and 230 of these were below the limit of reliable measurement. Refinement was by full-matrix least squares to a final R index of 0.095.

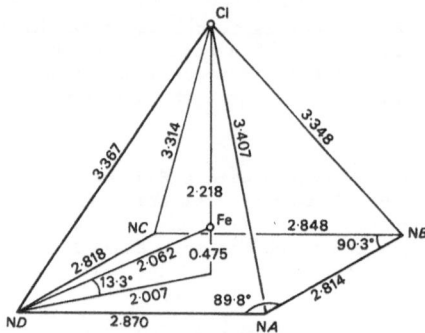

Fig. 126. The iron coordination in α-chlorohemin. Standard deviations are: Fe–Cl,
0.006; other distances 0.01 Å; angles, 0.3°.

METHOXYIRON(III) MESOPORPHYRIN-IX DIMETHYL ESTER

$C_{37}H_{43}FeN_4O_5$ F.W. = 680

I. The crystal structure and molecular stereochemistry of methoxy-iron(III)
   mesoporphyrin-IX dimethyl ester. J.L. HOARD, M.J. HAMOR, T.A. HAMOR and
   W.S. CAUGHEY, 1965. *J. Amer. Chem. Soc.*, <u>87</u>, 2312-2319.

Monoclinic, a = 11.55±2, b = 24.15±3, c = 12.51±1 Å, β = 98.4±1°, [U = 3452 Å³],
D_m = 1.31, Z = 4, D_x = 1.31.

Space group $I2$ (C_2^3), Im (C_s^3), or $I2/m$ (C_{2h}^3). $I2/m$ is confirmed by analysis.
Molecular symmetry, mirror plane. (The space group requires that the structure be
disordered. The alternative space groups were considered, but gave unsatisfactory
refinement.)

Atomic positions

	x	y	z
C(1)	0.7139±4	0.3949±3	0.4848±4
C(2)	0.6700±4	0.3533±3	0.5413±4
C(3)	0.4226±4	0.3555±2	0.8419±3
C(4)	0.3772±3	0.3964±2	0.8991±3
C(5)	0.7125±6	0.5000±0	0.4882±5
C(6)	0.5506±4	0.3565±2	0.6940±3
C(7)	0.3842±4	0.5000±0	0.9012±4
C(8)	0.6096±4	0.3839±2	0.6208±3
C(9)	0.4881±3	0.3827±2	0.7672±3
C(10)	0.4142±3	0.4498±2	0.8610±3
C(11)	0.7858±4	0.3885±4	0.3898±4
C(12)	0.6744±5	0.2910±3	0.5283±5
C(13)	0.4106±5	0.2926±2	0.8492±4
C(14)	0.3012±3	0.3890±2	0.9870±3

	x	y	z
C(15)	0.1738±2	0.4097±2	0.9504±3
C(16)	0.1040±4	0.3665±2	0.8829±4
C(17)	0.6849±3	0.4497±2	0.5243±3
½C(18)	0.7628±10	0.2684±6	0.6009±11
½C(19)	0.9199±9	0.4006±9	0.4357±9
C(20)	−0.0830±6	0.3474±4	0.7718±7
O(1)	0.1286±4	0.3179±2	0.8682±4
O(2)	0.0029±3	0.3866±2	0.8351±3
Fe	0.5845±1	0.5000±0	0.7193±0
N(1)	0.6199±3	0.4404±2	0.6092±2
N(2)	0.4825±2	0.4405±1	0.7803±2
O(3)	0.7185±3	0.5000±0	0.8193±3
C(21)	0.8307±7	0.5000±0	0.7965±7

The numbering scheme is arbitrary, and differs from that of the original.
The half atoms are the terminal methyl carbons of the disordered ethyl groups
Isotropic temperature factors are given also.

Structure

The structure is disordered. The average molecular site is occupied by a
superposition of enantiomorphic D and L isomers and thus appears to have mirror
symmetry. The overlap is fairly precise except for the ethyl and methyl groups
remote from the propionate chains, and the more important structural features are
clearly delineated. Bond lengths and angles are given in full. Some average
values (assuming 4mm symmetry for the porphine nucleus) are:

a = 1.368±12 Å   ab = 107.4±8°   cd = 125.9±8
b = 1.466±12   bc = 109.1±8   dd' = 124.1±8
c = 1.395±12   cc' = 107.0±8   ce = 126.4±8
d = 1.377±12
e = 2.073±6

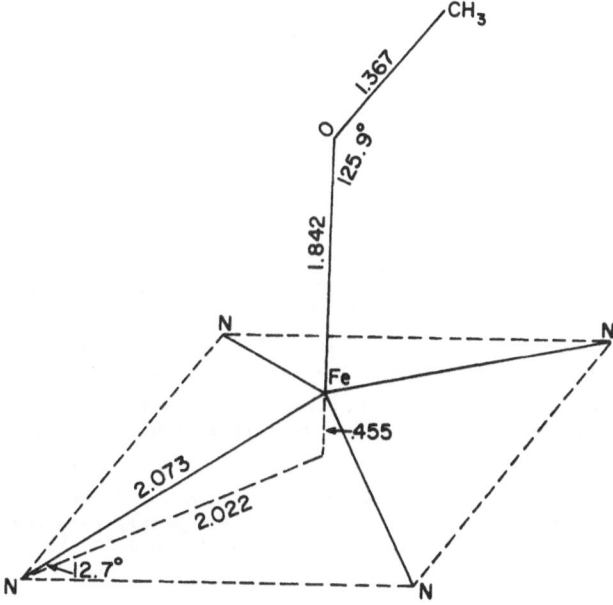

Fig. 127. The coordination of the iron atom in $C_{37}H_{43}FeN_4O_5$.

The porphine nucleus is somewhat ruffled, and the iron atom lies 0.49 Å from its mean plane. As a result of rotation about single bonds, the propionate chains lie almost normal to the porphine plane. The square-pyramidal coordination of the iron atom is illustrated in Fig. 127.

Details of analysis

The intensity data were recorded with a four-circle diffractometer and scintillation counter, using CoK_{α} radiation. The stationary-crystal, stationary-counter mode was used within the limiting sphere $\sin\theta/\lambda < 0.50$, and 1702 of a possible 1941 reflexions were observed above background. Refinement was by full-matrix least squares to a final R index of 0.119.

NICKEL(II) 2,4-DIACETYLDEUTEROPORPHYRIN-IX DIMETHYL ESTER

$C_{36}(H,D)_{36}NiN_4O_6 \cdot \frac{1}{2}C_6H_6$ F.W. = 718

, $\frac{1}{2}(C_6 H_6)$

I. The crystal and molecular structure of nickel(II) 2,4-diacetyldeuteroporph-
    yrin-IX dimethyl ester. T.A. HAMOR, W.S. CAUGHEY and J.L. HOARD, 1965.
    J. Amer. Chem. Soc., **87**, 2305-2312.

Triclinic, a = 15.615±20, *b* = 8.74±1, *c* = 13.57±2 Å, α = 101.8±2, β = 112.4±2,
γ = 86.4±2°, [*U* = 1676 Å³], D_m = 1.415, z = 2, D_x = 1.424.

Space group P1 (C_1^1) or P$\bar{1}$ (C_i^1). P$\bar{1}$ is confirmed by analysis. Molecular symmetry, centre for benzene molecule of crystallization.

Atomic positions

Atomic positions and isotropic temperature factors are given for the 50 non-hydrogen atoms. Apart from the nickel atom, and the poorly-resolved benzene of crystallization, standard deviations range from 0.004 to 0.009 Å.

Structure

Bond lengths and angles are given in full, and are consistent with expectation. Some mean values, (assuming *4mm* symmetry for the porphine nucleus) are:

a = 1.350±10 Å   d = 1.375±10   bc = 111.0±7   ce = 127.8±7
b = 1.446±10   e = 1.960±7   cc'= 104.4±7   dd'= 123.9±7
c = 1.383±10   ab ≐ 106.8±7°   cd = 125.3±7

The porphine skeleton is somewhat ruffled, with individual atoms lying as much as 0.07 Å from the mean plane. This ruffling is irregular, following no discernible symmetry.

Details of analysis

The intensity data were measured with a four-circle diffractometer and scintillation counter in the stationary-crystal stationary-counter mode, and MoK$_\alpha$ radiation. 3122 of a possible 4400 reflexions were observed in the sphere sinθ/λ<0.54. Refinement was by least squares to a final R index of 0.099.

TERRAMYCIN HYDROCHLORIDE

$C_{22}H_{24}N_2O_4$.HCl   F.W. = 496.7

I. The crystal structure of terramycin hydrochloride, $C_{22}H_{24}N_2O_4$.HCl. H. CID-DRESDNER, 1965. *Z. Kristallogr.*, 121, 170-189. (For earlier work, see 1, 2.)

Orthorhombic, a = 11.19, b = 12.49, c = 15.68 Å, [U = 2192 Å³], D_m = 1.51, Z = 4, D_x = 1.51.

Space group $P2_12_12_1$ (D_2^4)

Atomic positions

Atomic positions for the 34 non-hydrogen atoms are given. Average standard deviations are: Cl, 0.004 Å; O, 0.011 Å; N, 0.012 Å; C, 0.014 Å. Isotropic temperature factors are given also, as are the positions of 14 of the hydrogen atoms.

Structure

The structure is illustrated in Fig. 128. Bond lengths and angles are given in full, and generally support the formulation indicated above. However, the C-O distance in the amide group (1.31±2 Å) suggests that this oxygen atom is protonated. In this and other respects terramycin is very similar to aureomycin (3). As indicated above, extensive intramolecular hydrogen bonding is reasonably inferred. In addition, the structure is held together by hydrogen bonds to the chlorine atom; each molecule forms five such bonds, and each chlorine atom is coordinated to five different molecules. However, there appears to be no direct hydrogen bonding between terramycin molecules.

Details of analysis

The intensity data used were those from the original analysis (1). They were recorded on equi-inclination Weissenberg photographs about a and b. 1767 of 2100 accessible reflexions were observed and measured visually. Refinement was by least squares to a final R index of 0.142.

1. *Structure Reports*, <u>16</u>, 555.
2. *Ibid*, <u>24</u>, 720.
3. J. DONOHUE, J.D. DUNITZ, K.N. TRUEBLOOD and M.S. WEBBER, 1964. *J. Amer. Chem. Soc.*, <u>86</u>, 851.

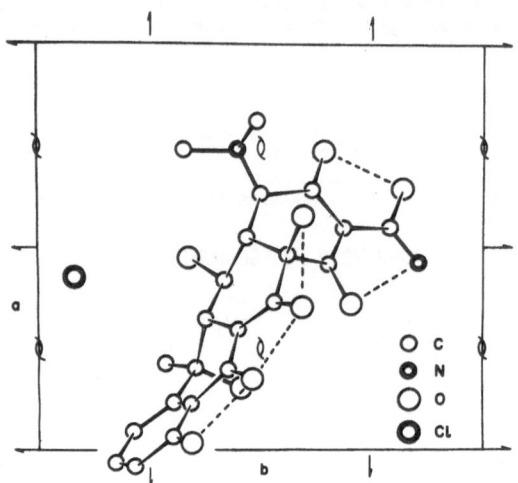

Fig. 128. The terramycin hydrochloride structure viewed along *c*.

ECDYSON

$C_{27}H_{44}O_6$ F.W. = 464.6

I.   Zur Chemie des Ecdysons.   II.   Bestimmung des Sterin-Skeletts und seiner Orientierung mit diffuser Roentgenstreuung in Kristallen von Ecdyson. W. HOPPE and R. HUBER, 1965. *Chem. Ber.*, <u>98</u>, 2353-2360.

II. Zur Chemie des Ecdysons. VII. Die Kristall- und Molekülstrukturanalyse des Insektenverpuppungshormons Ecdyson mit der automatisierten Faltmolekul-methode. R. HUBER and W. HOPPE, 1965. *Ibid.*, <u>98</u>, 2403-2424.

Orthorhombic, $a = 35.56 \pm 2$, $b = 9.92 \pm 2$, $c = 7.73 \pm 2$ Å, $U = 2725$ Å^3, $D_m = 1.127$, $Z = 4$, $D_x = 1.131$.

Space group $P2_12_12_1$ (D_2^4)

Atomic positions

Atomic positions are given for the 33 non-hydrogen atoms, and for 28 of the 44 hydrogen atoms. The standard deviations of the former range from 0.006 to 0.026 Å. Anisotropic temperature factors are given for the non-hydrogen atoms.

Structure

The analysis shows that the structure and relative configuration of the molecule can be described as: 2β, 3β, 14α, $22\beta_F$, 25-pentahydroxy-5β-choles-7-en-6-one (II)- (2β, 3β, 14α, 22β_F, 25-pentahydroxy-Δ^7-5β- cholestenon-(6)). Bond lengths and angles are given in full, and are consistent with this description. The crystal structure is held together by a system of intermolecular O-H..O bonds (of length 2.63 to 2.83 Å) involving all the oxygen atoms.

Details of analysis

The intensity data were recorded on equi-inclination Weissenberg photographs ($hk0$ to $hk6$; $h0l$ to $h7l$) using CuK_α radiation. About 3400 independent reflexions were observed, and measured with an integrating photometer. Refinement was by full-matrix least squares to a final R index of 0.15.

SUPRASTERYL II 4-IODO-5-NITROBENZOATE

$C_{35}H_{46}O_4NI$ F.W. = 671

I. The crystal and molecular structure of suprasteryl II 4-iodo-5-nitrobenzoate.
   C.P. SAUNDERSON, 1965. *Acta Cryst.*, <u>19</u>, 187-192.
   (For a preliminary report, see <u>1</u>.)

Monoclinic, a = 39.14±10, b = 10.34±3, c = 8.21±2 Å, β = 94°51', [U = 3311 Å³],
D_m = 1.33, Z = 4, D_x = 1.35.

Space group $C2$ (C_2^3) Atomic positions and isotropic temperature factors are given for the 41 non-hydrogen atoms in the structure. An anisotropic temperature factor is given for iodine. Assumed parameters are given for the hydrogen atoms.

Structure

The shape of the molecule is illustrated in Fig. 129. The bond lengths and angles (with standard deviations 0.04 to 0.07 Å, and 2 to 3°) are consistent with the formulation given above. The distance between the iodine atom and the carbonyl oxygen of an adjacent molecule is 3.1 Å. Other intermolecular distances are consistent with van der Waals interaction.

Details of analysis

Intensity data were recorded on equi-inclination Weissenberg photographs ($h0l$ to $h8l$, $hk0$ to $hk4$) using CuK_α radiation. 1996 reflexions were observed and estimated visually. Refinement was by block-diagonal least squares to a final R index of 0.12.

1. C.P. SAUNDERSON and D.C. HODGKIN, 1961. *Tetrahedron Letters*, No. 16, 573.

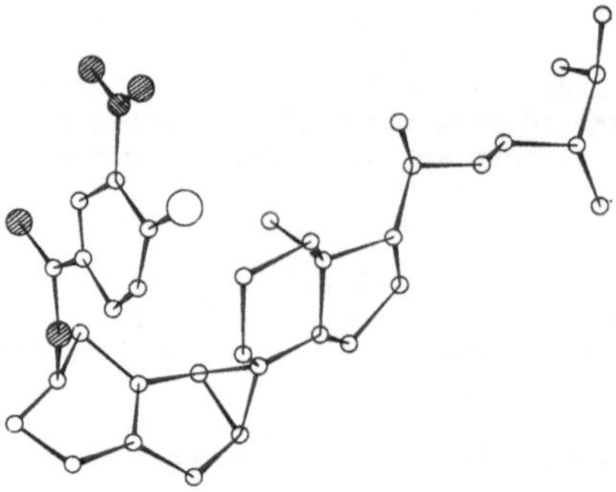

Fig. 129. A view of the $C_{35}H_{46}O_4NI$ molecule. Circles representing carbon, nitrogen, oxygen and iodine are in ascending order of size.

anti-7-NORBORNENYL p-BROMOBENZOATE

$C_{14}H_{13}O_2Br$ F.W. = 293.2

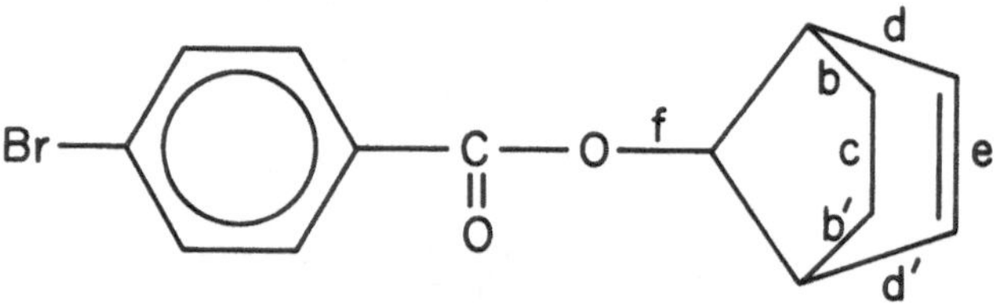

I. The crystal and molecular structure of anti-7-norbornenyl p-bromobenzoate.
A.C. MACDONALD and J. TROTTER, 1965. Acta Cryst., 19, 456-463.

Monoclinic, a = 8.81±2, b = 10.17±2, c = 14.10±3 Å, β = 99°51±5', U = 1245 Å³,
D_m = 1.54, Z = 4, D_x = 1.56, (CuK$_\alpha$, λ = 1.5418 Å).

Space group $P2_1/a$ (C_{2h}^5)

Atomic positions

	x	y	z
Br	0.2540	0.0733	0.2343
O(1)	0.1719	0.0652	0.7065
O(2)	0.0241	0.2393	0.6569
C(1')	0.2085	0.0945	0.3600
C(2')	0.2817	0.0171	0.4352
C(3')	0.2472	0.0349	0.5252
C(4')	0.1400	0.1264	0.5450
C(5')	0.0589	0.1976	0.4673
C(6')	0.0977	0.1841	0.3761
C(7')	0.1050	0.1501	0.6399
C(1)	0.2679	0.1716	0.8630
C(2)	0.2373	0.1383	0.9628
C(3)	0.1848	0.0149	0.9609
C(4)	0.1912	-0.0389	0.8620
C(5)	0.3645	-0.0501	0.8532
C(6)	0.4216	0.0951	0.8561
C(7)	0.1483	0.0861	0.8038

Standard deviations are given in full. Mean values are: Br, 0.002; O, 0.012; C, 0.016 Å. Anisotropic temperature factors are given also.

Structure

A view of the molecule, showing the numbering scheme, is given in Fig. 130. Bond lengths and angles are given in full, and are consistent with the formulation shown above. The *p*-bromobenzoate group is essentially planar. The norbornene nucleus has mirror symmetry within experimental error. Mean bond lengths and angles for the nucleus are:

a	=	1.51 Å		ab	=	101°
b	=	1.57		bc	=	103
c	=	1.56		ad	=	99
d	=	1.52		de	=	107
e	=	1.34		cd	=	105
f	=	1.44		aa'	=	96
				fa	=	112

Standard deviations are 0.02 to 0.03 Å, and 0.9 to 1.6°.

Details of analysis

Intensity data were measured with a four-circle diffractometer and scintillation counter using CuK_α radiation, in the range $2\theta < 145°$. 2059 of a possible 2506 reflexions were observed above background. Refinement was by block-diagonal least squares to a final *R* index of 0.18.

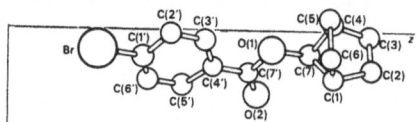

Fig. 130. A perspective drawing of the $C_{14}H_{13}O_2Br$ molecule, viewed along a*.

LONGIFOLENE HYDROCHLORIDE

$C_{15}H_{25}Cl$ F.W. = 240.8

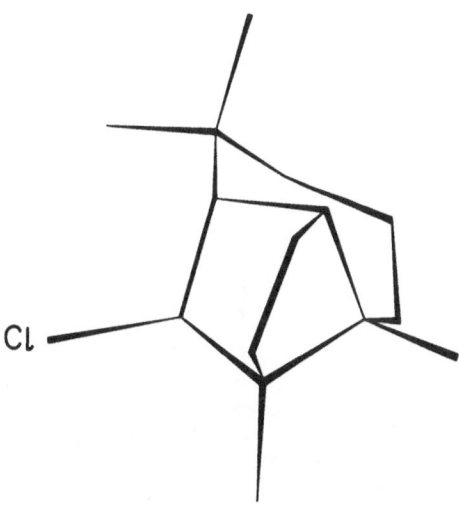

I.   The crystal structure of longifolene hydrochloride. A.F. CESUR and D.F. GRANT, 1965. *Acta Cryst.*, <u>18</u>, 55–59. (For an earlier report of two-dimensional work, see <u>1</u>.)

Orthorhombic, a = 8.505±3, b = 9.760±2, c = 16.674±3 Å, [U = 1384.1 Å³], D_m = 1.157, Z = 4, D_x = 1.154. (Cell dimensions for the isomorphous hydrobromide are given also. They are a = 8.53, b = 9.52, c = 16.75 Å.)

Space group $P2_12_12_1$ (D_2^4)

Atomic positions

	x	y	z
Cl	0.3364	−0.0694	0.2037
C(1)	0.3004	0.1092	0.3334
C(2)	0.1492	0.1953	0.3341
C(3)	0.0131	0.1387	0.3865
C(4)	0.0549	0.0800	0.4668
C(5)	0.1452	−0.0549	0.4694
C(6)	0.3259	−0.0480	0.4461
C(7)	0.3775	−0.1272	0.3673
C(8)	0,2730	−0.0406	0.3065
C(9)	0.5496	−0.0905	0.3594
C(10)	0.5539	0.0545	0.3937
C(11)	0.3872	0.0923	0.4159
C(12)	0.1902	0.3425	0.3628
C(13)	0.0847	0.2050	0.2494
C(14)	0.4232	−0.1027	0.5187
C(15)	0.3243	−0.2829	0.3619

(Original in Å). Mean standard deviations are: Cl, 0.004 Å; C, 0.010 to 0.019 Å. Temperature factors are given, as well as the positions of the hydrogen atoms.

Structure

The configuration of the molecule is as indicated above. Bond lengths and angles are given in full; these are consistent with a single-bonded but somewhat strained system. Intermolecular distances have normal van der Waals values.

Details of analysis

The intensities from a spherical crystal were recorded on equi-inclination Weissenberg photographs (0kl to 7kl; $h0l$) using CuK_α radiation. 1219 reflexions, constituting about 77% of those accessible, were observed. Refinement was by least squares to a final R index of 0.13.

1.   *Structure Reports*, <u>17</u>, 774.

2-BROMO-α-SANTONIN

$C_{15}H_{17}BrO_3$ F.W. = 325.2

I. Sesquiterpenoids. Part III. The stereochemistry of santonin: X-ray analysis of 2-bromo-α-santonin. J.D.M. ASHER and G.A. SIM, 1965. *J. Chem. Soc.*, 6041-6055. (For a preliminary account, see <u>1</u>.)

Orthorhombic, a = 7.34, b = 23.34, c = 8.28 Å, U = 1418 Å^3, D_m = 1.525, z = 4, D_x = 1.522.

Space group $P2_12_12_1$ (D_2^4)

Atomic positions

	x	y	z
C(1)	0.2283	0.2590	0.1818
C(2)	0.2292	0.2188	0.2920
C(3)	0.2309	0.2332	0.4720

	x	y	z
C(4)	0.2326	0.2944	0.5043
C(5)	0.2391	0.3349	0.3926
C(6)	0.2668	0.3980	0.4132
C(7)	0.4452	0.4146	0.3400
C(8)	0.4485	0.4067	0.1568
C(9)	0.4087	0.3444	0.1238
C(10)	0.2322	0.3191	0.2096
C(11)	0.4672	0.4730	0.4175
C(12)	0.3948	0.4633	0.5779
C(13)	0.6763	0.4966	0.4267
C(14)	0.2214	0.3045	0.6881
C(15)	0.0552	0.3468	0.1346
O(16)	0.2234	0.1986	0.5682
O(17)	0.2712	0.4196	0.5775
O(18)	0.4189	0.4876	0.7097
Br	0.2255	0.1381	0.2453

Standard deviations (in Å) are given in full. Mean values are: Br, 0.003; O, 0.025; C, 0.03 Å. Anisotropic temperature factors are given also.

Structure

The analysis establishes the structure and relative stereochemistry shown above. The methyl group attached to C(11) is α to the lactone ring, which has the envelope conformation with C(7) displaced 0.61 Å from the plane of C(6), C(11), C(12), O(17), and O(18). Bond lengths and angles are given in full.

Details of analysis

Intensity data were recorded on equi-inclination Weissenberg photographs ($hk0$ to $hk6$, CuK$_\alpha$ radiation), and precession photographs ($0kl$, MoK$_\alpha$ radiation). 1277 reflexions were observed and estimated visually. Refinement was by least squares to a final R index of 0.152.

1. J.D.M. ASHER and G.A. SIM, 1962. *Proc. Chem. Soc.*, 335.

2-BROMODIHYDROISOPHOTO-α-SANTONIC LACTONE ACETATE

$C_{17}H_{23}BrO_5$

F.W. = 387.3

M.P. = 117 – 118° (decomp.)

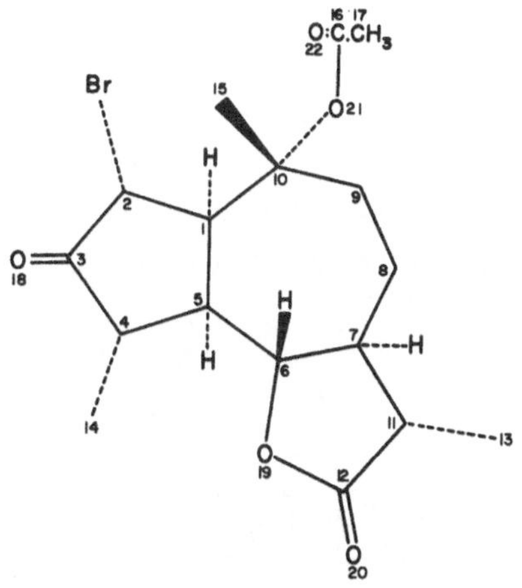

I. Sesquiterpenoids. Part II. The stereochemistry of isophotosantonic lactone: X-ray analysis of 2-bromo-dihydroisophoto-α-santonic lactone acetate. J.D.M. ASHER and G.A. SIM, 1965. *J. Chem. Soc.*, 1584-1594. (For a preliminary report, see 1.)

Orthorhombic, a = 11.05, b = 19.23, c = 7.93 Å, U = 1685 Å^3, z = 4, D_x = 1.526.

Space group $P2_12_12_1$ (D_2^4)

Atomic positions

	x	y	z
C(1)	−0.1819	0.2011	0.2631
C(2)	−0.2895	0.2463	0.2763
C(3)	−0.3729	0.2225	0.1280
C(4)	−0.3373	0.1510	0.0764
C(5)	−0.2003	0.1534	0.1048
C(6)	−0.1416	0.0812	0.0943
C(7)	−0.0084	0.0781	0.1596
C(8)	0.0001	0.0731	0.3521
C(9)	−0.0316	0.1378	0.4447
C(10)	−0.1662	0.1655	0.4252
C(11)	0.0378	0.0155	0.0842
C(12)	−0.0310	0.0175	−0.0866
C(13)	0.1728	0.0094	0.0556
C(14)	−0.3852	0.1280	−0.0883
C(15)	−0.2572	0.1115	0.4872
C(16)	−0.1140	0.2754	−0.4104
C(17)	−0.1712	0.3240	−0.2675
O(18)	−0.4549	0.2587	0.0461
O(19)	−0.1291	0.0605	−0.0723
O(20)	−0.0061	0.0056	−0.2318
O(21)	−0.1841	0.2188	−0.4292
O(22)	−0.0242	0.2911	−0.4832
Br	−0.2576	0.3462	0.2386

Standard deviations are given in full, in Å. Mean values are: Br, 0.003; O, 0.020; C, 0.028 Å. Anisotropic temperature factors are given also.

Structure

The analysis establishes the relative stereochemistry depicted above. (The absolute configuration, known from chemical considerations, is also correctly represented.) Bond lengths and angles are given in full. The cycloheptane ring adopts a 'flattened chair' conformation, with average valence angles of 115°. The α-lactone ring adopts an 'envelope' conformation, with C(7) displaced 0.60 Å from the plane of the remaining ring atoms.

Details of analysis

Intensity data were recorded on equi-inclination Weissenberg photographs ($hk0$ to $hk4$, CuK$_\alpha$ radiation), and precession photographs ($0kl$ and $h0l$, MoK$_\alpha$ radiation). 865 reflexions were observed and estimated visually. Refinement was by least squares to a final R index of 0.129.

1.   J.D.M. ASHER and G.A. SIM, 1962. *Proc. Chem. Soc.*, 111.

METHYL MALALEUCATE IODOACETATE

$C_{34}H_{51}O_6I$ F.W. = 682.7

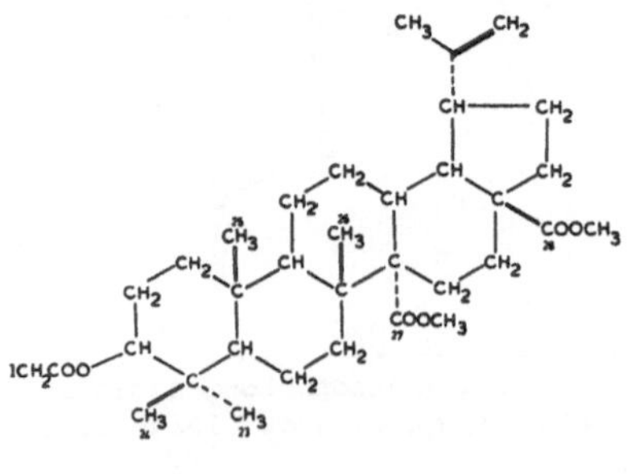

I. The determination of the crystal structure of methyl malaleucate iodoacetate. S.R. HALL and E.N. MASLEN, 1965. *Acta Cryst.*, 18, 265–279.

Orthorhombic, $a = 15.719 \pm 4$, $b = 24.533 \pm 7$, $c = 8.618 \pm 5$ Å, $U = 3323.3$ Å^3, $D_m = 1.37$, $Z = 4$, $D_x = 1.366$, (CuK$_\alpha$, $\lambda = 1.5418$ Å).

Space group P2$_1$2$_1$2$_1$ (D_2^4)

Atomic positions

Atomic positions are given in full. Temperature factors (some of them aniso-tropic) are given for the non-hydrogen atoms. [The thermal motion of some atoms appears to be very severe and anisotropic.]

Structure

Bond lengths and angles are given in full, and confirm the formulation given above. Some C-C distances are rather long, and possible reasons for this are discussed. Intermolecular distances are consistent with van der Waals inter-action.

Details of analysis

Intensity data were recorded on equi-inclination Weissenberg photographs (*hk*0 to *hk*7; 0*kl* to 2*kl*) using CuK$_\alpha$ radiation. Of 3795 independent reflexions in the copper sphere 1342 were observed, and estimated visually. *hkl* and $\overline{hkl}$ reflexions were measured separately. The structure was determined directly, using phases calculated from anomalous scattering considerations. Refinement was by block-diagonal least squares to a final *R* index of 0.079.

N(a)-ACETYL-7-ETHYL-5-DESETHYL-
ASPIDOSPERMIDINE N(b)-METHIODIDE

C$_{22}$H$_{31}$N$_2$OI   F.W. = 466.3

I. Cyclization in the cleavamine series: X-ray analysis of N(a)-Acetyl-7-ethyl-5-desethyl-aspidospermidine N(b)-methiodide. A. CAMERMAN, N. CAMERMAN and J. TROTTER, 1965. *Acta Cryst.*, 19, 314–320.

Orthorhombic, $a = 9.32 \pm 3$, $b = 11.28 \pm 3$, $c = 19.70 \pm 4$ Å, $U = 2071$ Å^3, $D_m = 1.493$, $Z = 4$, $D_x = 1.495$. (CuK$_\alpha$, $\lambda = 1.5418$ Å; MoK$_\alpha$, $\lambda = 0.7107$ Å).

Space group P2$_1$2$_1$2$_1$ (D_2^4)

Atomic positions

	x	y	z
I	0.9391±2	0.0074±2	0.4443±1
N(1)	0.3433±21	−0.0846±17	0.2379±10
C(2)	0.3339±22	−0.0836±18	0.3142±10
C(3)	0.3615±27	−0.2122±22	0.3395±13
C(4)	0.3769±26	−0.2101±22	0.4188±12
C(5)	0.5082±24	−0.1370±19	0.4376±13
C(6)	0.5321±24	−0.1291±20	0.5158±11
C(7)	0.4236±20	−0.0541±16	0.5518±12
C(8)	0.3330±24	0.0290±20	0.5059±11
N(9)	0.4277±20	0.0828±14	0.4519±9
C(10)	0.3203±26	0.1501±20	0.4048±11
C(11)	0.4011±24	0.1370±19	0.3333±12
C(12)	0.4526±23	0.0057±23	0.3331±10
C(13)	0.5599+21	−0.0132±18	0.2780±9
C(14)	0.7029±25	0.0211±22	0.2759±11

	x	y	z
C(15)	0.7765±26	0.0089±32	0.2121±11
C(16)	0.7039±27	−0.0283±21	0.1559±12
C(17)	0.5580±30	−0.0586±21	0.1563±13
C(18)	0.4878±26	−0.0555±20	0.2193±12
C(19)	0.5141±21	−0.0092±23	0.4071±10
C(20)	0.2999±29	−0.1334±23	0.5903±13
C(21)	0.3630±30	−0.1924±24	0.6554±14
C(22)	0.2477±27	−0.1410±21	0.1958±13
C(23)	0.1015±29	−0.1700±23	0.2230±14
O(24)	0.2829±20	−0.1627±16	0.1364±9
C(25)	0.5372±30	0.1700±22	0.4829±13

Anisotropic temperature factors are given also.

Structure

The stereochemistry and correct absolute configuration of the molecule are illustrated in Fig. 131. Bond lengths and angles are given in full, and are consistent with the name and formulation. The packing is illustrated in Fig. 132. There are no unusually short intermolecular contacts, and the shortest iodine-molecule distance (I---C) is 3.87 Å.

Details of analysis

The intensity data were measured with a four-circle diffractometer and scintillation counter, using MoK_α radiation. 1018 independent reflexions (about 80% of those accessible in the range $2\theta \leqslant 40°$) were observed above background. Refinement was by block-diagonal least squares to a final R index of 0.066. The absolute configuration was deduced by studying the effect of the anomalous scattering of the iodine atom on thirteen sets of symmetry-related reflexions.

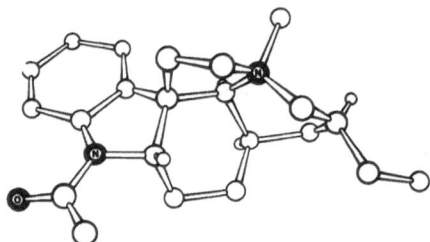

Fig. 131. Perspective drawing of $C_{22}H_{31}N_2OI$; the correct absolute configuration is shown.

CARACURINE-II DIMETHIODIDE

$C_{40}H_{44}I_2N_4O_2$ (Formula on p. 231)   F.W. = 866.6

I. The structure of caracurine-II: X-ray analysis of caracurine-II dimethiodide.
   A.T. McPHAIL and G.A. SIM, 1965. *J. Chem. Soc.*, 1663-1675.
   (For a preliminary report, see 1.)

Orthorhombic, $a = 18.59$, $b = 27.44$, $c = 7.52$ Å, $U = 3836$ Å^3, $D_m = 1.505$, $Z = 4$, $D_x = 1.502$.

Space group $P2_12_12_1$ (D_2^4)

Atomic positions

Atomic positions and temperature factors (anisotropic for iodine) are given for the 48 non-hydrogen atoms. Mean standard deviations are: I, 0.008; other, 0.07 Å.

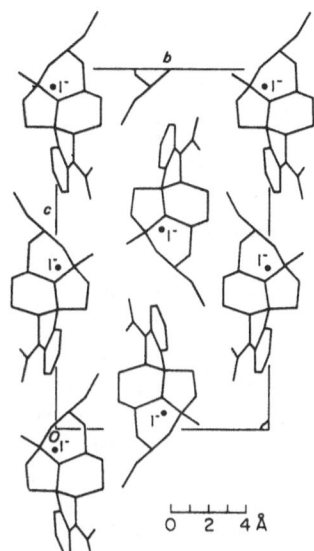

Fig. 132. Projection of the $C_{22}H_{31}N_2OI$ structure along [100].

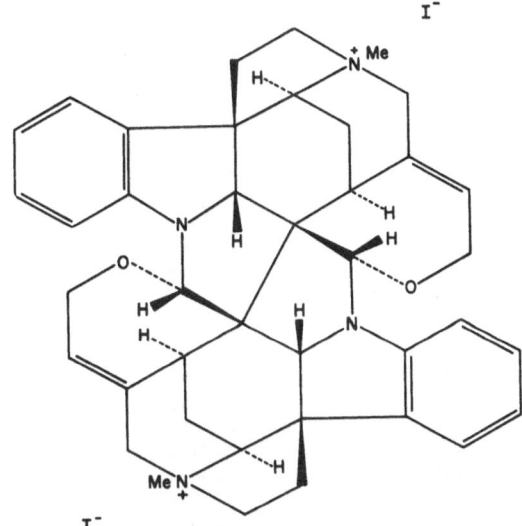

Formula for caracurine-II dimethiodide.

Structure

The analysis establishes the constitution and relative stereochemistry of the compound to be as depicted above. The molecule appears to have two-fold symmetry. The conformation is discussed in detail.

Details of analysis

Intensity data were recorded on equi-inclination Weissenberg photographs ($hk0$ to $hk5$) using CuK_α radiation. 1285 reflexions were observed and estimated visually. Refinement was by Fourier and least-squares methods to a final R index of 0.181.

1.   A.T. McPHAIL and G.A. SIM, 1961. *Proc. Chem. Soc.*, 416.

CHIMONANTHINE DIHYDROBROMIDE

$C_{22}H_{28}Br_2N_4$

F.W. = 508.3
M.P. = 188-189°C

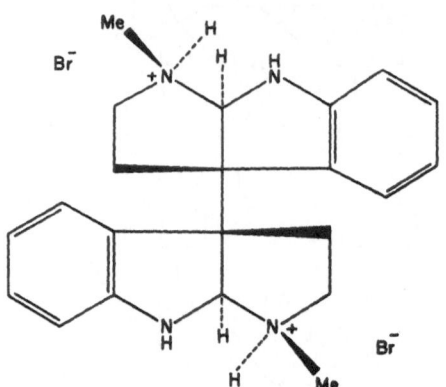

I. The structure of chimonanthine: X-ray analysis of chimonanthine dihydrobromide.
I.J. GRANT, T.A. HAMOR, J. M. ROBERTSON and G.A. SIM, 1965.   *J. Chem. Soc.*,
5678-5696. (For a preliminary report, see 1.)

Tetragonal, a = 13.95, c = 26.67 Å, U = 5190 Å^3, D_m = 1.35, z = 8, D_x = 1.31.

Space group $P4_12_12$ (D_4^4) or $P4_32_12$ (D_4^8)

Atomic positions

	x	y	z
C(1)	0.7967	0.4854	0.4351
C(2)	0.7670	0.5813	0.4280
C(3)	0.6668	0.6053	0.4330
C(4)	0.6007	0.5361	0.4435
C(5)	0.6302	0.4449	0.4527
C(7)	0.6303	0.2788	0.4689
C(9)	0.6774	0.2363	0.3850
C(10)	0.7690	0.2581	0.4115
C(11)	0.7340	0.3136	0.4600
C(12)	0.7232	0.4192	0.4479
C(13)	0.5073	0.1977	0.4127
C(1')	0.7016	0.3649	0.5779
C(2')	0.6986	0.4311	0.6161
C(3')	0.7823	0.4912	0.6248
C(4')	0.8641	0.4812	0.5988
C(5')	0.8582	0.4132	0.5578
C(7')	0.9096	0.3051	0.5048
C(9')	0.8899	0.1697	0.5449
C(10')	0.7932	0.1830	0.5187
C(11')	0.8022	0.2895	0.5057
C(12')	0.7815	0.3584	0.5487
C(13')	0.0587	0.2324	0.5380
N(6)	0.5710	0.3692	0.4647
N(8)	0.6039	0.2065	0.4268
N(6')	0.9423	0.3873	0.5262
N(8')	0.9607	0.2134	0.5122
Br(I)	0.4868	0.1069	0.5537
Br(II)	0.9715	0.0758	0.4173

Standard deviations are given in full. Mean values are: Br, 0.006; other
atoms, 0.05 Å.

Structure

The analysis establishes the relative stereochemistry indicated above.
(However, in the crystal the molecule is closer to the *cis* conformation than to
the *trans* conformation shown.) Bond lengths and angles are given in full, but
are not very accurate.

Details of analysis

Intensity data were recorded on equi-inclination Weissenberg photographs (*0kl*
to 9*kl*) using Cu*K*$_\alpha$ radiation. 2093 reflexions were observed and estimated visually.
Refinement was by least squares to a final *R* index of 0.149.

1.   I.J. GRANT, T.A. HAMOR, J.M. ROBERTSON and G.A. SIM, 1962. *Proc. Chem. Soc.*,
     148.

CORYMINE HYDROBROMIDE MONOHYDRATE

$C_{22}H_{27}N_2O_4Br.H_2O$ F.W. = 481.4

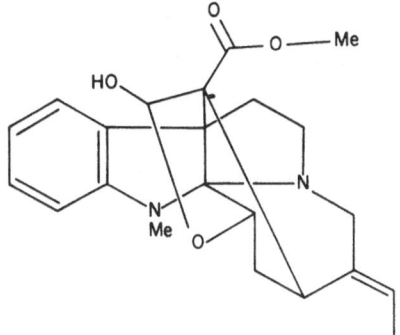

I.   The seed alkaloids of *Hunteria umbellata:* the X-ray crystal structure of
     corymine hydrobromide monohydrate. C.W.L. BEVAN, M.B. PATEL, A.H. REES,
     D.R. HARRIS, M.L. MARSHAK and H.H. MILLS, 1965. *Chem. and Ind., G.B.*,
     <u>14</u>, 2006-2008.

Orthorhombic, a = 13.99, *b* = 16.84, *c* = 9.56 Å, [*U* = 2253 Å^3], *Z* = 4,
[*D$_x$* = 1.42].

Space group *P*2$_1$2$_1$2$_1$ (*D*$_2^4$)

Atomic positions

Atomic positions are not given.

Structure

The analysis establishes that the structure and absolute configuration of
the corymine molecule are as represented above.

Details of analysis

The intensity data were measured with a four-circle diffractometer and scintillation counter, using Cu*K*$_\alpha$ radiation. 1460 of 1528 accessible reflexions were observed. The relative phases were assigned by measuring the differences between
I_{hkl} and $I_{h\bar{k}l}$; this procedure leads to the correct absolute configuration of the
molecule. Refinement was by least squares to a final *R* index of 0.083.

HETERATISINE HYDROBROMIDE MONOHYDRATE

$C_{22}H_{33}O_5N.HBr.H_2O$ F.W. = 490.4

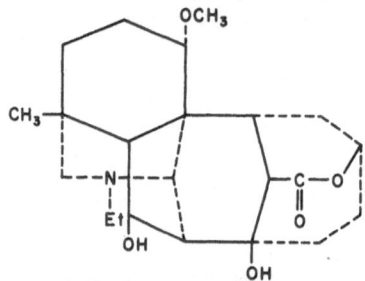

I. The crystal structure of heteratisine hydrobromide monohydrate. M. PRZYBYLSKA,
   1965. *Acta Cryst.*, <u>18</u>, 536–540. (For a preliminary account, see <u>1</u>.)

Monoclinic, a = 8.97±2, *b* = 14.54±3, *c* = 8.55±2 Å, β = 90°40±10', *U* = 1114 Å^3,
D_m = 1.459, *Z* = 2, D_x = 1.462.

Space group P2$_1$ (C_2^2) (The material is optically active.)

Atomic positions

   Atomic positions and temperature factors are given for the 30 non-hydrogen
atoms. For the bromine atom the temperature factor is anisotropic.

Structure

   The analysis confirms the formulation given above. Bond lengths and angles
are given in full, with standard errors ranging from 0.04 to 0.08 Å, and 3 to 5°.
The structure is stabilized by intra- and intermolecular hydrogen bonds as indicated in Fig. 133. The absolute configuration has not been determined, but that
represented in the Figure is consistent with that found for closely related diter-
penoid alkaloids.

Details of analysis

   The intensity data were measured with a linear diffractometer (*h*0*l* to *h*12*l*;
0*kl* to 2*kl*) using MoK$_\alpha$ radiation. 1345 reflexions were observed; this is about
52% of those obtainable with copper radiation. Refinement was by block-diagonal
least squares to a final *R* index of 0.12.

1. M. PRZYBYLSKA, 1963. *Canad. J. Chem.*, <u>41</u>, 2911.

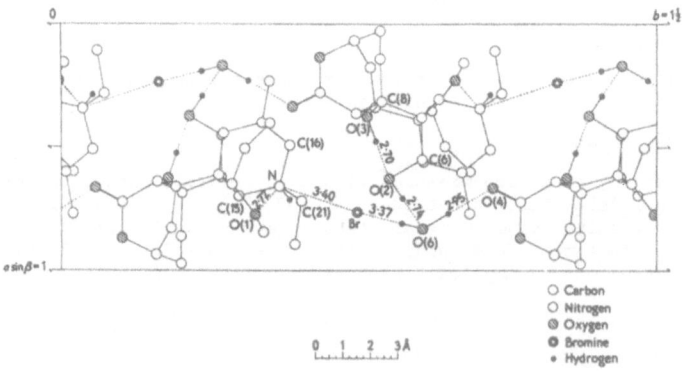

Fig. 133. The heteratisine hydrobromide monohydrate structure projected along *c*,
           showing hydrogen bonds and bromine contacts.

HUNTERBURNINE β-METHIODIDE

$C_{20}H_{27}IN_2O_2$ HO

I⁻   F.W. = 454.3
      M.P. = 277–280°C

I. The structure of hunterburnine: X-ray analysis of hunterburnine β-methiodide.
J.D.M. ASHER, J.M. ROBERTSON and G.A. SIM, 1965. *J. Chem. Soc.*, 6355-6367.
(For a preliminary report, see 1.)

Orthorhombic, a = 10.96, b = 18.83, c = 9.20 Å, U = 1899 Å³, D_m = 1.602, z = 4,
D_x = 1.589.

Space group $P2_12_12_1$ (D_2^4)

Atomic positions

	x	y	z
N(1)	0.8259	-0.0522	0.1763
C(2)	0.7180	-0.0311	0.1392
C(3)	0.6798	0.0541	0.1397
N(4)	0.5446	0.0560	0.1726
C(5)	0.4756	0.0107	0.0597
C(6)	0.4951	-0.0678	0.0867
C(7)	0.6347	-0.0776	0.1195
C(8)	0.7008	-0.1499	0.1147
C(9)	0.6679	-0.2153	0.1034
C(10)	0.7605	-0.2675	0.1277
C(11)	0.8765	-0.2466	0.1508
C(12)	0.9158	-0.1793	0.1878
C(13)	0.8274	-0.1312	0.1623
C(14)	0.7663	0.1015	0.2492
C(15)	0.7237	0.1764	0.2116
C(16)	0.5833	0.1809	0.2387
C(17)	0.5051	0.1360	0.1667
C(18)	0.7490	0.3298	0.1722
C(19)	0.7482	0.2923	0.0660
C(20)	0.7679	0.2065	0.0582
C(21)	0.9061	0.1945	0.0238
C(22)	0.5176	0.0317	0.3401
O(23)	0.9134	0.1184	-0.0264
O(24)	0.7213	-0.3397	0.1020
I	0.1278	0.0357	0.1841

Standard deviations are given in full, (in Å), and range from 0.02 to 0.06 Å
for carbon and oxygen atoms. Anisotropic temperature factors are given also.

Structure

The analysis establishes the structure and relative stereochemistry shown
above. Bond lengths and angles are given in full. The structure is stabilized
by a system of OH..I and NH..I bonds.

Details of analysis

Intensity data were recorded on equi-inclination Weissenberg photographs (*hk*0 to *hk*7, CuK$_\alpha$ radiation), and precession photographs (0*kl*, MoK$_\alpha$ radiation). 1622 reflexions were observed and estimated visually. Refinement was by least squares to a final *R* index of 0.154.

1.   *Structure Reports*, 27, 1008.

LUNARINE HYDROBROMIDE MONOHYDRATE
LUNARINE HYDROIODIDE MONOHYDRATE

$C_{25}H_{32}N_3O_4X.H_2O$

   X = Br
   X = I

                                        F.W. = 536.5
                                        F.W. = 583.5

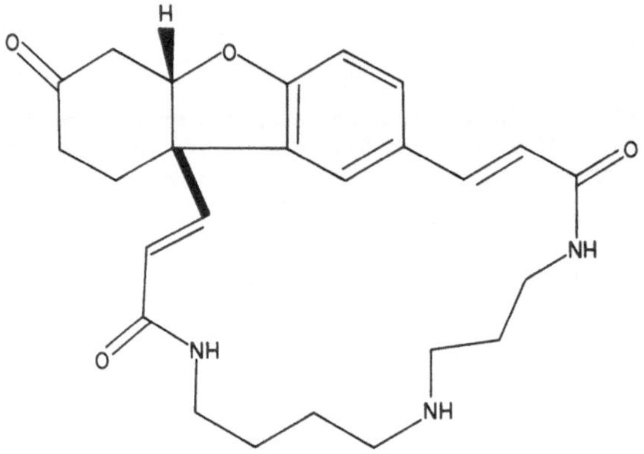

I. Lunarine. C. TAMURA, G.A. SIM, J.A.D. JEFFREYS, P. BLADON and G. FERGUSON, 1965. *Chem. Communic., G.B.*, 20, 485–486.

LUNARINE HYDROBROMIDE

Monoclinic, a = 11.37, *b* = 10.96, *c* = 12.41 Å, β = 120.3°, [*U* = 1336 Å^3], *Z* = 2, D_x = 1.34.

Space group $P2_1$ (C_2^2)

LUNARINE HYDROIODIDE

Orthorhombic, a = 11.60, *b* = 16.53, *c* = 14.01 Å, [*U* = 2686 Å^3], *Z* = 4, D_x = 1.44.

Space group $P2_12_12_1$ (D_2^4)

Atomic positions

Atomic positions are not given for either compound.

Structure

The analyses confirm that lunarine has the constitution and relative stereo-chemistry depicted above.

Details of analysis

For both compounds three-dimensional intensity data were recorded on equi-inclination Weissenberg photographs and estimated visually. Refinement was by least squares; the final *R* index for the hydrobromide is 0.137.

o-METHYLLYTHRINE HYDROBROMIDE

$C_{27}H_{32}BrNO_5 \cdot CH_3OH$ F.W. = 562.2

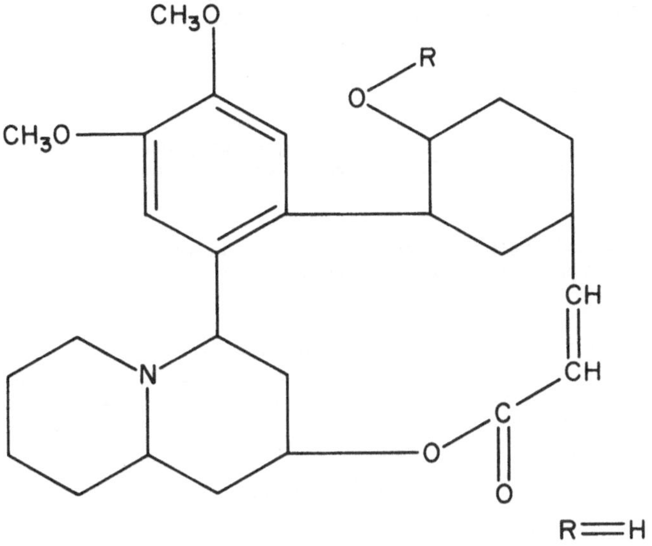

R══H

I. The structure of o-methyllythrine hydrobromide. D.E. ZACHARIAS, G.A. JEFFREY, B. DOUGLAS, J.A. WEISBACH, J.L. KIRKPATRICK, J.P. FERRIS, C.B. BOYCE and R.C. BRINER, 1965. *Experientia, Suisse*, <u>21</u>, 247-248.

Orthorhombic, $a = 11.58$, $b = 9.64$, $c = 22.86$ Å, $U = 2545$ Å^3, $z = 4$, $D_x = 1.47$.

Space group $P2_12_12_1$ (D_2^4)

Atomic positions

Atomic positions are not given.

Structure

The configuration of the non-hydrogen skeleton of the molecule is illustrated in Fig. 134. Oxidative studies have shown that the parent compound lythrine is as above, but with $R = H$.

Details of analysis

The intensity data were recorded on equi-inclination Weissenberg photographs about the a and b axes, using CuK_α radiation. 1226 reflexions were observed and measured visually. Refinement was by Fourier methods to a final R index of 0.22. The solvent molecule was disordered, and only its oxygen atom was resolved.

MITRAGYNINE HYDROIODIDE

$C_{23}H_{31}IN_2O_4$ F.W. = 526.4

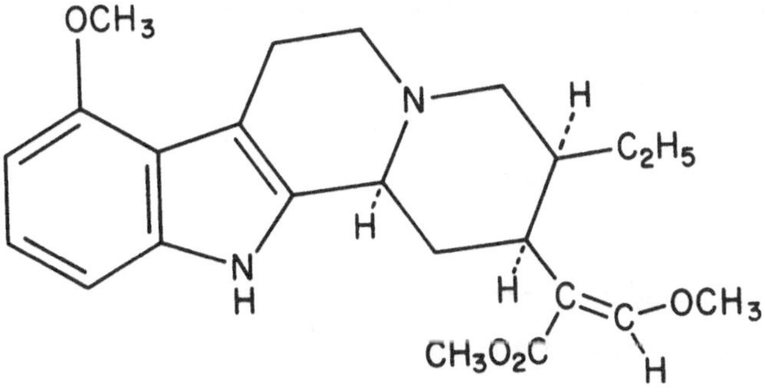

I. The structure of mitragynine hydroiodide. D.E. ZACHARIAS, R.D. ROSENSTEIN, and G.A. JEFFREY, 1965. *Acta Cryst.*, <u>18</u>, 1039-1043.

Orthorhombic, a = 11.5±5, b = 7.87±5, c = 26.69±5 Å, U = 2418 Å^3, D_m = 1.445, Z = 4, D_x = 1.445. (Corresponding values for the isotypic hydrobromide are: a = 11.41, b = 7.48, c = 27.42 Å, U = 2339 Å^3, D_m = 1.368, Z = 4, D_x = 1.360.)

Space group $P2_12_12_1$ (D_2^4)

Atomic positions

Atomic positions are given for the 30 non-hydrogen atoms. Standard deviations are 0.001 Å for iodine, and 0.013 to 0.027 Å for the other atoms.

Structure

Bond lengths and angles are given in full, and are in reasonable accord with standard values. The configuration is illustrated in Fig. 135.

Details of analysis

The intensity data were recorded on equi-inclination Weissenberg photographs ($0kl$ to $6kl$; $h0l$ to $h4l$) using CuK$_\alpha$ radiation. 1967 independent reflexions were observed and measured by visual estimation. Refinement was by successive differential syntheses to a final R index of 0.124.

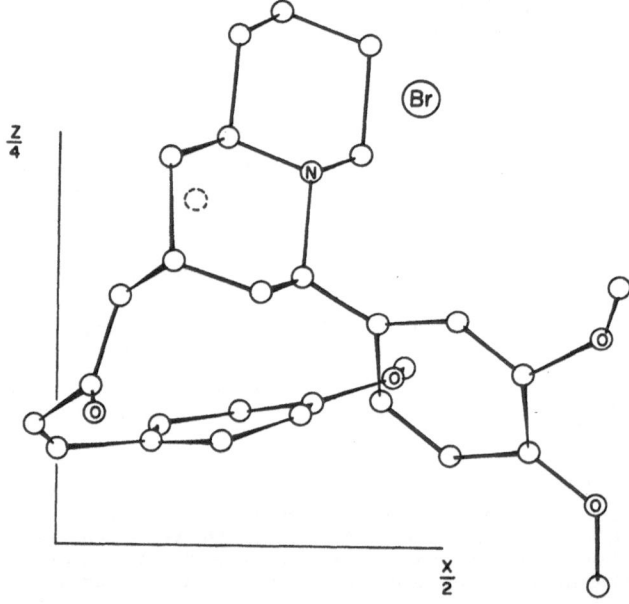

Fig. 134. The structure of *o*-methyllythrine hydrobromide, viewed along b. The broken circle represents the oxygen atom of the disordered solvent molecule.

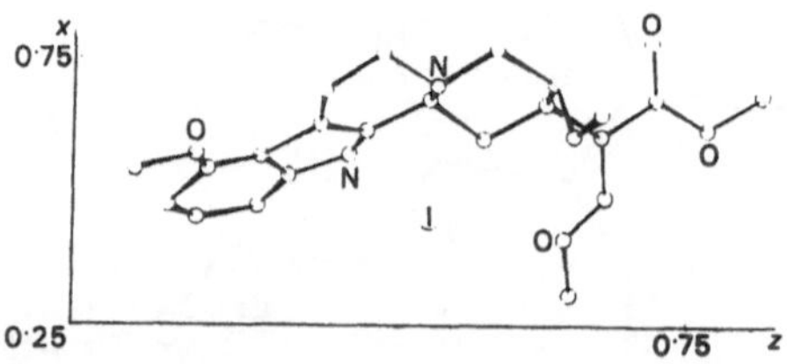

Fig. 135. The molecule mitragynine hydroiodide viewed along b.

SECURININE HYDROBROMIDE DIHYDRATE

$C_{13}H_{15}O_2N.HBr.2H_2O$ F.W. = 332.2

I. The crystal structure of securinine hydrobromide dihydrate and the molecular structure of securinine. S. IMADO, M. SHIRO and Z.-I. HORII, 1965. *Chem. Pharm. Bull., Tokyo*, <u>13</u>, 643–651.

Monoclinic, a = 7.71, b = 14.83, c = 7.04 Å, γ = 112.13°, [U = 745.6 Å^3], D_m = 1.492, Z = 2, D_x = 1.488. (Crystal data for the anhydrous material, and for the monohydrate are given also, and are reported in <u>1</u>.)

Space group $P2_1$ (C_2^2)   Note that c is the unique axis.

Atomic positions

	x	y	z
Br	0.3950	0.2114	0.9963
C(7)	0.0056	0.1173	0.5936
C(8)	0.8997	0.0777	0.7805
C(9)	0.8294	0.1561	0.8845
C(10)	0.8504	0.2393	0.7449
C(10a)	0.0541	0.2912	0.6893
C(5a)	0.2458	0.2650	0.4453
C(11)	0.1144	0.3033	0.3503
C(10b)	0.0890	0.3623	0.5068
C(3a)	0.2721	0.4509	0.5377
C(4)	0.4396	0.4295	0.5401
C(5)	0.4313	0.3379	0.4828
C(3)	0.2351	0.5311	0.5443
C(2)	0.0288	0.5009	0.5203
N(6)	0.1407	0.2182	0.6259
O	0.9362	0.5496	0.5271
O(1)	0.9499	0.4013	0.5089
O(H_2O)	0.6035	0.1126	0.3254
O(H_2O)	0.3837	0.0530	0.6482

The mean standard deviations are: Br, 0.0016 Å; O, 0.015 Å; N, 0.021 Å; C, 0.024 Å.

Structure

The analysis confirms that the chemical structure and absolute configuration of the securinine molecule are as represented above. The piperidine ring is boat-shaped, and the two double bonds C(3)=C(3a) and C(4)=C(5) are not coplanar. The bromine atom lies 3.29 Å from the nitrogen atom; the arrangement of N-C bonds and the N...Br vector is approximately tetrahedral, and N-H-Br bonding is inferred. The bromine atom and the water molecules are held together by hydrogen bonds to form spiral chains extended along c.

Details of analysis

The intensity data were recorded on equi-inclination Weissenberg photographs (*hk*0 to *hk*5) using CuK_α radiation. 908 of 1162 accessible reflexions were observed and measured visually. Refinement was by least squares to a final R index of about

0.12. The absolute configuration was deduced by comparing the effect of the anomalous scattering of the bromine atom on some pairs of symmetry-related reflexions.

1.   This volume, p. 427.

SPORIDESMIN

(METHYLENE DIBROMIDE ADDUCT OF SPORIDESMIN)

$C_{18}H_{20}O_6N_3S_2Cl.(0.65CH_2Br_2)$ F.W. = 587

I. The structure of the methylene dibromide adduct of sporidesmin at -150°C.
   J. FRIDRICHSONS and A.McL. MATHIESON, 1965. *Acta Cryst.*, <u>18</u>, 1043-1052.
   (For a preliminary report, see <u>1</u>.)

Orthorhombic, a = 9.681, b = 10.629, c = 23.358 Å [U = 2404 Å³], Z = 4, [D_x = 1.62], (all at -150°C). (The values at room temperature are: a = 9.64, b = 10.58, c = 23.88 Å, D_m = 1.65.)

Space group $P2_12_12_1$ (D_2^4)

Atomic positions

     Atomic positions are given for the 33 non-hydrogen atoms, but there is no indication of their accuracy.

Structure

     The shape of the molecule is indicated in Fig. 136 and 137. Bond lengths and angles are given in full. They are believed to be sufficiently accurate to establish the chemical formulation given above. Intermolecular distances are given. One of these, between an oxygen atom and a hydroxyl group, is 2.89 Å, and suggests a hydrogen bond.

Details of analysis

     Intensity data were recorded on equi-inclination Weissenberg photographs (0kl to 7kl; h0l) at -150°C, using CuK_α radiation. 1925 reflexions, about 80% of those accessible, were observed. During the structure determination (which was by Patterson methods) it was discovered that the site occupancy of the methylene bromide molecule was only 0.65. Refinement was by full-matrix least squares to a final R index of 0.144.

1.   J. FRIDRICHSONS and A. McL. MATHIESON, 1962. *Tetrahedron Letters, No. 26,* 1265.

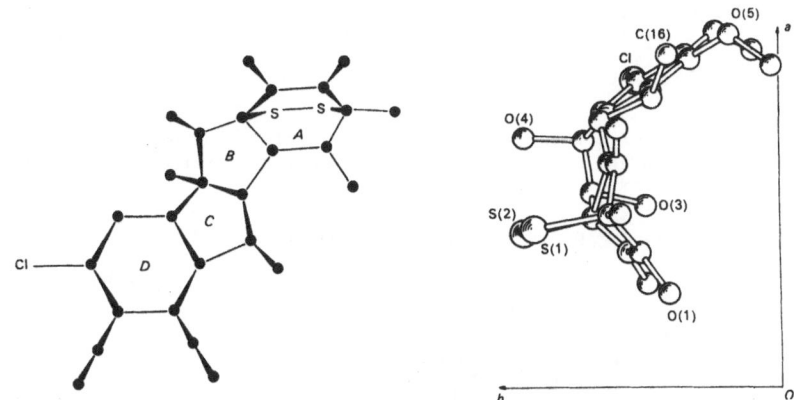

Fig. 136. Left: Sterical diagram of sporidesmin molecule.
Fig. 136. Right: A view of the sporidesmin molecule down the c axis.

VILLALSTONINE

$C_{41}H_{48}O_4N_4$ F.W. = 661

I. The structure of villalstonine.   C.E. NORDMAN and S.K. KUMRA, 1965.   *J. Amer. Chem. Soc.*, **87**, 2059-2060.

Monoclinic, a = 13.756, b = 13.645, c = 10.045 Å, β = 101°41', [U = 1846 Å^3], Z = 2, D_x = 1.25 (The unit cell also contains two molecules of methanol.)

Space group $P2_1$ (C_2^2)

Atomic positions

   Atomic positions are not given.

Structure

The molecular structure is as depicted above. From chemical considerations it is concluded that the absolute configuration of the molecule is correctly represented.

Details of analysis

The intensity data were recorded on integrated oscillation photographs. 3337 reflexions were observed and measured with a densitometer. Refinement was by least squares to a final R index of 0.059.

N(1')-CARBOXYBIOTIN-DI-(p-BROMOANILIDE)

$C_{23}H_{24}O_3N_4SBr_2$ F.W. = 596

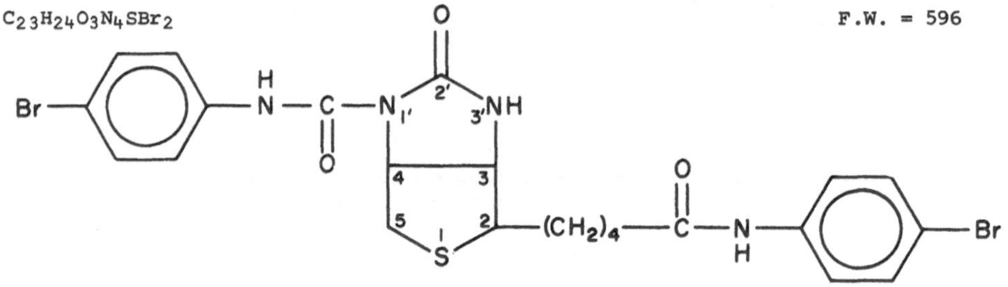

I. Structure of the bis-p-bromoanilide of carbon dioxide biotin. C. BONNEMERE, J.A. HAMILTON, L.K. STEINRAUF and J. KNAPPE, 1965. *Biochemistry, U.S.A.*, 4, 240-245.

Orthorhombic, $a = 19.34 \pm 1$, $b = 17.23 \pm 2$, $c = 7.04 \pm 1$ Å, $[U = 2346$ Å$^3]$, $D_m = 1.65$, $Z = 4$, $D_x = 1.688$.

Space group $P2_1 2_1 2_1$ (D_2^4)

Atomic positions

Atomic positions are given for the 33 non-hydrogen atoms. Standard deviations are not given explicitly, but are about 0.02 to 0.03 Å for the carbon, nitrogen and oxygen atoms. Temperature factors (anisotropic for bromine and sulphur) are given also.

Structure

The analysis confirms that the title compound has the formula indicated above. It is inferred that carbon dioxide attaches to N(1') of biotin in biological reactions. Bond lengths and angles are given in full, and are consistent with expectation. The structure is stabilized by one intramolecular and two intermolecular N-H..O bonds for each molecule.

Details of analysis

The intensity data were recorded on Weissenberg photographs ($hk0$ to $hk4$; $0kl$ to $6kl$) using CuKα radiation. 2254 of a possible 2700 reflexions were observed and measured visually. Refinement was by least squares to a final R index of 0.165.

ATROVENETIN ORANGE TRIMETHYL ETHER FERRICHLORIDE

$C_{22}H_{25}Cl_4FeO_6$ (Formula on p. 243)   F.W. = 582.8

I. Fungal metabolites. Part III. The structure of atrovenetin: X-ray analysis of atrovenetin orange trimethyl ether ferrichloride. I.C. PAUL and G.A. SIM, 1965. *J. Chem. Soc.*, 1097-1112.
   (For a preliminary report, see 1.)

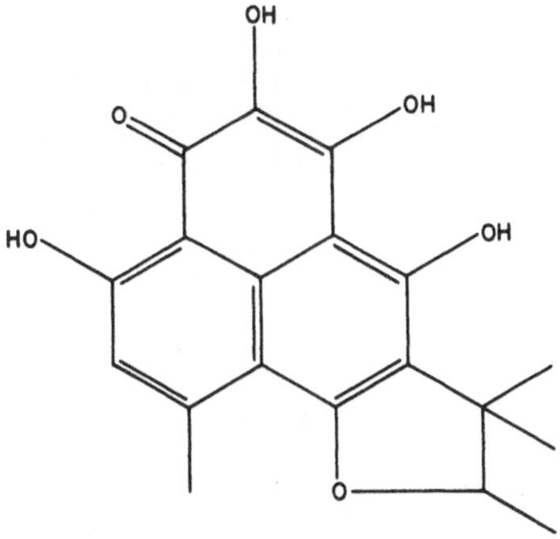

Formula

Monoclinic, $a = 17.04$, $b = 9.69$, $c = 15.66$ Å, $\beta = 96°35'$, $U = 2669$ Å^3, $D_m = 1.526$, $Z = 4$, $D_x = 1.510$.

Space group $P2_1$ (C_2^2)

Atomic positions

    Atomic positions and isotropic temperature factors of the 66 non-hydrogen atoms are given in the original.

Structure

    The analysis demonstrates that the molecular structure of atrovenetin is as shown in Fig. 138. Bond lengths and angles, (which are given in full) are consistent with this formulation, but are not very accurate.

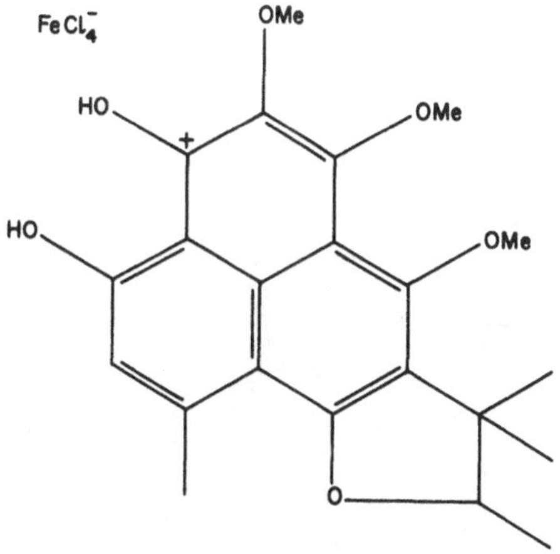

Fig. 138. The molecular structure of atrovenitin.

Details of analysis

Intensity data were recorded on *b*-axis equi-inclination Weissenberg photographs, using CuK_α radiation. 2360 reflexions were observed and estimated visually. Refinement was by Fourier methods, but was greatly hampered by a pseudo glide plane relating the two independent molecules. Refinement was terminated when the structure seemed to be unambiguously established; the final *R* index was 0.214.

1.   I.C. PAUL, G.A. SIM and G. A. MORRISON, 1962. *Proc. Chem. Soc.*, 352.

OPHIOBOLIN METHOXY BROMIDE

$C_{26}H_{39}O_5Br$ F.W. = 511.5

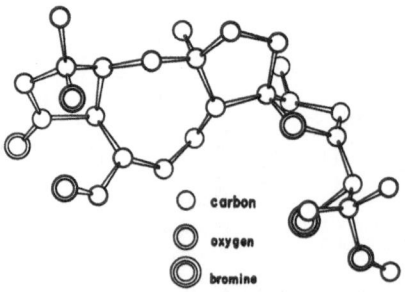

I. The structure of ophiobolin, a C_{25} terpenoid having a novel skeleton. S. NOZOE, M. MORISAKI, K. TSUDA, Y. IITAKI, N. TAKAHASHI, S. TAMURA, K. ISHIBASHI and M. SHIRASAKA, 1965. *J. Amer. Chem. Soc.*, <u>87</u>, 4968-4970.

Orthorhombic, *a* = 13.19, *b* = 22.27, *c* = 8.46 Å, [*U* = 2485 Å³], D_m = 1.40, *Z* = 4, D_x = 1.37.

Space group $P2_12_12_1$ (D_2^4)

Atomic positions

Atomic positions are not given.

Structure

The analysis establishes the stereochemistry and absolute configuration of ophiobolin as represented in the formula. Fig. 139 shows the molecular structure.

Details of analysis

The intensity data were recorded on equi-inclination Weissenberg photographs about *a* and *c*, using CuK_α radiation. 1499 reflexions were observed and measured visually. Refinement was by full-matrix least squares to a final *R* index of 0.114. The absolute configuration was determined from the anomalous scattering of the bromine atom.

Fig. 139. Molecular structure of ophiobolin methoxy bromide.

α-BROMO-ISOTUTINONE

C₁₅H₁₅O₆Br

$C_{15}H_{15}O_6Br$

F.W. = 371.2

I. The crystal structure of α-bromo-isotutinone at -150°C. M.F. MACKAY and A.McL. MATHIESON, 1965. *Acta Cryst.*, *19*, 417-425. (For a preliminary report, see *1*.)

Orthorhombic, a = 7.35, b = 14.06, c = 28.06 Å, U = 2910 Å³, D_m = 1.69 (at room temperature), Z = 8, D_x = 1.69. (Cell and intensities measured at -150°C. Calibration with Ag, a = 4.0776 Å.)

Space group $P2_12_12_1$ (D_2^4)

Atomic positions

Atomic positions for the non-hydrogen atoms of the two independent molecules (44 atoms) are given in full.

Structure

Bond lengths and angles for the two independent molecules are given in full. Average values, and the shape of the molecule, are shown in Fig. 140. [There is no indication of accuracy.] Intermolecular distances are consistent with van der Waals interaction.

Details of analysis

Partial three-dimensional intensity data were recorded on equi-inclination Weissenberg photographs (0*kl* to 3*kl*; *h0l*) with CuK$_\alpha$ radiation. 1630 of 1860 accessible reflexions were observed. Refinement was by Fourier methods to a final R index of 0.16. An anomaly of electron density was attributed to incomplete oxidation of the parent compound (α-bromo-isotutin) which thus replaced the title compound to the extent of 25% at one site.

1.   M.F. MACKAY and A. McL. MATHIESON, 1963. *Tetrahedron Letters*, *21*, 1399.

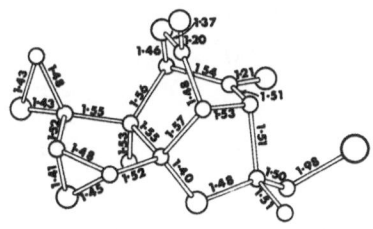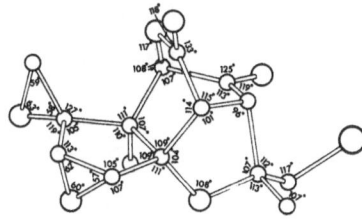

Fig. 140. Bond lengths and angles of the average α-bromo-isotutinone molecule. Carbon, oxygen and bromine atoms are represented by successively larger spheres.

DINITROGEN TETROXIDE, 1,4-DIOXAN (1;1 COMPLEX)

$C_4H_8N_2O_6$ F.W. = 180.1

N_2O_4 ,

I. Atomic arrangement in an addition compound in which dinitrogen tetroxide
    is the acceptor molecule. P. GROTH and O. HASSEL, 1965. *Acta Chem. Scand.*,
    19, 120-128. (For a preliminary report, see **1**.)

Triclinic, a = 5.46, b = 6.47, c = 6.81 Å, α = 108.5, β = 91.8, γ = 108.4°,
[U = 214 Å^3], Z = 1, D_x = 1.30 [1.40].

Space group P1 (C_1^1) or P$\bar{1}$ (C_i^1)
P$\bar{1}$ confirmed by analysis. Molecular symmetry, centre.

Atomic positions

	x	y	z
O(1)	0.29866	0.34929	0.34256
O(2)	0.74019	0.99492	0.13600
O(3)	0.10155	0.79781	0.05934
N	0.88179	0.06689	0.02812
C(1)	0.41893	0.59656	0.36295
C(2)	0.71944	0.64074	0.43833
C(3)	0.50237	0.53260	0.30439
C(4)	0.81660	0.81660	0.44090

Anisotropic temperature factors are given also.

Structure

     Endless chains of donor and acceptor molecules are extended in the [111]
direction. The arrangement of the molecules is shown in Fig. 141. The structure
appears to be disordered, with the chair-shaped 1,4-dioxan molecule adopting
the alternative orientations shown. The N...O distance shown is 2.76 Å, which
is somewhat less than the van der Waals separation. [However, there is no
indication of the accuracy attained.]

Details of analysis

     The intensity data were recorded on integrated Weissenberg photographs,
using both CuK_α and MoK_α radiation. About 440 reflexions were observed, and
measured photometrically. Refinement was by block-diagonal least squares to a
final R index of 0.11.

1. *Structure Reports*, **27**, 963.

1,4-DIOXAN, OXALYL CHLORIDE (1:1 COMPLEX)

1,4-DIOXAN, OXALYL BROMIDE (1:1 COMPLEX)

$C_6H_8Cl_2O_4$ F.W. = 215.0

$C_6H_8Br_2O_4$ F.W. = 303.9

I. X-ray analysis of the (1:1) addition compounds of 1,4-dioxan with oxalyl

chloride resp. oxalyl bromide. E. DAMM, O. HASSEL and CHR. RØMMING,
1965. *Acta Chem. Scand.*, <u>19</u>, 1159-1165.
(For a preliminary report, see <u>1</u>.)

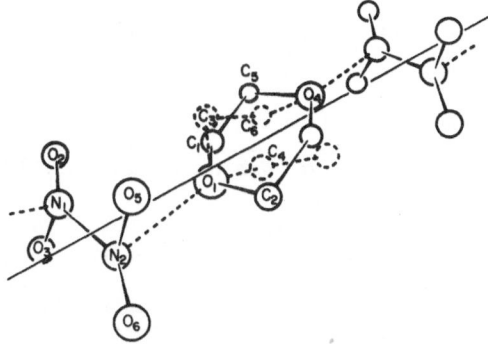

Fig. 141. Part of the chain of molecules. The alternative orientations of
the 1,4-dioxan molecule are shown.

CHLORIDE

Monoclinic, a = 7.58, b = 10.30, c = 7.09 Å, β = 126.5°, [U = 445 Å³], z = 2,
[D_x = 1.60], (all at -20 to -40°C).

Space group $P2_1/c$ (C_{2h}^5) Molecular symmetry, centre.

Atomic positions

	x	y	z
Cl	0.030	0.1920	0.128
O(1)	0.309	0.438	0.437
O(2)	-0.208	0.071	-0.263
C(1)	0.309	0.505	0.252
C(2)	0.484	0.606	0.348
C(3)	-0.077	0.061	-0.063

Standard deviations are: Cl, 0.006 Å; O, 0.013 Å; C, 0.021 Å. Temperature
factors are given also.

Structure

    The centrosymmetrical oxalyl chloride and 1,4-dioxan molecules occupy
centres of symmetry at 000 and $\frac{1}{2}\frac{1}{2}\frac{1}{2}$ respectively. The structure consists of
chains of alternating molecules, as illustrated in Fig. 142, extended in the
[111] direction.   The C-Cl and Cl-O distances are 1.70 ± 3 Å and 3.18 ± 2 Å
respectively. The latter distance does not differ significantly from the sum of
van der Waals radii (3.20 Å).

BROMIDE

Monoclinic, a = 7.67, b = 10.90, c = 7.14 Å, β = 127.2°, [U = 475 Å³],
z = 2, [D_x = 2.12] (all at room temperature).

Space group $P2_1/c$ (C_{2h}^5) Molecular symmetry, centre.

Atomic positions

	x	y	z
Br	0.043	0.1955	0.147
O(1)	0.297	0.442	0.432
O(2)	-0.219	0.068	-0.253
C(1)	0.297	0.507	0.241

	x	y	z
C(2)	0.503	0.600	0.370
C(3)	-0.068	0.055	-0.068

Standard deviations are: Br, 0.009 Å; O, 0.037 Å; C, 0.05 Å. Temperature factors are given also.

Structure

The structure resembles that of the chloride. The C–Br and Br–O distances are 1.96 ± 6 Å and 3.21 ± 4 Å respectively. The latter distance is slightly shorter than the sum of van der Waals radii, (3.35 Å), and weak bonding is inferred.

Details of analysis (both derivatives)

The intensity data for the zones [100] and [001] were recorded on Weissenberg (CuK$_\alpha$ radiation) and precession (MoK$_\alpha$ radiation) photographs. For the chloride derivative the temperature was maintained at -20 to -40°C. Refinement was by least squares to final R indices as follows:
Chloride: (0kl), 0.108; (hk0), 0.122.
Bromide: (0kl), 0.159; (hk0), 0.110.

1. O. HASSEL, 1961. Tidsskr. Kjemi, Bergvesen Met., 21, 60.

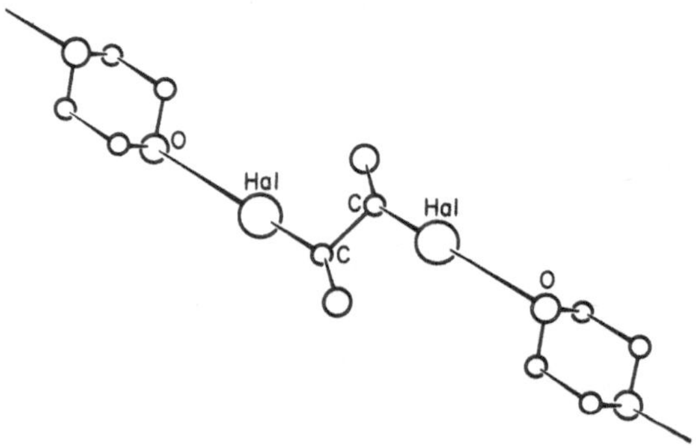

Fig. 142. A view of part of the structural chain.

CYCLOHEXANE-1,4-DIONE, DIIODOACETYLENE (1:1 COMPLEX)

$C_8H_8I_2O_2$ F.W. = 390.0

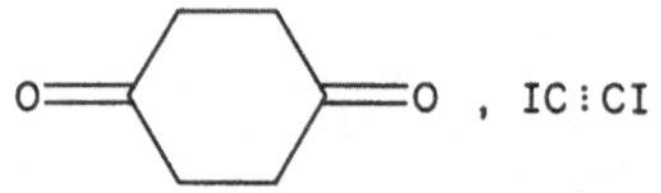

I. The crystal structure of the (1:1) addition compound cyclohexane-1,4-dione – diiodoacetylene. P. GROTH and O. HASSEL, 1965. Acta Chem. Scand., 19, 1733-1740.

Monoclinic, a = 9.61, b = 7.59, c = 8.72 Å, β = 117.3°, [U = 565 Å^3], Z = 2, D_x = 2.29.

Space group P2$_1$/c (C$^5_{2h}$) Molecular symmetry, centre.

Atomic positions are given for part of the structure.

Structure

The space group requires that each molecule occupy a centre of symmetry. It is shown, however, that the individual cyclohexane-1,4-dione molecules do not have the required symmetry, and simulate it only by being disordered. The electron-density distribution is consistent with a twisted boat conformation. The diiodoacetylene molecule is linear, with C–C = 1.17 Å and C–I = 1.98 Å. The mean intermolecular I...O distance is 2.95 Å.

Details of analysis

Three-dimensional intensity data were recorded on Weissenberg photographs taken at -20°C, using CuK$_\alpha$ radiation. About 500 reflexions were observed, and measured photometrically. Refinement of the disordered structure was by full-matrix least squares to a final R index of 0.082.

1,4-DISELENANE-TETRAIODOETHYLENE

$C_6H_8I_4Se_2$

F.W. = 745.7

I. Bonds connecting group VI donor atoms and halogen atoms in ethylene derivatives. TOR DAHL and O. HASSEL, 1965. *Acta Chem. Scand.*, <u>19</u>, 2000-2001.

Monoclinic, a = 6.50, b = 12.80, c = 9.21 Å, β = 99.6°, [U = 756 Å^3],
Z = 2, D_x = 3.36.

Space group $P2_1/c$ (C_{2h}^5) Molecular symmetry, centre.

Atomic positions are not given.

Structure

The structure consists of alternating acceptor and donor molecules, linked by selenium-iodine bonds to form chains extended along [101]. These chains are cross-linked by further Se-I bonds. Se-I distances are 3.40 to 3.43 Å.

Details of analysis

Intensity data were recorded for the 0*kl* and *h*0*l* zones, the former on Weissenberg photographs, with photometric measurement, and the latter with a diffractometer and counter. MoK$_\alpha$ radiation was used for both zones. The method of refinement is not stated, but a Fourier projection along *a* is given. The final R indices were: 0*kl*, 0.055; *h*0*l*, 0.077.

7,7,8,8-TETRACYANOQUINODIMETHAN, N,N,N',N'-TETRAMETHYL-

p-PHENYLENE DIAMINE (1:1 COMPLEX)

$C_{22}H_{20}N_6$

F.W. = 368.4

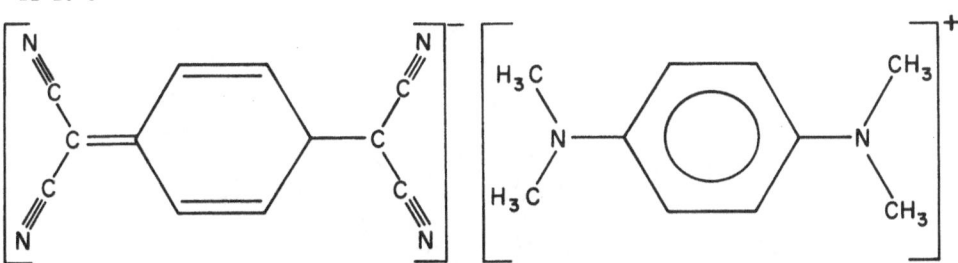

I. The crystal structure of the 1:1 complex of 7,7,8,8-tetracyanoquinodimethan and N,N,N',N'-tetramethyl-*p*-phenylene diamine. A. W. HANSON, 1965. *Acta Cryst.*, <u>19</u>, 610-613.

Monoclinic, a = 9.88 ± 3, b = 12.71 ± 4, c = 7.72 ± 3 Å, β = 97.34 ± 5°,
U = 961.5 Å^3, D_m = 1.25, Z = 2, D_x = 1.27.

Space group $C2$ (C_2^3), C_m (C_s^3) or $C2/m$ (C_{2h}^3) $C2/m$ is consistent with deduced structure. Molecular symmetry, centre.

Atomic positions

	x	y	z
C(1)	0.0590	0.0952	−0.0383
C(2)	0.1224	0	−0.0779
C(3)	0.2448	0	−0.1529
C(4)	0.3118	0.0941	−0.1890
N(5)	0.3683	0.1705	−0.2199
C(6)	0.0593	0.0951	0.4622
C(7)	0.1229	0	0.4214
N(8)	0.2407	0	0.3463
C(9)	0.3127	0.0987	0.3168

The standard deviations range from 0.004 to 0.007 Å. Anisotropic temperature factors are given also.

Structure

The dimensions of both molecule-ions are given in Fig. 143. (It is shown that thermal motion could have an appreciable effect on the apparent lengths of C(4)–N(5) and N(8)–C(9), but corrections have not been applied.) The distribution of ring bond lengths is, within experimental error, the same for both molecules. It appears that each ring has, as a result of complexing, acquired a character intermediate between aromatic and quinonoidal. Both molecules are planar except for the terminal groups (cyano and methyl) which are significantly displaced from the molecular planes. C(3) and N(8) are thus slightly pyramidal.

The structure consists of stacks, parallel to c, of equally-spaced, parallel, alternating molecules. The angle between c and the molecular planes is 58°; adjacent molecules in a stack overlap as shown in Fig. 144, with an

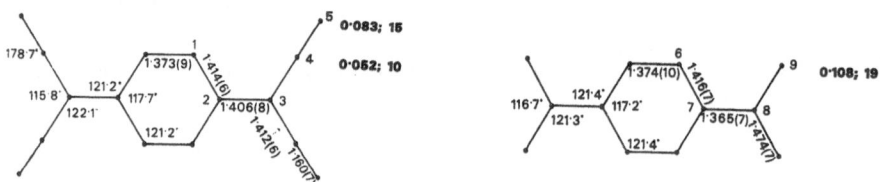

Fig. 143. Bond lengths and their e.s.d.'s (Å), and bond angles. Bold figures are the distances of certain atoms from the molecular planes, in Å, and in multiples of the corresponding e.s.d. of coordinates.

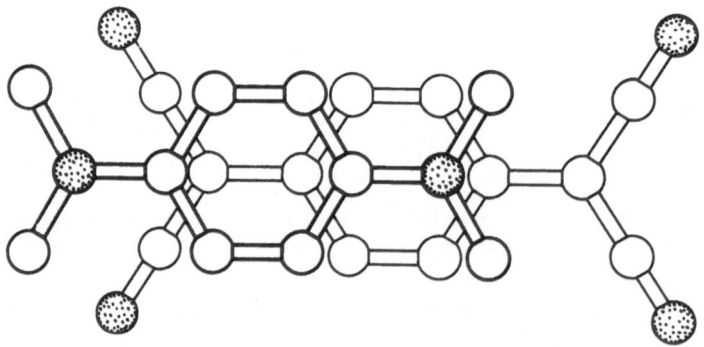

Fig. 144. Overlapping molecules, viewed normal to their plane.

interplanar spacing of 3.27 Å. Distances between stacks are consistent with van der Waals interaction.

Details of analysis

The intensity data were measured with a four-circle diffractometer and scintillation counter, using CuK_α radiation. 588 of 1122 accessible reflexions in the range $2\theta < 165°$ were observed. Refinement was by block-diagonal least squares to a final R index of 0.083.

AZULENE, *s*-TRINITROBENZENE (1:1 COMPLEX)

$C_{16}H_{11}N_3O_6$ F.W. = 341.3

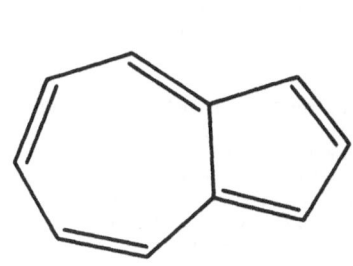
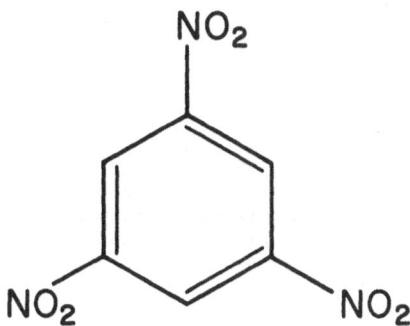

I.   The crystal structure of the azulene, *s*-trinitrobenzene complex. A. W. HANSON, 1965. *Acta Cryst.*, <u>19</u>, 19-26.

II. Molecular complexes exhibiting polarization bonding. V. The crystal structure of the azulene-*s*-trinitrobenzene complex. D. S. BROWN and S. C. WALLWORK, 1965. *Acta Cryst.*, <u>19</u>, 149.

I.   (Cell and intensity data at -95°C.)

Monoclinic, a = 16.39 ± 3, b = 6.66 ± 1, c = 13.77 ± 2 Å, β = 96.10 ± 5°, U = 1500 Å^3, Z = 4, D_x = 1.51.

(Cell data were also measured at room temperature, for which a = 16.39, b = 6.80, c = 13.99 Å, β = 95.86°, U = 1551 Å^3, D_m = 1.46, D_x = 1.46.)

Space group $P2_1/a$ (C_{2h}^5)

Atomic positions

	x	y	z
	s-Trinitrobenzene		
C(1)	0.1246	0.1236	0.1700
C(2)	0.1787	0.1201	0.2543
C(3)	0.1443	0.1233	0.3415
C(4)	0.0607	0.1316	0.3472
C(5)	0.0108	0.1371	0.2596
C(6)	0.0403	0.1316	0.1703
N(7)	0.1594	0.1179	0.0758
N(8)	0.1997	0.1149	0.4328
N(9)	-0.0787	0.1493	0.2627
O(10)	0.1112	0.1288	0.0011
O(11)	0.2332	0.1015	0.0771
O(12)	0.2715	0.0819	0.4277
O(13)	0.1694	0.1365	0.5083
O(14)	-0.1040	0.1455	0.3430
O(15)	-0.1220	0.1645	0.1849

Azulene

C(16)	0.2957	0.1222	0.8514
C(17)	0.3704	0.1286	0.9101
C(18)	0.4500	0.1340	0.8825
C(19)	0.4744	0.1343	0.7890
C(20)	0.4210	0.1314	0.6942
C(21)	0.3360	0.1255	0.6809
C(22)	0.2805	0.1219	0.7504
C(23)	0.5549	0.1383	0.7655
C(24)	0.5542	0.1372	0.6649
C(25)	0.4737	0.1342	0.6206

Standard deviations are stated to range from 0.002 to 0.004 Å. Also given are anisotropic temperature factors for the above atoms, and assumed positions for the hydrogen atoms. The occupancy of the azulene site is estimated to be about 0.93. Parameters are given also for an azulene molecule in an alternative orientation, with an occupancy of 0.07, but these have not been refined.

Structure

Bond lengths (corrected for thermal motion) and angles are given in full. Of the peripheral bonds in azulene none differs significantly in length from the mean value of 1.395 ± 4 Å. The length of the bridging bond is 1.498 ± 4 Å. The molecule is essentially planar. The benzene ring of the *s*-trinitrobenzene molecule is planar, but the NO_2 groups are bent and rotated out of this plane by varying amounts. Mean values of the bond lengths are: C–C, 1.383 ± 4; C–N, 1.480 ± 3, N–O, 1.243 ± 3 Å. (The latter value is considered to be less reliable because of uncertainty concerning the corrections for thermal motion.) The interior angles of the benzene ring are 123.3 ± 3° for the substituted positions, and 116.7 ± 3° for the non-substituted positions.

The molecules overlap each other as shown in Fig. 145, with an interplanar spacing of 3.33 Å.

Details of analysis

The intensity data were measured with a four-circle diffractometer and scintillation counter, using CuK_α radiation. The specimen was maintained at a temperature of -95°C during measurement. Within the range 2θ < 165°, 2409 of a possible 3333 reflexions were observed above background. The analysis was complicated by the presence of disorder, as about 7% of the azulene molecules were found to adopt an alternative orientation. Refinement was by block-diagonal least squares to a final *R* index of 0.060.

II.   (Cell and intensity data at room temperature.)

Monoclinic, a = 14.05 ± 2, b = 6.76 ± 2, c = 16.4 ± 2 Å, β = 96.0 ± 5°, U = 1548 Å^3, D_m = 1.45, Z = 4, D_x = 1.46, (CuK_α, λ = 1.542 Å).

Space group P2$_1$/c (C_{2h}^5)

[Note that the choice of *a* and *c* axes differs from that in I.] The azulene position was found to be disordered to a degree which prevented meaningful refinement, and no atomic coordinates or molecular dimensions are reported.

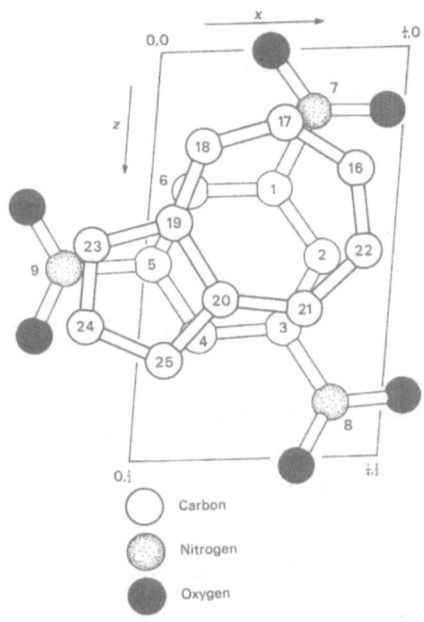

Fig. 145. The asymmetric unit of $C_{16}H_{11}N_3O_6$ projected along b.

1,3,7,9-TETRAMETHYLURIC ACID, PYRENE (1:1 COMPLEX)

$C_{25}H_{22}N_4O$ F.W. = 426.5

I. The crystal structure of the 1:1 molecular complex between 1,3,7,9-
    tetramethyluric acid and pyrene. A. DAMIANI, P. DE SANTIS, E. GIGLIO,
    A. M. LIQUORI, R. PULITI and O. RIPAMONTI, 1965. *Acta Cryst.*, 19,
    340–348.

(For a preliminary account, see 1.)

Monoclinic, $a = 9.71 \pm 2$, $b = 8.00 \pm 2$, $c = 15.0 \pm 2$ Å, $\beta = 117°00 \pm 12'$,
$U = 1041$ Å^3, [$Z = 2$], $D_x = 1.351$, (CuK_α, $\lambda = 1.542$ Å).

Space group Pc (C_s^2) or P2/c (C_{2h}^4)
Pc indicated by positive piezoelectric test.

Atomic positions

Positions and anisotropic temperature factors are given for the 32 non-hydrogen atoms. [However, the temperature coefficients b_{12} and b_{23} have arbitrarily been set to zero.] Standard deviations of position are given in full, and range from 0.01 to 0.03 Å.

Structure

Bond lengths and angles are given in full, with standard deviations which range from 0.010 to 0.035 Å, and 0.7 to 2.2°. [However, a comparison of chemically equivalent values in the pyrene molecule suggests that these estimates may be optimistic.] Both molecules are essentially planar, and their mean planes are inclined to one another by 3.6°. The plane-to-plane stacking of alternate molecules is illustrated in Fig. 146. The interplanar spacing is 3.48 Å. Intermolecular distances are generally consistent with van der Waals interaction.

Details of analysis

Intensity data were recorded on integrated equi-inclination Weissenberg photographs ($h0l$ to $h7l$) using CuK_α radiation. Of a possible 2400 reflexions, about 71% were observed, and estimated visually. Refinement was by differential syntheses to a final R index of 0.174.

1. *Structure Reports*, <u>26</u>, 754.

ANTHRACENE, PYROMELLITIC DIANHYDRIDE (1:1 COMPLEX)

PERYLENE, PYROMELLITIC DIANHYDRIDE (1:1 COMPLEX)

$C_{24}H_{12}O_6$ F.W. = 396.5

$C_{30}H_{14}O_6$ F.W. = 470.5

anthracene

pyromellitic dianhydride   perylene

I. Molecular compounds and complexes. III. The crystal structures of the equimolar π-molecular compounds of anthracene and perylene with pyromellitic dianhydride. J. C. A. BOEYENS and F. H. HERBSTEIN, 1965. *J. Phys. Chem.*, <u>69</u>, 2160-2176.

ANTHRACENE, PYROMELLITIC DIANHYDRIDE

Triclinic, a = 7.6, b = 10.0, c = 7.3 Å, α = 105, β = 115.5, γ = 101°, [U = 454 Å^3], D_m = 1.49, Z = 1, D_x = 1.48.

Space group $P1$ (C_1^1) or $P\bar{1}$ (C_i^1)
$P\bar{1}$ confirmed by analysis. Molecular symmetry, centre.

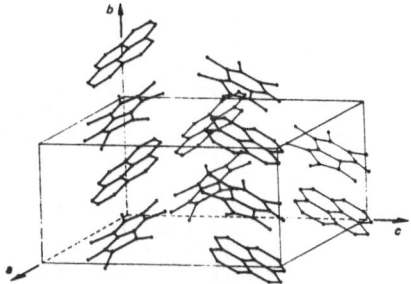

Fig. 146.   A clinographic view of the $C_{25}H_{22}N_4O$ structure, showing plane-to-
plane stacking.

Atomic positions

	x	y	z
C(D)	-0.163	0.055	-0.098
C(E)	0.033	0.146	0.134
C(F)	0.084	0.301	0.231
C(G)	0.272	0.382	0.485
C(A')	0.440	0.327	0.549
C(B')	0.405	0.185	0.440
C(C')	0.204	0.094	0.215
C(1)	-0.174	0.049	0.395
C(2)	0.022	0.143	0.610
C(3)	0.189	0.093	0.738
C(4)	0.104	0.299	0.752
C(5)	0.365	0.216	0.923
O(6)	0.005	0.381	0.756
O(7)	0.305	0.335	0.955
O(8)	0.545	0.226	0.080

Letter subscripts refer to the anthracene molecule.

Structure

The structure consists of stacks of alternating molecules, extended in the
c direction. Adjacent molecules in a stack overlap each other as illustrated
in Fig. 147. The molecules were assumed to be planar; the distance between
their planes is 3.23 Å. There is, however, no indication of the accuracy,
either of cell parameters or coordinates.

Details of analysis

Partial three-dimensional intensity data were recorded on Weissenberg
photographs (hk0 and hk1) using CuK_α radiation. 267 of a possible 336 reflexions
were observed and measured visually. Refinement was by Fourier methods to a
final R index of 0.185.

PERYLENE, PYROMELLITIC DIANHYDRIDE

Monoclinic, a = 14.613 ± 3, b = 7.16 ± 1, c = 10.1309 ± 4 Å, β = 94.67 ± 3°,
[U = 1056 Å^3], D_m = 1.50, Z = 2, D_x = 1.504, ($CuK_{\alpha1}$, λ = 1.54050 Å;
$CuK_{\alpha2}$, λ = 1.54436 Å).

Space group $P2_1/n$ (C_{2h}^5) Molecular symmetry, centre.

Atomic positions

	x	y	z
C(A)	-0.07988 ± 43	0.06877 ± 104	0.24042 ± 48
C(B)	-0.03015 ± 56	0.14132 ± 126	0.35060 ± 49

	x	y	z
C(C)	0.06104 ± 50	0.18038 ± 119	0.34605 ± 47
C(D)	0.10546 ± 42	0.15103 ± 113	0.22546 ± 52
C(E)	0.19630 ± 46	0.19225 ± 132	0.21931 ± 71
C(F)	0.23840 ± 37	0.16038 ± 121	0.10607 ± 73
C(G')	-0.18976 ± 41	-0.08623 ± 128	0.00755 ± 67
C(H')	-0.09669 ± 29	-0.04461 ± 93	0.00688 ± 43
C(I)	0.05274 ± 32	0.07395 ± 91	0.11168 ± 41
C(J')	-0.04054 ± 32	0.03455 ± 96	0.12336 ± 38
C(1')	0.07774 ± 30	0.57178 ± 93	0.08204 ± 43
C(2)	-0.08267 ± 26	0.48231 ± 86	0.04879 ± 39
C(3)	-0.00894 ± 31	0.55310 ± 92	0.12774 ± 40
C(4)	-0.16138 ± 37	0.47598 ± 116	0.12832 ± 66
C(5)	-0.04079 ± 49	0.58933 ± 118	0.25856 ± 46
O(6)	-0.23931 ± 28	0.41709 ± 97	0.09923 ± 61
O(7)	-0.13537 ± 34	0.54484 ± 78	0.25094 ± 41
O(8)	-0.00210 ± 44	0.64979 ± 85	0.35565 ± 37

Anisotropic temperature factors are given also. These have been analysed and used to correct the molecular parameters for thermal libration.

Structure

The structure consists of stacks of alternating molecules, extended in the b direction. The perylene molecule is planar, and the pyromellitic dianhydride molecule approximately so, although the carbonyl groups are bent somewhat out of the mean plane. Adjacent molecules in a stack overlap each other as illustrated in Fig. 148; The interplanar distance is 3.33 Å. There is no evidence of interaction between stacks. Bond lengths and angles are given in full. These are normal for the perylene molecule, but significant differences between chemically equivalent distances in pyromellitic dianhydride are noted and discussed.

Details of analysis

The intensity data were recorded on equi-inclination Weissenberg photographs, using CuKα radiation. 1187 of a possible 1545 reflexions were observed and measured visually. Refinement was by least-squares to a final R index of 0.144.

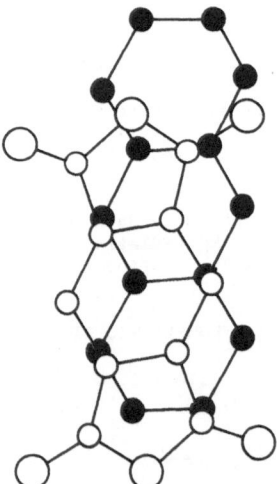

Fig. 147. Overlap of anthracene and pyromellitic dianhydride.

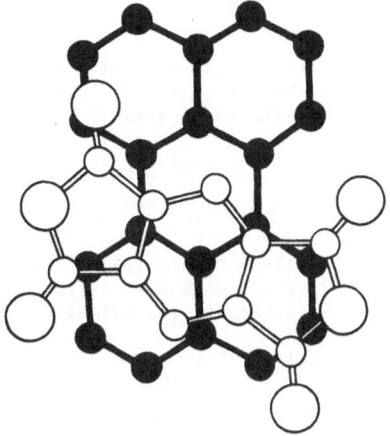

Fig. 148. Overlap of perylene and pyromellitic dianhydride.

POTASSIUM IODIDE, POTASSIUM TRIIODIDE, N-METHYLACETAMIDE COMPLEX

$C_6H_{21}K_2I_4N_6O_6$ $KI \cdot KI_3 \cdot 6(CH_3CONHCH_3)$ F.W. = 859

I. The crystal structure of a polyiodide complex with N-methylacetamide-
    $KI \cdot KI_3 \cdot 6(CH_3CONHCH_3)$. K. TOMAN, J. HONZL and J. JEČNÝ, 1965. *Acta Cryst.*,
    18, 673–677.

Trigonal, $a = 12.2$, $c = 15.2$ Å, $U = 1958$ Å^3, $Z = 2$, $[D_x = 3.0]$.

Space group $P31c$ (C_{3v}^4) or $P\bar{3}1c$ (D_{3d}^2) $P31c$ is consistent with analysis. Molecular
symmetry, $\bar{3}$.

Atomic positions

	x	y	z
I(1)	1/3	2/3	1/4
I(2)	1/3	2/3	3/4
I(3)	1/3	2/3	0.0559 ± 4
K(1)	0	0	0
K(2)	0	0	1/4
C(1)	0.360	0.345	0.352
C(2)	0.295	0.203	0.376
C(3)	0.329	0.035	0.403
N	0.373	0.163	0.388
O	0.176	0.137	0.377

Temperature factors are given also.

Structure

The inorganic framework of the structure consists of lines of alternating
I^- and I_3^- ions, extended in the c direction, interspersed with parallel lines
of equally-spaced K^+ ions. The mean I–I distance in the I_3^- ion is 2.94 Å;
crystal symmetry requires the ion to be linear and symmetrical, but it is
recognized that the structure is probably disordered, and the ion bent. Iodine
and potassium lines are separated by N-methylacetamide molecules, which lie in
planes normal to c. The NH groups are coordinated to the I^- ions at a distance
of 3.74 Å. The coordination is six-fold, slightly-staggered trigonal prismatic.
The oxygen atoms are coordinated to the K^+ ions at a distance of 2.74 Å. The
coordination is also six-fold, somewhat staggered trigonal prismatic, with
adjacent potassium ions sharing the trigonal faces. There appear to be no interactions, other than van der Waals, involving the I_3^- ions.

Details of analysis

The intensity data were measured with a diffractometer, using MoK$_\alpha$ radiation. 433 of 823 accessible reflexions were observed above background. Refinement was by Fourier and least-squares methods to a final R index of 0.126.

GUANIDINIUM CHLORIDE: N,N-DIMETHYLACETAMIDE (3:1 COMPLEX)

C$_7$H$_{27}$N$_{10}$Cl$_3$O   3[C(NH$_2$)$_3$Cl$^-$] : [CH$_3$CON(CH$_3$)$_2$]   F.W. = 373.7

I. The structure of the crystalline intermolecular complex: guanidinium chloride-disordered N,N-dimethylacetamide (3:1). D. J. HAAS, D. R. HARRIS and H. H. MILLS, 1965. *Acta Cryst.*, <u>18</u>, 623–627.

Monoclinic, a = 23.47 ± 1, b = 14.90 ± 1, c = 11.95 ± 1 Å, β = 90.95 ± 5°, [U = 4178 Å^3], D_m = 1.18, Z = 8, D_x = 1.188.

Space group Ia (C_5^4) or I2/a (C_{2h}^6). I2/a is consistent with analysis.

Atomic positions

	x	y	z
Cl(1)	0.0115	0.0927	0.1121
Cl(2)	0.1966	0.1385	0.1162
Cl(3)	0.5294	0.3440	0.0798
C(1)	0.9749	0.3707	0.2349
N(1)	0.9803	0.4446	0.1727
N(2)	0.9684	0.2914	0.1857
N(3)	0.9756	0.3757	0.3475
C(2)	0.8544	0.0287	0.1274
N(4)	0.8852	-0.0051	0.0470
N(5)	0.7989	0.0084	0.1347
N(6)	0.8793	0.0788	0.2091
C(3)	0.3724	0.2830	0.1185
N(7)	0.3990	0.2534	0.0236
N(8)	0.3201	0.2558	0.1390
N(9)	0.4021	0.3276	0.1930

Isotropic temperature factors are given also. Atomic positions are not reported for the disordered dimethylacetamide molecule.

Structure

The molecule packing is illustrated in Fig. 149. The crystal structure is a framework of guanidinium and chloride ions held together by electrostatic attraction, leaving cylindrical cavities parallel to c. These cavities (having an average diameter of 6 Å) are occupied by dimethylacetamide molecules in random orientations. Bond lengths and angles are given in full for the guanidinium ions. The mean C–N distance is 1.33 ± 3 Å.

Details of analysis

Intensity data were measured with a four-circle diffractometer, using CuK$_\alpha$ radiation, to a resolution of 1 Å. Refinement was by least squares, and in order to reduce the effect of the disordered dimethylacetamide molecules, reflexions for which 2θ was less than 30° were not used. The final R index was 0.16 for the reflexions used.

2,6-LUTIDINE, UREA (1:1 COMPLEX)

$C_8H_{13}N_3O$ F.W. = 167.2

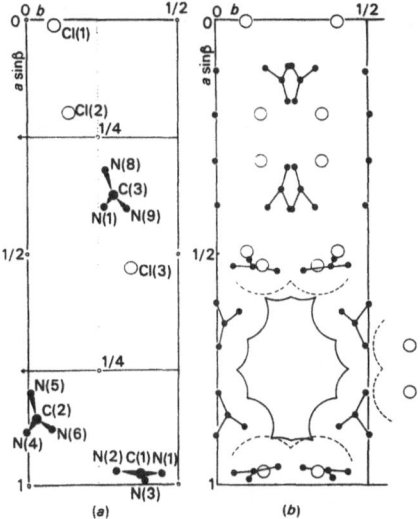

I. The crystal structure of the 1:1 complex formed by 2,6-lutidine and urea.
    J. D. LEE and S. C. WALLWORK, 1965. *Acta Cryst.*, <u>19</u>, 311–313.

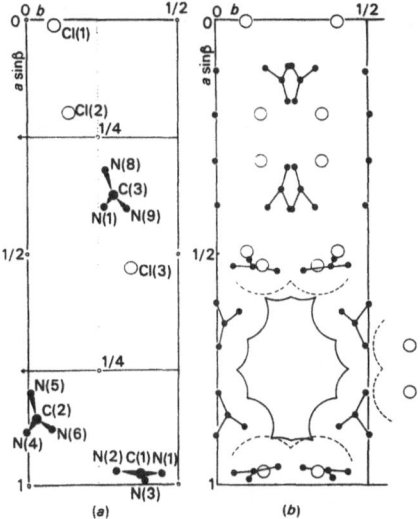

Fig. 149.   (a) Atoms in the asymmetric unit of the $C_7H_{27}N_{10}Cl_3O$ structure
            (z = 0 to 0.25) projected along c.   (b) Half of the unit cell
            projected along c, showing the cylindrical cavity parallel to c.

Monoclinic, C_c, a = 11.34 ± 3, b = 11.38 ± 3, c = 7.48 ± 2 Å, β = 99 ± 1°,
U = 955 Å³, D_m = 1.16, Z = 4, D_x = 1.165, (CuK$_\alpha$, λ = 1.542 Å).

Space group Cc (C_s^4) or C2/c (C_{2h}^6). C2/c consistent with refined structure.
Molecular symmetry, two-fold axis.

Atomic positions

	x	y	z
C(1)	0.103	−0.027	0.232
C(2)	0.104	0.097	0.232
C(3)	0.000	0.158	0.250
C(4)	0.214	−0.097	0.207
C(5)	0.000	0.595	0.250
N(1)	0.000	−0.088	0.250
N(2)	0.028	0.645	0.400
O	0.000	0.483	0.250

Isotropic temperature factors are given.

Structure

Bond lengths and angles are given in full, but are recognized to be of limited accuracy. A view of the structure, showing the assumed hydrogen bonding scheme, is given in Fig. 150.

Details of analysis

Intensity data for the *hk0* and *h0l* zones were recorded on oscillation and Weissenberg photographs, using CuK_α radiation. 104 reflexions were observed and measured photometrically. Refinement was by Fourier projections to a final *R* index of 0.08.

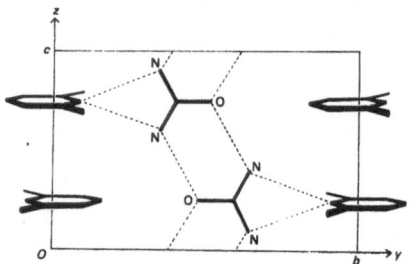

Fig. 150. A view along the *a* axis of one hydrogen-bonded layer of the 2,6-lutidine-urea (1:1 complex) structure. Hydrogen bonds are represented by broken lines; distances are: N-H...O, 3.05 Å; N-H...N, 3.24 Å.

PICRYL AZIDE, BIS-8-HYDROXYQUINOLINATOCOPPER(II), (2:1 COMPLEX)

$C_{30}H_{16}CuN_{14}O_{14}$ F.W. = 860.1

I. Molecular complexes. Part III. The crystal and molecular structure of the 1:2 molecular compound of bis-8-hydroxyquinolinatocopper(II) and picryl azide. A. S. BAILEY and C. K. PROUT, 1965. *J. Chem. Soc.*, 4867-4881.

Monoclinic, *a* = 16.14 ± 4, *b* = 30.93 ± 8, *c* = 6.90 ± 2 Å, γ = 105.6 ± 2°, *U* = 3321 Å^3, D_m = 1.709, *z* = 4, D_x = 1.718.

Space group Aa (C_s^4) or *A*2/*a* (C_{2h}^6) [first setting]. *A*2/*a* is consistent with deduced structure. Molecular symmetry, two-fold axis.

Atomic positions

	x	*y*	*z*
Cu	0.2500 ± 0	0	0.0878 ± 5
C(1)	0.2661 ± 7	0.0899 ± 3	0.0811 ± 21
C(2)	0.2529 ± 7	0.1326 ± 4	0.0739 ± 27

	x	y	z
C(3)	0.3224 ± 7	0.1720 ± 3	0.0740 ± 22
C(4)	0.4081 ± 6	0.1699 ± 3	0.0948 ± 26
C(5)	0.4218 ± 7	0.1275 ± 4	0.0913 ± 23
C(6)	0.5049 ± 6	0.1195 ± 3	0.1104 ± 24
C(7)	0.5143 ± 7	0.0768 ± 4	0.1193 ± 24
C(8)	0.4394 ± 6	0.0389 ± 3	0.1041 ± 22
C(9)	0.3508 ± 5	0.0883 ± 3	0.0878 ± 20
N(1)	0.3621 ± 4	0.0447 ± 2	0.0927 ± 17
O(1)	0.2018 ± 3	0.0514 ± 2	0.0740 ± 15
C(10)	0.8220 ± 6	0.0874 ± 3	0.4228 ± 24
C(11)	0.9040 ± 5	0.0799 ± 3	0.4038 ± 22
C(12)	0.9778 ± 6	0.1145 ± 3	0.3930 ± 23
C(13)	0.9687 ± 6	0.1579 ± 3	0.4127 ± 25
C(14)	0.8903 ± 6	0.1679 ± 3	0.4299 ± 24
C(15)	0.8194 ± 5	0.1321 ± 3	0.4328 ± 23
N(2)	0.9084 ± 5	0.0335 ± 3	0.3802 ± 20
N(3)	0.0536 ± 5	0.1941 ± 3	0.4165 ± 20
N(4)	0.8999 ± 6	0.2157 ± 3	0.4265 ± 20
N(5)	0.8450 ± 6	0.2309 ± 3	0.5078 ± 18
N(6)	0.8036 ± 6	0.2513 ± 3	0.5696 ± 25
N(7)	0.7299 ± 6	0.1382 ± 3	0.4264 ± 21
O(2)	0.8461 ± 5	0.0035 ± 3	0.4116 ± 20
O(3)	0.9782 ± 6	0.0273 ± 3	0.3455 ± 18
O(4)	0.5664 ± 5	0.2182 ± 3	0.4515 ± 19
O(5)	0.0988 ± 5	0.1943 ± 3	0.2757 ± 18
O(6)	0.7191 ± 5	0.1723 ± 3	0.3703 ± 19
O(7)	0.6707 ± 5	0.1067 ± 3	0.4714 ± 22

Anisotropic temperature factors are given also.

Structure

Both molecules are as illustrated, and the bond lengths and angles (given in full) are consistent with expectation. The Cu-O and Cu-N distances are 1.95 ± 1 and 1.96 ± 1 Å respectively. The picryl azide molecule departs from planarity in typical fashion, by rotation of the nitro and azido groups about the C-N bonds. Each 8-hydroxyquinoline residue is planar, but the angle between the planes of adjacent residues is 7°, so that the bis-8-hydroxyquino-linatocopper(II) molecule is slightly propeller-shaped.

The picryl azide molecule and each 8-hydroxyquinoline residue act as acceptor and donor, respectively, in a charge-transfer complex. As shown in Fig. 151, the structure consists of stacks, parallel to *c*, of alternating donors and acceptors, overlapping efficiently in the manner characteristic of charge-transfer interaction. The mean acceptor-donor distance is 3.45 Å. Pairs of stacks are covalently joined by the copper atoms.

Details of analysis

The intensity data were recorded on *c*-axis Weissenberg photographs, using CuK_α radiation, and measured visually. Refinement was by block-diagonal least squares to a final *R* index of 0.134.

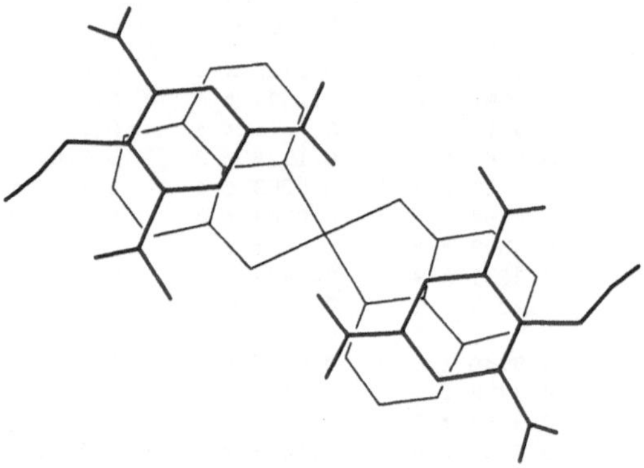

Fig. 151. Overlap of the picryl azide and bis-8-hydroxyquinolinatocopper(II) molecules.

BIS-8-HYDROXYQUINOLINATOPALLADIUM(II), CHLORANIL (1:1 COMPLEX)

$C_{24}H_{12}Cl_4N_2O_4Pd$ F.W. = 640.6

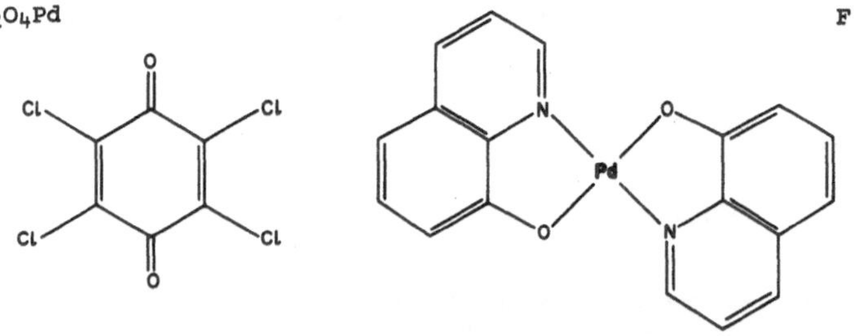

I. Molecular complexes. Part II. The crystal structure of the 1:1 complex. of bis-8-hydroxyquinolinatopalladium(II) and chloranil. B. KAMENAR, C. K. PROUT and J. D. WRIGHT, 1965. *J. Chem. Soc.*, 4851-4867.

Triclinic, a = 8.17 ± 2, b = 8.18 ± 2, c = 9.69 ± 2 Å, α = 99.5 ± 2, β = 77.8 ± 2, γ = 66.0 ± 2°, U = 546.5 Å^3, D_m = 1.931, Z = 1, D_x = 1.946.

Space group P1 (C_1^1) or P$\bar{1}$ (C_i^1). P$\bar{1}$ confirmed by analysis. Molecular symmetry, centre.

Atomic positions

	x	y	z
Pd	0	0	0
C(1)	0.7662 ± 25	0.3669 ± 19	0.2407 ± 14
C(2)	0.7186 ± 31	0.4792 ± 21	0.3859 ± 18
C(3)	0.8340 ± 30	0.4219 ± 22	0.4687 ± 15
C(4)	0.9948 ± 27	0.2364 ± 20	0.4178 ± 14
C(5)	0.1091 ± 29	0.1634 ± 23	0.4945 ± 15
C(6)	0.2568 ± 33	0.9817 ± 27	0.4408 ± 17
C(7)	0.2920 ± 28	0.8722 ± 21	0.2939 ± 16
C(8)	0.1767 ± 23	0.9408 ± 17	0.2101 ± 13
C(9)	0.0284 ± 23	0.1318 ± 16	0.2735 ± 14
N(1)	0.9112 ± 26	0.1931 ± 19	0.1895 ± 15
O(1)	0.2052 ± 26	0.8532 ± 21	0.0743 ± 16
C(10)	0.3811 ± 26	0.1964 ± 19	0.0462 ± 14
C(11)	0.6209 ± 20	0.8994 ± 15	0.0767 ± 12

	x	y	z
C(12)	0.4896 ± 19	0.1132 ± 14	0.1366 ± 10
O(2)	0.4869 ± 25	0.2007 ± 19	0.2522 ± 13
Cl(1)	0.2527 ± 10	0.4357 ± 6	0.1062 ± 5
Cl(2)	0.7501 ± 8	0.8058 ± 7	0.1820 ± 5

Anisotropic temperature factors are given also.

Structure

Both molecules are as illustrated, and are approximately planar.
The Pd–O and Pd–N distances are 1.98 ± 2 Å and 1.97 ± 2 Å respectively. Other
bond lengths and angles are given in full, and are consistent with expectation.
The structure consists of stacks (parallel to *a*) of alternating donor (bis-
8—hydroxyquinolinatopalladium(II)) and acceptor (chloranil) molecules. Adjacent
molecules in a stack are inclined at about 15° to each other, and the overlap
(illustrated in Fig. 152) is such that there can be little interaction between
π-systems. This result is attributed to packing forces and to a specific
interaction between chlorine and palladium atoms.

Details of analysis

Intensity data were recorded on *a*-axis Weissenberg photographs, using
CuK_α radiation. 1880 independent reflexions were observed and measured visually.
Refinement was by block-diagonal least squares to a final *R* index of 0.107.

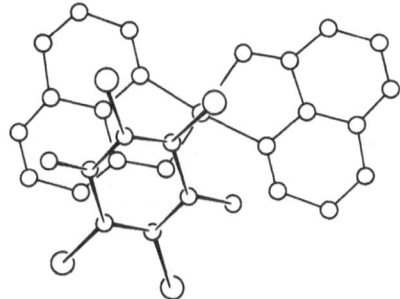

Fig. 152. Overlap of the chloranil and bis-8-hydroxyquinolinatopalladium(II)
           molecule.

BENZOTRIFUROXAN, BIS-8-HYDROXYQUINOLINATOCOPPER(II) (1:1 COMPLEX)

$C_{24}H_{12}CuN_8O_8$ F.W. = 604.2

I. Molecular complexes. Part IV. Observations on the structure of the 1:1
    molecular compound of bis-8-hydroxyquinolinatocopper(II) and benzotri-
    furoxan. C. K. PROUT and H. M. POWELL, 1965. *J. Chem. Soc.*, 4882–4887.

Monoclinic, a = 9.28 ± 2, b = 9.11 ± 2, c = 14.17 ± 3 Å, γ = 104.33 ± 20°, U = 1160.6 Å^3, D_m = 1.719, Z = 2, D_x = 1.725.

Space group $P2_1/b$ (C_{2h}^5) Molecular symmetry, centre.

Atomic positions

	x	y	z
Cu	0	0	0
N(1)	0.0391	-0.1762	0.0685
O(1)	0.1224	-0.0520	-0.0955
C(1)	-0.0075	-0.2279	0.1568
C(2)	0.0363	-0.3483	0.1857
C(3)	0.1246	-0.4283	0.1330
C(4)	0.1665	-0.3732	0.0461
C(5)	0.2496	-0.4324	-0.0221
C(6)	0.2978	-0.3680	-0.1153
C(7)	0.2493	-0.2407	-0.1385
C(8)	0.1584	-0.1715	-0.7341
C(9)	0.1173	0.2386	0.0114
C(11)	0.4445	-0.0219	0.0897
C(12)	0.5320	-0.1150	0.0553
C(13)	0.5989	-0.0892	-0.0355

For the benzotrifuroxan system only the benzene-ring carbon atoms (C(11), C(12), C(13)) are included. The positions of the other atoms of this molecule are uncertain.

Structure

Both molecules appear to be planar. The benzotrifuroxan molecule, which lacks, yet occupies a centre of symmetry, is disordered. The structure consists of stacks of alternating donor (bis-8-hydroxyquinolinatocopper(II)) and acceptor (benzotrifuroxan) molecules. Adjacent molecules in a stack are inclined at about 10° to each other, and the overlap is such that there can be little interaction between π-systems. The structure is remarkably similar to that of the complex of chloranil with bis-8-hydroxyquinolinatopalladium(II), reported elsewhere in this volume (1).

Details of analysis

The intensity data were recorded on b and c-axis equi-inclination Weissenberg photographs, using CuK$_\alpha$ radiation. Refinement, (greatly hampered by the disorder of the benzotrifuroxan molecule) was by least squares to a final R index of 0.195.

1. This volume, p. 262.

SODIUM IODIDE, METHANOL (1:3 COMPLEX)

$C_3H_{12}O_3INa$ $Na^+(CH_3OH)_3I^-$ F.W. = 246.0

I. Étude par diffraction de rayons X de complexes d'halogénures alcalins et des molecules organiques. III. Structure du complexe NaI.3CH$_3$OH.
P. PIRET and C. MESUREUR, 1965.   *J. Chim. Phys., Phys. Chim. Biol., Fr.*, 62, 287-290.

Hexagonal, a = 8.64 ± 2, c = 6.90 ± 2 Å, [U = 446 Å^3], Z = 2, D_x = 1.831.

Space group $P\bar{6}$ (C_{3h}^1), $P6_3$ (C_6^6) or $P6_3/m$ (C_{6h}^2). $P6_3/m$ confirmed by analysis. Molecular symmetry, $\bar{3}$ for Na, $\bar{6}$ for I, m for CH$_3$OH.

Atomic positions

	x	y	z
I	2/3	1/3	3/4
Na	0	0	0

	x	y	z
O	0.207	0.202	1/4
C	0.372	0.216	1/4

Isotropic temperature factors are given also.

Structure

The sodium atoms lie on the c axis, spaced at intervals of c/2 or 3.45 Å. Each sodium atom is coordinated to six oxygen atoms, distant 2.47 Å, in a distorted octahedral array. Adjacent octahedra share faces, thus forming columns extended along c. The columns are somewhat stretched; O...O distances are 3.06 Å for shared edges, and 3.88 Å otherwise. Parallel columns are held together by iodine ions, at distances slightly less than van der Waals.

Details of analysis

The intensity data were recorded on integrated Weissenberg photographs (hk0 to hk6) using MoKα radiation. The intensities for hk0 and hk1 were measured photometrically. Refinement was by Fourier projection (hk0); the parameters were confirmed by calculating structure amplitudes for hk1. The final R indices were: hk0, 0.099; hk1, 0.075.

MESITALDEHYDE, PERCHLORIC ACID (2:1 ADDUCT)

$C_{20}H_{25}ClO_6$ F.W. = 396.9

I. The crystal structure of mesitaldehyde-perchloric acid 2:1 adduct. C. D. FISHER, L. H. JENSEN and W. M. SCHUBERT, 1965. J. Amer. Chem. Soc., 87, 33-37.

Monoclinic, a = 8.43 ± 1, b = 6.91 ± 1, c = 17.39 ± 3 Å, β = 97°57 ± 9', [U = 1003 Å^3], D_m = 1.326, Z = 2, D_x = 1.315 (CuKα, λ = 1.5418 Å).

Space group P2$_1$ (C$_2^2$) or P2$_1$/m (C$_{2h}^2$). P2$_1$ inferred from intensity statistics and confirmed by analysis. Molecular symmetry, mirror plane.

Atomic positions

	x	y	z
Cl(1)	0.0249 ± 4	0.2500	0.2196 ± 2
O(2)	0.9848 ± 14	0.2500	0.1376 ± 5
O(3)	0.1956 ± 12	0.2500	0.2384 ± 7
O(4)	0.9549 ± 11	0.0857 ± 9	0.2503 ± 4
O(5)	0.9549 ± 11	0.4143 ± 9	0.2503 ± 4
O(6)	0.3532 ± 9	0.2500	0.7690 ± 4
O(7)	0.1869 ± 9	0.2500	0.6411 ± 4
C(8)	0.3226 ± 12	0.2500	0.0430 ± 6
C(9)	0.4892 ± 12	0.2500	0.0632 ± 6
C(10)	0.5888 ± 13	0.2500	0.0069 ± 6
C(11)	0.5276 ± 12	0.2500	0.9273 ± 6
C(12)	0.3605 ± 13	0.2500	0.9076 ± 6
C(13)	0.2548 ± 12	0.2500	0.9649 ± 6
C(14)	0.8490 ± 13	0.2500	0.4327 ± 6
C(15)	0.9687 ± 11	0.2500	0.4947 ± 5
C(16)	0.9220 ± 11	0.2500	0.5718 ± 5

	x	y	z
C(17)	0.7606 ± 12	0.2500	0.5800 ± 6
C(18)	0.6445 ± 12	0.2500	0.5153 ± 6
C(19)	0.6878 ± 13	0.2500	0.4404 ± 6
C(20)	0.5572 ± 16	0.2500	0.1485 ± 6
C(21)	0.6477 ± 12	0.2500	0.8707 ± 6
C(22)	0.0765 ± 14	0.2500	0.9464 ± 7
C(23)	0.2837 ± 12	0.2500	0.8281 ± 6
C(24)	0.0380 ± 12	0.2500	0.6382 ± 6
C(25)	0.5572 ± 18	0.2500	0.3704 ± 7
C(26)	0.7055 ± 15	0.2500	0.6611 ± 7
C(27)	0.1402 ± 11	0.2500	0.4787 ± 5

The above positions are not given in the original, but were taken from
Document No. 8079, obtained from A.D.I. Auxiliary Publications Project, Library
of Congress, Washington 25, D.C. Anisotropic temperature factors are given also.

Structure

Bond lengths and angles are given in full, and are consistent with expectation. The two crystallographically different mesitaldehyde moieties appear to
be hydrogen bonded to each other, *via* their carbonyl oxygens (O...O = 2.46 ± 1 Å).
The bridging hydrogen atom, however, was the only hydrogen atom not clearly
located, and reasons for this are discussed. The packing is illustrated in
Fig. 153; the mesitaldehyde moieties lie in adjacent parallel planes separated
by 3.46 Å, but the manner of their overlap does not appear to be determined by
specific interaction between them.

Details of analysis

The intensity data were recorded on uni-dimensionally integrated equi-
inclination Weissenberg photographs (0kl; h0l to h6l) using CuKα radiation. 881
of a possible 2050 reflexions were observed and measured with a microdensitometer.
Refinement was by full-matrix least squares to a final R index of 0.080.

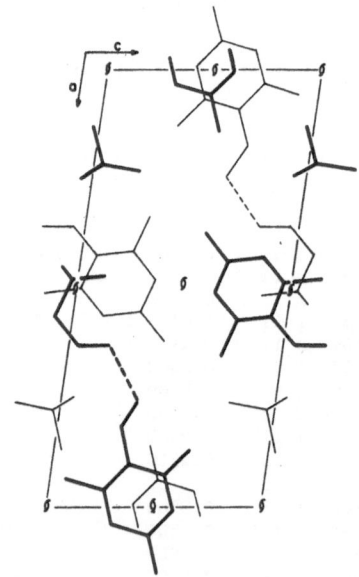

Fig. 153. The mesitaldehyde-perchloric acid (2:1 adduct) structure viewed along
       b. The dashed lines are presumed hydrogen bonds.

BENZOTRIFUROXAN, 13,14-DITHIATRICYCLO[8,2,1,1^{4,7}]TETRA-
DECA-4,6,10,12-TETRAENE, (1:1 COMPLEX)

$C_{18}H_{12}N_6O_6S_2$ F.W. = 472.5

I. Molecular complexes. Part I. The crystal and molecular structure of
the 1:1 adduct of benzotrifuroxan and 13,14-dithiatricyclo[8,2,1,1^{4,7}]-
tetradeca-4,6,10,12-tetraene. B. KAMENAR and C. K. PROUT, 1965. *J.
Chem. Soc.*, 4838-4851.

Triclinic, a = 9.71 ± 3, b = 8.01 ± 3, c = 15.28 ± 3 Å, α = 102.2 ± 2,
β = 96.2 ± 2, γ = 117.9 ± 2°, U = 996.7 Å^3, D_m = 1.566, z = 2, D_x = 1.574.
Space group P1 (C_1^1) or P$\bar{1}$ (C_i^1). P$\bar{1}$ is consistent with the deduced structure.

Atomic positions

Positions for the 32 non-hydrogen atoms are given. The standard deviations
are 0.005 Å for sulphur, and 0.02 to 0.04 Å for the other atoms. Anisotropic
temperature factors are given also.

Structure

The analysis confirms the formulations given above. The benzotrifuroxan
molecule is not strictly planar, as each furoxan ring is twisted slightly out
of the plane of the benzene nucleus. The molecule of 13,14-dithiatricyclo-
[8,2,1,1^{4,7}]tetradeca-4,6,10,12-tetraene (DTTD) has the centrosymmetrical *trans*
or step form, with parallel thiophen rings. Bond lengths and angles are given in
full, but are not very accurate. The structure consists of plane-to-plane
stacks of molecules, parallel to *a*, in which pairs of thiophen donor groups of
the DTTD molecules alternate with the acceptor molecules of benzotrifuroxan.
The dihedral angle between donor and acceptor planes is 9½°.

Details of analysis

The intensity data were recorded on equi-inclination Weissenberg photographs about *a*, using CuK_α radiation, 1192 reflexions were observed and measured
visually. Refinement was by block-diagonal least squares to a final R index of
0.137.

ETHYLENE OXIDE HYDRATE

6.4C_2H_4O.46H_2O F.W. = 1111

I. Polyhedral clathrate hydrates. IX. Structure of ethylene oxide hydrate.
R. K. McMULLAN and G. A. JEFFREY, 1965. *J. Chem. Phys.*, <u>42</u>, 2725-2732.

Cubic, a = 12.03 ± 1 Å (at -25°C), [U = 1741 Å^3], D_m = 1.059 ± 4 (at -20°C),
z = 1, D_x = 1.059.

Space group Pm3n (O_h^3) or P43n (T_d^4).   Pm3n confirmed by analysis.

Atomic positions

Atom	Set	Multi-plicity	Site Occupancy	x	y	z
Water Structure						
O(1)	(i)	16	1	0.18362	0.18362	0.18362
O(2)	(k)	24	1	0.00000	0.30710	0.11819
O(3)	(c)	6	1	0.00000	0.50000	0.25000
H(7)	(i)	16	½	0.2350	0.2350	0.2350
H(8)	(k)	24	½	0.0000	0.4233	0.1967
H(9)	(k)	24	½	0.0000	0.3767	0.1667
H(10)	(k)	24	½	0.0000	0.3350	0.0350
H(11)	(l)	48	½	0.0750	0.2483	0.1350
H(12)	(l)	48	½	0.1033	0.2217	0.1633
Guest Molecule						
C(4)	(k)	24	1/4	0.0000	0.2552	0.4388
C(4A)	(l)	48	1/8	0.0591	0.2448	0.4842
O(5)	(h)	12	1/6	0.1527	0.5000	0.0000
O(5A)	(k)	24	1/6	0.0000	0.3438	0.4735
EtO(6)	(a)	2	1/5	0.0000	0.0000	0.0000

Structure

This is a gas hydrate structure (type I), in which the water molecules of
the unit cube form a hydrogen-bonded framework defining two pentagonal dodecahedral and six tetrakaidecahedral voids. The latter are occupied by ethylene
oxide molecules, which are disordered, and appear to be axial hindered rotators.
The two dodecahedral voids are partially occupied by unidentified molecules
which are probably ethylene oxide also.

Details of analysis

The intensity data were recorded at -30°C on equi-inclination Weissenberg
photographs (hk0 to hk7) using CuK_α radiation. 300 of a possible 395 independent
reflexions were observed and measured visually. Refinement was by full-matrix
least squares, with variable site-occupancy parameters, to a final R index of 0.081.

TETRAHYDROFURAN, HYDROGEN SULPHIDE DOUBLE HYDRATE

$8C_4H_8O.7.33H_2S.136H_2O$ F.W. = 3271

I.   Polyhedral clathrate hydrates. X. Structure of the double hydrate of
     tetrahydrofuran and hydrogen sulphide. T.C.W. MAK and R. K. McMULLAN,
     1965. *J. Chem. Phys.*, <u>42</u>, 2732-2737.

Cubic, a = 17.31   1 Å, $[U$ = 5187 Å^3], D_m = 1.051, Z = 1, D_x = 1.049, (measurements at -20°C).

Space group Fd3m (O_h^7)

Atomic positions

Atom	Set	Point Symmetry	Multi- plicity	Site Occupancy	x	y	z
Water Structure							
O	(a)	$\bar{4}3m$	8	1	0.00000	0.00000	0.00000
O	(e)	3m	32	1	-0.09228	-0.09228	-0.09228
O	(g)	m	96	1	-0.05744	-0.05744	-0.24487
H	(e)'	3m	32	½	-0.060	-0.060	-0.060
H	(e)"	3m	32	½	-0.033	-0.033	-0.033
H	(g)'	m	96	½	-0.080	-0.080	-0.146
H	(g)"	m	96	½	-0.070	-0.070	-0.191
H	(g)'"	m	96	½	-0.024	-0.024	-0.245
H	(i)	1	192	½	-0.024	0.091	0.278
Guest molecules							
S	(c)	3m	16	0.458	0.12500	0.12500	0.12500
THF	(b)	$\bar{4}3m$	8	1	0.50000	0.50000	0.50000

Structure

    This is a gas hydrate structure (type II), in which the 136 water molecules of the unit cube form a hydrogen-bonded framework defining 16 pentagonal dodecahedral and 8 hexakaidecahedral voids. The latter voids enclose tetrahydrofuran molecules, which appear to undergo free rotation. The pentagonal dodecahedral voids are occupied by H_2S molecules, with a statistical occupancy of 46%..

Details of analysis

    The intensity data were recorded on equi-inclination Weissenberg photographs (hk0 to hk7) using CuK$_\alpha$ radiation. 258 of a possible 320 independent reflexions were observed and measured visually. Refinement was by full-matrix least squares, with the site occupancy of the sulphur atom treated as a variable parameter, to a final R index of 0.116.

HEXAMETHYLENETETRAMINE HEXAHYDRATE

$C_6H_{12}N_4.6H_2O$ $(CH_2)_6N_4.6H_2O$ F.W. = 248.3

    I.   Hexamethylenetetramine hexahydrate: a new type of clathrate hydrate.
         T.C.W. MAK, 1965. *J. Chem. Phys.*, 43, 2799-2805.

    II. Hexamethylenetetramine hexahydrate: a new type of clathrate hydrate.
         G. A. JEFFREY and T. C. W. MAK, 1965. *Science*, 149, 178-179.

Rhombohedral, a = 7.30 ± 1 Å, α = 105.4 ± 2°, $[U$ = 337.1 Å^3], Z = 1, D_x = 1.223, (measured at -20°C).

The structure is reported in terms of a *hexagonal* cell, for which $a_H = 11.62$, $c_H = 8.67$ Å.

Space group R32 (D_3^7), R3m (C_{3v}^5), or $R\bar{3}m$ (D_{3d}^5). R3m is confirmed by analysis.

Atomic positions

Atom	Equivalent set	Point symmetry	x	y	z
Water Structure					
O(1)	9(b)	m	0.1288 ± 6	0.2576	-0.0249 ± 18
O(2)	9(b)	m	-0.1418 ± 4	-0.2836	0.0249 ± 16
H(4)	9(b)	m	-0.185	-0.370	-0.035
H(5)	9(b)	m	0.147	0.294	-0.138
Guest molecule					
N(1)	3(a)	3m	0	0	0.2994 ± 23
N(2)	9(b)	m	0.0697 ± 5	0.1394	0.5316
C(1)	9(b)	m	0.0673 ± 6	0.1346	0.3600 ± 14
C(2)	9(b)	m	-0.0683 ± 6	-0.1366	0.5861 ± 17
H(1)	18(c)	1	0.018	0.184	0.317
H(2)	9(b)	m	-0.069	-0.138	0.712
H(3)	9(b)	m	-0.122	-0.244	0.547

Anisotropic temperature factors are given also.

Structure

This is a new type of clathrate hydrate. The water framework consists of stacks of slightly-puckered $(H_2O)_6$ rings extended in the c_H direction. Adjacent stacks are displaced with respect to one another in the stack direction, and are cross-linked by hydrogen bonds. The hexamethylenetetramine molecules occupy voids in the framework, and are hydrogen-bonded to it *via* three of the four nitrogen atoms. A portion of the structure is shown in Fig. 154.

Details of analysis

The intensity data were recorded at -20°C on equi-inclination Weissenberg photographs (8 layers about a_H) using CuK_α radiation. 292 of a possible 318 reflexions were observed, and measured visually. Refinement was by full-matrix least squares to a final R index of 0.096.

HEXAETHYL BORAZINE

$C_{12}H_{30}B_3N_3$ F.W. = 248.8

I. Molecular and crystal structure of hexaethylborazine $B_3(C_2H_5)_3N_3(C_2H_5)_3$. M. A. VISWAMITRA and S. N. VAIDYA, 1965. *Z. Kristallogr.*, **121**, 472-475.

Triclinic, a = 9.910 ± 15, b = 9.300 ± 15, c = 6.080 ± 15 Å, α = 106°28', β = 100°25', γ = 120°28' (all ± 1°), [U = 425.7 Å³], D_m = 1.00, z = 1, D_x = 0.98.

Space group P1 (C_1^1) or P$\bar{1}$ (C_i^1). P$\bar{1}$ is assumed, although this choice requires the structure to be disordered, with boron and nitrogen regarded as equivalent.

Atomic positions

	x	y	z
(B,N)(1)	0.07	0.188	0.049
(B,N)(2)	0.155	0.108	-0.02
(B,N)(3)	0.091	-0.078	-0.058
C(4)	0.138	0.387	0.092
C(5)	0.318	0.22	-0.046
C(6)	0.18	-0.164	-0.122
C(7)	0.255	0.515	0.371
C(8)	0.288	0.20	-0.314
C(9)	0.308	-0.128	0.11

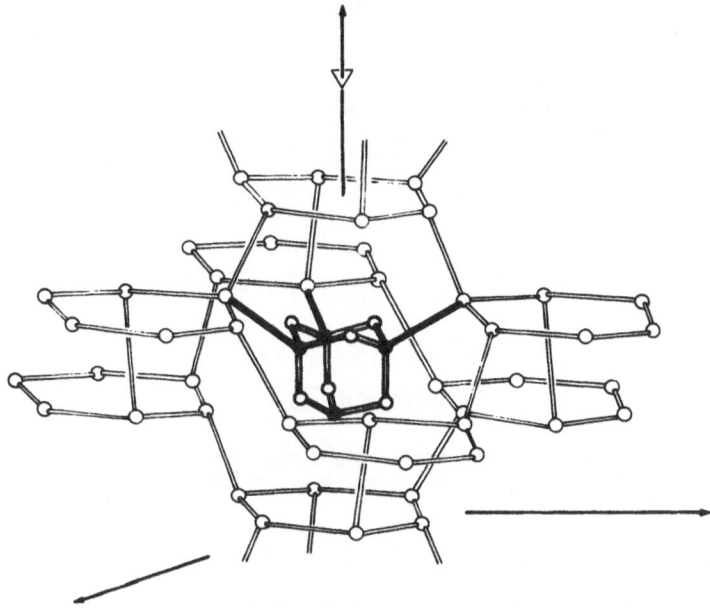

Fig. 154. A part of the hexamethylenetetramine hexahydrate structure.

Structure

The borazine ring is planar. The terminal methyl groups lie alternately above and below the ring plane. The ethyl groups appear to lie in planes which are normal to the ring plane. The mean B–N distance is 1.423 ± 15 Å, and the mean (B,N)–C–C angle is 110°.

Details of analysis

The intensity data were recorded on equi-inclination Weissenberg photographs ($hk0$ to $hk3$) using CuK_α radiation. 383 of a possible 763 reflexions were observed. Refinement was by least squares to a final R index of 0.19.

TETRAMETHYLAMMONIUM HEXAHYDROHEXABORATE

$C_8H_{30}N_2B_6$ $[N(CH_3)_4]_2B_6H_6$ F.W. = 219.3

I. The crystal and molecular structure of tetramethylammonium hexahydrohexaborate. R. SCHAEFFER, Q. JOHNSON and G. S. SMITH, 1965. *Inorg. Chem.*, *4*, 917-918.

Cubic, a = 11.84 ± 1 Å, U = 1660 Å^3, D_m = 0.88, Z = 4, D_x = 0.89.
(Crystal data for the corresponding caesium salt are given also.)

Space group F432 (O^3) F$\bar{4}$3m (T_d^2) or Fm3m (O_h^5). Fm3m is consistent with the analysis.

Atomic positions

	Position	x	z
B	24(e)	0.1007 ± 8	0
C	32(f)	0.3219 ± 4	0.3219 ± 4
N	8(c)	¼	¼
H(B)	24(e)	0.194 ± 6	0
H(N)	96(m)	0.368 ± 3	0.269 ± 4

Temperature factors (anisotropic for boron and carbon) are given also.

Structure

The hexahydrohexaborate ion has octahedral symmetry and is situated at the origin. B-B and B-H distances are 1.69 ± 1 Å and 1.11 ± 7 Å, respectively. The tetramethylammonium ion is located at ¼,¼,¼. The length of the tetrahedrally-oriented C-N bonds is 1.48 ± 1 Å.

Details of analysis

The intensity data were measured with a four-circle diffractometer, using MoK$_\alpha$ radiation. 59 independent reflexions were observed. Refinement was by full-matrix least squares to a final R index of 0.053.

BIS(o-DODECACARBORANE)

$C_4H_{20}B_{22}$ $H_{11}B_{10}C_2-C_2B_{10}H_{11}$ F.W. = 286.4

I.   Crystal and molecular structure of $C_4B_{20}H_{22}$, bis(o-dodecacarborane).
     L. H. HALL, A. PERLOFF, F. A. MAUER and S. BLOCK, 1965. *J. Chem. Phys.*, *43*, 3911-3917.

Monoclinic, a = 7.014 ± 1, b = 9.862 ± 1, c = 12.360 ± 2 Å, β = 90°31 ± 2',
[U = 855.0 Å^3], D_m = 1.11, Z = 2, D_x = 1.11.

Space group P2$_1$/n (C_{2h}^5) Molecular symmetry, centre.

Atomic positions

	x	y	z
C(1)	-0.00498 ± 13	0.05870 ± 10	0.46009 ± 7
Hb(4)	0.15600 ± 16	0.05912 ± 12	0.35900 ± 9
Hb(5)	-0.07810 ± 16	0.02166 ± 13	0.33283 ± 9
B(2)	-0.06849 ± 18	0.22025 ± 13	0.49684 ± 11
B(3)	0.17004 ± 18	0.18181 ± 13	0.46020 ± 10
B(6)	-0.22419 ± 18	0.11823 ± 14	0.41693 ± 10
B(7)	-0.18438 ± 21	0.28836 ± 15	0.38082 ± 12
B(8)	0.05929 ± 21	0.32804 ± 15	0.40774 ± 12
B(9)	0.20171 ± 20	0.22504 ± 15	0.32316 ± 11
B(10)	0.04706 ± 20	0.12198 ± 15	0.24345 ± 11
B(11)	-0.19226 ± 20	0.16070 ± 15	0.27906 ± 11
B(12)	-0.01786 ± 21	0.29167 ± 16	0.27365 ± 12

Atoms designated Hb ("hybrid") are assumed to consist of 50% each of carbon and boron. Anisotropic temperature factors are given for the above atoms. The positions and isotropic temperature factors of the hydrogen atoms are given also.

Structure

The molecule consists of two $C_2B_{10}H_{11}$ units, joined by a C–C bond across a centre of symmetry. The carbon and boron atoms of each unit lie on the corners of a slightly-distorted icosahedron. A view of the structure is given in Fig. 155. The carbon atom .in each unit which is not involved in the bridging bond is disordered, and appears to share with boron two of the five icosahedral sites adjacent to the bridgehead carbon. Bond lengths and angles are given in full. Some mean bond lengths are:

$$
\begin{array}{lr}
\text{C–C' (bridging)} & 1.522 \pm 2 \text{ Å} \\
\text{B–C} & 1.722 \pm 2 \\
\text{B–B} & 1.772 \pm 2
\end{array}
$$

The errors quoted are mean e.s.d.'s from the least-squares analysis. The standard deviations derived from comparison of chemically equivalent bonds are 0.005 to 0.012 Å. The values given here are not corrected for thermal motion. [In the original, corrections have been applied assuming: (a), riding motion; (b), independent motion. Neither of these assumptions is realistic. In some cases the "corrections" attributed to (a) are negative.]

Details of analysis

The intensity data were recorded with a four-circle diffractometer and proportional counter, using CuK_α radiation. The θ–2θ scan method was used, and 1749 of a possible 1899 reflexions were observed. Refinement was by full-matrix least squares to a final R index of 0.064.

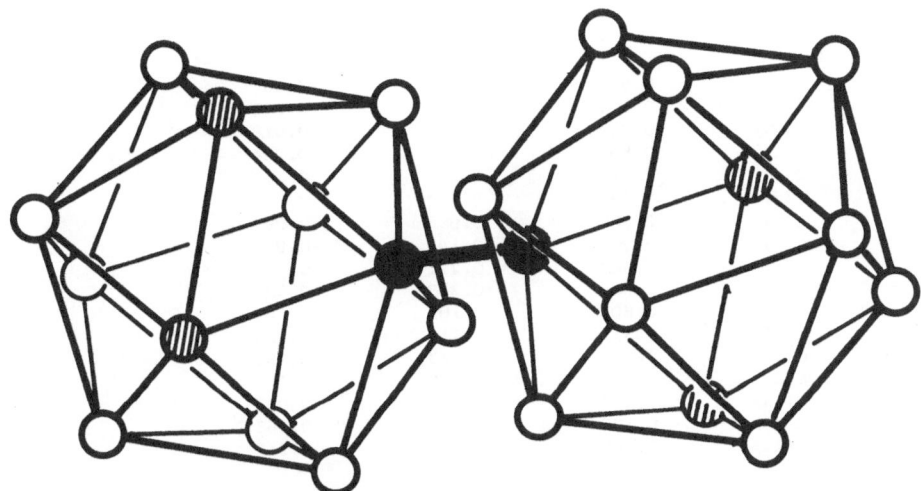

Fig. 155. The configuration of the bis(o-dodecacarborane) molecule. Solid circles represent carbon atoms, open circles boron atoms, and the shaded circles the hybrid boron-carbon atoms, as explained in the text.

1,1,4,4-TETRAMETHYL-2,3,5,6-TETRAPHENYL-1,4-
DISILICOCYCLOHEXADIENE-2,5

F.W. = 472.8

$C_{32}H_{32}Si_2$

H_5C_6 C_6H_5

H_3C C=C CH_3

Si   Si

H_3C C=C CH_3

H_5C_6 C_6H_5

I.   [The crystal structure of 1,1,4,4-tetramethyl-2,3,5,6-tetraphenyl-1,4-disilicocyclohexadiene-2,5.]   N. G. BOKIJ and JU. T. STRUČKOV, 1965. Ž. Strukt. Khim. SSSR, **6**, 571-578 [J. Struct. Chem., **6**, 543-549].

(This appears to be a preliminary communication.)

I.   Triclinic, $a = 12.63 \pm 2$, $b = 9.18 \pm 6$, $c = 6.33 \pm 2$ Å, $\alpha = 109.0 \pm 1$, $\beta = 94.5 \pm 1$, $\gamma = 93.0 \pm 1°$, $U = 689$ Å^3, $D_m = 1.14$, $Z = 1$, $D_x = 1.14$. (β is incorrectly given as $54.5 \pm 1°$ in J. Struct. Chem., English translation.)

Space group $P1$ (C_1^1) or $P\bar{1}$ (C_i^1). $P\bar{1}$ confirmed by analysis.   Molecular symmetry, centre.

Atomic positions

	x	y	z
Si	0.540	0.183	-0.001
C(1)	0.634	0.041	0.042
C(2)	0.396	0.101	-0.018
C(3)	0.556	0.217	-0.264
C(4)	0.572	0.376	0.249
C(5)	0.752	0.100	0.073
C(6)	0.809	0.083	-0.097
C(7)	0.920	0.145	-0.051
C(8)	0.967	0.221	0.173
C(9)	0.906	0.238	0.338
C(10)	0.801	0.181	0.307
C(11)	0.317	0.218	-0.034
C(12)	0.295	0.333	0.166
C(13)	0.222	0.439	0.156
C(14)	0.173	0.431	-0.067
C(15)	0.192	0.326	-0.263
C(16)	0.265	0.215	-0.247

Structure

The analysis confirms the formulation given above. The central ring is non-planar, as the silicon atoms lie ± 0.20 Å from the plane of the four carbon atoms. The planes of the phenyl rings and of the group CH_3-Si-CH_3 are approximately perpendicular to the mean plane of the central ring. Bond lengths and angles are given in full, but are not very accurate (the range of aromatic C-C distances is 1.32 to 1.47 Å). Some mean bond lengths are:

Si-C   1.88 Å

C=C   1.30

C-C_6H_5 1.52

The bonds to the silicon atoms are arranged tetrahedrally. Intermolecular distances appear to be van der Waals contacts.

Details of analysis

Intensity data were recorded on equi-inclination Weissenberg photographs ($hk0$ to $hk4$) and on reciprocal-lattice photographs ($0kl$ to $1kl$) using MoK_α radiation. 1100 independent reflexions were observed and estimated visually. Refinement was by least squares to an R index of 0.161, and is continuing.

BENZOYL(TRIPHENYLPHOSPHORANYLIDENE)METHYL CHLORIDE

BENZOYL(TRIPHENYLPHOSPHORANYLIDENE)METHYL BROMIDE

BENZOYL(TRIPHENYLPHOSPHORANYLIDENE)METHYL IODIDE

$C_{26}H_{20}OPCl$ F.W. = 414.9

$C_{26}H_{20}OPBr$ F.W. = 459.3

$C_{26}H_{20}OPI$ F.W. = 506.3

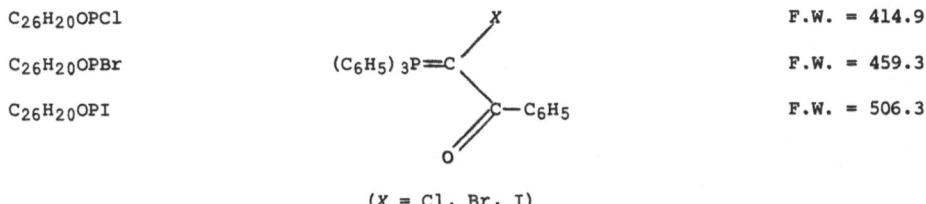

(X = Cl, Br, I)

I. The crystal and molecular structure of benzoyl(triphenylphosphoranylidene)-methyl chloride. F. S. STEPHENS, 1965. *J. Chem. Soc.*, 5658-5678.

II. The crystal and molecular structure of benzoyl(triphenylphosphoranylidene)-methyl iodide. *Idem*, 1965. *Ibid.*, 5640-5650.

I. CHLORIDE

Monoclinic, a = 11.13 ± 2, b = 12.64 ± 3, c = 15.47 ± 3 Å, β = 97°32 ± 15', U = 2158 Å^3, D_m = 1.279, Z = 4, D_x = 1.277.

Space group $P2_1/a$ (C_{2h}^5)

Atomic positions

	x	y	z
Cl	0.3726	0.4844	0.0776
P	0.3412	0.4394	0.2638
O	0.4199	0.6525	0.2913
C(1)	0.2070	0.4794	0.3115
C(2)	0.1502	0.4017	0.3591
C(3)	0.0463	0.4339	0.3970
C(4)	0.0042	0.5362	0.3846
C(5)	0.0544	0.6099	0.3357
C(6)	0.1613	0.5815	0.2972
C(7)	0.3042	0.3126	0.2146
C(8)	0.3770	0.2272	0.2297
C(9)	0.3487	0.1297	0.1915
C(10)	0.2382	0.1192	0.1340
C(11)	0.1633	0.2049	0.1189
C(12)	0.1927	0.3031	0.1568
C(13)	0.4634	0.4171	0.3501
C(14)	0.4426	0.3827	0.4317
C(15)	0.5485	0.3576	0.4956
C(16)	0.6636	0.3677	0.4742
C(17)	0.6819	0.4034	0.3939
C(18)	0.5824	0.4256	0.3295
C(19)	0.3802	0.5278	0.1862

	x	y	z
C(20)	0.4255	0.6248	0.2109
C(21)	0.4709	0.7051	0.1527
C(22)	0.4180	0.8007	0.1445
C(23)	0.4620	0.8789	0.0902
C(24)	0.5597	0.8555	0.0476
C(25)	0.6135	0.7585	0.0564
C(26)	0.5686	0.6792	0.1091

(These are given in Å, in the original.) Standard deviations are:
Cl, 0.004; O, 0.012; C, 0.013 to 0.019 Å. Isotropic temperature factors are
given also.

Structure

Bond lengths and angles are given in full. The phosphorus-carbon double
bond length is 1.736 ± 14 Å. The carbonyl group is almost coplanar with the
group P=C-Cl. The benzoyl ring is rotated 58° from the plane containing the
carbonyl group and the two adjacent carbon atoms.

Details of analysis

The intensity data were measured with a linear diffractometer ($0kl$ to
$12,kl$) using MoK_α radiation and balanced filters. 2212 of 4759 accessible
reflexions were observed and of these, 2189 were used in the analysis.
Refinement was by least squares to a final R index of 0.167.

I. BROMIDE

Monoclinic, a = 11.149, b = 12.663, c = 15.686 Å, β = 97°27', U = 2196 Å³,
D_m =1.380, Z = 4, D_x = 1.389.

Space group $P2_1/a$ (C_{2h}^5)

The bromide is isotypic with the chloride.

II. IODIDE

Monoclinic, a = 8.248 ± 10, b = 20.544 ± 40, c = 13.488 ± 38 Å, β = 101°21 ±
15', U = 2234 Å³, D_m = 1.51, Z = 4, D_x = 1.505.

Space group $P2_1/c$ (C_{2h}^5)

Atomic positions

	x	y	z
I	-0.0187	0.2423	0.1266
P	0.1359	0.1260	0.2891
O	-0.061	0.174	0.418
C(1)	0.001	0.059	0.308
C(2)	-0.159	0.058	0.252
C(3)	-0.257	0.002	0.258
C(4)	-0.200	-0.045	0.320
C(5)	-0.046	-0.050	0.374
C(6)	0.062	0.006	0.362
C(7)	0.224	0.106	0.183
C(8)	0.159	0.056	0.118
C(9)	0.250	0.049	0.033
C(10)	0.365	0.086	0.024
C(11)	0.436	0.133	0.082
C(12)	0.355	0.143	0.168
C(13)	0.299	0.126	0.397
C(14)	0.321	0.172	0.480
C(15)	0.484	0.175	0.552
C(16)	0.603	0.124	0.538
C(17)	0.575	0.083	0.467

	x	y	z
C(18)	0.440	0.080	0.395
C(19)	0.027	0.198	0.277
C(20)	-0.066	0.214	0.344
C(21)	-0.193	0.272	0.338
C(22)	-0.343	0.258	0.350
C(23)	-0.447	0.317	0.341
C(24)	-0.370	0.373	0.331
C(25)	-0.225	0.390	0.319
C(26)	-0.095	0.330	0.329

(These are given in Å, in the original.) Standard deviations are: I, 0.004; P, 0.016; O, 0.037; C, 0.05 to 0.09 Å. Isotropic temperature factors are given also.

Structure

Bond lengths and angles are given in full. The length of the phosphorus-carbon double bond is 1.71 ± 5 Å. The plane containing the carbonyl group and adjacent atoms is rotated about 12° from the plane containing the group P=C-I, and the benzoyl ring is rotated 63° from the plane containing the carbonyl group and adjacent atoms. Both rotations are in the same sense.

Details of analysis

The intensity data were measured with a linear diffractometer ($0kl$ to $8kl$), using MoK_α radiation with balanced filters. 938 of a possible 6150 reflexions were observed, and of these, 914 were used in the analysis. Refinement was by least squares to a final R index of 0.157.

p-TOLYL TRIPHENYLPHOSPHORANYLIDENEMETHYL SULPHONE

$C_{26}H_{23}O_2PS$ F.W. = 430.5

M.P. = 185-186°C

I. An X-ray determination of the molecular structure of a Wittig reagent: p-tolyl triphenylphosphoranylidenemethyl sulphone. P. J. WHEATLEY, 1965. *J. Chem. Soc.*, 5785-5800.

Monoclinic, a = 25.633 ± 87, b = 8.981 ± 8, c = 20.733 ± 71 Å, β = 111°54 ± 10', U = 4428.5 Å^3, D_m = 1.30, Z = 8, D_x = 1.291 (CuK_α, λ = 1.542 Å).

Space group Cc (C_s^4) or $C2/c$ (C_{2h}^6). $C2/c$ confirmed by analysis.

Atomic positions

Positions (in Å) and isotropic temperature factors are given for the 30 non-hydrogen atoms. Standard deviations are 0.005 Å for sulphur and phosphorus, and 0.01 to 0.03 Å for the remaining atoms.

Structure

Bond lengths and angles are given in full. Some bond lengths are: P-C, 1.71 ± 2; P-C$_6$H$_5$ (mean), 1.81 ± 2; C-S, 1.69 ± 2; S-C$_7$H$_7$, 1.77 ± 2 Å. The phenyl rings bonded to the phosphorus atom are so oriented that the group C-P-(C$_6$H$_5$)$_3$ resembles a three-bladed propeller. Intermolecular distances are consistent with van der Waals interaction.

Details of analysis

The intensity data were recorded with a linear diffractometer ($h0l$ to $h,12,l$) using MoK_α radiation. 1750 of a possible 6249 reflexions were observed above background. Refinement was by least squares to a final R index of 0.164.

N-p-BROMOPHENYLTRIPHENYLPHOSPHINE IMIDE, DIMETHYLACETYL-
ENEDICARBOXYLATE ADDUCT

C$_{30}$H$_{25}$NO$_4$PBr   (C$_6$H$_5$)$_3$P=C—C—(C$_6$H$_4$)Br   F.W. = 574.4

$$\text{H}_3\text{CO}_2\text{C} \quad \text{CO}_2\text{CH}_3$$

I. The structure of the adduct of Ph$_3$P=N.Ph and MeO$_2$C.C=C.CO$_2$Me. T. C. W. MAK and J. TROTTER, 1965. *Acta Cryst.*, **18**, 81-88.

Monoclinic, a = 11.83 ± 2, b = 9.24 ± 2, c = 25.43 ± 4 Å, β = 104°15 ± 5', U = 2694 Å^3, D_m = 1.4, Z = 4, D_x = 1.416 (CuK_α, λ = 1.5418 Å; MoK_α, λ = 0.7107 Å).

Space group $P2_1/c$ (C_{2h}^5)

Atomic positions

Positions and temperature factors (anisotropic for the bromine atom) are given for the 37 non-hydrogen atoms.

Structure

The molecular configuration is illustrated in Fig. 156. Bond lengths and angles are given in some detail, and are consistent with expectation. The arrangement of bonds to the phosphorus atom is tetrahedral; the phosphorus-carbon bonds are: double bond, 1.70 ± 3 Å; single bond (mean of three), 1.83 ± 3 Å. The intermolecular distances are consistent with van der Waals interaction.

Details of analysis

The intensity data were recorded with a four-circle diffractometer and scintillation counter (MoK_α radiation, Θ-2Θ scan, 2Θ < 41°), and on Weissenberg photographs (CuK_α radiation, $h0l$ to $h4l$, visual estimation). 1940 reflexions were observed and used in the refinement, which was by block-diagonal least squares to a final R index of 0.21.

BIS(DIPHENYLPHOSPHINO)ETHYLAMINE ETHYL IODIDE

C$_{28}$H$_{30}$INP$_2$ [(C$_6$H$_5$)$_2$P.N(C$_2$H$_5$).P(C$_2$H$_5$)(C$_6$H$_5$)$_2$]$^+$I$^-$ F.W. = 569.4

I. X-ray studies of aminophosphine complexes of molybdenum and palladium, and of an aminophosphonium iodide. D. S. PAYNE, J. A. A. MOKUOLU and J. C. SPEAKMAN, 1965. *Chem. Communic.*, 599.

Orthorhombic, a = 13.90, b = 21.22, c = 10.04 Å, [U = 2961 Å^3], Z = 4, [D_x = 1.28].

Space group $P2_12_12_1$ (D_2^4)

Atomic positions are not given.

Structure

The analysis confirms the formulation given above. Mean bond lengths and angles are given. For the four-coordinated phosphorus atom, P-N = 1.75 ± 3 Å and N-P-N = 109.5 ± 7°. For the three-coordinated phosphorus atom, P-N = 1.88 ± 3 Å and N-P-N = 102.0 ± 7°.

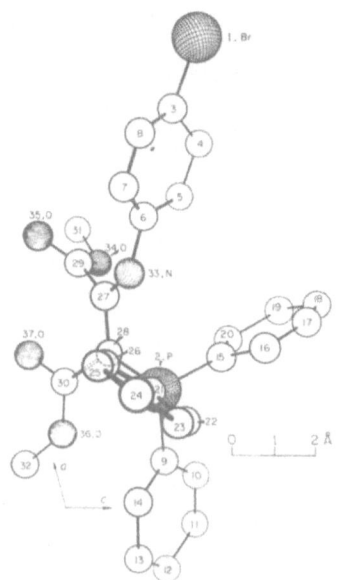

Fig. 156. The $C_{30}H_{25}NO_4PBr$ molecule, viewed along b.

Details of analysis

The intensity data were recorded photographically, using CuK_α radiation. 1221 reflexions were observed and measured visually. Refinement was by least squares to a final R index of 0.13.

PHOSPHOBENZENE B

$C_{36}H_{30}P_6$ $(P.C_6H_5)_6$ F.W. = 648.5

I. The crystal and molecular structure of a trigonal form of phosphobenzene B, $(PC_6H_5)_6$. J. J. DALY, 1965. *J. Chem. Soc.*, 4789–4799.

Trigonal, a = 13.026, c = 11.547 Å, U = 1696.8 Å^3, D_m = 1.265, Z = 2, D_x = 1.269.

Space group $P3c1$ (C_{3v}^4) or $P\bar{3}c1$ (D_{3d}^4). $P\bar{3}c1$ suggested by statistics and confirmed by analysis. Molecular symmetry, $\bar{3}$

Atomic positions

	x	y	z
P	0.1127	0.1645	0.0513
C(1)	0.2150	0.3067	-0.0190
C(2)	0.2160	0.3251	-0.1390
C(3)	0.2996	0.4374	-0.1847
C(4)	0.3772	0.5276	-0.1048
C(5)	0.3728	0.5096	0.0125
C(6)	0.2902	0.3976	0.0564

The positions are given in Å in the original. Mean standard deviations are: P, 0.0016; C, 0.01 Å. Anisotropic temperature factors are given also.

Structure

   The nucleus of the molecule is a ring of six phosphorus atoms in the chair
form. Adjacent phosphorus atoms lie 0.592 Å above and below the equatorial
plane. To each phosphorus atom is attached a phenyl group in what might (by
analogy with cyclohexane) be described as the equatorial position. The phenyl
rings are inclined 5.6° to *c*. Bond lengths and angles are given in full: normal
aromatic values are found for the phenyl groups, and other values are given
in Fig. 157.

Details of analysis

   Intensity data were recorded with a linear diffractometer, using MoK_α
radiation. 879 of a possible 1397 reflexions were observed above background.
Refinement was by least squares to a final R index of 0.072.

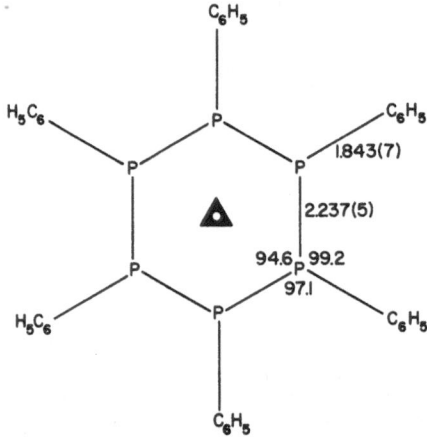

Fig. 157. Some bond lengths and angles in phosphobenzene B. Standard deviations
           are: P–P, 0.005; P–C, 0.007 Å; P–P–P, 0.1; P–P–C, 0.02°.

2,2-DIPHENYL-4,4,6,6-TETRACHLOROCYCLOTRIPHOSPHAZATRIENE

$C_{12}H_{10}Cl_4N_3P_3$ $(C_6H_5)_2Cl_4P_3N_3$ F.W. = 431.0

   I. X-ray crystallography of the phenyltriphosphonitriles. I. The crystal
   structure of 2,2-diphenyl-4,4,6,6-tetrachlorocyclotriphosphazatriene.
   N. V. MANI, F. R. AHMED and W. H. BARNES, 1965. *Acta Cryst.*, 19,
   693–698.

Monoclinic, *a* = 10.653 ± 2, *b* = 13.223 ± 2, *c* = 12.776 ± 2 Å, β = 94°54 ± 3',
U = 1793.2 Å^3, D_m = 1.598, Z = 4, D_x = 1.596 (Cu$K_{\alpha 1}$ λ = 1.54050 Å;
Cu$K_{\alpha 2}$, λ = 1.54434 Å).

Space group $P2_1/n$ (C_{2h}^5)

Atomic positions

	x	*y*	*z*
N(1)	0.0388	0.2016	0.0693
P(2)	0.1269	0.2806	0.1386
N(3)	0.0760	0.3125	0.2493
P(4)	−0.0426	0.2634	0.2910
N(5)	−0.1263	0.1884	0.2186
P(6)	−0.0747	0.1479	0.1144
Cl(1)	0.0049	0.1969	0.4284

	x	y	z
Cl(2)	-0.1583	0.3715	0.3357
Cl(3)	-0.2176	0.1428	0.0036
Cl(4)	-0.0404	-0.0006	0.1346
C(1)	0.2811	0.2265	0.1608
C(2)	0.3075	0.1330	0.1201
C(3)	0.4303	0.0954	0.1341
C(4)	0.5228	0.1492	0.1878
C(5)	0.4965	0.2416	0.2285
C(6)	0.3757	0.2814	0.2137
C(7)	0.1503	0.3931	0.0657
C(8)	0.1924	0.3862	-0.0347
C(9)	0.2221	0.4719	-0.0876
C(10)	0.2095	0.5662	-0.0429
C(11)	0.1646	0.5734	0.0529
C(12)	0.1361	0.4878	0.1087

Standard deviations are given in full, in Å. Average values are: P, 0.0015; Cl, 0.002; N, 0.005; C, 0.005 to 0.010 Å. Hydrogen positions are given, and anisotropic temperature factors for the non-hydrogen atoms.

Structure

The shape of the molecule is illustrated in Fig. 158. Bond lengths and angles are given in full; mean values of those in and adjacent to the cyclophosphazene ring are as follows (refer to Fig. 159.):

$a = 1.578 \pm 5$ Å $\Theta(1) = 119.2 \pm 3°$
$b = 1.555 \pm 5$ $\Theta(2) = 119.7 \pm 3$
$c = 1.615 \pm 5$ $\Theta(3) = 122.0 \pm 3$
$d = 1.998 \pm 2$ $\Theta(4) = 115.2 \pm 2$
$e = 1.788 \pm 6$ $\Theta(5) = 100.3 \pm 1$
                     $\Theta(6) = 104.4 \pm 3$

Intermolecular distances are consistent with van der Waals contacts.

Details of analysis

Intensity data were collected with a four-circle diffractometer and scintillation counter, using MoK_α radiation. Within the range $\sin \theta/\lambda < 0.68$, 2529 of a possible 4472 reflexions were observed. Refinement was by block-diagonal least squares to a final R index of 0.048.

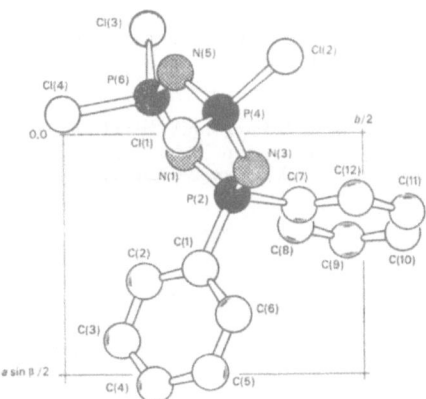

Fig. 158. The $C_{12}H_{10}Cl_4N_3P_3$ molecule viewed along c.

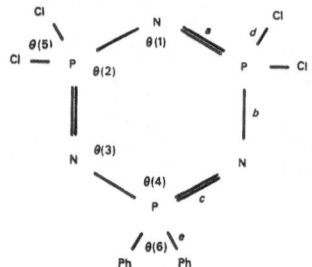

Fig. 159. The cyclophosphazene ring and adjacent atoms.

CACODYLIC ACID

$C_2H_7O_2As$ $(CH_3)_2AsO.OH$ F.W. = 138.0

    I. Stereochemistry of arsenic. Part XVI. Cacodylic acid. J. TROTTER and
T. ZOBEL, 1965. *J. Chem. Soc.*, 4466-4471.

Triclinic, a = 6.53 ± 1, b = 6.82 ± 1, c = 6.61 ± 1 Å, α = 77°30 ± 5,
β = 78°45 ± 5, γ = 55°9 ± 5', U = 234.9 Å^3, D_m = 1.95, Z = 2, D_x = 1.95.
Space group $P1$ (C_1^1) or $P\bar{1}$ (C_i^1). $P\bar{1}$ is consistent with the deduced structure.
Atomic positions

	x	y	z
As(1)	0.1667	0.0602	0.1785
C(2)	0.4072	-0.2732	0.2794
C(3)	0.1953	0.2660	0.3049
O(4)	0.1879	0.1290	-0.0726
O(5)	-0.1089	0.1157	0.2543

    Standard deviations (in Å) are given in full. Anisotropic temperature
factors are given also.

Structure

    The configuration of bonds around the arsenic atom is tetrahedral, with
bond angles ranging from 106 to 115°. The mean As-C distance is 1.91 ± 4 Å.
The As-O distances are, within experimental error, equal, with the mean value
1.62 ± 3 Å. Adjacent molecules are linked by pairs of O-H...O bonds (of length
2.57 Å) to form centrosymmetrical dimers of the carboxylic acid type.

Details of analysis

    The intensity data were recorded on equi-inclination Weissenberg photo-
graphs ($0kl$ to $5kl$) using CuK_α radiation. 806 reflexions were observed and
measured visually. Refinement was by block-diagonal least squares to a final
R index of 0.149.

10-CHLORO-5,10-DIHYDROPHENARSAZINE

$C_{12}H_9NAsCl$ F.W. = 277.5

I. Stereochemistry of arsenic. Part XIII. 10-Chloro-5,10-dihydrophenarsazine.
A. CAMERMAN and J. TROTTER, 1965. *J. Chem. Soc.*, 730-738.

Orthorhombic, a = 5.47 ± 1, b = 13.91 ± 2, c = 14.30 ± 2 Å, U = 1088 Å³,
D_m = 1.693, Z = 4, D_x = 1.694.

(Crystal data of a monoclinic form containing solvent of crystallization
are also given in the paper.)

Space group $P2_12_12_1$ (D_4^4)

Atomic positions

	x	y	z
As(1)	0.1305	0.4818	0.1659
Cl(2)	-0.1196	0.5858	0.2465
N(3)	-0.2282	0.4773	-0.0164
C(4)	-0.1678	0.3165	0.1912
C(5)	-0.3629	0.2545	0.1754
C(6)	-0.5215	0.2698	0.1003
C(7)	-0.4750	0.3461	0.0372
C(8)	-0.2700	0.4079	0.0499
C(9)	-0.1248	0.3935	0.1312
C(10)	0.3277	0.6197	0.0395
C(11)	0.3495	0.6747	-0.0411
C(12)	0.1799	0.6622	-0.1132
C(13)	-0.0110	0.5969	-0.1030
C(14)	-0.0374	0.5409	-0.0213
C(15)	0.1366	0.5526	0.0516

Anisotropic temperature factors are given also.

Structure

The molecule is slightly folded about the As-N axis, the angle between the
two *o*-phenylene groups being 169°, with the chlorine atom lying outside the
angle. Bond lengths and angles are given in full; the more important values
are: As-Cl, 2.301 ± 4; As-C, 1.917 ± 7; C-N, 1.371 ± 9; C-C (mean),
1.406 ± 5 Å; Cl-As-C, 96.1 ± 2; C-As-C, 97.0 ± 4°; C-N-C, 128.0 ± 8°. Intermolecular distances are consistent with van der Waals interaction.

Details of analysis

The intensity data were measured with a four-circle diffractometer, using
CuK_α radiation. In the range 2θ < 148°, 1033 reflexions (79% of the accessible
total), were observed above background. Refinement was by block-diagonal least
squares to a final R index of 0.056.

DIETHYLMAGNESIUM

$C_4H_{10}Mg$

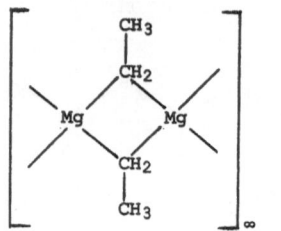

F.W. = 82.4

I. Die Kristallstruktur des diathylmagnesiums. E. WEISS, 1965. *J. Organo-metal. Chem.*, **4**, 101-108.

Tetragonal, a = 7.29 ± 2, c = 5.34 ± 2 Å, U = 284.0 Å^3, z = 2, D_x = 0.964.

Space group not determined, but assumed to be $P4_2/mcm$ (D_{4h}^{10}). Molecular symmetry, $\overline{4}2m$ for a disordered structure.

Atomic positions

		x	y	z
Mg	2(*b*)	0	0	¼
C(1)	8(*n*)	0.23	0.13	0
C(2)	8(*n*)	0.24	0.33	0

C(1) and C(2) are half-atoms, representing a disordered structure. Assumed hydrogen positions are given also.

Structure

Each magnesium atom is tetrahedrally coordinated to four methylene groups. Adjacent magnesium atoms are joined by pairs of ethyl groups, as shown above, to form polymeric chains extended along c.

Details of analysis

Powder intensity data were recorded with a counter diffractometer, using CuK_α radiation. It is shown that these data are consistent with the proposed structure.

TELLURIUM DIMETHANETHIOSULPHONATE, THIOUREA (1:1 COMPLEX)

$C_4H_{14}N_4O_4S_6Te$ $Te(CS(NH_2)_2)_2(S_2O_2CH_3)_2$ F.W. = 502.2

I. The crystal and molecular structure of a *trans* square-planar complex of tellurium dimethanethiosulphonate with thiourea. O. FOSS, K. MAROY and S. HUSEBYE, 1965. *Acta Chem. Scand.*, **19**, 2361-2368.

Monoclinic, a = 12.50, b = 5.60, c = 12.80 Å, β = 98°, [U = 887 Å^3], z = 2, [D_x = 1.88].

Space group $P2_1/n$ (C_{2h}^5) Molecular symmetry, centre.

Atomic positions

	x	y	z
Te	0	0	0
S(1)	0.0355	0.2543	0.1776
C(1)	0.0840	0.0312	0.2685
N(1)	0.1602	-0.1208	0.2478

	x	*y*	*z*
N(2)	0.0440	0.0225	0.3595
S(2)	0.1459	0.2618	−0.0848
S(3)	0.2837	0.0821	−0.0337
C(1)	0.2910	−0.1620	−0.1225
O(1)	0.3740	0.2430	−0.0465
O(2)	0.2879	−0.0024	0.0730

Mean standard deviations are: S, 0.01 Å; O, 0.04 Å; N, 0.05 Å; C, 0.06 Å. Temperature factors are given also.

Structure

The shape and dimensions of the molecule are given in Fig. 160. (Bond lengths and angles are given in full.) The structure is held together by a system of N-H...O bonds of length 2.90 to 3.14 Å.

Details of analysis

The intensity data were recorded on zero-level Weissenberg photographs about the principal axes, using CuK_α radiation. 310 of a possible 376 reflexions were observed, and measured visually. Refinement was by least squares to a final *R* index of 0.098.

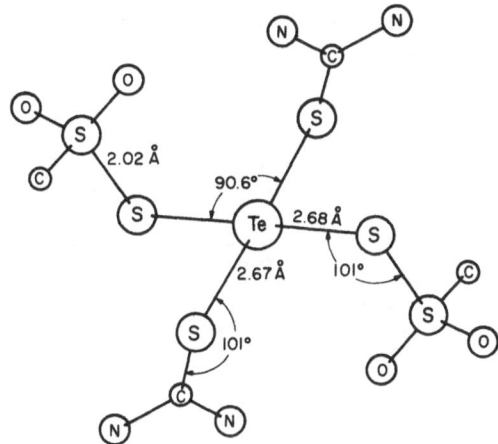

Fig. 160. The $C_4H_{14}N_4O_4S_6Te$ molecule as seen along the *b* axis. Standard deviations are: Te-S, 0.015 Å; S-S, 0.02 Å; S-Te-S, 0.5°; Te-S-S, 0.7°; Te-S-C, 2°.

TETRATHIOUREATELLURIUM(II) DICHLORIDE

TETRATHIOUREATELLURIUM(II) DICHLORIDE DIHYDRATE

$C_4H_{16}Cl_2N_8S_4Te$ F.W. = 503.0

$$Te[SC(NH_2)_2]_4Cl_2 \cdot nH_2O \quad (n = 0, 2)$$

$C_4H_{16}Cl_2N_8S_4Te \cdot 2H_2O$ F.W. = 539.0

I. The crystal structures of tetrathioureatellurium(II) dichloride and its dihydrate. K. FOSHEIM, O. FOSS, A. SCHEIE and S. SOLHEIMSNES, 1965. *Acta Chem. Scand.*, 19, 2336–2348.

ANHYDROUS MATERIAL

Triclinic, *a* = 5.83, *b* = 7.77, *c* = 10.86 Å, α = 95.5, β = 90, γ = 119.5°, [*U* = 425.6 Å³], *z* = 1, D_x = 1.6.

Space group P1 (C_1^1) or P$\bar{1}$ (C_i^1). P$\bar{1}$ is consistent with analysis. Molecular symmetry, centre.

Atomic positions

	x	y	z
Te	0	0	0
S(1)	-0.205	-0.347	0.0983
C(1)	0.052	-0.287	0.208
N(1)	0.251	-0.320	0.174
N(2)	0.065	-0.190	0.316
S(2)	-0.212	0.141	0.1740
C(2)	0.028	0.253	0.296
N(3)	0.276	0.300	0.277
N(4)	-0.028	0.313	0.403
Cl	0.547	-0.233	0.4510

DIHYDRATE

Monoclinic, a = 6.01, b = 16.49, c = 9.95 Å, β = 98.5°, [U = 975.3 Å^3], Z = 2, [D_x = 1.83].

Space group P2$_1$/c (C_{2h}^5) Molecular symmetry, centre.

Atomic positions

	x	y	z
Te	0	0	0
S(1)	0.121	0.1565	0.033
C(1)	-0.076	0.200	0.122
N(1)	-0.285	0.175	0.100
N(2)	-0.016	0.265	0.199
S(2)	0.102	-0.0245	0.268
C(2)	-0.096	0.034	0.333
N(3)	-0.021	0.096	0.410
N(4)	-0.316	0.015	0.310
Cl	-0.512	0.3240	0.269
H$_2$O	-0.511	0.151	0.442

Structure

The coordination of the tellurium in both compounds is square planar, with a mean Te-S distance of 2.685 ± 10 Å. The thiourea groups are variously oriented. Both structures are stabilized by hydrogen bonds involving the nitrogen and chlorine atoms and, for the dihydrate, the water molecule.

Details of analysis

The intensity data were recorded on zero-level Weissenberg photographs, using CuK$_\alpha$ radiation. The intensities were measured visually and, in part, photometrically. Refinement was by Fourier methods to final R indices ranging from 0.08 to 0.10 for the several projections.

TRITHIOUREATELLURIUM(II) HYDROGEN DIFLUORIDE

C$_3$H$_{14}$F$_4$N$_6$S$_3$Te   Te(CS(NH$_2$)$_2$)$_3$(HF$_2$)$_2$ F.W. = 434.0

I. The crystal structure of trithioureatellurium(II) hydrogendifluoride.
    O. FOSS and S. HAUGE, 1965. *Acta Chem. Scand.*, 19, 2395-2403.

Monoclinic, a = 5.91, b = 20.68, c = 11.47 Å, β = 95°, [U = 1397 Å^3], Z = 4, [D_x = 2.06].

Space group P2$_1$/c (C_{2h}^5)

Atomic positions

	x	y	z
Te	-0.1274	0.0953	0.0500
S(1)	-0.2870	0.1352	0.2348
C(1)	-0.061	0.142	0.341
N(1)	-0.109	0.131	0.449
N(2)	0.160	0.151	0.313
S(2)	-0.1522	0.2008	-0.0510
C(2)	-0.385	0.192	-0.158
N(3)	-0.596	0.185	-0.127
N(4)	-0.330	0.185	-0.268
S(3)	0.0700	0.0401	-0.1458
C(3)	-0.156	0.024	-0.251
N(5)	-0.370	0.016	-0.226
N(6)	-0.110	0.019	-0.362
F(1)	0.133	0.187	-0.339
F(2)	0.278	0.283	-0.397
F(3)	0.476	0.128	-0.484
F(4)	0.327	0.032	-0.424

Structure

Pairs of tellurium atoms are linked by sulphur bridges to form binuclear cations as illustrated in Fig. 161. The tellurium and sulphur atoms in a cation are coplanar. Adjacent cations are held together by HF_2 anions *via* N-H...F bonds of length 2.63 to 2.92 Å (standard deviation, 0.07 Å).

Details of analysis

The intensity data were recorded on *a* and *c* axis Weissenberg photographs, using CuK_α radiation, and measured visually. 304 of a possible 456 reflexions were observed. Refinement was by Fourier methods to a final *R* index of 0.10.

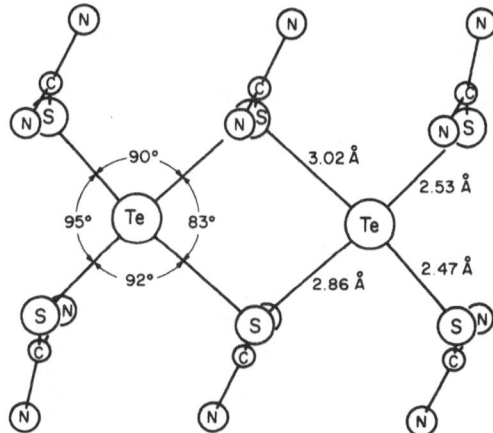

Fig. 161. The binuclear cation. Standard deviations are 0.015 Å and 0.4°.

TELLURIUM DIBROMIDE, ETHYLENETHIOUREA (1:1 COMPLEX)

TELLURIUM DIIODIDE, ETHYLENETHIOUREA (1:1 COMPLEX)

$C_6H_{12}N_4Br_2S_2Te$ F.W. = 491.7

$C_6H_{12}N_4I_2S_2Te$ F.W. = 585.7

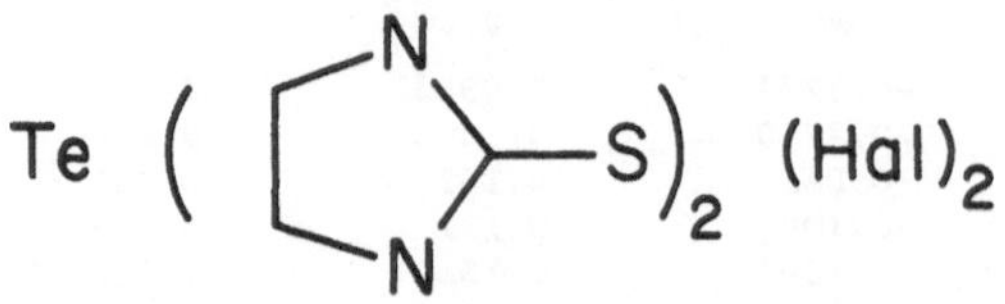

I. The crystal and molecular structures of *trans* square-planar complexes of tellurium dibromide and diiodide with ethylenethiourea. O. FOSS, H. M. KJOGE and K. MAROY, 1965. *Acta Chem. Scand.*, 19, 2349-2360.

DIBROMIDE

Monoclinic, a = 17.30, b = 5.89, c = 15.57 Å, β = 120°, [U = 1374 Å³], Z = 4, [D_X = 2.38].

Space group Cc (C_s^4) or $C2/c$ (C_{2h}^6). $C2/c$ confirmed by analysis. Molecular symmetry, centre.

Atomic positions

	x	y	z
Te	¼	¼	0
Br	0.3775	-0.079	0.1070
S	0.2224	0.301	0.1540
C(1)	0.152	0.075	0.145
N(1)	0.138	-0.109	0.092
C(2)	0.067	-0.252	0.089
C(3)	0.045	-0.134	0.164
N(2)	0.099	0.077	0.184

     Mean standard deviations are: Br, 0.005 Å; S, 0.013 Å. The positions of the carbon and nitrogen atoms were not refined, but were adjusted to conform to known molecular geometry.

DIIODIDE

Monoclinic, a = 17.17, b = 6.24, c = 15.97 Å, β = 119½°, [U = 1489 Å³], Z = 4, [D_X = 2.61].

Space group, same as dibromide.

Atomic positions

	x	y	z
Te	¼	¼	0
I	0.3835	-0.086	0.1116
S	0.2190	0.302	0.1480
C(1)	0.151	0.086	0.142
N(1)	0.138	-0.091	0.091
C(2)	0.070	-0.231	0.093
C(3)	0.046	-0.110	0.162
N(2)	0.101	0.086	0.184

     Mean standard deviations are: I, 0.004 Å; S, 0.016 Å. As for the dibromide, the positions of the carbon and nitrogen atoms were not refined.

Structure (both derivatives)

     The structures are isotypic. The coordination of the tellurium atom is square planar. Bond lengths and angles are as follows:

Dibromide: Te-Br, 2.78 ± 1 Å;   Te-S, 2.690 ± 15 Å; Br-Te-S, 89.2 ± 3°.
Diiodide: Te-I, 2.97 ± 1 Å;   Te-S, 2.69 ± 2 Å;   I-Te-S, 89.2 ± 4°.

Details of analysis

The intensity data were recorded on zero-level Weissenberg photographs (integrated for the diiodide) about the b axis and the ab diagonal, using CuK$_\alpha$ radiation. The intensities were measured visually. For the dibromide, 264 of a possible 323 reflexions were observed, and for the diiodide, 259 of 342. Refinement was by Fourier methods to final R indices of 0.10 for both compounds.

trans-DITHIOCYANATODIAMMINECOPPER

$C_2H_6CuN_4S_2$ F.W. = 213.8

$$\begin{array}{c} NH_3 \\ | \\ S-C-N-Cu-N-C-S \\ | \\ NH_3 \end{array}$$

I. Crystal structure of *trans*–dithiocyanatodiamminecopper. CHIN-LING HUANG and YUAN CHU CHEN, 1965. *Sci. Sinica*, <u>14</u>, 924–927.

Orthothombic, a = 14.02, b = 8.88, c = 6.10 Å, $[U = 746 \text{ Å}^3]$, D_m = 1.87, Z = 4, D_x = 1.87.

Space group Pna2$_1$ (C_{2v}^9) or Pnam (D_{2h}^{16}). Pnam is consistent with the analysis. Molecular symmetry, mirror plane.

Atomic positions

	x	y	z
Cu	0.176	0.102	¼
S(1)	0.324	0.602	¼
S(2)	0.033	−0.391	¼
C(1)	0.274	0.438	¼
C(2)	0.079	−0.225	¼
N(1)	0.236	0.312	¼
N(2)	0.114	−0.101	¼
N(3) (NH$_3$)	0.044	0.199	¼
N(4) (NH$_3$)	0.316	0.025	¼

Structure

The copper atom is bonded to four nitrogen atoms, at an average distance of 2.04 Å, in a square planar array. The thiocyanate groups are linear, as is the thiocyanate-copper-thiocyanate chain. Octahedral coordination is completed by the sulphur atoms of neighbouring atoms (Cu...S = 3.05 Å).

Details of analysis

The structure was determined from Fourier projections along b and c. The final agreement index was 0.19.

SILVER FULMINATE

CAgNO   AgCNO   F.W. = 149.9

I. The crystal structure of silver fulminate. D. BRITTON and J. D. DUNITZ, 1965. *Acta Cryst.*, <u>19</u>, 662–668.

   (The structure of an orthorhombic and of a trigonal form are reported.)

Orthorhombic form: a = 3.864 ± 6, b = 10.722 ± 18, c = 5.851 ± 10 Å, U = 242.4 Å^3, Z = 4, D_x = 4.107.

Space group $Cmc2_1$ (C_{2v}^{12}), $C2cm$ (C_{2v}^{16}) or $Cmcm$ (D_{2h}^{17}) $Cmcm$ is confirmed by the analysis. Molecular symmetry; 2/m.

Atomic positions

	x	y	z
Ag	0	0	0
O	0	0.3753 ± 17	¼
N	0	0.2584 ± 25	¼
C	0	0.1564 ± 15	¼

The quoted standard deviations are said to be optimistic. Isotropic temperature factors are given also.

Structure

As illustrated in Fig. 162, the structure consists of endless zigzag chains -C-Ag-C-Ag-C- extended in the c direction. The C-Ag-C linkage is linear, and the angle Ag-C-Ag is 82°. The distance Ag-C is 2.23 ± 2 Å, and the fulminate ion is linear, with C-N = 1.09 ± 3 Å, and N-O = 1.25 ± 3 Å. There is an inter-chain Ag-O distance of 2.77 Å.

Details of analysis

The intensity data were recorded with a linear diffractometer and scintillation counter ($hk0$ to $hk6$, $\sin\theta < 0.5$) using MoK_α radiation. 127 independent reflexions were observed. Refinement was by full-matrix least squares to a final R index of 0.134.

Trigonal form: Rhombohedral axes, $a = 9.109 \pm 15$ Å, $\alpha = 115°44'$, $U = 393.3$ Å^3, $Z = 6$, $D_x = 3.796$. Hexagonal axes, $a = 15.427 \pm 26$, $c = 5.726 \pm 10$ Å, $Z = 18$. *Space group* $R3$ (C_3^4) or $R\bar{3}$ (C_{3i}^2). $R\bar{3}$ is confirmed by analysis.

Atomic positions

	x	y	z
Ag	0.0635 ± 2	0.1493 ± 2	−0.2017 ± 2
O	0.8521 ± 25	0.7081 ± 24	0.3578 ± 24
N	0.6398 ± 17	0.5060 ± 17	0.2106 ± 17
C	0.4349 ± 20	0.3285 ± 20	0.0701 ± 20

The quoted standard deviations are said to be optimistic. Isotropic temperature factors are given also.

Structure

The structural unit consists of hexameric puckered rings, as illustrated in Fig. 163. Some bond lengths and angles are: Ag-C, 2.17 ± 3 Å; C-N, 1.12 ± 3 Å; N-O, 1.20 ± 3 Å; C-Ag-C, 163°; Ag-C-Ag, 81°. The fulminate ions are not quite linear (C-N-O = 172°). There is an inter-ring Ag-O distance of 2.45 Å. The intra-ring Ag-Ag separation is 2.82 Å, shorter than the value of 2.89 Å found for metallic silver.

Details of analysis

The intensity data were recorded on Weissenberg photographs ($hk0$ to $hk4$) using MoK_α radiation. 321 of a possible 420 reflexions were observed and measured visually. Refinement was by least squares to a final R index of 0.151.

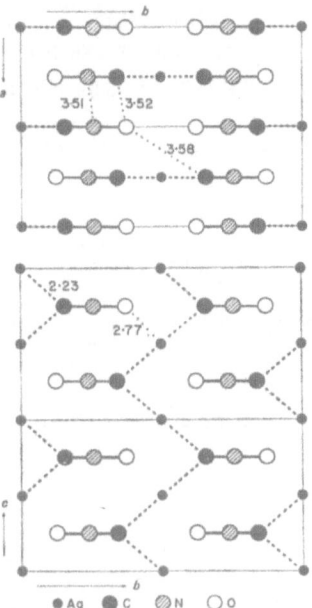

Fig. 162. Orthorhombic silver fulminate. Top: view along *c* axis. Bottom: view along *a* axis. Distances are given in Å. All Ag-C distances are shown as dashed lines.

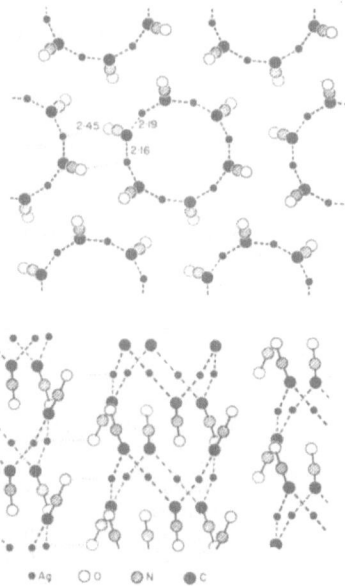

Fig. 163. Rhombohedral silver fulminate. Top: view down hexagonal *c* axis. Bottom: side view of the central line of hexamer units shown in the top view. Distances are given in Å. All Ag-C distances are shown as dashed lines.

ZEISE SALTS
(POTASSIUM TRICHLOROETHYLENE PLATINOATE)
(POTASSIUM TRIBROMOETHYLENE PLATINOATE)

$C_2H_4Cl_3KPt.H_2O$ $K(PtC_2H_4X_3).H_2O$ (X = Cl, Br)   F.W. = 386.6

$C_2H_4Br_3KPt.H_2O$ F.W. = 520.0

  I. [X-ray investigation of potassium trichloro(ethylene) platinoate(II).]
      G. B. BOKIJ and G. A. KUKINA, 1965. Ž. Strukt. Khim. SSSR, 6, 706-715
      [J. Struct. Chem., 6, 670-677].

      (These structures have been redetermined. For earlier work, see 1, 2.
      For subsequent three-dimensional work, see 3, 4.)

X = Cl:

Monoclinic, a = 10.85 ± 2, b = 8.53 ± 1, c = 4.81 ± 1 Å, β = 97°, [U =
442 Å^3], Z = 2, [D_x = 2.90].

Space group $P2_1$ (C_2^2)

Atomic positions

	x	y	z
Pt	0.215	0.000	0.328
Cl(1)	0.424	0.000	0.532
Cl(2)	0.215	0.269	0.328
Cl(3)	0.209	0.733	0.328
K	0.436	0.304	0.931
O(H$_2$O)	0.344	0.579	0.798
C(1)	0.007	0.000	0.285
C(2)	0.045	0.000	0.025

Standard deviations are: Pt, 0.0025; Cl, 0.018; K, 0.015; O, 0.047
C, 0.066 Å.

Structure

     The coordination of the platinum ion is essentially square planar. Chlorine
atoms occupy three corners of the square, at distances 2.27 to 2.39 Å, and
the ethylenic double bond occupies the fourth, with a mean Pt...C distance of
2.25 Å. The axis of the ethylene molecule is normal to the plane of the other
atoms of the array. The potassium atom is coordinated to six chlorine atoms
at the corners of a trigonal prism, and to two water molecules.

X = Br

Monoclinic, a = 11.38 ± 2, b = 8.78 ± 2, c = 5.01 ± 1 Å, β = 97°,
[U = 497 Å^3], Z = 2, [D_x = 3.47].

Space group $P2_1$ (C_2^2)

Atomic positions

	x	y	z
Pt	0.211	0.000	0.329
Br(1)	0.424	0.000	0.519
Br(2)	0.211	0.282	0.329
Br(3)	0.211	0.719	0.329

	x	y	z
K	0.438	0.275	0.938
O(H_2O)	(0.33)	0.592	(0.81)
C(1)	(-0.003)	0.000	(0.144)
C(2)	(0.067)	0.000	(0.012)

Standard deviations are: Pt, 0.0028; Br, 0.008; K, 0.017; O, 0.049; C, 0.070 Å. Values in parentheses 'were not completely accurately determined'.

Structure

The structure is isotypic with that of the chlorine derivative. The platinum-bromine coordination distances range from 2.42 to 2.52 Å.

Details of analysis

Intensities of $h0l$ and $hk0$ were recorded photographically and estimated visually. Refinement was by Fourier methods. Reliability indices range from 0.15 to 0.21 for the several projections.

1. *Structure Reports*, <u>21</u>, 503.

2. *Structure Reports*, <u>18</u>, 633.

3. M. BLACK, R. H. B. MAIS and P. G. OWSTON, 1969. *Acta Cryst.*, <u>B25</u>, 1753.

4. J. A. J. JARVIS, B. T. KILBOURN and P. G. OWSTON, 1970. *Ibid.*, <u>B26</u>, 876.

PYRIDINIUM TETRABROMORHENATE(II)

$C_5H_7Br_4NRe$ $(C_5H_5NH)^+(ReBr_4H)^-$ F.W. = 587

    I.   [The crystal structure of (PyH)HRe^{II}Br$_4$.] P. A. KOZ'MIN, V. G. KUZNECOV and Z. V. POPOVA, 1965. *Ž. Strukt. Khim. SSSR*, <u>6</u>, 651-652 [*J. Struct. Chem.*, <u>6</u>, 624-625].

α-FORM

Tetragonal, a = 11.908 ± 5, c = 7.605 ± 3 Å, [U = 1078 Å^3], D_m = 3.63, Z = 4, D_x = 3.58.

Space group $I4/mmmm$ (D_{4h}^{17}) (confirmed by analysis.) Molecular symmetry, $4mm$ for tetrabromorhenate anion, m for pyridinium cation.

Atomic positions

		x	y	z
Re	4(e)	0	0	0.1447
Br	16(m)	0.142	0.142	0.231
(C,N)	16(l)	0.418	0.082	0
C(2)	16(l)	0.387	0.030	0
C(3)	16(l)	0.470	0.113	0

(The C,N parameters represent a pyridine molecule which suffers two-fold rotational disorder in its own plane.)

β-FORM

Orthorhombic, a = 7.76 ± 2, b = 16.76 ± 4, c = 16.83 ± 4 Å, [U = 2189 Å^3], Z = 8, D_x = 3.56.

Space group $Cccm$ (D_{2h}^{20}) or $Ccc2$ (C_{2v}^{13}). $Cccm$ confirmed by analysis. Molecular symmetry, mirror plane for tetrabromorhenate anion.

Atomic positions

	x	y	z
Re	0.104	0.251	0
Br(1)	0.020	0.396	0
Br(2)	0.020	0.108	0
Br(3)	0.020	0.254	0.144
(N,C)(1)	0.220	0.000	0.167
C(2)	0.220	0.071	0.208
C(3)	0.220	0.071	0.292

Structure

In both forms the rhenium atom occurs in the dimeric anion $(Br_4Re-ReBr_4)^{4-}$ (or $(HBr_4Re-ReBr_4H)^{2-}$). The bromine atoms lie at the corners of a square prism, and the rhenium atoms lie on its axis. Some bond lengths are: Re–Re, 2.21 and 2.27 Å; Re–Br, 2.48 and 2.50 Å.

Details of analysis

There are few details. Fourier and least-squares methods were used. The final R index for the α-form was 0.142.

CHROMIUM CARBONYL ACETOPHENONE COMPLEX

WOLFRAM CARBONYL ACETOPHENONE COMPLEX

$C_{13}H_8O_6Cr$ $(CO)_5M(C_6H_5COCH_3)$ F.W. = 312.2

$C_{13}H_8O_6W$ F.W. = 444.1

I. [On the existence of metal carbonyl carbene complexes.] O. S. MILLS and A. D. REDHOUSE, 1965. *Angew. Chem.*, 77, 1142-1143 [*Internat. Edit.*, 4, 1082].

CHROMIUM COMPLEX

Monoclinic, a = 9.91, b = 21.70, c = 6.34 Å, β = 96°35', [U = 1350 Å^3], z = 4, D_x = 1.54.

WOLFRAM COMPLEX

Monoclinic, a = 10.00, b = 21.90, c = 6.45 Å, β = 96°12', [U = 1404 Å^3], z = 4, D_x = 2.10.

Space group Cc (C_s^4) or $C2/c$ (C_{2h}^6). Cc confirmed by analysis.

Atomic positions are not given.

Structure

The structure of the chromium complex has been determined, and that of the wolfram complex is isomorphous. The molecule, illustrated in Fig. 164, corresponds to a carbene-metal complex. The phenyl group is perpendicular to the plane of CH_3OCCCr; the latter plane approximately bisects the equatorial (CO–Cr–CO) axes of the chromium octahedron. The Cr–carbene distance of 2.05 Å is significantly longer than the Cr–CO distances.

Details of analysis

The intensity data for the chromium complex were recorded on precession photographs, using MoK_α radiation. 565 reflexions were observed. The method of refinement is not stated.

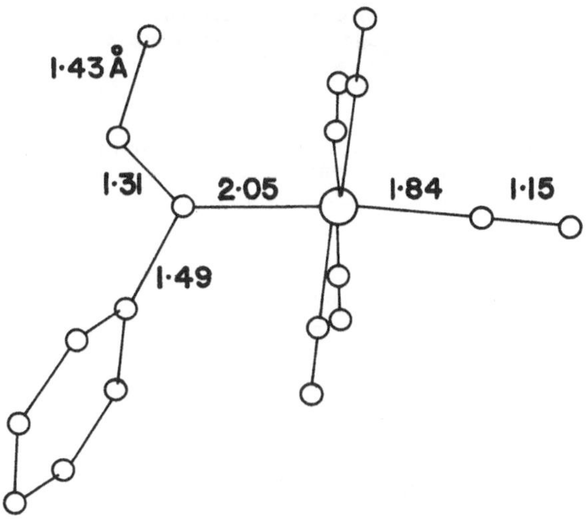

Fig. 164. The molecule of the chromium complex $C_{13}H_8O_6Cr$. Distances not shown are: Cr-CO (equat.) 1.84 to 1.92 Å; C-O (equat.) 1.12 to 1.15 Å. Standard deviations are 0.02 to 0.03 Å.

PENTAKIS(METHYLISONITRILE)COBALT(I) PERCHLORATE

$C_{10}H_{15}ClCoN_5O_4$ $[Co(CNCH_3)_5]^+[ClO_4]^-$ F.W. = 363.6

I. The crystal and molecular structure of pentakis(methylisonitrile)cobalt-(I)perchlorate. F. A. COTTON, T. G. DUNNE and J. S. WOOD, 1965. *Inorg. Chem.*, <u>87</u>, 318-325.

Trigonal, a = 11.61 ± 2, c = 10.75 ± 2 Å, $[U = 1255$ Å$^3]$, D_m = 1.48, Z = 3, D_x = 1.45.

Space group $P3_121$ (D_3^4) or enantiomorph. Molecular symmetry, two-fold axis.

Atomic positions

	x	y	z
Co	0.2455 ± 2	0.2455 ± 2	½
C(1)	0.083 ± 2	0.083 ± 2	½
N(1)	−0.015 ± 1	−0.015 ± 1	½
C(2)	−0.140 ± 3	−0.140) ± 3	½
C(3)	0.190 ± 1	0.296 ± 2	0.640 ± 2
N(2)	0.157 ± 1	0.330 ± 1	0.725 ± 1
C(4)	0.109 ± 2	0.365 ± 2	0.834 ± 2
C(5)	0.247 ± 2	0.385 ± 2	0.412 ± 2
N(3)	0.247 ± 1	0.471 ± 2	0.358 ± 2
C(6)	0.231 ± 2	0.570 ± 2	0.292 ± 3
Cl	0.4110 ± 6	0.4110 ± 6	0
O(1)	0.312 ± 3	0.383 ± 2	0.089 3
O(2)	0.431 ± 1	0.529 ± 1	−0.063 ± 2

Isotropic temperature factors are given also.

Structure

The coordination of the cobalt atom is essentially trigonal bipyramidal. The methylisonitrile groups are linear, and colinear with the cobalt atom. Minor deviations from the ideal configuration are discussed. The coordination of the chlorine atom is tetrahedral. Average interatomic distances are: Co-C, 1.87 ± 2Å; C-N, 1.14 ± 2 Å; N-CH$_3$, 1.44 ± 3 Å; Cl-O, 1.43 ± 2 Å.

Details of analysis

The intensity data were recorded on equi-inclination Weissenberg photographs (hk0 to hk,10) using MoK$_\alpha$ radiation. 700 reflexions were observed in the range sinθ<0.4. Refinement was by full-matrix least squares to a final R index of 0.091.

1,5-DI(IRON TRICARBONYL)-3-METHYLENEPENTA-1,4-DIENE

(RED-ORANGE ISOMER)

C$_{12}$H$_6$Fe$_2$O$_6$ F.W. = 357.9

I. Complexes du fer-carbonyle avec l'acétylène et ses dérivés. I. Structure du 1,5-di(fer-tricarbonyle)-3-méthylène-1,4-pentadiène, Fe$_2$(CO)$_6$(C$_2$H$_2$)$_3$, isomère rouge-orange (p.f. 62°C). P. PIRET, J. MEUNIER-PIRET and M. VAN MEERSSCHE, 1965. Acta Cryst., 19, 78-84.

Monoclinic, a = 14.75 ± 2, b = 13.26 ± 2, c = 7.036 ± 10 Å, β = 94.6 ± 2°, [U = 1375.5 Å^3], D$_m$ = 1.724, Z = 4, D$_x$ = 1.733, (calibration with silver, a = 4.086 Å).

Space group Cc (C$_s^4$) or C2/c (C$_{2h}^6$). Cc indicated by statistics and confirmed by analysis.

Atomic positions

	x	y	z
Fe(1)	0.5000	0.3295	0.5000
Fe(2)	0.6291	0.2047	0.4901
O(1)	0.6296	0.5004	0.5090
O(2)	0.4466	0.3727	0.8963
O(3)	0.3481	0.4109	0.2543
O(4)	0.7037	0.2586	0.8694
O(5)	0.7890	0.2865	0.3214
O(6)	0.6677	-0.0121	0.4797
C(1)	0.5749	0.4286	0.5149
C(2)	0.4636	0.3614	0.7365
C(3)	0.4069	0.3785	0.3624
C(4)	0.6684	0.2329	0.7037
C(5)	0.7250	0.2502	0.3882
C(6)	0.6570	0.0763	0.4840
C(7)	0.5566	0.2776	0.2662
C(8)	0.5404	0.1825	0.2166
C(9)	0.4535	0.1412	0.2768
C(10)	0.4365	0.1751	0.4658
C(11)	0.5130	0.1801	0.5995
C(12)	0.4011	0.0777	0.1650

Standard deviations (in Å) are given in full. Mean values are: Fe, 0.005 Å; O, 0.025 Å; C, 0.03 Å. Isotropic temperature factors are given also.

Structure

The molecules (the configuration of which is illustrated in Fig. 168) has a non-crystallographic two-fold axis joining the terminal methylene group and the mid-point of the iron atoms. Bond lengths and angles are given in full, and are consistent with the formula given above. The Fe-Fe distances is

2.527 ± 6 Å. The Fe – CH distances range from 1.96 to 2.26 Å (±0.03 Å), and
the distances of the iron atoms from the mid-points of the double bonds are both
2.06 ± 3 Å. The intermolecular distances are consistent with van der Waals
interaction.

Details of analysis

The intensity data were recorded on integrated equi-inclination Weissenberg
photographs (*hk*0 to *hk*5) using CoK_α radiation. 765 reflexions were observed
and measured visually. Refinement was by least squares to a final *R* index of
0.123.

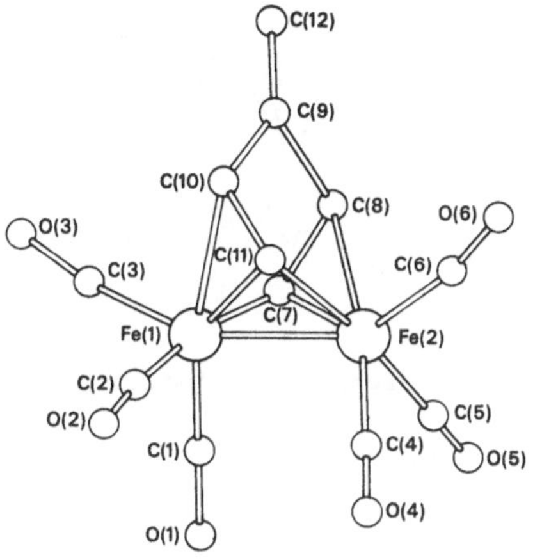

Fig. 165. Molecular configuration and the numbering scheme in $C_{12}H_6Fe_2O_6$.

IRON CARBONYL DIPHENYLACETYLENE COMPLEX

$C_{36}H_{20}Fe_3O_8$ $Fe_3(CO)_8(C_6H_5C_2C_6H_5)_2$ F.W. = 748

    I. The molecular structures of two isomers of $Fe_3(CO)_8(C_6H_5C_2C_6H_5)_2$.
       R. P. DODGE and V. SCHOMAKER, 1965. *J. Organometal. Chem. Netherl.*,
       <u>3</u>, 274–284.

BLACK ISOMER

Monoclinic,   *a* = 9.39, *b* = 18.45, *c* = 18.29 Å, β = 96.8°, *U* = 3148 Å^3,
D_m = 1.578,   z = 4, D_x = 1.57.
Space group $P2_1/c$ (C_{2h}^5)

VIOLET ISOMER

Monoclinic,   *a* = 38.49, *b* = 8.31, *c* = 21.75 Å, β = 115.1°, *U* = 6304 Å^3,
D_m = 1.575,   z = 8, D_x = 1.576.
Space group Cc (C_s^4) or $C2/c$ (C_{2h}^6). C2/c is confirmed by analysis.

Atomic positions

Positions and isotropic temperature factors for the 47 non-hydrogen atoms
of each of the isomers are given in the original.

Structure

Structural details of both isomers are represented schematically in Figs.
166 and 167. In the black isomer, the two diphenylacetylene molecules have fused
to a tetraphenylbutadiene unit, which combines with an iron atom to form a
planar ferracyclopentadiene ring, as shown. The remaining two iron atoms are

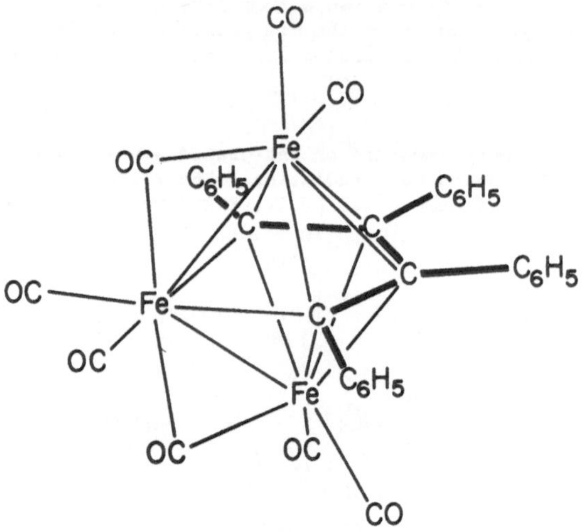

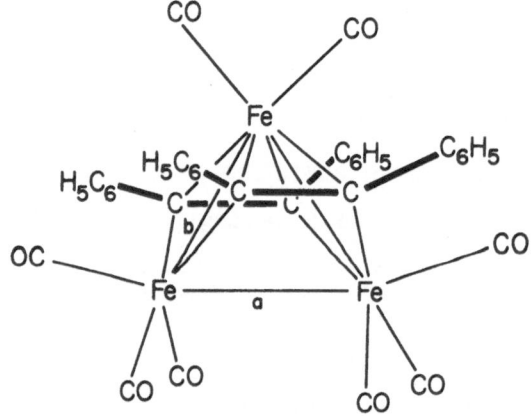

Fig. 166. Details of the black isomer of $C_{36}H_{20}Fe_3O_8$.

Fig. 167. Details of the violet isomer of $C_{36}H_{20}Fe_3O_8$.

π-bonded to the ring, one on each side. Some mean interatomic distances are:

    Fe – Fe   = 2.431 ± 3 Å
    Fe – CO   = 1.80 ± 2
    Fe – C   = 2.13 ± 2
    C – C₆H₅   = 1.51 ± 2
    C – C(ring)   = 1.45 ± 2

  The phenyl rings are twisted by about 55° (all in the same sense) out of the plane of the ferracyclopentadiene ring. Otherwise the molecule has approximate *mm*2 symmetry.

  In the violet isomer, the diphenylacetylene molecules remain separate, being located on opposite sides of a triangle of iron atoms. Some mean interatomic distances are:

    Fe – Fe (a)   = 2.592 ± 5 Å   Fe – C   = 2.00 ± 2 Å
    Fe – Fe (b)   = 2.463 ± 5   C – C₆H₅   = 1.48 ± 3
    Fe – CO   = 1.76 ± 3   C≡C   = 1.38 ± 3

Apart from the twists of the phenyl rings, this isomer too has approximate *mm*2 symmetry.

Details of analysis

The intensity data for both forms were measured with a four-circle diffractometer in the stationary-crystal stationary-counter mode, using Zr-filtered MoK_α
radiation. 1510 of a possible 1811 reflexions were observed for the black form,
and 1649 of 1969 for the violet form. Refinement was by full-matrix least
squares to final R indices of 0.087 and 0.127, respectively, for the black
and violet forms.

BIS-METHALLYLNICKEL

$C_8H_{14}Ni$

F.W. = 168.9

I. Kristall- und Molekülstruktur von *bis*-Methallylnickel $Ni[(CH_2)_2C.CH_3]_2$.
    R. UTTECH and H. DIETRICH, 1965. *Z. Kristallogr.*, **122**, 60–72.

Monoclinic, $a = 6.05$, $b = 13.28$, $c = 5.83$ Å, $\beta = 117.1°$, $U = 423$ Å^3,
$Z = 2$, $D_x = 1.32$.

Space group $P2_1/c$ (C_{2h}^5) Molecular symmetry, centre.

Atomic positions

	x	y	z
Ni	½	0	½
C(1)	0.4388 ± 13	0.1427 ± 5	0.5527 ± 14
C(2)	0.6702 ± 11	0.1297 ± 4	0.5534 ± 11
C(3)	0.6771 ± 11	0.0813 ± 5	0.3410 ± 13
C(4)	0.9039 ± 12	0.1441 ± 5	0.7969 ± 13

Isotropic temperature factors are given, as are assumed positions for the
hydrogen atoms.

Structure

The molecular configuration is indicated schematically in Fig. 168. Two
centro-symmetrically related methallyl groups are coordinated, *via* their partial
double bonds, to the nickel atom. In addition to a crystallographic centre of
symmetry, the molecule has an approximate mirror plane, containing the nickel
atom and the C-CH$_3$ bonds. Relevant distances are: Ni-CH$_2$, 2.02 ± 1 Å;
Ni-C, 1.98 ± 1 Å; C-CH$_2$, 1.41 ± 1 Å; C-CH$_3$, 1.49 ± 2 Å. The methallyl group
is not planar, the C-CH$_3$ bond being bent out of the allyl plane by about 12°.
The bend is towards the nickel atom, and does not appear to be related to
steric forces.

Details of analysis

The intensity data were recorded on Weissenberg photographs about various
axes, using CuK_α radiation. 715 of 839 accessible reflexions were observed
and measured visually. Refinement was by Fourier methods to a final R index of
0.079. (Least-squares refinement was attempted, but led to anomalous results
attributed to systematic errors in the intensity data.)

ALLYLPALLADIUM CHLORIDE COMPLEX

$C_6H_{10}Pd_2Cl_2$ $[(C_3H_5)PdCl]_2$ F.W. = 366.4

I. The structure of the allylpalladium chloride complex $(C_3H_5PdCl)_2$ at
    −140°C. A. E. SMITH, 1965. *Acta Cryst.*, **18**, 331–340.

II.   Structure of and bonding in $[(C_3H_5)PdCl]_2$.   W. E. OBERHANSLI and L. F.
DAHL, 1965.   *J. Organometal. Chem.*, $\underline{3}$, 43-54.

(For reports of two-dimensional work see 1.)

I.   *Monoclinic*, $a = 7.46_2$, $b = 7.42_7$, $c = 8.60_5$Å, $\beta = 93.5°$, $[U = 476.9 \text{ Å}^3]$,
$Z = 2$, $D_x = 2.556$, (Cell data at room temperature;   intensity data at $-140°C$).

Space group $P2_1/n$ (C_{2h}^5) Molecular symmetry, centre.

Atomic positions

	x	y	z
Pd	0.62600 ± 6	0.31440 ± 6	0.44990 ± 5
Cl	0.3238 ± 3	0.4201 ± 2	0.3899 ± 2
C(1)	0.6355 ± 12	0.0773 ± 11	0.3130 ± 10
C(2)	0.8817 ± 11	0.1895 ± 11	0.4738 ± 9
C(3)	0.7494 ± 16	0.0596 ± 15	0.4411 ± 13

Anisotropic temperature factors are given for these atoms.   The hydrogen
atoms have been located but their positions are not given explicitly.

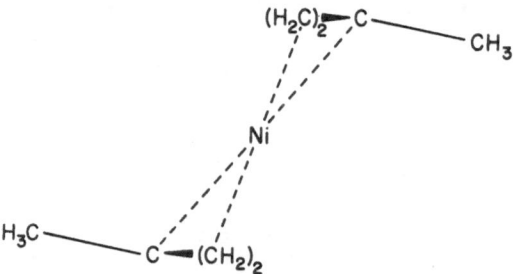

Fig. 168.   Schematic representation of the (bis-methallylnickel) molecule,
viewed along the CH_2-CH_2 vector.

Structure

The molecular packing is illustrated in Fig. 169.   Intermolecular distances
appear to be consistent with van der Waals interaction.   A further view of the
dimer is given in Fig. 170.   Bond lengths and angles are given in full;   some
average values are:

Pd–Cl	2.413 ± 2 Å
C–Pd	2.117 ± 8
C–C	1.376 ± 15
Cl–Pd–Cl	$88.3 \pm 2°$ [88.1]
Pd–Cl–Pd	92.1 ± 2 [91.9]
C(1)–C(3)–C(2)	119.8 ± 9

The hydrogen atoms are stated to be in the plane of the allyl group.

Details of analysis

The intensity data were recorded on equi-inclination Weissenberg photographs
($0kl$ to $7kl$; $h0l$ to $h31$) using MoK$_\alpha$ radiation.   1890 independent reflexions
were observed and measured visually.   Refinement was by full-matrix least squares
to a final R index of 0.055.

II.   *Monoclinic*, $a = 7.46 \pm 2$, $b = 7.43 \pm 2$, $c = 8.61 \pm 2$ Å, $\beta = 93.6 \pm 1°$,
$U = 475.9 \text{ Å}^3$, $D_m = 2.51$, $Z = 2$, $D_x = 2.56$ (MoK$_\alpha$, $\lambda = 0.7107$ Å).

Space group as in I

Atomic positions

	x	y	z
Pd	0.6264 ± 2	0.3159 ± 2	0.4504 ± 2
Cl	0.3248 ± 8	0.4204 ± 8	0.3921 ± 8
C(1)	0.6391 ± 39	0.0711 ± 38	0.3130 ± 31
C(2)	0.8866 ± 32	0.1960 ± 35	0.4796 ± 30
C(3)	0.7540 ± 57	0.0767 ± 48	0.4430 ± 42

Isotropic temperature factors are given for the carbon atoms, and isotropic temperature factors for palladium and chlorine. The origin, and the numbering system of the carbon atoms, have been changed from the original to correspond to I.

Structure

The structure is as described in I, but is considered to be less accurate. Bond lengths and angles are given in full.

Details of analysis

The intensity data were recorded on equi-inclination Weissenberg photographs ($h0l$ to $h8l$) using MoK$_\alpha$ radiation. 535 independent reflexions were observed and estimated visually. Refinement was by full-matrix least squares to a final R index of 0.069.

1. *Structure Reports*, <u>27</u>, 820.

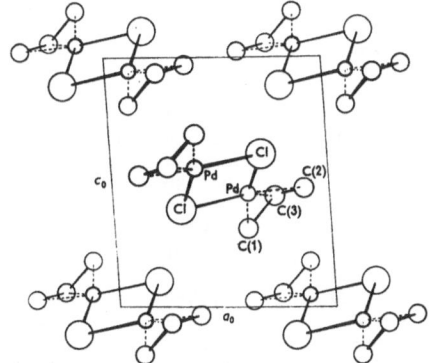

Fig. 169. The structure of the allylpalladium chloride complex viewed along b.

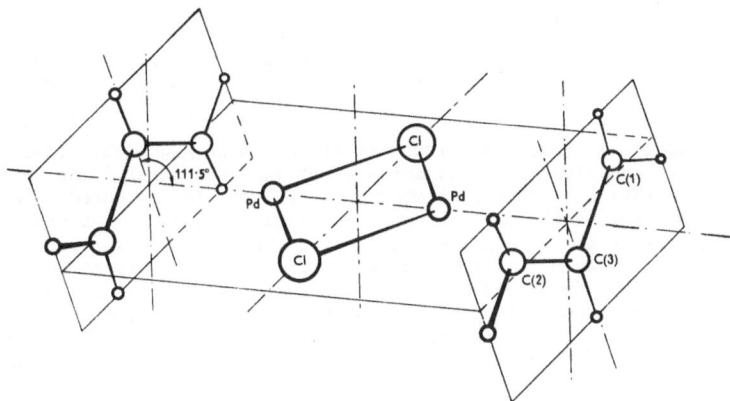

Fig. 170. Schematic drawing of the dimer.

2,4-BIS-(π-CYCLOPENTADIENYLCOBALT-CYCLOPENTADIEN)YL-

CYCLOPENTADIENE-2,4.

$C_{25}H_{24}Co_2$ $(C_5H_5)Co(C_5H_5C_5H_4C_5H_5)Co(C_5H_5)$ F.W. = 442

I. [X-ray structural study of the compound $Co_2C_{25}H_{24}$.]   O. V. STAROVSKIJ
    and J. T. STRUČKOV, 1965. *Ž. Strukt. Khim. SSSR*, <u>6</u>, 248-261 [*J. Struct.
    Chem.*, <u>6</u>, 228-240].

    For preliminary results, see <u>1</u>, <u>2</u>.

Orthorhombic, a = 5.83, b = 13.38, c = 23.74 Å, (all ± 1%), U = 1851 Å^3,
Z = 4, D_x = 1.593.

Space group Pbc2$_1$ (C$_{2v}^5$) or Pbcm (D$_{2h}^{11}$). Pbc2$_1$ confirmed by analysis.

Atomic positions

	x	y	z
Co(1)	0.7737	0.6729	0.5003
Co(2)	0.7121	0.1280	0.7879
C(1)	1.0152	0.4512	0.6299
C(2)	0.7910	0.4585	0.6109
C(3)	0.6613	0.3798	0.6295
C(4)	0.8181	0.3166	0.6667
C(5)	1.0215	0.3597	0.6687
C(11)	0.8551	0.7816	0.4410
C(12)	0.6988	0.8186	0.4860
C(13)	0.5124	0.7709	0.4863
C(14)	0.5215	0.7018	0.4448
C(15)	0.7116	0.7091	0.4166
C(11')	0.6852	0.5324	0.5726
C(12')	0.8086	0.5194	0.5138
C(13')	1.0033	0.5730	0.5130
C(14')	0.9929	0.6500	0.5624
C(15')	0.7822	0.6363	0.5838
C(21)	0.5736	-0.0101	0.7661
C(22)	0.4514	0.0473	0.8143
C(23)	0.5756	0.0562	0.8628
C(24)	0.7878	0.0218	0.8492
C(25)	0.8052	-0.0207	0.7958
C(21')	0.7470	0.2271	0.6967
C(22')	0.5936	0.2492	0.7473
C(23')	0.7272	0.2730	0.7991
C(24')	0.9436	0.2306	0.7907
C(25')	0.9189	0.1817	0.7924

    (Numbering scheme as in original) Also given are isotropic temperature
factors for the above atoms, and assumed positions for the hydrogen atoms.

Structure

    Some details of the molecular structure are given in Fig. 171. (C–C bond
lengths and angles are given in full, but are not very accurate). The structure
consists of two sandwich fragments bridged by a central five-membered ring.
In each sandwich fragment the outer ring is essentially planar. For each inner
ring four of the atoms lie in a plane parallel to that of the outer ring, but
the fifth, bridgehead, atom is displaced from this plane, away from the cobalt
atom, by 0.4 to 0.5 Å. The atoms of the central ring lie in a plane which is
approximately normal to the principal planes of the sandwich fragments. The
respective conformations of the two sandwiches differ. That about Co(1) is
antiprismatic, while that about Co(2) is prismatic. Intermolecular distances
are given, and appear to be normal van der Waals contacts.

Details of analysis

Intensity data were recorded on equi-inclination Weissenberg photographs ($0kl$ to $4kl$) using CoK_α radiation. 1100 independent reflexions were observed and measured visually. Refinement was by least squares to a final R index of 0.19.

1. O. V. STAROVSKIJ and J. T. STRUČKOV, 1960. *Ž. Strukt. Khim. SSSR*, **3**, 612.

2. *Idem*, 1965. *Ibid.*, **5**, 144.

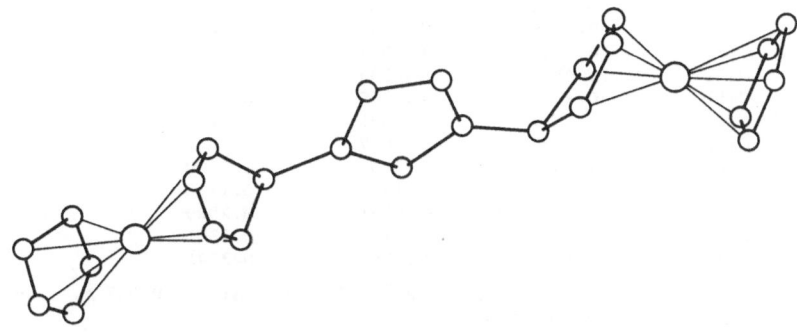

Fig. 171. The molecular configuration in $C_{25}H_{24}Co_2$. The Fe–C distances indicated range from 1.91 to 2.17 Å, with a mean value of 2.02 Å.

1,5-DI(IRON TRICARBONYL)-3-METHYLENEPENTA-1,4-DIENE

(DARK-RED ISOMER)

$C_{12}H_6Fe_2O_6$ F.W. = 357.9

III

I. Complexes du fer-carbonyle avec l'acétylène et ses dérivés. II. Structure du difer hexacarbonyle-méthylène-cyclopentadiényle, $Fe_2(CO)_6(C_2H_2)_3$, isomère rouge fonce (p. déc. 140°C). J. MEUNIER-PIRET, P. PIRET and M. VAN MEERSSCHE, 1965. *Acta Cryst.*, **19**, 85-91.

Monoclinic, $a = 13.26 \pm 2$, $b = 7.18 \pm 2$, $c = 13.72 \pm 2$ Å, $\beta = 102.5 \pm 5°$, $[U = 1275.3$ Å$^3]$, $Z = 4$, $D_x = 1.86$, (calibration with silver, $a = 4.086$ Å).

Space group $P2_1/c$ (C_{2h}^5)

Atomic positions

	x	y	z
Fe(1)	0.3148	-0.0119	0.1012
Fe(2)	0.1842	0.1052	0.2124
O(1)	0.1282	-0.0467	-0.0544
O(2)	0.3441	0.3676	0.0381
O(3)	0.4463	-0.2113	-0.0090
O(4)	0.4744	0.0175	0.2857
O(5)	0.3163	0.3899	0.3246
O(6)	0.0722	0.3566	0.0636
C(1)	0.1996	-0.0282	0.0066
C(2)	0.3328	0.2255	0.0641
C(3)	0.3951	-0.1307	0.0351
C(4)	0.4100	0.0133	0.2161
C(5)	0.2668	0.2760	0.2770
C(6)	0.1179	0.2635	0.1223
C(7)	0.2689	-0.2667	0.1572
C(8)	0.2023	-0.1802	0.2160
C(9)	0.0959	-0.1247	0.1776
C(10)	0.0584	-0.0231	0.2529
C(11)	0.1406	-0.0166	0.3381
C(12)	0.2284	-0.1054	0.3141

Standard deviations are: Fe, 0.002 Å; O, 0.012 Å; C, 0.015 Å. Isotropic temperature factors are given also.

Structure

The molecule (the configuration of which is illustrated in Fig 172) has a non-crystallographic mirror plane normal to the ring plane, and containing the iron atoms. Bond lengths and angles are given in full, and are consistent with the formula given above. Some mean distances are: $CH-CH_2$, 1.46 ± 2 Å; $CH-CH$(ring), 1.43 ± 2 Å; Fe(2)$-CH$(ring), 2.068 ± 15 Å; Fe(1)$-CH_2$, 2.123 ± 15 Å.
Intermolecular distances are consistent with van der Waals interaction.

Details of analysis

The intensity data were recorded on integrated equi-inclination Weissenberg photographs ($h0l$ to $h5l$) using CoK_α radiation. The intensities were measured photometrically. Refinement was by least squares to a final R index of 0.150.

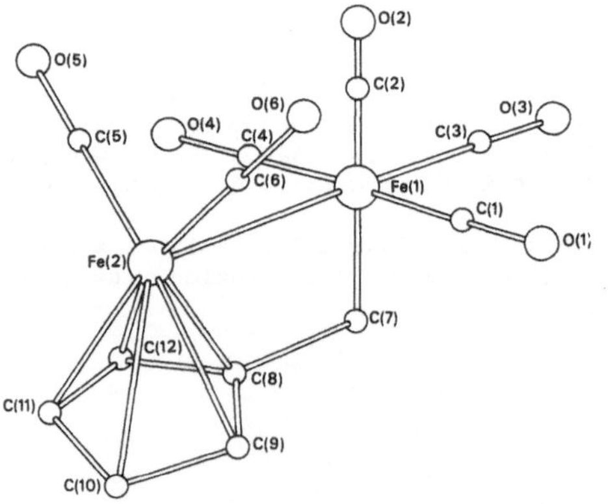

Fig. 172. Molecular configuration and the numbering scheme in $C_{12}H_6Fe_2O_6$.

π-CYCLOPENTADIENYL-π-(1)-2,3-DICARBOLLYLIRON(III)

C$_7$H$_{16}$B$_9$Fe   F.W. = 253.4

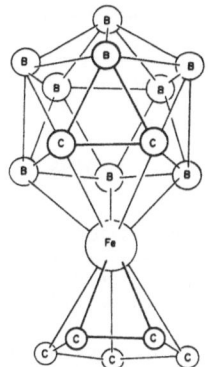

I. The crystal and molecular structure of C$_5$H$_5$FeB$_9$C$_2$H$_{11}$. A. ZALKIN, D. H. TEMPLETON and T. E. HOPKINS, 1965. *J. Amer. Chem. Soc.*, **87**, 3988-3990.

Monoclinic, a = 11.470, b = 6.629, c = 16.808 Å, β = 99.86°, [U = 1259 Å^3], z = 4, D_x = 1.34.

Space group $P2_1/c$ (C_{2h}^5)

Atomic positions

	x	y	z
Fe	0.2244	-0.0331	0.1897
C(1R)	0.0917	-0.0331	0.0909
C(2R)	0.1817	0.0862	0.0767
C(3R)	0.2779	-0.0361	0.0780
C(4R)	0.2433	-0.2285	0.0942
C(5R)	0.1278	-0.2228	0.1026
C(2)	0.3345	-0.1476	0.2871
C(3)	0.2008	-0.1537	0.2978
B(4)	0.1354	0.0749	0.2794
B(5)	0.2502	0.2337	0.2580
B(6)	0.3764	0.0795	0.2621
B(7)	0.3035	-0.1630	0.3822
B(8)	0.1786	-0.0172	0.3790
B(9)	0.2127	0.2301	0.3555
B(10)	0.3607	0.2328	0.3447
B(11)	0.4159	-0.0112	0.3609
B(12)	0.3173	0.0825	0.4194

(R indicates an atom in the cyclopentadienyl ring.)

The standard deviations are about 0.001 Å for the iron atom, and about 0.01Å for the others. The root-mean-square amplitudes of thermal motion are given also.

Structure

The non-hydrogen skeleton of the molecule is shown above. The atoms of the carborane part (B$_9$C$_2$H$_{11}$) lie at eleven of the twelve corners of a nearly-regular icosahedron, and the iron atom lies at the twelfth corner, but a little farther from the centre. The cyclopentadienyl ring eclipses the adjacent five-membered boron-carbon ring; however, the cyclopentadienyl ring appears to undergo severe twisting motion, and the relative orientations may be of little significance. Some mean bond lengths are as follows:

Fe - B	= 2.09 Å
Fe - C (carborane)	= 2.04
Fe - C (cyclopentadienyl)	= 2.07
B - B	= 1.75
B - C	= 1.68
C - C (carborane)	= 1.58
C - C (cyclopentadienyl)	= 1.36

These distances are not corrected for severe thermal motion; it is estimated that appropriate corrections could be as great as 0.05 Å.

Details of analysis

The intensity data were recorded with a four-circle diffractometer and scintillation counter, in the stationary-crystal stationary-counter mode, using MoK_α radiation. 1639 reflexions were observed above background. Refinement was by least squares to a final R index of 0.074.

1,1'-TETRAMETHYLETHYLENEFERROCENE

$C_{16}H_{20}Fe$ F.W. = 268.2

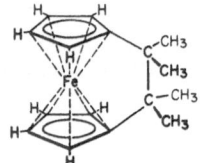

I. The crystal and molecular structure of 1,1'-tetramethylethyleneferrocene. M. B. LAING and K. N. TRUEBLOOD, 1965. *Acta Cryst.*, 19, 373-381.

Monoclinic, a = 7.756, b = 10.97, c = 15.41 Å, (all ± 0.1%), β = 92.63 ± 10°, [U = 1310 Å], D_m = 1.35, Z = 4, D_x = 1.36 (Calibration with ceric oxide, a = 5.411 Å).

Space group $P2_1/c$ (C_{2h}^5)

Atomic positions

	x	y	z
C(1)	0.1419 ± 12	0.3877 ± 10	0.2661 ± 5
C(2)	0.0363 ± 13	0.4252 ± 12	0.3366 ± 6
C(3)	0.1059 ± 15	0.3765 ± 11	0.4165 ± 11
C(4)	0.2433 ± 16	0.2998 ± 11	0.3970 ± 7
C(5)	0.2693 ± 15	0.3059 ± 9	0.3048 ± 6
C(6)	0.1416 ± 13	0.4444 ± 9	0.1741 ± 6
C(7)	0.3075 ± 13	0.5266 ± 9	0.1643 ± 6
C(8)	0.3712 ± 16	0.5716 ± 11	0.2550 ± 6
C(9)	0.2773 ± 20	0.6554 ± 11	0.3096 ± 8
C(10)	0.3620 ± 27	0.6468 ± 14	0.3931 ± 10
C(11)	0.5127 ± 16	0.5724 ± 14	0.3938 ± 8
C(12)	0.5167 ± 18	0.5267 ± 13	0.3059 ± 7
C(13)	0.1313 ± 17	0.3383 ± 13	0.1074 ± 7
C(14)	-0.0323 ± 15	0.5153 ± 15	0.1583 ± 8
C(15)	0.4572 ± 17	0.4523 ± 15	0.1264 ± 8
C(16)	0.2741 ± 17	0.6377 ± 12	0.1036 ± 8
Fe	0.2889 ± 2	0.4768 ± 1	0.3531 ± 1

Anisotropic temperature factors and assumed hydrogen positions are given.

Structure

A view of the structure is given in Fig. 173, and the bond lengths and angles are given in Fig. 174. The angle between the planes of the rings is 23°, and these rings are staggered with respect to one another by about 9°. The eclipsed configuration about the bond C(6)–C(7) is avoided by a rotation about this bond of about 26°. Iron to ring-carbon distances range from 1.96 to 2.10 Å, with standard deviation 0.009 to 0.013 Å. Intermolecular distances are consistent with van der Waals interaction.

Details of analysis

Intensity data were recorded on integrated and unintegrated Weissenberg photographs (0*kl* to 3*kl*, *h*0*l* to *h*5*l*) using CuK$_\alpha$ radiation. Of 2149 accessible reflexions, 1219 were observed and measured by photometric and visual methods. Refinement was by block-diagonal least squares to a final *R* index of 0.079.

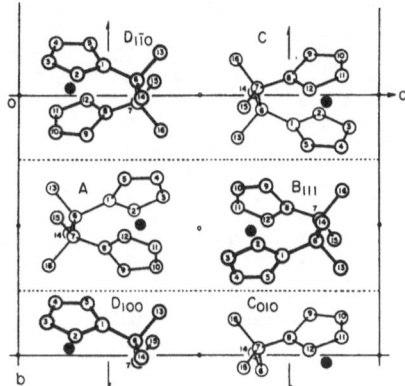

Fig. 173. Part of the 1,1'-tetramethylethyleneferrocene structure viewed down *a**.

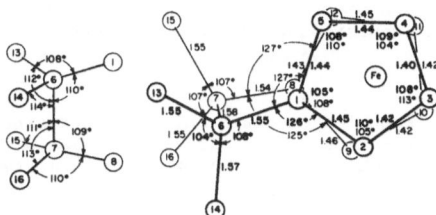

Fig. 174. Bond lengths and angles in the 1,1'-tetramethylethyleneferrocene structure. Standard deviations vary from 0.01 to 0.02 Å and 1 to 2°. The molecule is viewed along the average to the normals of the planes of the rings. Darker numbers refer to the nearer ring. Some bridge angles are shown to the left.

α-KETO-1,1'-TRIMETHYLENEFERROCENE

$C_{13}H_{12}OFe$ F.W. = 240.1

I. The crystal structure of α-keto-1,1'-trimethyleneferrocene. N. D. JONES, R. E. MARSH and J. H. RICHARDS, 1965. *Acta Cryst.*, 19, 330–336.

Monoclinic, a = 22.981 ± 2, b = 7.381 ± 1, c = 5.833 ± 1 Å, β = 93.38 ± 2°,
$[U$ = 987.6 Å³], D_m = 1.60, Z = 4, D_X = 1.61 ($CuK_{\alpha 1}$, λ = 1.54051 Å;
$CuK_{\alpha 2}$, λ = 1.54433 Å).

Space group $P2_1/a$ (C_{2h}^5)

Atomic positions

	x	y	z
Fe	0.15150± 4	0.1259 ± 1	0.1913 ± 1
O	0.0955 ± 3	0.4880 ± 7	0.6090 ± 9
C(1)	0.1584 ± 3	0.3702 ± 8	0.3418 ± 12
C(2)	0.1925 ± 3	0.2429 ± 9	0.4727 ± 11
C(3)	0.2350 ± 3	0.1701 ± 10	0.3344 ± 12
C(4)	0.2278 ± 3	0.2544 ± 9	0.1141 ± 14
C(5)	0.1794 ± 3	0.3767 ± 8	0.1161 ± 12
C(6)	0.0634 ± 3	0.0985 ± 8	0.1599 ± 13
C(7)	0.0892 ± 3	−0.0438 ± 9	0.3029 ± 12
C(8)	0.1285 ± 3	−0.1426 ± 8	0.1736 ± 14
C(9)	0.1275 ± 4	−0.0637 ± 9	−0.0508 ± 13
C(10)	0.0873 ± 3	0.0815 ± 8	−0.0580 ± 12
C(11)	0.1014 ± 3	0.4421 ± 8	0.4123 ± 12
C(12)	0.0501 ± 3	0.4343 ± 9	0.2391 ± 12
C(13)	0.0224 ± 3	0.2460 ± 9	0.2396 ± 13

    Also given are anisotropic temperature factors and assumed hydrogen para-
meters.

Structure

    Bond angles and intramolecular distances are given in Fig. 175. The rings
are nearly planar, and the dihedral angle between them is 8.8°. The two rings,
if projected on the average plane of both of them, would be 'staggered', or
rotated with respect to each other, by 11.8°. The molecular packing is illustrated in Fig. 176. Intermolecular distances are consistent with van der Waals
interaction.

Details of analysis

    Intensity data were recorded on equi-inclination Weissenberg photographs
($hk0$ to $hk5$) using CuK_{α} radiation. 1196 visually estimated reflexions were
included in the refinement, which was by block-diagonal least squares to a final
R index of 0.067.

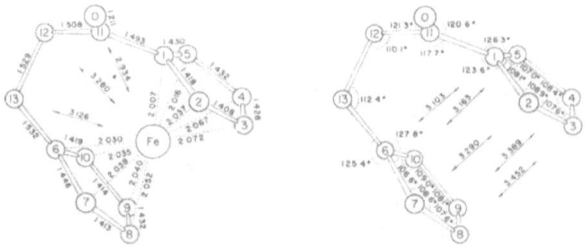

Fig. 175. Bond angles and interatomic distances for $C_{13}H_{12}OFe$. Standard
            deviations of distances are 0.007 Å for Fe–C, and 0.010 for the others.

BIS-[1'-(1-ETHYLFERROCENYL)]

$C_{24}H_{26}Fe_2$ $(C_5H_4CH_2CH_3)Fe(C_5H_4C_5H_4)Fe(C_5H_4CH_2CH_3)$ F.W. = 426.1

    I.   [The crystal structure and molecular structure of diethyldiferrocenyl.]
         Z. L. KALUSKI and J. T. STRUČKOV, 1965. *Ž. Strukt. Khim. SSSR*, <u>6</u>,

104-112 [*J. Struct. Chem.* **6**, 90-97].

(This structure is also described in **1**.)

Monoclinic, $a = 19.14 \pm 6$, $b = 7.52 \pm 3$, $c = 16.29 \pm 6$ Å, $\beta = 127.5 \pm 3°$, $U = 1860$ Å^3, $D_m = 1.45$, $Z = 4$, $D_x = 1.52$.

Space group Cc (C_s^4) or $C2/c$ (C_{2h}^6). $C2/c$ confirmed by analysis. Molecular symmetry, centre.

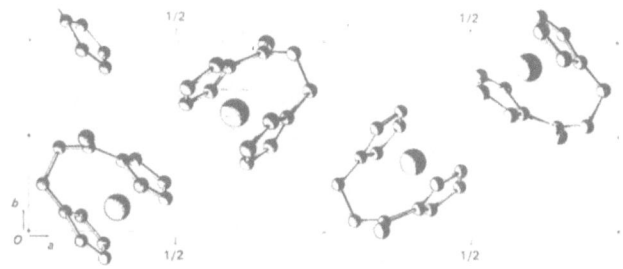

Fig. 176. The structure of $C_{13}H_{12}OFe$ viewed along c.

Atomic positions

	x	y	z
Fe	0.1649	0.5538	0.0372
C(1)	0.272	0.426	0.164
C(2)	0.222	0.514	0.192
C(3)	0.135	0.441	0.128
C(4)	1.131	0.308	0.061
C(5)	0.215	0.298	0.083
C(6)	0.372	0.477	0.218
C(7)	0.472	0.346	0.304
C(1')	0.207	0.739	-0.019
C(2')	0.162	0.826	0.017
C(3')	0.074	0.756	-0.042
C(4')	0.065	0.625	-0.113
C(5')	0.147	0.614	-0.098

Standard deviations are 0.001 Å for iron, and 0.015 to 0.029 Å for the carbon atoms. Also given are isotropic temperature factors for the atoms above, and assumed positions for the hydrogen atoms.

Structure

A view of the molecule is given in Fig. 177. Bond lengths and angles are given in full; some distances (averaged where appropriate) are:

C(1')-C(1')	=	1.38 Å
C-C (ring)	=	1.44
C-C (single)	=	1.55
Fe-C (ring)	=	2.07

The five membered rings are planar, but opposing rings in a ferrocenyl sandwich are inclined at 2.3° to each other. The average inter-ring spacing is 3.33 Å. The sandwich conformation is very nearly prismatic; one ring is rotated relative to the other by 3.6° about the line joining their centres. Intermolecular distances are consistent with van der Waals interaction.

Details of analysis

The intensity data were recorded photographically ($h0l$ to $h4l$; $hk0$ to $hk4$) using MoK_α radiation. 880 reflexions were observed and measured visually. Refinement was by diagonal least squares to a final R index of 0.16.

1. Z. KALUSKI, 1965. *Bull. Acad. Polon. Sci., Ser. Sci. Chim.*, **13**, 355.

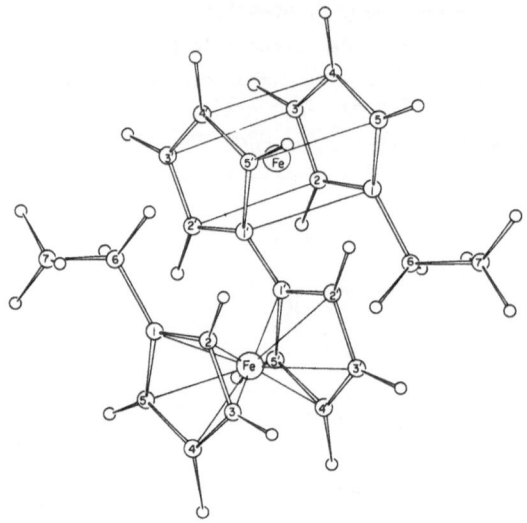

Fig. 177. A view of the molecule of bis-[1'-(1-ethylferrocenyl)].

BIS-[1-(2'-CHLORFERROCENYL)]

$C_{20}H_{16}Fe_2Cl_2$ $(C_5H_4Cl)Fe(C_5H_4C_5H_4)Fe(C_5H_4Cl)$ F.W. = 439.0

I. [Crystal and molecular structure of dichlordiferrocenyl.] Z. L. KALUSKI
    and J. T. STRUČKOV, 1965. Ž. Strukt. Khim. SSSR, 6, 745-754 [J. Struct.
    Chem., 6, 705-713].

    (For preliminary results, see 1.)

Monoclinic, a = 10.63 ± 3, b = 8.65 ± 3, c = 10.60 ± 3 Å, β = 121.5 ± 5°,
U = 855 Å³, D_m = 1.68, Z = 2, D_x = 1.70.

Space group $P2_1/c$ (C_{2h}^5) Molecular symmetry, centre.

Atomic positions

	x	y	z
Fe	0.2460	0.4610	0.4602
Cl	0.2894	0.0770	0.5028
C(1)	0.4528	0.5293	0.5290
C(2)	0.4426	0.4629	0.6546
C(3)	0.3367	0.5551	0.6725
C(4)	0.2834	0.6720	0.5549
C(5)	0.3659	0.6665	0.4797
C(1')	0.2088	0.3101	0.2926
C(2')	0.1848	0.2368	0.3971
C(3')	0.0791	0.3079	0.4282
C(4')	0.0219	0.4310	0.3185
C(5')	0.0921	0.4339	0.2351

    Also given are isotropic temperature factors for the above atoms, and
assumed positions for the hydrogen atoms.

Structure

    The molecular configuration is illustrated in Fig. 178. Bond lengths and
angles are given in full; some distances (averaged where appropriate) are:

C-C (inter-ring)	= 1.51 Å
C-C (ring)	= 1.46
C-Cl	= 1.76
Fe-C	= 2.06

All rings are planar; the connected rings are nearly coplanar, and the chlorine atom lies in the plane of the outer ring. Opposing rings in a ferrocenyl sandwich are not parallel, but are inclined at 4° to each other. The conformation of the sandwich is very nearly prismatic, with one ring turned relative to the other by 5.2° about the line joining their centres. Intermolecular distances are consistent with van der Waals interaction.

Details of analysis

The intensity data were recorded photographically (h0l to h4l; hk0 to hk4) using MoK$_\alpha$ radiation. 1100 reflexions were observed and measured visually. Refinement was by least squares to a final R index of 0.173.

1. Z. L. KALUSKI and J. T. STRUČKOV, 1965. *Ž. Strukt. Khim. SSSR,* 6, 475.

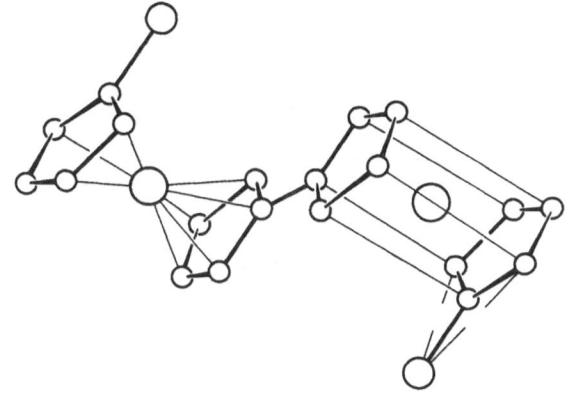

Fig. 178. A view of the bis-[1-(2'-chloroferrocenyl)] molecule.

DIHYDRIDODI-π-CYCLOPENTADIENYLMOLYBDENUM

$C_{10}H_{12}Mo$ $(\pi-C_5H_5)_2MoH_2$ F.W. = 228.1

I. The crystal and molecular structure of dihydridodi-π-cyclopentadienyl-molybdenum. M. GERLOCH and R. MASON, 1965. *J. Chem. Soc.,* 296-304.

(For a preliminary report, see 1.)

Monoclinic, a = 14.30 ± 4, b = 5.90 ± 2, c = 10.41 ± 3 Å, β = 104.0 ± 2°, U = 852 Å^3, Z = 4, D_x = 1.78.

Space group Cc (C_s^4) or C2/c (C_{2h}^6). C2/c is consistent with deduced structure. Molecular symmetry, two-fold axis.

Atomic positions

	x	y	z
Mo(1)	0	0.33188	¼
C(1)	-0.1027	0.0610	0.2712
C(2)	-0.1459	0.1796	0.1634
C(3)	-0.1667	0.4145	0.2027
C(4)	-0.1290	0.4187	0.3423
C(5)	-0.0897	0.1906	0.3849

The reported standard deviations are: Mo, 0.004 Å; C, 0.02 to 0.04 Å.
Also given are the positions of the hydrogen atoms and anisotropic temperature
factors for the non-hydrogen atoms.

Structure

The configuration of the molecule is illustrated in Fig. 179. The angle
between the cyclopentadienyl ring planes is 34 ± 1°. The Mo–C distances range
from 2.22 to 2.37 Å. Bond lengths (including Mo–H) and angles are given in
full. [A subsequent report (2) suggests that the work is not sufficiently
accurate either to reveal the position of the molybdenum-bonded hydrogen atom,
or to demonstrate significant differences in the C–C distances.]

Details of analysis

The intensity data were recorded on Weissenberg (*h0l*; *h1l*) and precession
(*hk0*; *h0l*) photographs using MoK_α radiation. 274 reflexions were observed
and measured visually. Refinement was by least squares to a final *R* index of
0.087.

1. *Structure Reports, 27*, 849.

2. S. C. ABRAHAMS and A. P. GINSBERG, 1966. *Inorg. Chem., 5*, 500.

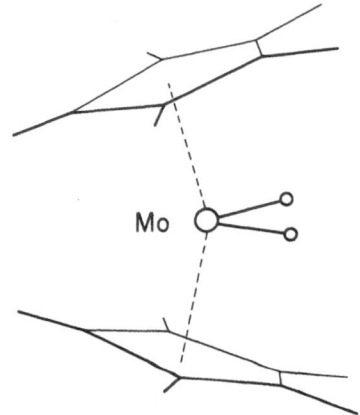

Fig. 179. The molecular configuration of $(\pi\text{–}C_5H_5)_2MoH_2$.

μ-(DIMETHYLPHOSPHIDO)HYDRIDODI-π-CYCLOPENTADIENYLTETRA-

CARBONYLDIMOLYBDENUM

$C_{16}H_{17}Mo_2O_4P$ F.W. = 496.2

I. Structure of and bonding in $[(C_5H_5)_2Mo_2H\{P(CH_3)_2\}(CO)_4]$. Evidence for a
    symmetrical, bent, three-center metal-hydrogen-metal bond. R. J. DOEDENS
    and L. F. DAHL, 1965. *J. Amer. Chem. Soc., 87*, 2576-2581.

Triclinic, *a* = 9.16 ± 2, *b* = 9.44 ± 2, *c* = 11.570 ± 25 Å, α = 84°01 ± 10',
β = 82°28 ± 10, γ = 61°32 ± 10', [*U* = 871 Å³], D_m = 1.89, Z = 2, D_x = 1.89.
(The results are, however, reported in terms of a *C*-centred cell for which a_C =
9.16, b_C = 16.59, c_C = 11.57 Å, α_C = 92°39', β_C = 97°32', γ_C = 90°34', Z = 4.)

Space group P1 (C_1^1) or P$\bar{1}$ (C_i^1). P$\bar{1}$ confirmed by analysis.

Atomic positions

	x	y	z
Mo(1)	0.48480 ± 9	0.33630 ± 5	0.24560 ± 8
Mo(2)	0.82560 ± 9	0.36860 ± 5	0.20440 ± 8

	x	y	z
P	0.6690 ± 3	0.4357 ± 2	0.3300 ± 2
C(1)	0.3742 ± 14	0.4350 ± 8	0.2417 ± 11
O(1)	0.3116 ± 11	0.4939 ± 6	0.2400 ± 9
C(2)	0.4281 ± 14	0.3468 ± 8	0.0815 ± 11
O(2)	0.3935 ± 13	0.3498 ± 7	-0.0189 ± 10
C(3)	0.9587 ± 14	0.3863 ± 7	0.3507 ± 11
O(3)	1.0386 ± 13	0.3979 ± 7	0.4325 ± 10
C(4)	0.8693 ± 14	0.2603 ± 8	0.2489 ± 11
O(4)	0.8963 ± 12	0.1948 ± 6	0.2770 ± 9
C(5)	0.7098 ± 16	0.4394 ± 8	0.4890 ± 12
C(6)	0.6502 ± 16	0.5460 ± 9	0.3127 ± 13
C(7)	0.2958 ± 22	0.2595 ± 11	0.2924 ± 17
C(8)	0.3859 ± 20	0.2042 ± 10	0.2416 ± 15
C(9)	0.5190 ± 18	0.2015 ± 10	0.3015 ± 14
C(10)	0.5167 ± 23	0.2524 ± 12	0.4068 ± 18
C(11)	0.3804 ± 21	0.2871 ± 11	0.3963 ± 16
C(12)	1.0046 ± 17	0.3881 ± 9	0.0870 ± 13
C(13)	0.8903 ± 16	0.3481 ± 9	0.0160 ± 13
C(14)	0.7688 ± 17	0.3977 ± 9	0.0034 ± 13
C(15)	0.8104 ± 18	0.4709 ± 9	0.0690 ± 14
C(16)	0.9507 ± 18	0.4638 ± 10	0.1215 ± 14

Temperature factors (anisotropic for molybdenum and phosphorus) are given also.

Structure

The configuration of the molecule is shown in Fig. 180; a non-crystallographic two-fold axis is defined by the phosphorus atom and the midpoint of the molybdenum atoms. Bond lengths and angles are given in full; the distances shown in Fig. 180 are average values. Selected values of some bond angles, averaged where appropriate, are: Mo-P-Mo, 84.7 ± 2°; C-P-C, 97.3 ± 7°; C-Mo-C, 77.6 ± 6°.

The positions of the hydrogen atoms were not determined, but it is reasonably inferred that the metal-coordinated hydrogen atom must be symmetrically located 1.8 Å from each molybdenum atom, and must therefore participate in a symmetrical, bent, three-centre metal-hydrogen-metal bond.

Details of analysis

The intensity data were recorded on equi-inclination Weissenberg (0kl to 12kl) and precession (hk0 to hk2) photographs, using MoK$_\alpha$ radiation. 4008 independent reflexions were observed and measured visually. Refinement was by full-matrix least squares to a final R index of 0.105.

IODOCARBONYL-π-CYCLOPENTADIENYL-PENTAFLUOROETHYLRHODIUM

C$_8$H$_5$OF$_5$IRh   π-C$_5$H$_5$Rh(CO)(C$_2$F$_5$)I   F.W. = 441.9

I. The crystal and molecular structure of racemic iodocarbonyl-π-cyclopentadienyl-pentafluoroethylrhodium. M. R. CHURCHILL, 1965. *Inorg. Chem.*, **4**, 1734-1739.
(For a preliminary account, see **1**.)

Monoclinic, a = 12.410 ± 10, b = 7.823 ± 8, c = 12.632 ± 10 Å, β = 109.86 ± 5°, U = 1153.2 Å^3, D$_m$ = 2.50, Z = 4, D$_x$ = 2.543.

Space group P2$_1$/c (C$_{2h}^5$)

Atomic positions

	x	y	z
Rh	0.23249 ± 15	0.42302 ± 27	-0.04362 ± 15
I	0.10381 ± 15	0.16244 ± 27	-0.02388 ± 15
F(1)	0.2154 ± 13	0.5173 ± 25	0.1678 ± 12

	x	y	z
F(2)	0.3289 ± 13	0.3066 ± 24	0.1859 ± 11
F(3)	0.3730 ± 17	0.7281 ± 25	0.1415 ± 16
F(4)	0.4868 ± 12	0.5164 ± 25	0.1485 ± 13
F(5)	0.4312 ± 14	0.5753 ± 28	0.2868 ± 12
O	0.4231 ± 22	0.1720 ± 31	−0.0217 ± 18
C(1)	0.2960 ± 22	0.4522 ± 33	0.1305 ± 16
C(2)	0.4005 ± 22	0.5758 ± 52	0.1787 ± 23
C(3)	0.3598 ± 26	0.2634 ± 42	−0.0271 ± 20
C(4)	0.2279 ± 28	0.5389 ± 38	−0.2093 ± 20
C(5)	0.1174 ± 22	0.4785 ± 42	−0.2215 ± 20
C(6)	0.0796 ± 21	0.5759 ± 38	−0.1378 ± 23
C(7)	0.1684 ± 29	0.6913 ± 40	−0.0790 ± 22
C(8)	0.2628 ± 23	0.6724 ± 42	−0.1187 ± 26

Anisotropic temperature factors for the above atoms and assumed positions for the hydrogen atoms are given also.

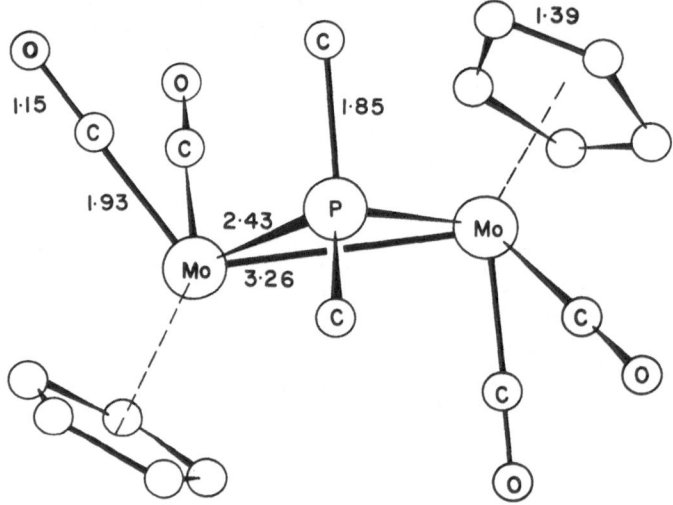

Fig. 180. The molecular configuration of $C_{16}H_{17}Mo_2O_4P$. The standard deviations of distances range from 0.006 to 0.026 Å.

Structure

The molecular structure is illustrated in Fig. 181. (Bond lengths and angles are given in full.) The molecule can be regarded as an octahedral complex of rhodium; the angles between the carbonyl, iodide and perfluoroethyl ligands are close to 90°, and the six-fold coordination is completed by the formally tridentate π-cyclopentadienyl group. The orientation of the ring with respect to the three ligands opposite is fixed by the large size of the iodine atom, which lies almost equally distant (3.56 and 3.51 Å) from C(5) and C(6). The distances of the rhodium atom from the ring carbon atoms range from 2.22 to 2.26 Å.

Details of analysis

The intensity data were recorded on Weissenberg photographs (*h0l* to *h4l*; *hk0* to *hk2*) using MoK_α radiation. 960 reflexions were observed and measured visually. Refinement was by full-matrix least squares to a final *R* index of 0.07.

1. M. R. CHURCHILL, 1965. *Chem. Comm.*, 86.

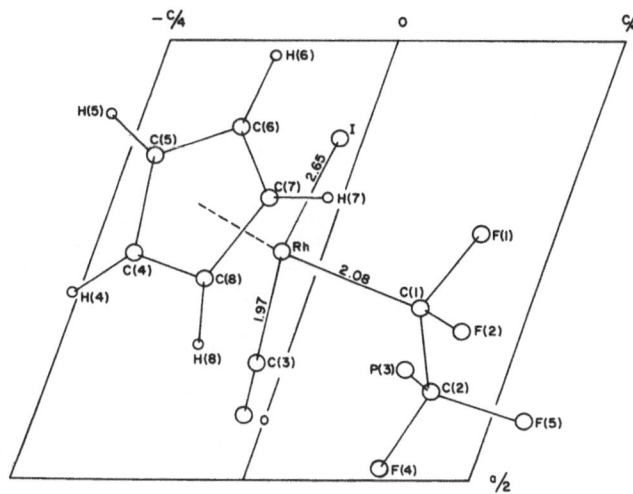

Fig. 181. The structure of π-C₅H₅Rh(CO)(C₂F₅)I viewed along *b*. Mean standard
deviations of the distances shown are: Rh-I, 0.003 Å; Rh-C,
0.025 Å.

URANIUM CHLORIDE π-TRICYCLOPENTADIENYL

$C_{15}H_{18}ClU$ $(C_5H_6)_3UCl$ F.W. = 471.8

I. The crystal structure of uranium chloride π-tricyclopentadienyl. CHI-
HSIANG WONG, TUNG-MOU YEN and TSENG YUH LEE, 1965. *Acta Cryst.*, <u>18</u>,
340-345.

Monoclinic, a = 8.26 ± 3, *b* = 12.50 ± 3, *c* = 13.81 ± 3 Å, β = 90.6 ± 5°,
[*U* = 1426 Å³], D_m = 2.19, *Z* = 4, D_x = 2.184, (calibration with gold, *a* =
4.0702 Å).

Space group P2₁/n (C_{2h}^5)

Atomic positions

	x	*y*	*z*
U	0.5379 ± 2	0.3442 ± 1	0.7189 ± 1
Cl	0.4476 ± 17	0.5365 ± 14	0.6859 ± 9
C(1)	0.368	0.179	0.638
C(2)	0.275	0.274	0.616
C(3)	0.368	0.334	0.549
C(4)	0.509	0.272	0.526
C(5)	0.510	0.178	0.583
C(6)	0.303	0.276	0.842
C(7)	0.315	0.380	0.860
C(8)	0.472	0.408	0.900
C(9)	0.550	0.304	0.910
C(10)	0.440	0.224	0.870
C(11)	0.822	0.370	0.615
C(12)	0.845	0.278	0.673
C(13)	0.854	0.294	0.772
C(14)	0.837	0.408	0.783
C(15)	0.817	0.455	0.688

Anisotropic temperature factors are given for uranium and chlorine. Carbon atoms were not resolved; the positions given correspond to idealized five-membered rings fitted to regions of positive electron density.

Structure

Two views of the proposed structure are given in Fig. 182. The bond length U—Cl is 2.559 ± 16 Å, and the mean U—C distance is 2.74 Å.

Details of analysis

The intensity data were recorded on equi-inclination Weissenberg photographs ($h0l$ to $h5l$, $0kl$ to $1kl$) using MoK$_\alpha$ radiation. 687 of a possible 1051 reflexions were observed and measured visually. Refinement was by block-diagonal least squares to a final R index of 0.12.

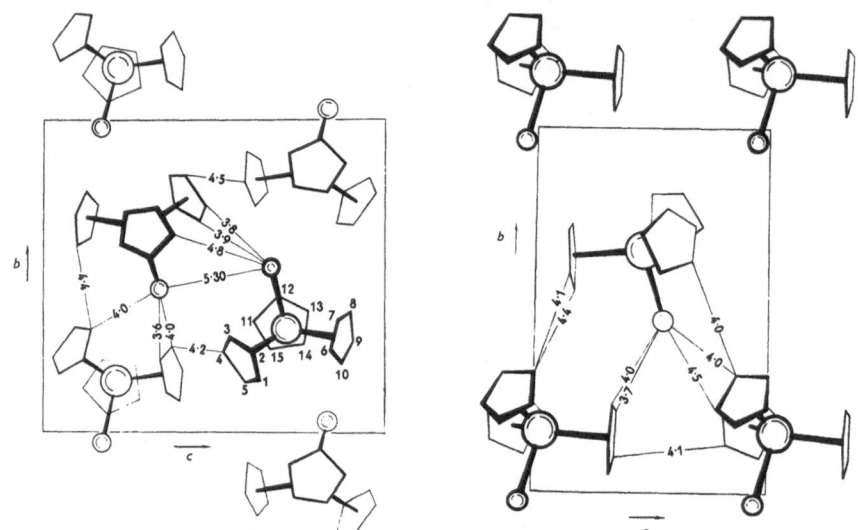

Fig. 182. The structure of (C$_5$H$_6$)$_3$UCl viewed along (a) a, and (b) c.

BENZENECHROMIUM TRICARBONYL

C$_9$H$_6$O$_3$Cr   C$_6$H$_6$Cr(CO)$_3$ F.W. = 214.2

I. Three-dimensional crystal structure of benzenechromium tricarbonyl with further comments on the dibenzenechromium structure. M.F. BAILEY and L.F. DAHL, 1965. *Inorg. Chem.*, *4*, 1314–1319.

(For earlier, two-dimensional work, see *1*, *2*.)

Monoclinic, a = 6.17 ± 2, b = 11.07 ± 4, c = 6.57 ± 2 Å, β = 101.5 ± 1°, [U = 440 Å^3], D_m = 1.64, Z = 2, D_x = 1.62.

Space group $P2_1$ (C_2^2) or $P2_1/m$ (C_{2h}^2). $P2_1/m$ confirmed by analysis. Molecular symmetry, mirror plane.

Atomic positions

	x	y	z
Cr	0.33190 ± 15	¼	0.02250 ± 13
C(1)	0.1804 ± 8	0.3119 ± 4	−0.2973 ± 6
C(2)	0.3761 ± 10	0.3769 ± 4	−0.2273 ± 6
C(3)	0.5753 ± 8	0.3142 ± 4	−0.1598 ± 7

	x	y	z
C(4)	0.5538 ± 10	¼	0.2557 ± 0
O(4)	0.6899 ± 9	¼	0.4002 ± 8
C(5)	0.1827 ± 8	0.3642 ± 4	0.1453 ± 7
O(5)	0.0894 ± 7	0.4341 ± 3	0.2248 ± 5

Anisotropic temperature factors for the above atoms and assumed positions for the hydrogen atoms are given also.

Structure

A view of the molecule is given in Fig. 183. The plane through the carbonyl groups is nearly parallel to the plane of the benzene ring, and the molecule as a whole has approximate 3m symmetry. Bond lengths and angles are given in full. Some mean values for the chromium tricarbonyl group are: Cr–C, 1.84 ± 1 Å, C–O, 1.14 ± 1Å; Cr–C–O, 179.2 ± 5°; C–Cr–C, 89.6 ± 4°. The chromium atom lies 1.72 Å from the plane of the benzene molecule, with a mean distance from the ring carbon atoms of 2.23 Å. There is no evidence of any three-fold distortion of the ring. Intermolecular distances are consistent with van der Waals interaction.

Details of analysis

The intensity data were recorded on equi-inclination Weissenberg photographs (0kl to 5kl; hk0 to hk6) using MoK$_\alpha$ radiation. 518 reflexions were observed and measured visually. Refinement was by full-matrix least squares to a final R index of 0.042.

1. *Structure Reports*, 23, 695.

2. *Ibid.*, 26, 685.

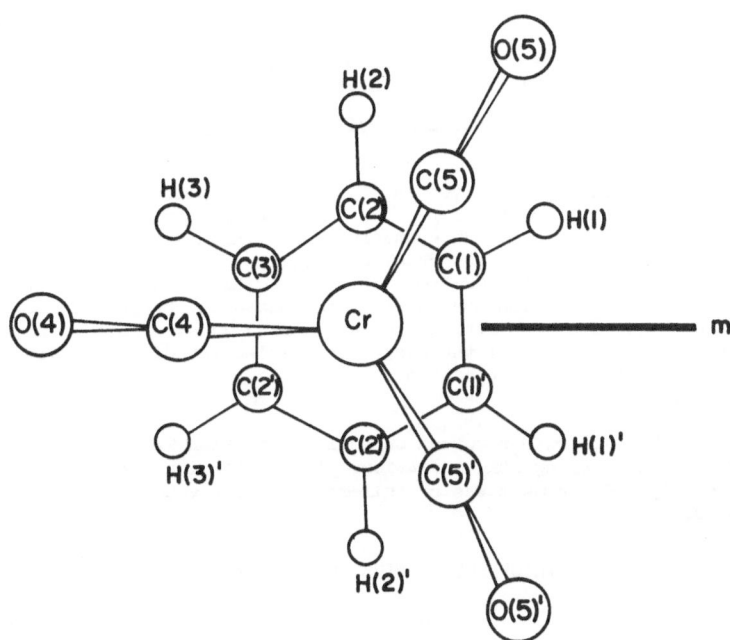

Fig. 183. The $C_6H_6Cr(CO)_3$ molecule, viewed along the non-crystallographic three-fold axis.

HEXAMETHYLBENZENECHROMIUM TRICARBONYL

$C_{15}H_{18}O_3Cr$ $(H_3C)_6C_6Cr(CO)_3$ F.W. = 298.3

I. The structure of hexamethylbenzenechromium tricarbonyl with comments on
the dibenzenechromium structure. M. F. BAILEY and L. F. DAHL, 1965. *Inorg.
Chem.*, *4*, 1298-1306.

Orthorhombic, a = 13.67 ± 3, b = 13.53 ± 3, c = 15.27 ± 3 Å, [U = 2824 Å^3],
D_m = 1.39, z = 8, D_x = 1.40.

Space group Pbca (D_{2h}^{15})

Atomic positions

	x	y	z
Cr	0.19270 ± 9	0.24310 ± 11	0.07700 ± 7
C(7)	0.1994 ± 10	0.3381 ± 8	-0.0074 ± 7
O(7)	0.2070 ± 8	0.3953 ± 7	-0.0615 ± 5
C(8)	0.2888 ± 7	0.3062 ± 10	0.1391 ± 7
O(8)	0.3477 ± 6	0.3474 ± 7	0.1790 ± 6
C(9)	0.1042 ± 8	0.3163 ± 9	0.1340 ± 7
O(9)	0.0421 ± 6	0.3645 ± 7	0.1696 ± 6
C(1)	0.1358 ± 7	0.1057 ± 7	0.1366 ± 6
C(2)	0.0816 ± 8	0.1223 ± 7	0.0564 ± 7
C(3)	0.1384 ± 8	0.1366 ± 8	-0.0238 ± 6
C(4)	0.2390 ± 10	0.1300 ± 8	-0.0204 ± 7
C(5)	0.2908 ± 6	0.1120 ± 7	0.0586 ± 7
C(6)	0.2358 ± 8	0.0989 ± 7	0.1370 ± 6
H$_3$C(1)	0.0745 ± 8	0.0908 ± 9	0.2220 ± 8
H$_3$C(2)	-0.0275 ± 8	0.1286 ± 9	0.0567 ± 9
H$_3$C(3)	0.0796 ± 10	0.1540 ± 9	-0.1062 ± 7
H$_3$C(4)	0.3014 ± 9	0.1389 ± 9	-0.1039 ± 7
H$_3$C(5)	0.3987 ± 9	0.0991 ± 9	0.0599 ± 7
H$_3$C(6)	0.2918 ± 8	0.0795 ± 10	0.2218 ± 8

Anisotropic temperature factors are given also.

Structure

A view of the molecule is given in Fig. 184. The plane through the carbonyl groups is parallel to the mean plane of the hexamethylbenzene moiety,
and the molecule as a whole has 3m symmetry. Bond lengths and angles are given
in full. Some mean values for the chromium tricarbonyl group are: Cr–C,
1.81 ± 1 Å; C–O, 1.16 ± 1 Å; Cr–C–O, 178 ± 1°; C–Cr–C, 89.6 ± 3°. The
chromium atom lies 1.73 Å from the plane of the hexamethylbenzene molecule,
with a mean distance from the ring carbon atoms of 2.23 Å. There is no evidence
of any three-fold distortion of the ring. Intermolecular distances are consistent with van der Waals interaction.

Details of analysis

The intensity data were recorded on equi-inclination Weissenberg photographs about *b* and *c*, using MoK$_\alpha$ radiation. 1447 reflexions were observed and
measured visually. Refinement was by full-matrix least squares to a final *R*
index of 0.105.

THIOPHENECHROMIUM TRICARBONYL

$C_7H_4O_3SCr$ $(C_4H_4S)Cr(CO)_3$ F.W. = 220.2

I. Structure of thiophenechromium tricarbonyl, $C_4H_4SCr(CO)_3$. M. F. BAILEY
and L. F. DAHL, 1965. *Inorg. Chem.*, *4*, 1306-1314.

Monoclinic, a = 6.06 ± 2, b = 10.79 ± 3, c = 6.65 ± 2 Å, β = 102.0 ± 1°,
[U = 425 Å^3], D_m = 1.74, z = 2, D_x = 1.72.

Space group $P2_1$ (C_2^2) or $P2_1/m$ (C_{2h}^2). $P2_1/m$ confirmed by analysis. Molecular symmetry, mirror plane.

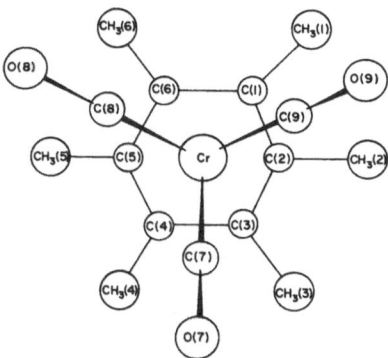

Fig. 184. The $(H_3C)_6C_6Cr(CO)_3$ molecule viewed along the non-crystallographic three-fold axis.

Atomic positions

Chromium tricarbonyl:

	x	y	z
Cr	0.3279 ± 5	¼	0.0372 ± 4
C(7)	0.5512 ± 42	¼	0.2560 ± 39
O(7)	0.6912 ± 40	¼	0.4151 ± 39
C(8)	0.1808 ± 24	0.3644 ± 16	0.1482 ± 24
O(8)	0.0823 ± 22	0.4375 ± 14	0.2366 ± 20

Thiophene:

	x	y	z
S(1)	0.1564	¼	-0.3058
C(1)	0.3528	0.3645	-0.2304
C(2)	0.5619	0.3160	-0.1502
S(2)	0.4898	0.3573	-0.1986
C(3)	0.5654	0.2045	-0.1625
C(4)	0.3816	0.1280	-0.2142
C(5)	0.1770	0.1948	-0.2840
C(6)	0.2103	0.3203	-0.2836

The positions specified for the thiophene molecule are used to describe a disordered structure, and are assigned occupancy factors of 1/3. Isotropic temperature factors are given also.

Structure

The molecular configuration is illustrated in Fig. 185. In the crystal structure the thiophene molecule suffers three-fold rotational disorder, with the alternative orientations achieved by rotations of 120° about the normal to the thiophene plane. The geometry of the thiophene molecule was assumed to be that of the free molecule. Bond lengths and angles are given for the chromium tricarbonyl group, and appear to be normal. The mean distance from the chromium to the sulphur atom is 2.31 Å, and to the ring carbon atoms, 2.21 Å. The coordination of the chromium atom is thus octahedral, with three carbonyl, one sulphur, and two C=C ligands.

Details of analysis

The intensity data were recorded on equi-inclination Weissenberg photographs ($h0l$ to $h6l$; $hk0$ to $hk6$) using CuK_α radiation. 653 reflexions were observed and measured visually. Refinement was by full-matrix least squares, with the disordered thiophene molecule constrained to its assumed (free-molecule) geometry, to a final R index of 0.132.

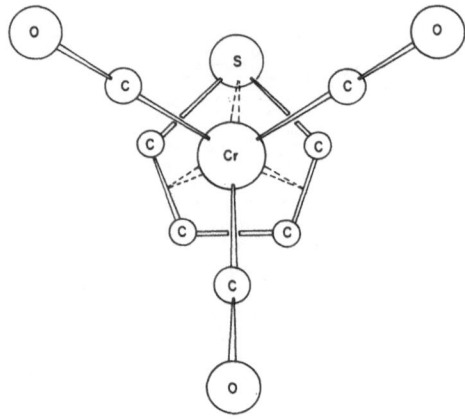

Fig. 185. The molecular configuration of $(C_4H_4S)Cr(CO)_3$.

TETRAPHENYLCYCLOBUTADIENE IRON TRICARBONYL

$C_{31}H_{20}FeO_3$ $Fe(CO)_3(C_6H_5C_2C_6H_5)_2$ F.W. = 496.4

   I. The crystal and molecular structure of $Fe(CO)_3(C_6H_5C_2C_6H_5)_2$. R. P. DODGE and V. SCHOMAKER, 1965. *Acta Cryst.*, <u>18</u>, 614-617.

     (For an earlier report by the same authors, see <u>1</u>.)

Monoclinic, a = 8.93, b = 18.72, c = 14.09 Å, β = 92.7°, $[U = 2355$ Å$^3]$, D_m = 1.39, Z = 4, D_x = 1.40.

Space group $P2_1/c$ (C_{2h}^5) Atomic positions and temperature factors are given for the 35 non-hydrogen atoms.

Structure

    The form of the molecule is illustrated in Fig. 186. Mean values of some bond lengths and angles are as follows:

Fe–CO	1.750 ± 13 Å
C–O	1.179 ± 17
Fe–CC_6H_5	2.067 ± 13
C_6H_5C–CC_6H_5	1.459 ± 17
C–C_6H_5	1.468 ± 17
C–C (aromatic)	1.406 ± 22
OC–Fe–CO	97.0 ± 6°
C_6H_5C–Fe–CC_6H_5	41.3 ± 5°

The cyclobutadiene ring is square planar. The phenyl groups are bent out of the plane of this ring, away from the iron atom an average of 10.8° (range 6.9 to 16.7°). In addition all phenyl groups are twisted (in the same sense) by angles of 32.4, 36.4, 28.6, and 60.8°.

Details of analysis

    Intensity data were collected with a four-circle diffractometer, using MoKα radiation. 1776 of a possible 2204 reflexions were observed. Refinement was by full-matrix least squares to a final R index of 0.094 (for observed reflexions).

1. *Structure Reports,* <u>24</u>, 601.

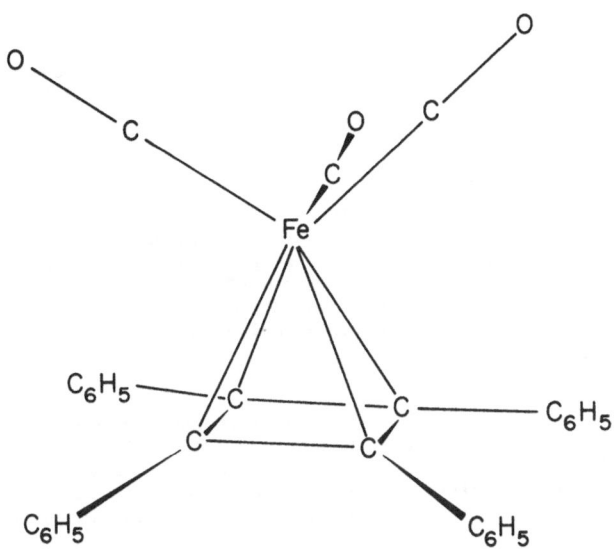

Fig. 186. The molecular configuration of $Fe(CO)_3(C_6H_5C_2C_6H_5)_2$.

BIS-CYCLOOCTA-1,5-DIENENICKEL(0)

$C_{16}H_{24}Ni$ F.W. = 275.1

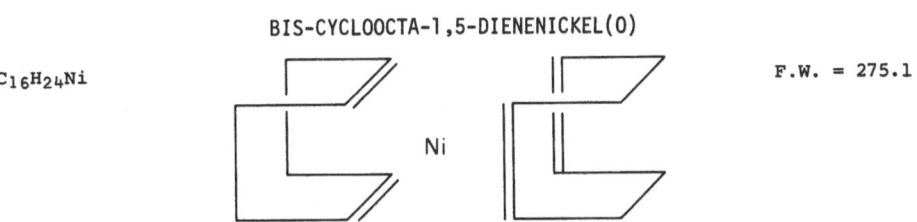

I. Die Kristallstruktur von *bis*-Cyclooctadien-(1,5)-Nickel(0). H. DIERKS
   and H. DIETRICH, 1965. *Z. Kristallogr.*, 122, 1-23.

Triclinic, a = 10.84 ± 4, b = 9.12 ± 4, c = 7.38 ± 3 Å, α = 109.95 ± 40°,
β = 84.72 ± 40°, γ = 108.13 ± 40°, U = 652 Å³, Z = 2, D_x = 1.399.
Space group $P1$ (C_1^1) or $P\bar{1}$ (C_i^1). $P\bar{1}$ is confirmed by the analysis.

Atomic positions

	x	y	z
Ni	0.2539 ± 1	0.0542 ± 1	0.2500 ± 1
C(1)	0.1960 ± 5	0.1570 ± 7	0.0660 ± 8
C(2)	0.1307 ± 5	0.1934 ± 7	0.2378 ± 7
C(3)	0.1666 ± 6	0.3587 ± 7	0.3975 ± 8
C(4)	0.2581 ± 6	0.3672 ± 7	0.5502 ± 8
C(5)	0.3506 ± 5	0.2694 ± 7	0.4720 ± 7
C(6)	0.4205 ± 5	0.2642 ± 6	0.3046 ± 7
C(7)	0.4192 ± 6	0.3620 ± 8	0 1765 ± 9
C(8)	0.3160 ± 6	0.2695 ± 8	0.0126 ± 9
C(9)	0.0969 ± 5	−0.1295 ± 7	0.3086 ± 8
C(10)	0.1913 ± 5	−0.0704 ± 7	0.4529 ± 8
C(11)	0.2910 ± 6	−0.1477 ± 8	0.4656 ± 9
C(12)	0.3506 ± 6	−0.2167 ± 8	0.2680 ± 8
C(13)	0.3657 ± 5	−0.1117 ± 7	0.1458 ± 8
C(14)	0.2779 ± 5	−0.1382 ± 6	0.0040 ± 7

	x	y	z
C(15)	0.1509 ± 6	−0.2714 ± 7	−0.0381 ± 8
C(16)	0.0734 ± 6	−0.2898 ± 8	0.1430 ± 9

Isotropic temperature factors are given also, as are assumed positions for the hydrogen atoms.

Structure

The configuration of the molecule is indicated schematically above, and more realistically in Fig. 187. The nickel atom is coordinated to four double bonds, as shown, with the ligands oriented approximately tetrahedrally. Bond lengths and angles are given in full. The Ni–C distances are all 2.12 ± 1 Å. The cycloocta-1,5-dienyl rings are puckered; mean bond lengths and angles are: C≡C, 1.39 ± 1 Å; C–C, 1.52 ± 1 Å (1.49 to 1.55 Å); C≡C–C, 125 ± 1°; C–C–C, 113 ± 1°.

Details of analysis

The intensity data were recorded on equi-inclination Weissenberg photographs about three axes, using CuK_α radiation. 2453 reflexions were observed and measured visually. Refinement was by Fourier methods to a final R index of 0.102. (Least-squares refinement was attempted, but abandoned because of anomalous results, attributed to systematic errors in the intensity data.)

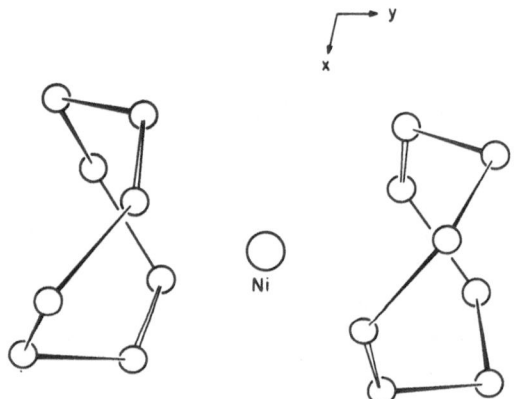

Fig. 187. The $C_{16}H_{24}Ni$ molecule viewed along c.

CYCLOPENTADIENYL[1-(CYCLOPENTADIENYL)-1,2,3,4-TETRA-

METHYL-π-CYCLOBUTENYL]NICKEL

$C_{18}H_{22}Ni$ $(CH_3)_4C_4C_5H_5\ NiC_5H_5$ F.W. = 297.0

I. The structure of a metal-cyclobutenyl complex, $(CH_3)_4C_4C_4H_5\ NiC_5H_5$. W. OBERHANSLI and L. F. DAHL, 1965. *Inorg. Chem.*, **4**, 150–157.

Orthorhombic, $a = 11.66 ± 3$, $b = 11.77 ± 3$, $c = 11.40 ± 3$ Å, $U = 1565$ Å^3, $D_m = 1.33$, $z = 4$, $D_x = 1.28$.

Space group $P2_12_12_1$ (D_2^4)

Atomic positions

	x	y	z
Ni	0.3351 ± 1	0.2068 ± 1	0.7557 ± 2
C(1)	0.4538 ± 15	0.0765 ± 16	0.7090 ± 16
C(2)	0.4781 ± 16	0.1132 ± 16	0.8242 ± 15

	x	y	z
C(3)	0.5083 ± 14	0.2278 ± 17	0.8211 ± 14
C(4)	0.4971 ± 14	0.2654 ± 14	0.7037 ± 14
C(5)	0.4677 ± 12	0.1708 ± 14	0.6348 ± 14
C(6)	0.1996 ± 11	0.2980 ± 14	0.7043 ± 11
C(7)	0.2133 ± 11	0.2854 ± 14	0.8294 ± 11
C(8)	0.1818 ± 10	0.1678 ± 11	0.8176 ± 10
C(9)	0.1255 ± 10	0.1894 ± 11	0.6931 ± 11
C(10)	−0.0041 ± 11	0.2156 ± 13	0.7124 ± 10
C(11)	−0.0866 ± 13	0.1628 ± 14	0.6515 ± 15
C(12)	−0.2037 ± 16	0.2066 ± 18	0.6959 ± 16
C(13)	−0.1761 ± 16	0.2942 ± 17	0.7755 ± 25
C(14)	−0.0487 ± 14	0.2996 ± 15	0.7882 ± 14
C(15)	0.1462 ± 11	0.1070 ± 13	0.5947 ± 12
C(16)	0.1518 ± 14	0.0769 ± 16	0.9055 ± 15
C(17)	0.2010 ± 15	0.4018 ± 15	0.6257 ± 15
C(18)	0.2347 ± 13	0.3614 ± 14	0.9318 ± 15

Isotropic temperature factors are given also.

Structure

The molecular configuration is illustrated in Fig. 188. A non-crystallographic, but fairly precise mirror plane contains the cyclopentadiene ring carbon
atoms C(10) to C(14), the nickel atom, C(7), C(9), C(15), C(18), C(3), and the
mid-point of C(1) and C(5). The nickel atom is 1.764 Å from the mean plane of
the cyclopentadienyl anion (mean Ni–C distance, 2.11 Å), and 1.700 Å from the
plane of the allylic fragment of the cyclobutenyl ring (mean Ni–C distance, 1.96
Å). The three methyl groups bonded to the latter fragment are displaced by
0.13 to 0.21 Å from its plane. Bond lengths and angles are given in full,
and are consistent with expectation.

Details of analysis

The intensity data were recorded on equi-inclination Weissenberg photographs
($hk0$ to $hk12$) using MoKα radiation. 970 reflexions were observed and measured
visually. Refinement was by full-matrix least squares to a final R index of
0.078.

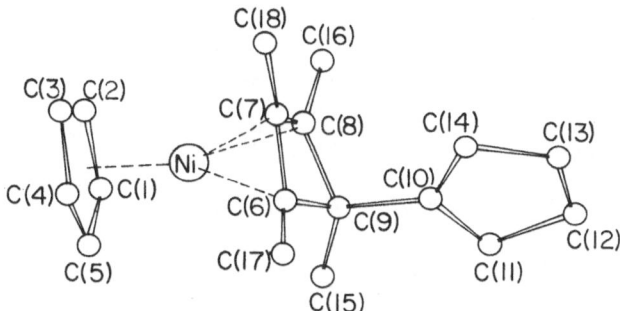

Fig. 188. The $C_{18}H_{22}Ni$ molecule viewed along b.

1,5-CYCLOOCTADIENE-DUROQUINONE-NICKEL

$C_{18}H_{24}O_2Ni$ F.W. = 331.1

I. Structure of and bonding in 1,5-cyclooctadiene-duroquinone-nickel. M. D.
    GLICK and L. F. DAHL, 1965. *J. Organometal. Chem.*, **3**, 200–221.

Monoclinic, a = 14.26 ± 5, b = 7.95 ± 3, c = 14.17 ± 5 Å, β = 94°27 ± 10',
[U = 1602 Å³], D_m = 1.44, z = 4, D_x = 1.37.

Space group Pn (C_s^2) or P2/n (C_{2h}^4). P2/n confirmed by analysis. Molecular symmetry, two-fold axis for each of two independent molecules.

	x	y	z
Ni(1)	0.2500	0.0594 ± 2	0.2500
Ni(2)	0.7500	0.2911 ± 2	0.2500
C(1)	0.1312 ± 11	0.3437 ± 24	0.1876 ± 12
C(2)	0.1998 ± 8	0.2278 ± 20	0.1441 ± 10
C(3)	0.2928 ± 9	0.2350 ± 19	0.1498 ± 9
C(4)	0.3480 ± 12	0.3718 ± 26	0.2036 ± 13
C(5)	0.0681 ± 9	−0.1495 ± 19	0.1358 ± 9
C(6)	0.1591 ± 7	−0.1544 ± 15	0.1953 ± 7
C(7)	0.2459 ± 7	−0.1664 ± 14	0.1470 ± 7
C(8)	0.3358 ± 7	−0.1527 ± 15	0.2048 ± 7
C(9)	0.4247 ± 8	−0.1461 ± 19	0.1522 ± 9
O(10)	0.2432 ± 6	−0.1789 ± 13	0.0603 ± 6
C(11)	0.6777 ± 11	0.0052 ± 23	0.1345 ± 11
C(12)	0.6434 ± 9	0.1236 ± 18	0.2076 ± 9
C(13)	0.6547 ± 9	0.1160 ± 19	0.3013 ± 10
C(14)	0.7133 ± 12	−0.0280 ± 25	0.3488 ± 12
C(15)	0.6378 ± 9	0.5012 ± 20	0.0668 ± 9
C(16)	0.6972 ± 7	0.5068 ± 15	0.1612 ± 7
C(17)	0.6487 ± 7	0.5192 ± 13	0.2468 ± 7
C(18)	0.7052 ± 7	0.5069 ± 16	0.3350 ± 7
C(19)	0.6515 ± 8	0.5029 ± 19	0.4255 ± 9
O(20)	0.5618 ± 6	0.5338 ± 12	0.2430 ± 6

Temperature factors (anisotropic for the nickel atoms) are given also.

Structure

Each discrete monomeric molecule consists of a nickel atom sandwiched between a boat-shaped 1,5-cyclooctadiene ring and a duroquinone ring. The two sets of parallel double bonds in the two rings are perpendicular, and these bonds are tetrahedrally disposed with respect to the nickel atom. The molecular configuration is illustrated in Fig. 189, together with average bond lengths and angles. The detailed geometry of the rings, and the nature and implications of the bonding are discussed. The intermolecular distances are consistent with van der Waals interaction.

Details of analysis

The intensity data were recorded on equi-inclination Weissenberg photographs (hk0 to hk15) and precession photographs (h0l to h2l; 0kl) using MoK_α radiation. 1757 reflexions were observed and measured visually. Refinement was by full-matrix least squares to a final R index of 0.101.

NORBORNADIENE PALLADIUM CHLORIDE

$C_7H_8PdCl_2$ F.W. = 269.8

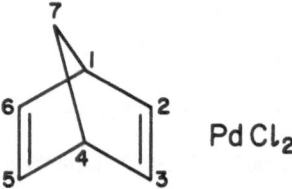

$PdCl_2$

I. The refinement of the crystal structure of norbornadiene palladium chloride. N. C. BAENZIGER, G. F. RICHARDS and J. R. DOYLE, 1965. *Acta Cryst.*, <u>18</u>, 924-926.

This is a redetermination, at liquid nitrogen temperature, of the structure reported in <u>1</u>.

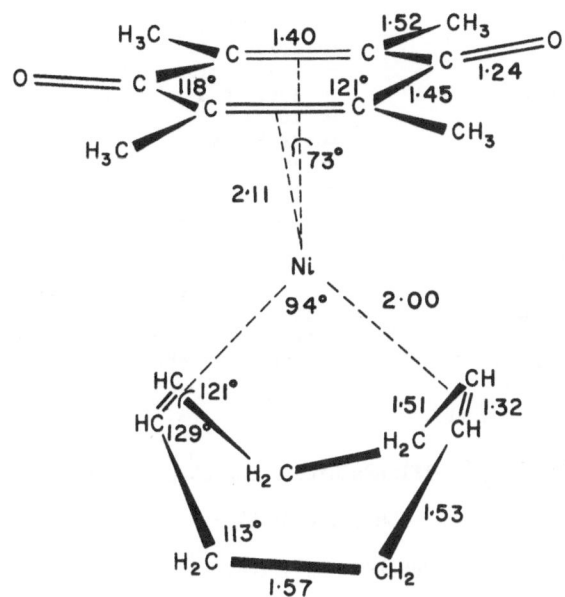

Fig. 189. Configuration, bond lengths, and angles in the average 1,5-cyclooct-
adiene-duroquinone-nickel molecule. Standard deviations range from
0.010 to 0.015 Å, and 0.6 to 1.1°.

Orthorhombic, a = 9.3160 ± 15, b = 7.085 ± 1, c = 11.685 ± 15 Å, [U =
771.3 Å³, z = 4, D_X = 2.32]. (Cell and intensity data at -195°C.)

Space group Pnma (D_{2h}^{16}) Molecular symmetry, mirror plane.

Atomic positions

	x	y	z
Pd	0.60144 ± 4	0.25	0.47763 ± 4
Cl(1)	0.53180 ± 15	0.25	0.28674 ±11
Cl(2)	0.36886 ± 12	0.25	0.54600 ± 12
C(1,4)	0.84450 ± 44	0.41036 ± 65	0.57929 ± 33
C(2,3)	0.70051 ± 41	0.34646 ± 68	0.63414 ± 30
C(5,6)	0.82012 ± 36	0.34644 ± 72	0.45379 ± 32
C(7)	0.94333 ± 63	0.25	0.62247 ± 52
H(1,4)	0.8692 ±54	0.5358 ±92	0.5944 ±54
H(2,3)	0.6367 ±50	0.4413 ±85	0.6701 ±51
H(5,6)	0.8156 ±58	0.4320 ±83	0.3906 ±50
H(7)	0.0589 ±85	0.25	0.5791 ±76
H(7')	0.9473 ±86	0.25	0.7106 ±69

Anisotropic temperature factors are given for the non-hydrogen atoms.

Structure

The structure is essentially as reported in 1, but has been more accurately
determined. Interatomic distances and angles are as follows:

Pd-Cl(1)	=	2.323 ± 1 Å
Pd-Cl(2)	=	2.310 ± 1
Pd-C(1,2)	=	2.159 ± 4
Pd-C(5,6)	=	2.166 ± 4
C(1)-C(2)	=	1.554 ± 6
C(1)-C(6)	=	1.552 ± 6

C(1)–C(7)	=	1.547 ± 6
C(2)–C(3)	=	1.366 ± 10
C(5)–C(6)	=	1.366 ± 10
C(1)–C(2)–C(3)	=	106.9 ± 4°
C(4)–C(5)–C(6)	=	107.0 ± 4
C(2)–C(1)–C(7)	=	99.5 ± 4
C(6)–C(1)–C(7)	=	100.4 ± 4
C(2)–C(1)–C(6)	=	100.3 ± 4

C–H distances and some non-bonded C–C distances are given also.

Details of analysis

Intensity data were measured with a diffractometer (of modified Weissen-berg design) and proportional counter, using $CuK\alpha$ radiation. Levels scanned were $h0l$ to $h5l$ and $0kl$ to $6kl$. 838 reflexions were measured and absorption corrections were applied. Refinement was by full-matrix least squares to a final R index of 0.0275.

1. *Structure Reports*, 26, 687.

1-ETHOXY-1,2,3,4-TETRAPHENYLCYCLOBUTENYL PALLADIUM(II)

CHLORIDE DIMER

$C_{60}H_{50}O_2Pd_2Cl_2$ $[(C_6H_5)_4C_4OC_2H_5]_2Pd_2Cl_2$ F.W. = 1087

I. Structures of two metal-cyclobutenyl isomers of $[(C_6H_5)_4C_4OC_2H_5]_2Pd_2Cl_2$. L. F. DAHL and W. E. OBERHANSLI, 1965. *Inorg. Chem.*, 4, 629–637.

Endo ISOMER

Monoclinic, a = 9.90 ± 2, b = 20.75 ± 4, c = 13.75 ± 3 Å, β = 118°40 ± 10', U = 2479 Å³, D_m = 1.46, Z = 2, D_x = 1.46.

Space group $P2_1/c$ (C_{2h}^5) Molecular symmetry, centre.

Exo ISOMER

Monoclinic, a = 13.17 ± 3, b = 13.57 ± 3, c = 14.05 ± 3 Å, β = 92°30 ± 10', U = 2506 Å³, D_m = 1.46, Z = 2, D_x = 1.45.

Space group $P2_1/n$ (C_{2h}^5) Molecular symmetry, centre.

Atomic positions

Atomic positions are given for the 33 non-hydrogen atoms of each isomer. Results are given both for conventional refinement and for rigid-body refine-ment, in which the phenyl rings were constrained to their known geometry. Mean standard deviations are: Pd, 0.001 Å; Cl, 0.005 Å; other, 0.01 to 0.04 Å.

Structure

The configuration of the *endo* isomer is indicated in Fig. 190. Adjacent palladium atoms share chlorine ligands, as shown, to form centrosymmetrical dimers. The configuration of the *exo* isomer is identical, except that the groups attached to the tetrahedral carbon atom of the cyclobutenyl ring are interchanged, so that the ethoxy group is remote from, instead of close to, the palladium atom. Each palladium atom is coordinated to two chlorine atoms, as shown, and to the allylic fragment of the cyclobutenyl ring. This fragment is symmetrically disposed with respect to the palladium-chlorine plane, and is inclined at 95° to it. The dihedral angle of the cyclobutenyl ring is 24°, with the tetrahedral carbon atom bent away from the palladium atom. Bond lengths and angles are given in some detail. The palladium atoms lie 2.43 Å from the chlorine atoms, and 2.09 to 2.18 Å from the carbon atoms of the allylic fragment.

Details of analysis

The intensity data were recorded on equi-inclination Weissenberg photo-graphs (0kl to 10,kl for the *endo* isomer; hk0 to hk,13 for the *exo* isomer) using CuK$_\alpha$ radiation. 2461 and 1764 reflexions were observed for the *endo* and *exo* isomers, respectively, and. were measured visually. Refinement was by full-matrix least squares, with the phenyl rings constrained to their known geometry. The final R indices were 0.136 and 0.133, respectively.

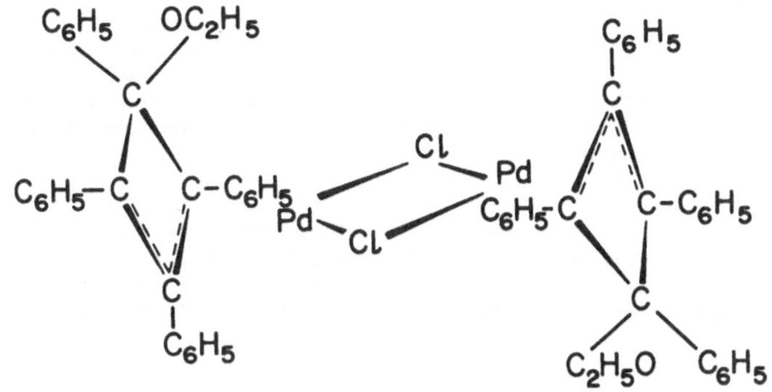

Fig. 190. Configuration of the *endo* isomer of $C_{60}H_{50}O_2Pd_2Cl_2$.

DIPENTENE PLATINUM (II) CHLORIDE

$C_{10}H_{16}Cl_2Pt$ F.W. = 402.2

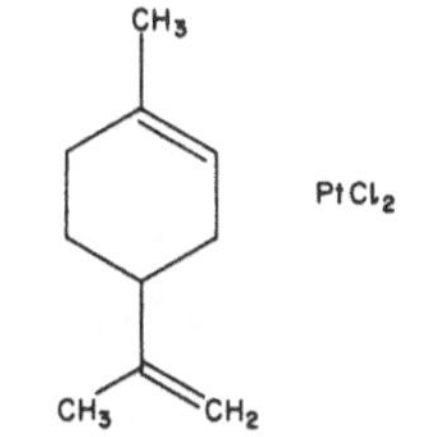

I. The crystal structure of dipentene platinum (II) chloride. N. C. BAENZIGER, R. C. MEDRUD and J. R. DOYLE, 1965. *Acta Cryst.*, <u>18</u>, 237–242.

Orthorhombic, a = 10.188 ± 5, b = 7.992 ± 2, c = 14.401 ± 9 Å, U = 1173 Å^3, D_m = 2.21, Z = 4, D_x = 2.28, (CuK$_\alpha$, λ = 1.5418 Å; MoK$_\alpha$, λ = 0.7107 Å).

Space group $P2_1cn$ (C_{2v}^9) or $Pmcn$ (D_{2h}^{16}) Structural considerations indicate $P2_1cn$.

Atomic positions

	x	y	z
Pt	0.25	0.09693 ± 16	0.01738 ± 9
Cl(1)	0.3360 ± 12	0.3415 ± 17	-0.0485 ± 8
Cl(2)	0.2486 ± 24	-0.0098 ± 12	-0.1321 ± 5
C(1)	0.1864 ± 31	0.2362 ± 44	0.1449 ± 21
C(2)	0.3291 ± 38	0.1800 ± 49	0.1559 ± 25
C(3)	0.3850 ± 31	0.0536 ± 41	0.2031 ± 20
C(4)	0.2837 ± 41	-0.0970 ± 57	0.1998 ± 24
C(5)	0.1532 ± 44	-0.0387 ± 61	0.2530 ± 28
C(6)	0.1217 ± 49	0.1278 ± 69	0.2147 ± 33
C(7)	0.2771 ± 36	-0.1320 ± 42	0.0973 ± 18

	x	y	z
C(8)	0.1458 ± 33	-0.1269 ± 48	0.0414 ± 20
C(9)	0.3827 ± 41	-0.2395 ± 59	0.0631 ± 27
C(10)	0.1388 ± 44	0.3910 ± 64	0.1238 ± 26

Temperature factors (anisotropic for the platinum atom) are given also.

Structure

The coordination of the platinum atom is square planar. Two adjacent coordination positions are occupied by chlorine atoms, and the remaining two by the double bonds of the dipentene molecule. The molecular configuration is illustrated in Fig. 191. [Note however, that the accuracy of the analysis does not permit the identification of the double bonds, other than by the proximity of the platinum atom.] Intermolecular distances are consistent with van der Waals interaction.

Details of analysis

The intensity data were recorded on equi-inclination Weissenberg photographs ($0kl$ to $8kl$, CuK_α radiation) and precession photographs ($h0l$ to $h4l$, MoK_α radiation). 932 of 1311 accessible reflexions were observed and measured visually. Refinement was by full-matrix least squares to a final R index of 0.071.

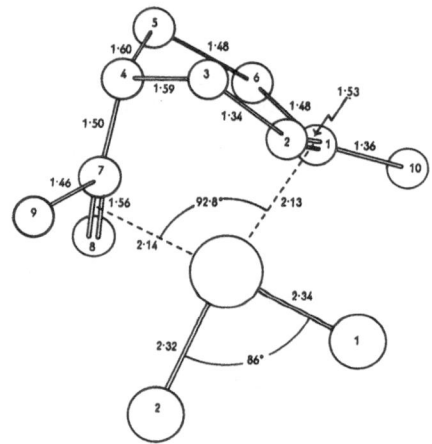

Fig. 191. The dipentene platinum (II) chloride molecule, viewed along a.
The standard deviations of the distances shown are: Pt-Cl, 0.01 Å;
Pt-C, 0.03 Å; C-C, 0.06 Å.

DIPEROXOAQUOETHYLENEDIAMINECHROMIUM(IV) MONOHYDRATE

$C_2H_8CrN_2O_4 \cdot 2H_2O$ $[Cr(O_2)_2(H_2O)(C_2H_8N_2)] \cdot H_2O$ F.W. = 212.1

I. The crystal and molecular structure of diperoxoaquaethylenediamine-
chromium(IV)monohydrate, $[Cr(O_2)_2(H_2O)en](H_2O)$. R. STOMBERG, 1965.
Ark. Kemi, <u>24</u>, 47–71.

Orthorhombic, a = 7.594 ± 4, b = 12.258 ± 8, c = 8.168 ± 5 Å, U = 760.3 Å^3,
D_m = 1.863, Z = 4, D_x = 1.853, (calibration with KCl, a = 6.2929 Å).

Space group Pbc2₁ (C_{2v}^5) or Pbcm (D_{2h}^{11}) Pbc2₁ assumed, but is shown that a disordered structure in Pbcm is also possible.

Atomic positions

	x	y	z
Cr	0.2410	0.0130	¼
O(1)	0.2366	0.0168	0.0108
O(2)	0.2828	0.9116	0.0826
O(1')	0.2378	0.0206	0.4785
O(2')	0.2792	0.9113	0.4119
O(3)	0.9774	0.9869	0.2442
O(4)	0.9550	0.2446	0.0054
N(1)	0.2054	0.1816	0.2482
N(2)	0.5019	0.0571	0.2459
C(1)	0.3671	0.2319	0.3030
C(2)	0.5202	0.1727	0.2204

For comparison, parameters resulting from a refinement of a disordered structure in Pbcm are given also.

Structure

[Cr(O₂)₂(H₂O)en] exists in the structure as a discrete molecule, with the configuration shown in Fig. 192. The coordination of the chromium atom is pentagonal bipyramidal; the two peroxo groups and one of the nitrogen atoms of the ethylenediamine molecule form the equatorial plane, while the apical positions are occupied by the other nitrogen atom and a water molecule. The molecule of [Cr(O₂)₂(H₂O)en] and the remaining water molecule are held together by a system of strong O-H...O and N-H...O bonds, for details of which the original should be consulted.

Details of analysis

The intensity data were recorded on equi-inclination Weissenberg photographs (hk0 to hk5, h0l to h2l, 0kl to 1kl) using CuK$_\alpha$ radiation. Because of crystal decomposition, a new specimen was used for each layer recorded. 417 independent reflexions were observed and measured visually. Refinement was by least squares to a final R index of 0.115.

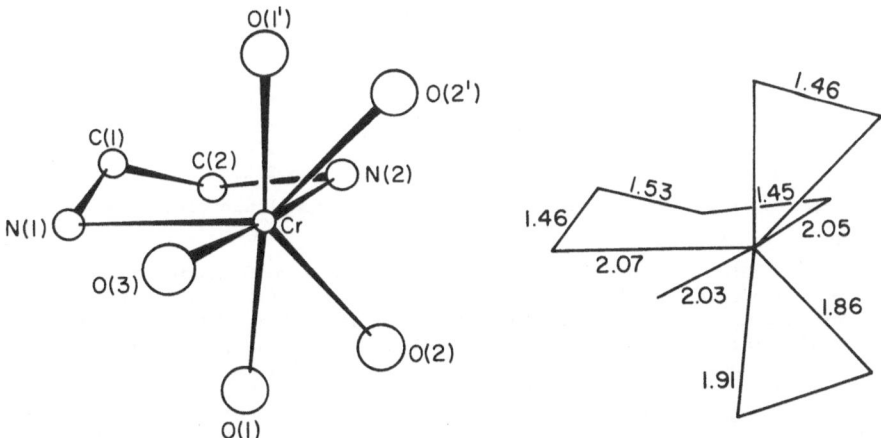

Fig. 192. The [Cr(O₂)₂(H₂O)en] molecule. Standard deviations of bond lengths
range from 0.01 to 0.03 Å.

trans-DICHLOROBIS(ETHYLENEDIAMINE)COBALT(III) HEXATHIONATE

MONOHYDRATE

$C_8H_{32}Cl_4Co_2N_8O_6S_6 \cdot H_2O$ F.W. = 806.5

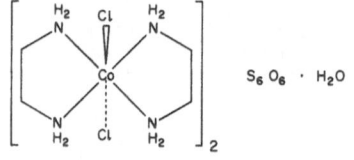

$S_6O_6 \cdot H_2O$

I. The crystal structure of *trans*-dichlorobis(ethylenediamine)cobalt(III)
    hexathionate monohydrate. O. FOSS and K. MAROY, 1965. *Acta Chem. Scand.*,
    <u>19</u>, 2219-2228.

    (For earlier work, see <u>1</u>.)

Orthorhombic, a = 12.12, b = 19.13, c = 6.43 Å, $[U = 1491$ Å$^3]$,Z = 2,
D_x = 1.80.

Space group Pba2 (C_{2v}^8) or Pbam (D_{2h}^9). Pba2 is confirmed by analysis. Molecular symmetry, two-fold axis for hexathionate ion and water molecule.

Atomic positions

	x	y	z
S(1)	0.2541	0.0919	0.4901
S(2)	0.1647	0.0046	0.3767
S(3)	0.0713	0.0290	0.6181
O(1)	0.1730	0.1373	0.5819
O(2)	0.2999	0.1151	0.2964
O(3)	0.3387	0.0645	0.6212
Co	0.0415	0.2762	−0.0193
Cl(1)	0.1690	0.2963	0.2219
Cl(2)	−0.0872	0.2567	−0.2614
N(1)	0.0831	0.1779	−0.0144
C(1)	−0.0156	0.1348	0.0780
C(2)	−0.0492	0.1797	0.2731
N(2)	−0.0681	0.2506	0.1963
N(3)	0.0016	0.3764	−0.0193
C(3)	0.0873	0.4161	−0.1355
C(4)	0.1433	0.3715	−0.3054
N(4)	0.1492	0.2981	−0.2307
H_2O	0.5000	0	0.2716

Temperature factors (anisotropic for sulphur atoms) are given also.

Structure

The hexathionate ion has the helical *trans-trans* conformation shown in
Fig. 193. The *trans*-dichlorobis(ethylenediamine)cobalt(III) cation is essentially centrosymmetrical, and the coordination of ligands to the cobalt atom is
octahedral. The cobalt and the nitrogen atoms are coplanar, and the Co-Cl
bonds lie 88° and 89° from this plane. Mean bond lengths for the coordination
octahedron are: Co-Cl, 2.24 ± 1 Å; Co-N, 1.96 ± 2 Å. The angle N-Co-N
(for both nitrogen atoms in the same ethylenediamine molecule) is 87°. The
C-C bonds are twisted somewhat out of the cobalt-nitrogen plane, in such a
way as to preserve the approximate centre of symmetry.

Details of analysis

The intensity data were recorded on equi-inclination Weissenberg photographs (*hk*0 to *hk*3; 0*kl*) using Fe*K*$_\alpha$ radiation. 564 of a possible 700 reflex-

ions were observed and measured visually. Refinement was by block-diagonal
least squares to a final R index of 0.085.

1. *Structure Reports*, <u>23</u>, 573.

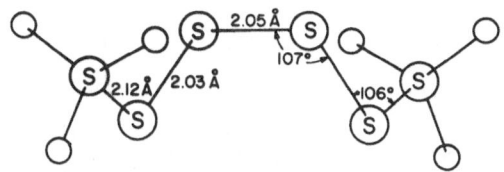

Fig. 193. Conformation of the hexathionate ion. Standard deviations of
distances and angles are 0.010 to 0.015 Å, and 1°.

NITRITOBIS(ETHYLENEDIAMINE)NICKEL TETRAFLUOROBORATE

$C_4H_{16}BF_4N_5O_2$ $[Ni(NH_2CH_2CH_2NH_2)_2(NO_2)]BF_4$ F.W. = 311.7

   I. The structure of $Ni(ethylenediamine)_2(NO_2)BF_4$. M. G. B. DREW, D. M.
   L. GOODGAME, M. A. HITCHMAN and D. ROGERS, 1965. *Chem. Communic. G.B.*,
   <u>14</u>, 477-479.

Orthorhombic, a = 15.142, b = 10.305, c = 8.153 Å, [U = 1272 Å^3], z = 4,
[D_x = 1.62].

Space group $Pn2_1a$ (C_{2v}^9) *or* $Pnma$ (D_{2h}^{16}). $Pn2_1a$ *is consistent with the deduced*
structure.

Atomic positions are not given.

Structure

    The nickel atom is coordinated to the nitrogen atoms of two bidentate
ethylenediamine ligands in an approximately square planar array. Distorted
octahedral coordination is completed by a nitrogen atom and an oxygen atom of
two different nitrite goups. The nickel-ligand distances range from 2.10 to
2.22 Å. The nitrite groups serve as bridges, joining the nickel atoms to
form endless polymeric chains, and are non-linear, with O-N-O = 111°.

Details of analysis

    811 independent reflexions were measured visually. Refinement was by
full-matrix least squares to a final R index of 0.138. The analysis was
complicated by disorder of the nitrite groups.

1,2-DICHLORODIETHYLENEDIAMINECOBALT(III) CHLORIDE MONOHYDRATE

(*cis* FORM)

$C_4H_{16}Cl_3CoN_4 \cdot H_2O$ $CoCl_2(NH_2CH_2CH_2NH_2)_2$ F.W. = 303.5

   I. Die Struktur des 1,2-Dichloro-diaethylendiamin-kobalt(III) chloridmono-
   hydrats. A. HUELLEN, K. PLIETH and G. RUBAN, 1965. *Naturwissenschaften*,
   <u>52</u>, 618.

Monoclinic, a = 12.00, b = 6.87, c = 16.48 Å, β = 122°, [U = 1152 Å^3],
D_m = 1.73, z = 4, D_x = 1.75.

Space group $P2_1/c$ (C_{2h}^5)

Atomic positions

	x	y	z
Co	0.176	0.395	0.194
Cl(1)	-0.027	0.237	0.619
Cl(2)	0.088	0.690	0.151
Cl(3)	0.434	0.650	0.110
N(1)	0.245	0.118	0.222
N(2)	0.357	0.517	0.265
N(3)	0.175	0.391	0.064
N(4)	0.175	0.391	0.316
C(1)	0.238	0.203	0.077
C(2)	0.274	0.049	0.153
C(3)	0.304	0.488	0.394
C(4)	0.356	0.602	0.345
O	0.245	0.939	0.400

Structure

The cobalt atom is coordinated to two chlorine atoms at distances of 2.24 ± 2 Å, and to four nitrogen atoms at distances ranging from 2.02 to 2.14 Å. The coordination is essentially octahedral; the two chlorine atoms occupy adjacent corners of the octahedron, as do the two nitrogen atoms of each of the bidentate ethylenediamine ligands. The ethylenediamine planes are therefore approximately at right angles to each other.

Details of analysis

The intensity data were recorded photographically, using CoK_α radiation. Refinement was by Fourier methods to final R indices as follows: $h0l$, 0.116; $hk0$, 0.17; $hk1$, 0.215.

TRISACETYLACETONATOCHROMIUM (III)

$C_{15}H_{21}O_6Cr$ $Cr(CH_3COCHCOCH_3)_3$ F.W. = 349.3

I. The crystal structure of trisacetylacetonatochromium (III). B. MOROSIN, 1965. *Acta Cryst.*, <u>19</u>, 131-137.

Monoclinic, a = 14.031 ± 9, b = 7.55 ± 5, c = 16.379 ± 11 Å, β = 99.06 ± 20°, [U = 1714 Å³], D_m = 1.374, Z = 4, D_x = 1.362 ($MoK_{\alpha1}$, λ = 0.70926 Å).

Space group $P2_1/c$ (C_{2h}^5) .

Atomic positions

	x	y	z
Cr	0.23771	0.26578	0.47071
O(1)	0.11258	0.33858	0.41152
O(2)	0.18055	0.18920	0.56564
O(3)	0.36548	0.19218	0.52550
O(4)	0.29698	0.34347	0.37596
O(5)	0.21717	0.02765	0.42352
O(6)	0.25741	0.50286	0.51967
C(4)	0.30395	0.25276	0.31081
C(5)	0.23428	-0.02536	0.35347
C(45)	0.27410	0.08116	0.29743
C(X4)	0.34770	0.35068	0.24573
C(X5)	0.21023	-0.21721	0.33426
C(6)	0.33308	0.56527	0.56060
C(3)	0.43110	0.29259	0.56316
C(36)	0.41881	0.47041	0.58000
C(X3)	0.52800	0.20577	0.59184
C(X6)	0.32767	0.75494	0.59008

	x	y	z
C(1)	0.03285	0.31731	0.43459
C(2)	0.09222	0.18624	0.57122
C(12)	0.01789	0.23933	0.50880
C(X1)	-0.05339	0.37613	0.37093
C(X2)	0.06938	0.11372	0.65401

The mean standard deviations of the coordinates are 0.0014, 0.0065, and 0.015 Å for chromium, oxygen, and carbon respectively. Anisotropic temperature factors are given for all atoms.

Structure

The octahedral coordination of the chromium atom and the molecular packing are illustrated in Fig. 194. The coordination octahedron is almost regular. The mean Cr-O distance is 1.951 ± 7 Å and the mean intrachelate O-O distance is 2.786 ± 13 Å. Other O-O distances in the octahedron range from 2.731 to 2.771 Å. Bond lengths and angles are given in full. Mean values for the chelate ring are given in Fig. 195.

Details of analysis

Intensity data were measured with a four-circle diffractometer and scintill-ation counter using MoK_α radiation. Of 1490 reflexions in the range $2\theta<45°$, 1266 were observed above background. Refinement was by differential syntheses to a final R index of 0.066. One of the chelate rings was found to undergo rather severe thermal motion, and this phenomenon was associated with diffuse reflexions. These reflexions were shown to be temperature dependent, and it was concluded that the apparent thermal motion of the chelate ring was real, and not a result of static disorder.

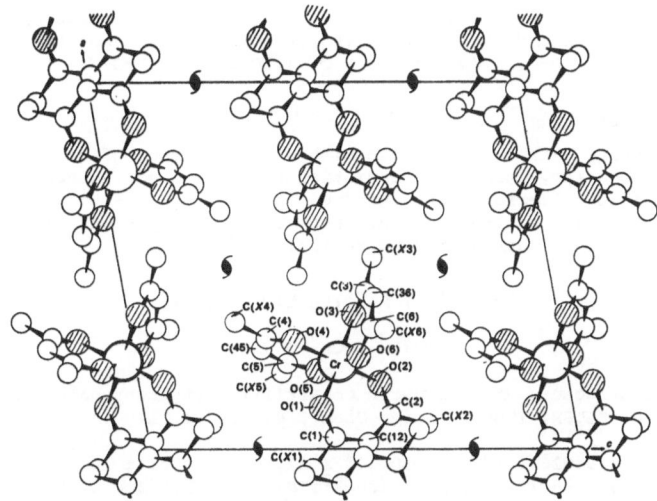

Fig. 194.   The structure of $C_{15}H_{21}O_6Cr$ viewed along b. The chromium atoms near $y = \frac{1}{4}$ are denoted by double circles; those near $y = 3/4$ by a single open circle.

DI-μ-DIPHENYLPHOSPHINATOACETYLACETONATOCHROMIUM(III)

$C_{44}H_{48}O_{12}Cr_2P_2$ $(C_5H_7O_2)_2Cr(OP(C_6H_5)_2O)_2Cr(C_5H_7O_2)$ F.W. = 934.8

I. The crystal structure of di-μ-diphenylphosphinatoacetylacetonatochromium-(III). C. E. WILKES and R. A. JACOBSON, 1965. *Inorg. Chem.*, <u>4</u>, 99-103.

Triclinic, $a = 12.64 \pm 5$, $b = 15.57 \pm 5$, $c = 13.35 \pm 1$ Å, $\alpha = 112.4 \pm 2$, $\beta = 112.5 \pm 2$, $\gamma = 84.2 \pm 1°$, $[U = 2241$ Å$^3]$, $D_m = 1.4$, $Z = 2$, $D_x = 1.383$.

Space group P1 (C_1^1) or P$\bar{1}$ (C_i^1). P$\bar{1}$ is confirmed by the analysis.

Atomic positions

   Atomic positions are given for the 60 non-hydrogen atoms. Average standard deviations are: Cr, 0.002 Å; P, 0.002 Å; O, 0.006 Å; C, 0.013 Å. Temperature factors (anisotropic for Cr, P and O) are given also.

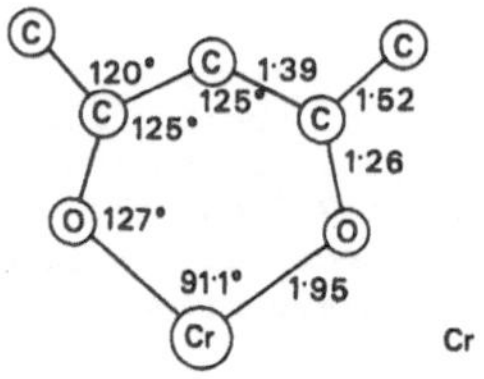

Fig. 195.   Mean bond lengths and angles for the chelate ring in $C_{15}H_{21}O_6Cr$. Standard deviations are: C-C, 0.021; C-O, 0.016; Cr-O, 0.007 Å, O-Cr-O, 0.6°; other angles 1°.

Structure

   The molecular configuration is indicated schematically in Fig. 196. Each chromium atom is octahedrally coordinated, being attached to two oxygen atoms of two different diphenyl phosphinate groups, and to the oxygen atoms of two bidentate acetylacetonate groups. Adjacent chromium atoms share diphenyl phosphinate groups to form a dimer, with an eight-membered puckered ring, as shown. Bond lengths and angles are given in full. Some mean values are: Cr-O, 1.955 ± 9 Å; P-C, 1.80 ± 1 Å; P-O, 1.50 ± 1 Å; P-O-Cr, 151° (141° to 161°).

Details of analysis

   The intensity data were recorded with a four-circle diffractometer and scintillation counter, using MoK_α radiation with balanced filters. 5100 of 10,300 accessible reflexions were observed in the range $2\theta < 56°$, and were measured by the peak-counting method. Refinement was by block-diagonal least-squares to a final R index (for 4400 reflexions) of 0.09.

COBALT(II) ACETYLACETOACETONATE

$C_{10}H_{16}CoO_4$ Co$(CH_3COCH_2COCH_3)_2$ F.W. = 259.2

   I. Crystal structure of tetrameric cobalt(II) acetylacetonate. F. A. COTTON and R. C. ELDER, 1965. *Inorg. Chem.*, **4**, 1145-1151.

Triclinic, $a = 8.516 \pm 12$, $b = 10.243 \pm 17$, $c = 13.781 \pm 20$ Å, $\alpha = 93.5 \pm 3$, $\beta = 90.4 \pm 3$, $\gamma = 98.7 \pm 3°$, $U = 1186$ Å^3, $D_m = 1.45$, $Z = 4$, $D_x = 1.45$.

Space group P1 (C_1^1) or P$\bar{1}$ (C_i^1) [Although the space group is nowhere mentioned, it is clear that P$\bar{1}$ is consistent with the analysis.]

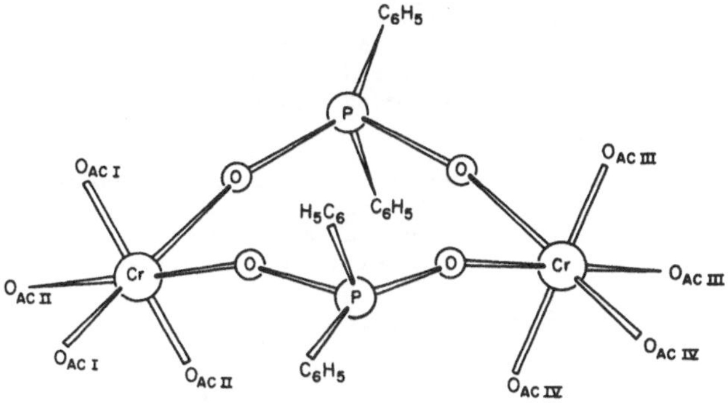

C$_6$H$_5$

P

O$_{AC\ I}$ O   O   O$_{AC\ III}$

H$_5$C$_6$ C$_6$H$_5$

O$_{AC\ II}$ Cr   O   P   O   Cr   O$_{AC\ III}$

O$_{AC\ I}$ O$_{AC\ II}$ C$_6$H$_5$ O$_{AC\ IV}$

O$_{AC\ IV}$

Fig. 196. The molecular configuration of $C_{44}H_{48}O_{12}Cr_2P_2$, indicated schematically.

Atomic positions

Atomic positions and isotropic temperature factors are given for the
thirty non-hydrogen atoms. Mean standard deviations are 0.01 Å for cobalt,
and 0.02 to 0.05 Å for the other atoms.

Structure

The structural unit consists of a centrosymmetrical tetramer, illustrated
schematically in Fig. 197. Essentially octahedral coordination is achieved
by the sharing of oxygen atoms between adjacent cobalt atoms; the terminal
cobalt atom is joined to its neighbour by sharing an octahedral *face*, and the

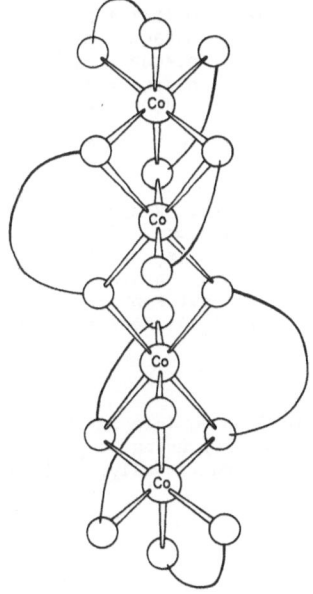

Fig. 197. A schematic drawing of tetrameric cobalt(II)acetylacetoacetonate.
Only the oxygen atoms of the acetylacetonate ligands are shown
explicitly, the remainder being indicated by curved lines.

central two cobalt atoms are joined by sharing an *edge*. There is necessarily some distortion of coordination angles, and the cobalt atoms lie as far as 0.47 Å from the ideal ligand planes. Bond lengths and angles are given in full. Some mean Co-O distances are: for unshared oxygen, 2.06 ± 4 Å; for edge-shared oxygen, 2.13 ± 3 Å; for face-shared oxygen, 2.24 ± 3 Å.

Details of analysis

The intensity data were recorded on zero- and upper-level precession photographs, using CuK_α radiation. The task was greatly complicated by twinning on the (100) plane. 880 reflexions were observed and measured visually. Refinement was by least squares to a final R index of 0.154.

COPPER(II) ETHYLACETOACETATE

$C_{12}H_{18}CuO_6$ $(C_2H_5.O.CO.CH.CO.CH_3)_2Cu$ F.W. = 321.7

  I. The crystal structure of copper (II) ethyl acetoacetate. G. A. BARCLAY and A. COOPER, 1965. *J. Chem. Soc.*, 3746-3751.

    (For earlier work, see 1.)

Monoclinic, a = 11.578 ± 6, b = 4.527 ± 2, c = 13.791 ± 6 Å, β = 104.5 ± 4°, U = 699.7 Å^3, D_m = 1.51, Z = 2, D_x = 1.53.

Space group $P2_1/n$ (C^5_{2h}) Molecular symmetry, centre.

Atomic positions

	x	y	z
Cu	0.0000	0.0000	0.0000
O(1)	0.1437 ± 10	−0.2143 ± 29	0.0149 ± 9
O(2)	−0.0323 ± 9	−0.2054 ± 29	0.1123 ± 8
O(3)	−0.0005 ± 10	−0.5236 ± 36	0.2417 ± 8
C(1)	0.2990 ± 15	−0.5469 ± 48	0.0725 ± 14
C(2)	0.1814 ± 16	−0.4241 ± 37	0.0818 ± 14
C(3)	0.1360 ± 13	−0.5305 ± 51	0.1525 ± 11
C(4)	0.0318 ± 16	−0.4049 ± 39	0.1662 ± 13
C(5)	−0.1112 ± 15	−0.4183 ± 41	0.2597 ± 14
C(6)	−0.1068 ± 16	−0.4980 ± 80	0.3628 ± 14

    Anisotropic temperature factors are given also.

Structure

The molecule is essentially planar except for C(6), which lies 0.51 Å from the mean plane of the remaining atoms. [It is implied that C(5) also lies significantly off this plane, but this is not so.] Bond lengths and angles are given in Fig. 198. The copper atom is coordinated to four oxygen atoms in a square planar arrangement, as shown, but approximate octahedral coordination is completed by the carbon atoms C(3) of adjacent molecules, at a distance of 3.12 Å from the copper atom.

Details of analysis

Intensity data were recorded on equi-inclination Weissenberg photographs ($h0l$ to $h2l$; $hk0$ to $hk4$; $0kl$ to $1kl$) using CuK_α radiation. 575 independent reflexions were observed and measured visually. Refinement was by least squares to a final R index of 0.12.

1. *Structure Reports*, 18, 641.

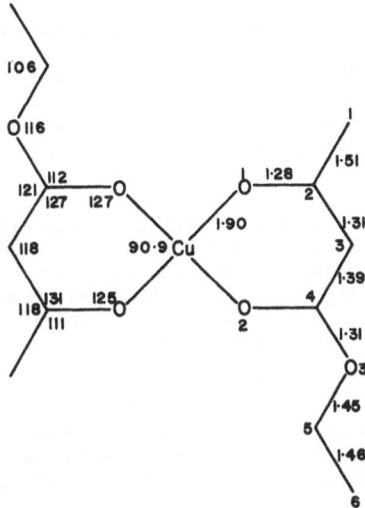

Fig. 198. Bond lengths and angles in copper(II) ethylacetoacetate. Standard
deviations are 0.01 to 0.03 Å and 0.5 to 1.2°.

BIS(3-PHENYL-2,4-PENTANEDIONATO)COPPER(II)

$C_{22}H_{22}CuO_4$ F.W. = 414.0

I. Bis(3-phenyl-2,4-pentanedionato)copper. I. Molecular and crystal
    structure. J. W. CARMICHAEL, JR., L. K. STEINRAUF and R. L. BELFORD,
    1965. *J. Chem. Phys.*, **43**, 3959-3966.

Monoclinic, a = 10.250 ± 10, b = 6.778 ± 6, c = 13.763 ± 13 Å, β = 93°33 ±5',
[U = 954.3 Å³], D_m = 1.442, z = 2, D_x = 1.440.

Space group $P2_1/c$ (C_{2h}^5) Molecular symmetry, centre.

Atomic positions

	x	y	z
Cu	0	0	0
O(1)	0.16340 ± 59	0.12258 ± 135	−0.02224 ± 44
O(2)	−0.04659 ± 56	0.21054 ± 128	0.08236 ± 45
C(1)	0.35484 ± 88	0.31299 ± 237	−0.00475 ± 77
C(2)	0.21682 ± 81	0.26458 ± 205	0.02443 ± 61
C(3)	0.15997 ± 82	0.38164 ± 195	0.09796 ± 59
C(4)	0.02454 ± 86	0.34472 ± 219	0.11811 ± 61
C(5)	−0.04090 ± 86	0.48993 ± 223	0.18631 ± 76
C(6)	0.23653 ± 77	0.53807 ± 207	0.15160 ± 60

	x	y	z
C(7)	0.27276 ± 90	0.71300 ± 223	0.10806 ± 68
C(8)	0.34826 ± 97	0.85530 ± 228	0.15871 ± 76
C(9)	0.38332 ± 88	0.82373 ± 233	0.25915 ± 76
C(10)	0.34386 ± 87	0.64648 ± 220	0.30331 ± 67
C(11)	0.27174 ± 84	0.50363 ± 205	0.25167 ± 61

Anisotropic temperature factors are given also.

Structure

The coordination of the copper atom is square planar, with a mean Cu–O distance of 1.906 ± 7 Å, and intra-ligand O–Cu–O angle $91.4 \pm 3°$. Other bond lengths and angles are given in full, and have their expected values. The acetyl carbon skeleton is tilted by $14°$ from the plane of the copper and oxygen atoms. The phenyl ring is twisted through a torsional angle of $70°$ from the mean plane of the chelate ring. The molecular packing is illustrated in Fig. 199.

Details of analysis

The intensity data were recorded on equi-inclination Weissenberg photographs ($h0l$ to $h3l$) using CuK_{α} radiation. 1082 of a possible 1232 reflexions were observed and measured visually. Refinement was by least squares to a final R index of 0.0834.

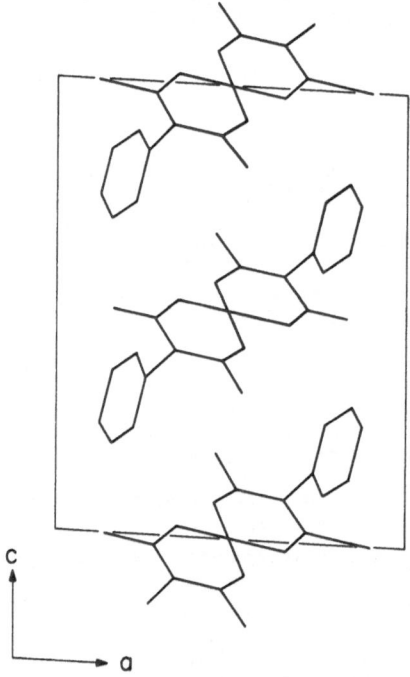

Fig. 199. The structure of $C_{22}H_{22}CuO_4$ viewed along b.

BIS(ACETYLACETONATO)NICKEL(II)

$C_{10}H_{16}NiO_4$ $Ni(CH_3COCH_2COCH_3)_2$ F.W. = 258.9

I. The crystal and molecular structure of bis(acetylacetonato)nickel(II).

G. J. BULLEN, R. MASON and P. PAULING, 1965. *Inorg. Chem.*, *4*, 456–462.

Orthorhombic, a = 23.23 ± 4, b = 9.64 ± 2, c = 15.65 ± 2 Å, [U = 3505 Å^3], D_m = 1.455, Z = 12, D_x = 1.460.
(Crystal data for a different orientation of axes are given in <u>1</u>.)

Space group $Pca2_1$ (C_{2v}^5) or $Pcam$ (D_{2h}^{11}). $Pca2_1$ is consistent with analysis.

Atomic positions

Atomic positions and anisotropic temperature factors are given for the 45 non-hydrogen atoms. Mean standard deviations are about 0.01 Å for nickel, and 0.04 to 0.10 Å for the other atoms.

Structure

The structural unit consists of a trimer, illustrated schematically in Fig. 200. Essentially octahedral coordination of the nickel atoms is achieved by the sharing of faces between adjacent oxygen octahedra. Average Ni–O distances are 2.01 Å for unshared, and 2.12 Å for shared, oxygen atoms.

Details of analysis

The intensity data were recorded on oscillation photographs about a and c, using CuK_α radiation. 1730 reflexions were measured visually. Refinement was by least squares to a final R index of 0.14.

1. *Structure Reports*, <u>20</u>, 473.

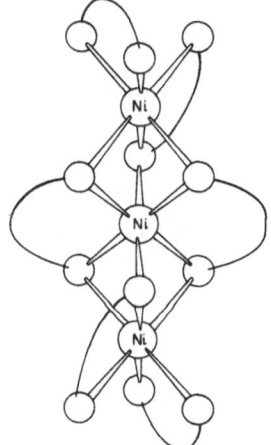

Fig. 200. A schematic drawing of the $(C_{10}H_{18}NiO_4)_3$ trimer. Only the oxygen atoms of the acetylacetonate ligands are shown explicitly, the remainder being indicated by curved lines.

μ-ETHYLENEDIAMINE-BIS[TRIMETHYL(ACETYLACETONATO)PLATINUM(IV)]

$C_{18}H_{40}N_2O_4Pt_2$ [(CH$_3$)$_3$.Pt.(CH$_3$.CO.CH.CO.CH$_3$)]$_2$(H$_2$N.CH$_2$.CH$_2$.CH$_2$.NH$_2$)   F.W. = 738.7

M.P. = 196–198°C
(decomp.)

I. The stereochemistry of β-diketo-complexes with trimethylplatinum(IV). Part IV. The crystal structure of μ-ethylenediamine-bis[trimethyl-(acetylacetonato)platinum(IV)]. A. ROBSON and M. R. TRUTER, 1965. *J. Chem. Soc.*, 630–637.

Monoclinic, $a = 23.19 \pm 3$, $b = 6.46 \pm 1$, $c = 16.42 \pm 2$ Å, $\beta = 103.0 \pm 3°$, $U = 2397$ Å^3, $D_m = 2.05$, $Z = 4$, $D_x = 2.05$.

Space group Ia (C_s^4) or $I2/a$ (C_{2h}^6). $I2/a$ is consistent with deduced structure. Molecular symmetry, centre.

Atomic positions

	x	y	z
Pt	0.0969	0.0916	-0.0745
O(1)	0.158	0.310	-0.001
O(2)	0.093	-0.047	0.045
N	0.017	0.295	0.064
C(1)	0.032	0.431	0.002
C(2)	0.107	0.260	-0.185
C(3)	0.167	-0.083	-0.090
C(4)	0.036	-0.094	-0.159
C(5)	0.136	-0.025	0.114
C(6)	0.173	0.103	0.128
C(7)	0.184	0.297	0.077
C(8)	0.232	0.434	0.103
C(9)	0.123	-0.163	0.185

Isotropic temperature factors are given also. Standard deviations (in Å) are given in full. Mean values are: Pt, 0.003; O, 0.04; N, 0.05; C, 0.08 Å.

Structure

Bond lengths and angles are given in full but are not very accurate. Some values (and the numbering system) are shown in Fig. 201. Each platinum atom is octahedrally coordinated by one nitrogen atom of an ethylenediamine group, by the two oxygen atoms of a chelate acetylacetone group, and three methyl groups in *cis* configuration. The ethylenediamine group acts as a bridge in the formation of a centrosymmetrical dimer. The acetylacetone group is planar, and the platinum atom lies 0.39 Å from the plane. Intermolecular distances are consistent with van der Waals interaction.

Details of analysis

Intensity data were recorded on equi-inclination Weissenberg photographs ($h0l$ to $h3l$: $0kl$) using CuK_α radiation. 533 reflexions were observed and estimated visually. Refinement was by least squares to a final R index of 0.09.

BIS(1-PHENYL-1,3-BUTANEDIONATO) VANADYL (*cis* FORM)

$C_{20}H_{18}O_5V$ F.W. = 389.3

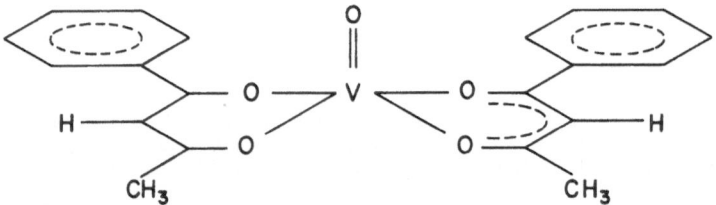

I. *Bis*(1-phenyl-1,3-butanedionato) vanadyl. I. Molecular and crystal structure of the *cis* form. PING-KAY HON, R. LINN BELFORD and C. E. PFLUGER, 1965. *J. Chem. Phys.*, **43**, 1323-1333.

Monoclinic, $a = 8.130 \pm 3$, $b = 22.599 \pm 10$, $c = 10.505 \pm 4$ Å, $\beta = 106°47 \pm 3'$, [$U = 1848$ Å^3], $D_m = 1.409$, $Z = 4$, $D_x = 1.404$.

Space group $P2_1/c$ (C_{2h}^5)

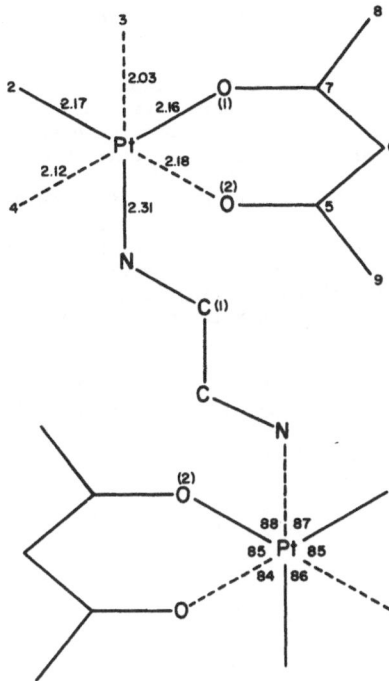

Fig. 201. Some bond lengths and angles in $C_{18}H_{40}N_2O_4Pt_2$. Standard deviations range from 0.04 to 0.08 Å and 1 to 3°.

Atomic positions

	x	y	z
V	0.18652 ± 32	0.09907 ± 8	0.12413 ± 19
O(1)	0.33553 ± 115	0.12052 ± 33	0.25381 ± 75
O(2)	0.02535 ± 106	0.05133 ± 28	0.18734 ± 70
O(3)	0.27692 ± 110	0.02396 ± 30	0.07242 ± 68
O(4)	0.24146 ± 112	0.13417 ± 30	-0.03174 ± 74
O(5)	0.00265 ± 110	0.15728 ± 29	0.09689 ± 73
C(1)	0.27942 ± 199	0.20219 ± 56	-0.19514 ± 125
C(2)	0.19133 ± 193	0.18385 ± 52	-0.09038 ± 108
C(3)	0.05546 ± 179	0.21917 ± 51	-0.06948 ± 116
C(4)	-0.02574 ± 173	0.20402 ± 45	0.02280 ± 114
C(5)	-0.17792 ± 161	0.24045 ± 44	0.03889 ± 98
C(6)	-0.21396 ± 185	0.29698 ± 50	-0.01923 ± 114
C(7)	-0.35880 ± 204	0.32776 ± 55	-0.00096 ± 130
C(8)	-0.46265 ± 212	0.30495 ± 60	0.07452 ± 140
C(9)	-0.42366 ± 195	0.24962 ± 58	0.13362 ± 127
C(10)	-0.27897 ± 200	0.21757 ± 49	0.11691 ± 114
C(11)	0.37764 ± 174	-0.07529 ± 49	0.07766 ± 122
C(12)	0.26216 ± 178	-0.02882 ± 44	0.11712 ± 114
C(13)	0.14951 ± 181	-0.04329 ± 51	0.19341 ± 111
C(14)	0.03480 ± 172	-0.00347 ± 48	0.22404 ± 98
C(15)	-0.08152 ± 173	-0.02084 ± 50	0.30204 ± 96
C(16)	-0.23266 ± 204	0.01309 ± 54	0.28944 ± 111
C(17)	-0.34650 ± 199	-0.00097 ± 58	0.36478 ± 126
C(18)	-0.30464 ± 229	-0.05006 ± 61	0.45253 ± 116
C(19)	-0.15377 ± 219	-0.08217 ± 56	0.46669 ± 124
C(20)	-0.04069 ± 182	-0.06957 ± 47	0.39038 ± 112

Anisotropic temperature factors are given also.

Structure

The molecule has the *cis* configuration shown above. The coordination of the vanadium atom is distorted square pyramidal, with V-O as the axis, and with the vanadium atom lying 0.56 Å above the base of the pyramid. Coordination distances and angles are: V-O, 1.612 ± 10 Å; V-O, 1.946 ± 8 to 1.986 ± 8 Å (mean 1.966 Å); O=V-O, 104.0 ± 4° to 106.9 ± 4° (mean 105.7°); O-V-O (adjacent oxygen atoms), 82.2 ± 3° to 87.5 ± 3° (mean 85.9°); O-V-O (non-adjacent oxygen atoms), 146.5 ± 3° and 149.9 ± 3° (mean 148.2°). The dimensions of the oxygen pyramid are given in Fig. 202. The chelate rings are planar, and inclined to each other at 162°. Each phenyl ring is twisted slightly out of the plane of the adjacent chelate ring, one by 6.5°, and the other by 18.8°. The molecular packing is illustrated in Fig. 203.

Details of analysis

The intensity data were recorded on equi-inclination Weissenberg photographs (0*kl* to 5*kl*; *hk*0) using CoK_α and CuK$_\alpha$ radiation. 1894 of a possible 2415 reflexions were observed and measured photometrically. Refinement was by least squares to a final R index of 0.088.

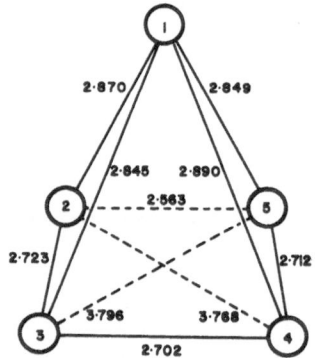

Fig. 202. The dimensions of the oxygen pyramid in $C_{20}H_{18}O_5V$.

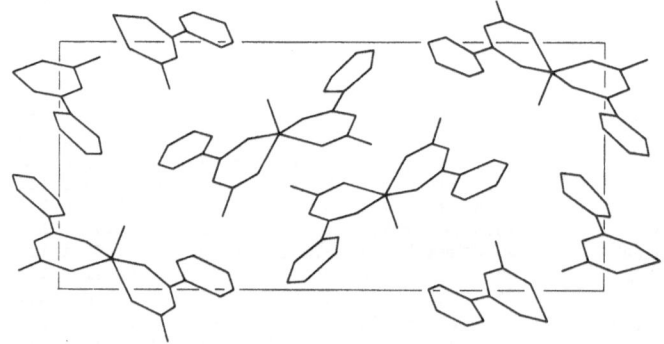

Fig. 203. The structure of $C_{20}H_{18}O_5V$ viewed along *a*.

MERCURY(II) TETRATHIOUREATETRATHIOCYANATOCOBALTATE(II)

$C_8H_{16}CoHgN_{12}S_8$ $Co(SCN)_4^{2-} \cdot Hg[SC(NH_2)_2]_4^{2+}$ F.W. = 796.3

I. The crystal structure of $Hg[SC(NH_2)_2]_4Co(SCN)$. A. KORCZYNSKI and

M. A. PORAJ-KOŠIC, 1965. *Rocz. Chem.*, **39**, 1567-1583.

Tetragonal, a = 17.27 ± 20, c = 4.27 ± 1 Å, $[U = 1274 \text{ Å}^3]$, D_m = 2.09, Z = 2, D_x = 1.91 [2.08].

Space group $I4$ (C_4^5), $I\bar{4}$ (S_4^2) or $I4/m$ (C_{4h}^5). $I\bar{4}$ is consistent with the analysis. *Molecular symmetry*, 4.

Atomic positions

	x	y	z
Hg	0	0	0
Co	0	1/2	3/4
S(1)	0.122	-0.012	0.35
S(2)	0.363	0.1935	0.862
C(1)	0.191	0.0503	0.199
N(A)	0.180	0.1205	0.145
N(B)	0.271	0.040	0.165
C(2)	0.404	0.125	0.595
N(2)	0.438	0.084	0.460

Mean standard deviations are about 0.01 Å for sulphur, and 0.02 Å for the other non-metallic atoms.

Structure

The cobalt atom is coordinated to four thiocyanate groups, *via* their nitrogen atoms, in a somewhat flattened tetrahedral arrangement. The Co-N-C-S linkage is approximately linear. The mercury atom is tetrahedrally coordinated to the sulphur atoms of four thiourea molecules. The coordination distances are: Co-N, 2.01 ± 2 Å; Hg-S, 2.59 ± 1 Å. A view of the structure is given in Fig. 204. The intermolecular N...N contacts shown have lengths of 3.10 and 3.20 Å, and are presumably hydrogen bonds.

Details of analysis

The intensity data were recorded photographically ($hk0$ to $hk4$; $0kl$ to $2kl$) using MoKα radiation. 750 reflexions were observed and measured visually. Refinement was by Fourier methods to a final R index of 0.142.

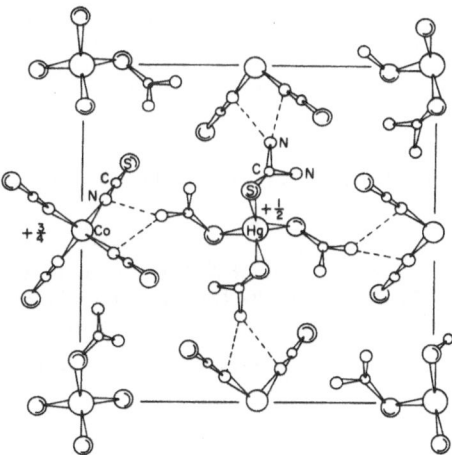

Fig. 204. The structure of $C_8H_{16}CoHgN_{12}S_8$ viewed along c.

BISTHIOUREA-CADMIUM FORMATE

$C_4H_{10}N_4O_4S_2Cd$ $Cd[SC(NH_2)_2]_2(HCOO)_2$ F.W. = 354.7

I.   The crystal and molecular structure of bisthiourea-cadmium formate.
     M. NARDELLI, G. F. GASPARRI and P. BOLDRINI, 1965. *Acta Cryst.*,
     18, 618-623.

     (For a preliminary report, see 1.)

Orthorhombic, a = 8.000 ± 9, b = 17.878 ± 5, c = 3.933 ± 7 Å, U = 562.5 Å³,
D_m = 2.091, Z = 2, D_x = 2.093.

Space group $P2_12_12$ (D_2^3) Molecular symmetry, two-fold axis.

Atomic positions

	x	y	z
Cd	0	0	0.0970
S	0.1904	0.0622	0.6025
O(1)	-0.1815	0.0969	0.1677
O(2)	-0.4148	0.1617	0.0896
N(1)	0.0237	0.1902	0.5579
N(2)	0.2688	0.1968	0.8349
C(1)	0.1577	0.1583	0.6709
C(2)	-0.3349	0.1045	0.0780

     Standard deviations are given in full, in Å. Mean values are: Cd,
0.003, S, 0.007; other atoms, 0.028 to 0.036 Å. Anisotropic temperature
factors are given also.

Structure

     Details of the structure are illustrated in Fig. 205. The cadmium ion is
coordinated octahedrally to four sulphur atoms (with which it is coplanar)
and two oxygen atoms from symmetry-related HCOO⁻ groups. The octahedra are
linked by S-S edges in endless chains along the two-fold axes parallel to c.
Bond lengths and angles are given in full. The more important and accurate
values are given in Fig. 205.

Details of analysis

     Intensity data were recorded on integrated and non-integrated equi-inclin-
ation Weissenberg photographs (hk0 to hk3; 0kl to 6kl) using CuK_α radiation.
The intensities were measured photometrically, and were corrected for absorption.
Refinement was by differential syntheses to a final R index of 0.128.

1. *Structure Reports*, 27, 835.

ACETATOPENTAAMMINECOBALT(III) CHLORIDE PERCHLORATE

$C_2H_{18}Cl_2CoN_5O_6$ F.W. = 338.0

I.   The structure of acetatopentaamminecobalt(III) in [Co(NH₃)₅(CH₃CO₂)]-
     (Cl)(ClO₄). E. B. FLEISCHER and R. FROST, 1965. *J. Amer. Chem. Soc.*,
     87, 3998.

Orthorhombic, a = 22.01, b = 9.75, c = 11.41 Å, [U = 2449 Å³], D_m = 1.84,
Z = 8, D_x = 1.83.

Space group $Pbca$ (D_{2h}^{15})

Atomic positions are not given.

Structure

     The configuration of the acetatopentaamminecobalt(III) ion is illustrated
in Fig. 206. Some bond lengths are: C-CH₃ = 1.58 Å; C=O = 1.23 Å;

C-O = 1.30 Å;   C-N (mean) = 2.00 Å.   (Standard deviations are 0.02 to 0.03 Å.)
[No other distances, and no angles are given.]

Details of analysis

The intensity data were measured with a four-circle diffractometer and
scintillation counter, using MoK$_\alpha$ radiation. 1143 reflexions were observed;
the method of refinement is not stated, but the final R factor was 0.14.

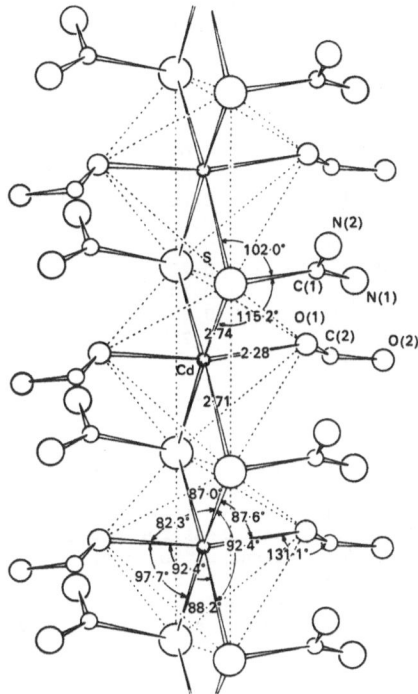

Fig. 205.   Clinographic projection of a bisthiourea-cadmium formate chain
            parallel to *c*. Standard deviations are: Cd-S, 0.01 Å; Cd-O,
            0.02 Å; S-Cd-S, 0.2°; S-Cd-O, 0.8°.

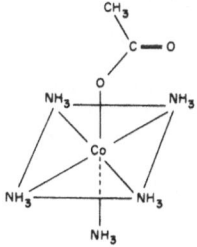

Fig. 206.   Configuration of the acetatopentaammine-cobalt(III) ion.

COPPER(II) FORMATE DIHYDRATE

$C_2H_2O_4Cu.2H_2O$ $Cu(HCOO)_2.2H_2O$ F.W. = 189.6

I. The crystal structure of copper(II) formate dihydrate. M. BUKOWSKA-
    STRZYZEWSKA, 1965. *Acta Cryst.*, <u>19</u>, 357–362.

Monoclinic, a = 8.54 ± 2, b = 7.15 ± 1, c = 9.50 ± 2 Å, β = 96°48',
U = 574.2 Å^3, D_m = 2.17, Z = 4, D_x = 2.19.

Space group $P2_1/c$ (C_{2h}^5) Molecular symmetry, centre for each of two crystallographically distinct copper atoms.

Atomic positions

	x	y	z
Cu(1)	0.500	0	0.500
Cu(2)	0	0	0
O(1)	0.413	0.267	0.083
O(2)	0.406	0.096	0.277
O(3)	0.296	0.644	0.007
O(4)	0.075	0.705	0.094
O(5) H$_2$O	0.089	0.103	0.184
O(6) H$_2$O	0.211	0.484	0.425
C(1)	0.467	0.217	0.198
C(2)	0.192	0.600	0.078

Standard deviations are 0.010 Å for oxygen and 0.014 Å for carbon.

Structure

   The structure consists of copper ions linked by a three-dimensional network of formate bridges as shown in Fig. 207. Each copper ion is coordinated
to six oxygen atoms forming a distorted octahedron. For one copper ion all six
oxygen atoms are contributed by different formate groups; four are at a
distance of 2.02 Å in a square planar arrangement, while the remaining two are
at the somewhat greater distance of 2.28 Å. The coordination of the other
copper ion is similar except that the four nearer oxygen atoms are contributed
by water molecules. The structure is stabilized by O–H...O bonds of length
2.79 to 2.99 Å.

Details of analysis

   The intensity data for the three principal zones were recorded on Weissenberg photographs, using CuK_α radiation. 195 of 254 accessible reflexions were
observed and measured visually. Refinement was by Fourier methods to a final R
index of 0.09.

BARIUM BIS-DIOXOMOLYBDENUM(V) OXALATE PENTAHYDRATE

$C_4BaMoO_{12}.5H_2O$ $BaMoO_4(C_2O_4)_2.5H_2O$ F.W. = 563.4

I. The molecular structure of a diamagnetic, doubly oxygen-bridged, binu-
    clear complex of molybdenum(V) containing a metal-metal bond. F. A.
    COTTON and S. M. MOREHOUSE, 1965. *Inorg. Chem.*, <u>4</u>, 1377–1381.

Trigonal, a = 10.63 ± 3, c = 11.65 ± 3 Å, [U = 1140 Å^3], D_m = 2.9, Z = 3,
D_x = 2.89 [2.46].

Space group $P3_121$ (D_3^4) or its enantiomorph $P3_221$ (D_3^6) Molecular symmetry,
two-fold axis.

Atomic positions

	x	y	z
Ba	0.1185 ± 4	0	1/3
Mo	0.9452 ± 4	0.5029 ± 4	0.0666 ± 4

	x	y	z
O(1)	0.087 ± 3	0.380 ± 3	0.296 ± 3
O(2)	0.058 ± 4	0.717 ± 4	0.291 ± 3
O(3)	0.277 ± 4	0.669 ± 4	0.339 ± 3
O(4)	0.136 ± 3	0.567 ± 3	0.117 ± 3
O(5)	0.972 ± 4	0.483 ± 4	0.436 ± 3
O(6)	0.306 ± 3	0.915 ± 3	0.438 ± 3
O(7)	0.044 ± 3	0.201 ± 3	0.430 ± 3
C(1)	0.049 ± 5	0.312 ± 5	0.396 ± 4
C(2)	0.038 ± 5	0.404 ± 5	0.865 ± 4
O(8)	0.885 ± 4	0.648 ± 4	0.586 ± 4
O(9)	0.0	0.097 ± 5	1/6

O(3), O(8) and O(9) belong to water molecules. Average standard deviations
are: Ba and Mo, 0.004 Å; C and O, 0.03 to 0.04 Å. Isotropic temperature
factors are given also.

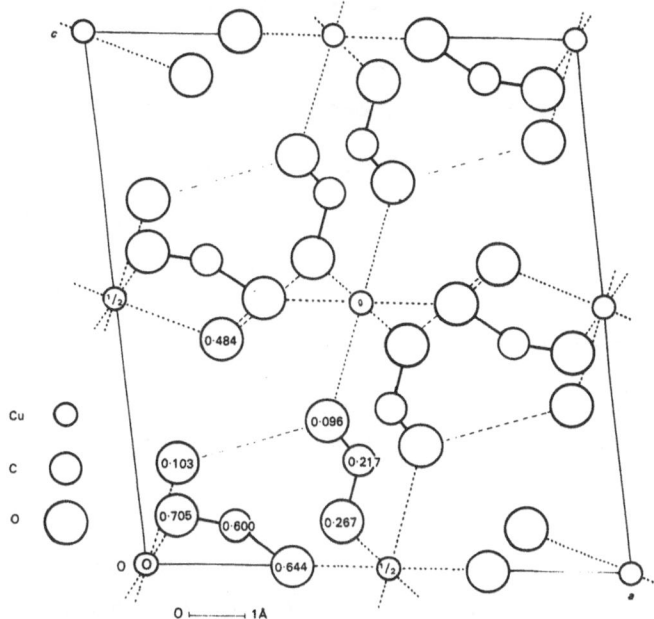

Fig. 207.   The structure of cupric formate dihydrate projected along b. y
            coordinates are given for the asymmetric unit. Hydrogen bonds
            are shown as dashed, and Cu-O bonds as dotted lines.

Structure

The structure consists of a barium cation, (highly coordinated to water
molecules) and a binuclear complex anion with a double oxygen bridge (Fig. 208).
Each molybdenum atom is octahedrally coordinated to six oxygen atoms as follows:
two bridging atoms at 1.90 ± 3 Å; two oxalate oxygen atoms at 2.13 ± 3 Å;
one terminal oxygen atom at 1.70 ± 3 Å; one aqueous oxygen atom at 2.22 ± 4 Å.
The octahedron is somewhat distorted, with angles between adjacent ligands
ranging from 75° to 108°. The Mo-Mo distance of 2.54 ± 1 Å is suggestive of
metal-metal bonding.

Details of analysis

The intensity data were recorded on equi-inclination Weissenberg photographs

(*hk*0 to *hk*,11) using MoK_α radiation. 500 reflexions were observed in the
range sinθ<0.4, and were measured visually. Refinement was by full-matrix
least squares to a final *R* index of 0.102.

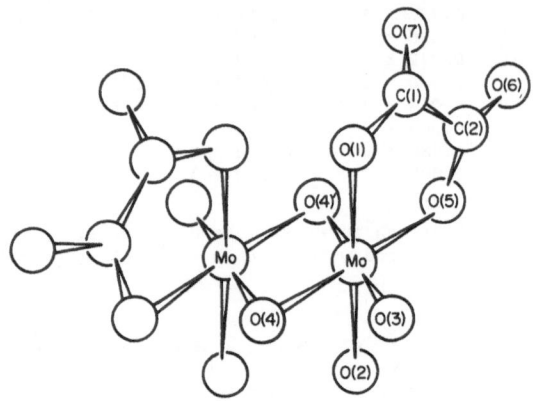

Fig. 208.   The binuclear anion in $C_4BaMoO_{12}.5H_2O$.

SODIUM GLYCYLGLYCYLGLYCINO CUPRATE (II) MONOHYDRATE

$C_6H_8N_3O_4CuNa.H_2O$ $Na[CuNH_2(CH_2CON)_2CH_2COO].H_2O$ F.W. = 290.7

   I.   Crystallographic studies of metal-peptide complexes.   II.   Sodium glycyl-
glycylglycino cuprate (II) monohydrate. H. C. FREEMAN, J. C. SCHOONE
and J. G. SIME, 1965. *Acta Cryst.*, <u>18</u>, 381-392.

   (For a preliminary account, see <u>1</u>.)

Monoclinic, a = 14.328 ± 6, b = 10.556 ± 6, c = 13.175 ± 6 Å,β = 92°58 ± 1',
U = 1990 Å^3, D_m = 1.92, Z = 8, D_x = 1.94.

Space group $I2/c$ (C_{2h}^6) or Ic (C_5^4). $I2/c$ is confirmed by analysis. Molecular
symmetry, two-fold axis.

Atomic positions

	x	y	z
Cu	0.55799 ± 10	0.56687 ± 12	0.65496 ± 10
Na	0.62645 ± 29	-0.01055 ± 34	0.59436 ± 27
O(1)	0.57219 ± 51	0.19439 ± 60	0.60997 ± 66
O(2)	0.28519 ± 48	0.49369 ± 66	0.59384 ± 56
O(3)	0.36608 ± 48	0.90904 ± 56	0.73979 ± 47
O(4)	0.40780 ± 56	0.74323 ± 61	0.82916 ± 49
O(5$_w$)	0.58434 ± 67	-0.05196 ± 73	0.42786 ± 55
N(1)	0.68921 ± 53	0.49954 ± 73	0.63523 ± 63
N(2)	0.52335 ± 60	0.39825 ± 71	0.62271 ± 61
N(3)	0.42031 ± 62	0.59701 ± 70	0.65182 ± 61
C(1)	0.68815 ± 65	0.35857 ± 84	0.63563 ± 80
C(2)	0.58489 ± 81	0.30908 ± 84	0.62172 ± 70
C(3)	0.42480 ± 67	0.37098 ± 79	0.61147 ± 73
C(4)	0.36989 ± 68	0.49457 ± 83	0.61790 ± 67
C(5)	0.37360 ± 75	0.71624 ± 87	0.64768 ± 67
C(6)	0.38205 ± 65	0.79511 ± 82	0.74491 ± 69

Anisotropic temperature factors are given also.

Structure

The copper-tripeptide dimer is illustrated schematically in Fig. 209.
The coordination of the copper atom is square pyramidal; the relatively weak
bond (2.568 ± 8 Å) to the apical nitrogen is shown dotted. The bonds to the
basal atoms (shown solid) are much stronger, with Cu-O = 1.933 ± 7 Å, Cu-N =
1.891 ± 8, 1.997 ± 8, and 2.039 ± 8 Å. The copper atom is displaced about
0.115 Å from the basal plane towards the apex. The copper coordination is
further illustrated in Fig. 210, which also shows the sodium coordination and
the hydrogen bonding system. Bond lengths and angles are given in full.
The original article should be studied for further details.

Details of analysis

Three-dimensional intensity data were recorded on Weissenberg photographs
about both a and b, using CuK_α radiation. 1502 of a possible 2100 reflexions
were observed and measured visually. Refinement was by full-matrix least squares
to a final R index of 0.132 for all reflexions.

1. *Structure Reports*, 27, 1049.

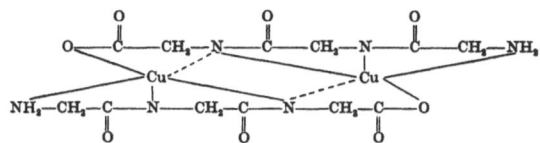

Fig. 209. The copper-tripeptide dimer.

DISODIUM GLYCYLGLYCYLGLYCYLGLYCINO CUPRATE (II) DECAHYDRATE

$C_8H_{10}N_4O_5CuNa_2 \cdot 10H_2O$ $Na_2[CuNH_2\{CH_2CON\}_3CH_2COO]10H_2O$ F.W. = 531.9

I. Crystallographic studies of metal-peptide complexes. III. Disodium
    glycylglycylglycylglycino cuprate (II) decahydrate. H. C. FREEMAN and
    M. R. TAYLOR, 1965. *Acta Cryst.*, 18, 939-952.

(For a preliminary report, see 1.)

Triclinic, a = 7.665 ± 6, b = 10.204 ± 9, c = 14.872 ± 10 Å, α = 93°48 ± 2',
β = 107°39 ± 3', γ = 94°17 ± 3', U = 1100.5 Å³, D_m = 1.67, Z = 2,
D_x = 1.600.

Space group P1 (C_1^1) or P$\bar{1}$ (C_i^1). P$\bar{1}$ was confirmed by the structure analysis.

Atomic positions

Atomic positions, anisotropic temperature factors and standard deviations
are given in full for the thirty non-hydrogen atoms.

Structure

The copper atom is coordinated to the four nitrogen atoms of the peptide
group in an approximately square planar arrangement. Discrete glycylglycylglycylglycino cuprate (II) ions are extensively bonded to water molecules.
Both sodium ions are octahedrally coordinated, with adjacent octahedra sharing
a face of three water molecules. A view of the structure, and many bond lengths
and angles are given in Figs. 211 and 212.

Details of analysis

The intensity data were recorded on equi-inclination Weissenberg photographs (0kl to 6kl; $h0l$ to $h1l$) using CuK_α radiation. Of 4319 accessible
reflexions 3690 were observed and estimated visually. Refinement was by full-
matrix least squares to a final R index of 0.092.

1. H.C. FREEMAN and M.R. TAYLOR, 1964. *Proc. Chem. Soc.*, 88.

350 ORGANIC COMPOUNDS

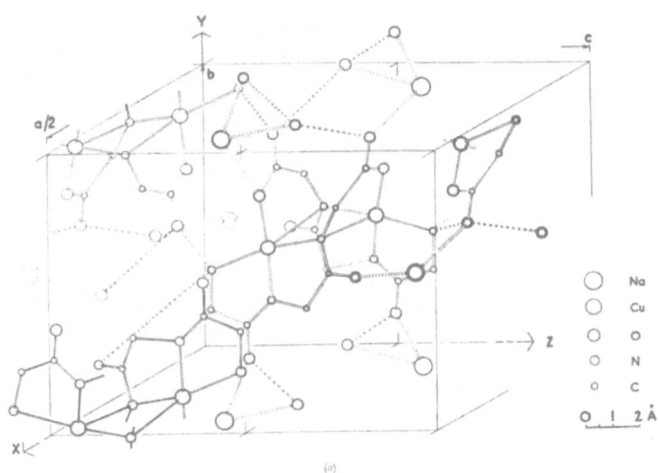

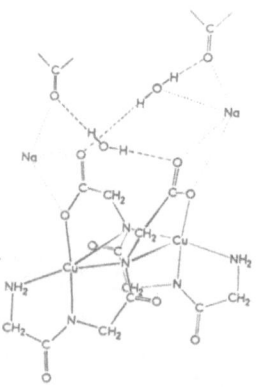

Fig. 210. (a) Part of unit cell contents of $C_6H_8N_3O_4CuNa.H_2O$. Dashed lines
show hydrogen bonds, dotted lines represent electrostatic interactions
with sodium ions. (b) One dimeric complex and its environment,
from (a).

OXODIPEROXO-1,10-PHENANTHROLINECHROMIUM(VI)

$C_{12}H_8CrN_2O_5$ $[CrO(O_2)_2(C_{12}H_8N_2)]$ F.W. = 312.2

I. The crystal structure of oxodiperoxo-1,10-phenanthrolinechromium(VI),
   $[CrO(O_2)_2(C_{12}H_8N_2)]$. R. STOMBERG, 1965. *Ark. Kemi*, <u>24</u>, 111-131.

Orthorhombic, a = 10.554 ± 7, b = 6.857 ± 3, c = 16.239 ± 10 Å, U = 1175.2 Å^3,
D_m = 1.74, Z = 4, D_x = 1.764 (calibration with lead nitrate, a = 7.8404 Å at
21°C).

Space group Pn2₁a (C_{2v}^9) or *Pnma* (D_{2h}^{16}). *Pnma* was suggested by intensity
statistics, and confirmed by analysis. Molecular symmetry, mirror plane.

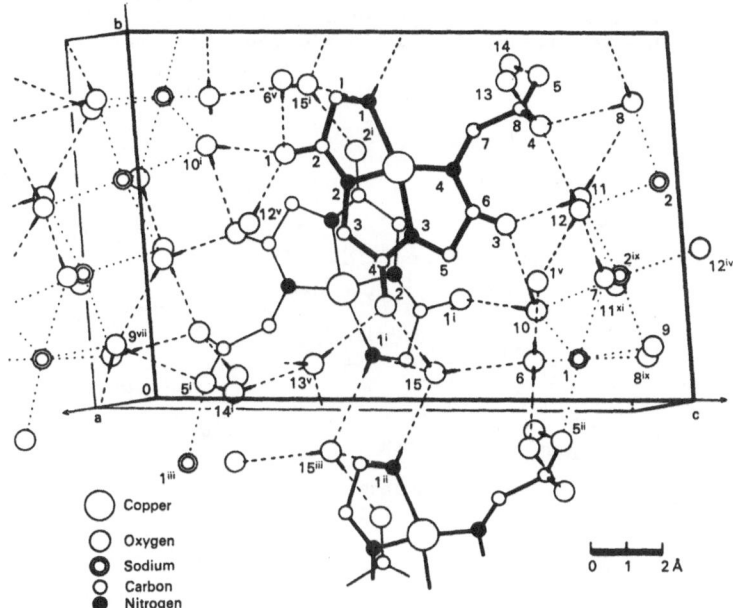

Fig. 211.   The structure of $C_8H_{10}N_4O_5CuNa_2 \cdot 10H_2O$ viewed along the normal to
(100).   Hydrogen bonds are dashed and sodium ion-water molecule
contacts are dotted.   Proton donation in a hydrogen bond is represented by a short full line at the appropriate end of the bond.

Atomic positions

	x	y	z
Cr	0.2467 ± 4	¼	0.1735 ± 2
O(1)	0.2389 ± 11	0.5200 ± 22	0.1571 ± 7
O(2)	0.1308 ± 12	0.4362 ± 23	0.1927 ± 8
O(3)	0.3273 ± 22	¼	0.2544 ± 14
N(1)	0.4118 ± 18	¼	0.1004 ± 8
N(2)	0.1758 ± 18	¼	0.0424 ± 11
C(1)	0.5308 ± 22	¼	0.1325 ± 13
C(2)	0.6313 ± 25	¼	0.0780 ± 16
C(3)	0.6230 ± 21	¼	0.9937 ± 14
C(4)	0.4989 ± 21	¼	0.9615 ± 13
C(5)	0.3925 ± 19	¼	0.0138 ± 11
C(6)	0.4730 ± 21	¼	0.8752 ± 12
C(7)	0.3513 ± 25	¼	0.8468 ± 15
C(8)	0.2419 ± 21	¼	0.9010 ± 13
C(9)	0.2677 ± 18	¼	0.9848 ± 12
C(10)	0.1107 ± 26	¼	0.8769 ± 17
C(11)	0.0258 ± 25	¼	0.9350 ± 15
C(12)	0.0463 ± 23	¼	0.0161 ± 14

Isotropic temperature factors are given also.

Structure

    The configuration of the molecule is shown in Fig. 213.   The coordination
of the chromium atom is pentagonal dipyramidal;   the equatorial plane contains

four peroxidic oxygen atoms and one nitrogen atom, whereas the oxide oxygen
atom and the other nitrogen atom occupy apical positions. The chromium atom is
displaced by 0.26 Å from the equatorial plane in the direction of the apical
oxygen atom.

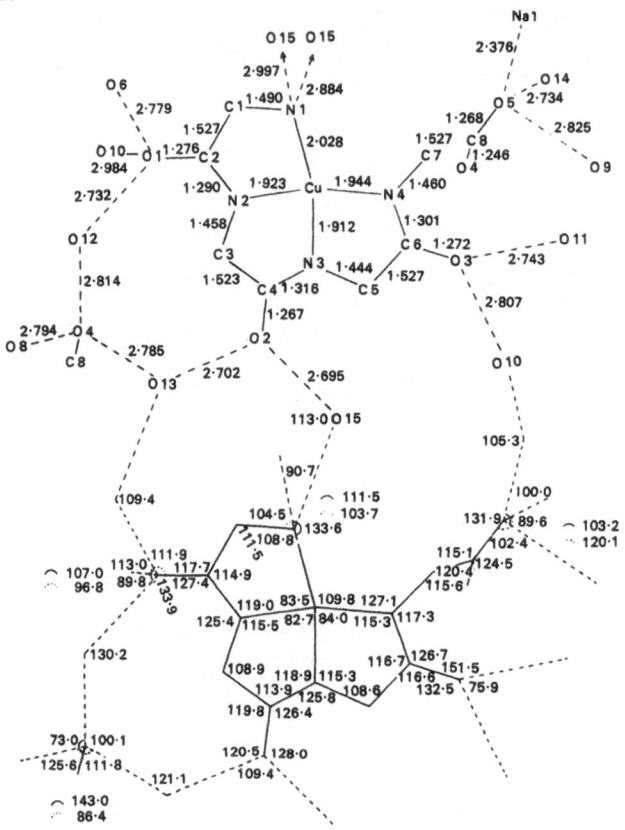

Fig. 212. Interatomic distances and angles in the glycylglycylglycylglycino
          cuprate (II) ion and its contacts. Standard deviations are from
          0.003 to 0.007 Å, and 0.3 to 0.7°.

Details of analysis

    The intensity data were recorded on equi-inclination Weissenberg photo-
graphs using CuK$_\alpha$ radiation for $h0l$ to $h4l$, $0kl$ to $2kl$, and $hk0$ to $hk1$, and
MoK$_\alpha$ radiation for $h5l$. 476 independent reflexions were observed and measured
visually. Refinement was by least squares to a final R index of 0.134.

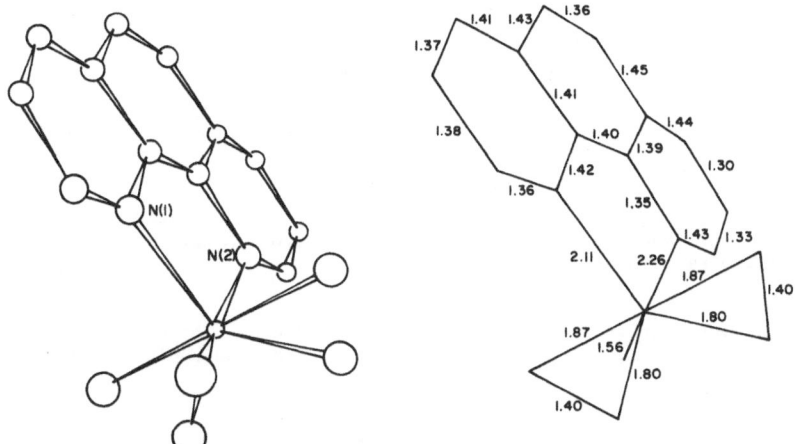

Fig. 213. The [CrO(O$_2$)$_2$(C$_{12}$H$_8$N$_2$)] molecule. The standard deviations of the distances shown range from 0.01 to 0.04 Å.

DIPOTASSIUM BIS(TRIMETHYLENEDINITRAMINE)NICKELATE(II) TETRAHYDRATE

C$_6$H$_6$N$_8$O$_8$K$_2$Ni.4H$_2$O   F.W. = 533.1

I. The crystal structure of dipotassium bis(trimethylenedinitramine)-nickelate(II) tetrahydrate. D. M. LIEBIG and J. H. ROBERTSON, 1965. *J. Chem. Soc.*, 5801-5809.

Orthorhombic, a = 8.91 ± 1, b = 10.86 ± 1, c = 10.22 ± 1 Å, U = 998 Å^3, D_m = 1.808, Z = 2, D_x = 1.818.

Space group Pba2 (C_{2v}^8) or Pbam (D_{2h}^9). Pbam confirmed by analysis. Molecular symmetry, 2/m.

Atomic positions

	x	y	z
Ni	0	0	0
K	0.0040	0.3017	½
C(1)	0.0976	0.2578	0

	x	y	z
C(2)	0.0090	0.2438	0.1246
N(1)	-0.0594	0.1180	0.1267
N(2)	-0.1485	0.0938	0.2223
O(1)	-0.2175	-0.0124	0.2205
O(2)	-0.1759	0.1708	0.3113
O(3)	0	$\frac{1}{2}$	0.3373
O(4)	0.1955	0.1015	$\frac{1}{2}$

The coordinates in the original are given in Å. Standard deviations are given in full, and range from 0.003 Å for K to 0.015 Å for C. Anisotropic temperature factors are given also.

Structure

The conformation of the anion is as shown above. The coordination of the nickel ion is square planar; some bond lengths and angles in the anion are:

$$
\begin{aligned}
a   &= 1.898 \pm 7 \text{ Å} \\
b   &= 1.50 \pm 1 \\
c   &= 1.51 \pm 1 \\
d   &= 1.29 \pm 1 \\
e   &= 1.26 \pm 1 \\
f   &= 1.31 \pm 1 \\
aa' &= 86.0 \pm 6° \\
ab &= 120 \pm 1 \\
bc &= 109 \pm 1 \\
cc' &= 115 \pm 1 \\
ad &= 124 \pm 1 \\
bd &= 117 \pm 1 \\
de &= 122 \pm 1 \\
df &= 117 \pm 1 \\
ef &= 120 \pm 1
\end{aligned}
$$

The bonds a,b,d,e,f are coplanar. Details of the oxygen-water and oxygen-potassium coordination are given. It is suggested that the approach of water molecules to the available octahedral sites of the nickel ion is blocked by the presence of the nitro groups.

Details of analysis

The intensity data were recorded on equi-inclination Weissenberg photographs ($0kl$ to $6kl$; $h0l$ to $h7l$) using CuK$_\alpha$ radiation. 771 of a possible 1005 reflexions were observed and measured visually. Refinement was by least squares to a final R index of 0.125.

BIS(2,2'-DIPYRIDYLIMINATO)PALLADIUM(II)

$C_{20}H_{16}N_6Pd$ F.W. = 446.7

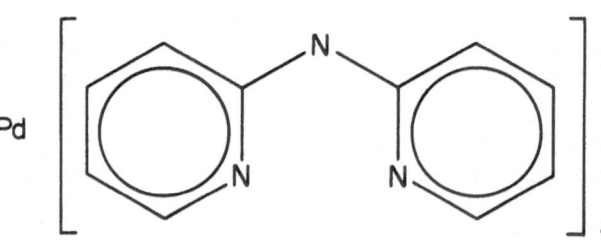

I. The crystal structure of bis-(2,2'-dipyridyliminato)palladium(II), $Pd(C_{10}H_8N_3)_2$. H. C. FREEMAN and M. R. SNOW, 1965. *Acta Cryst.*, <u>18</u>, 843-850.

(For a preliminary report, see <u>1</u>.)

Monoclinic, a = 15.405 ± 6, b = 12.770 ± 3, c = 9.046 ± 1 Å, β = 96.75 ± 8°,
U = 1767 Å^3, D_m = 1.70, Z = 4, D_x = 1.68, (calibration with platinum powder,
a = 3.9231 Å).

Space group Cc (C_s^4) or $C2/c$ (C_{2h}^6). $C2/c$ confirmed by structure analysis.
Molecular symmetry, centre.

Atomic positions

	x	y	z
Pd	0.5000	0.5000	0.0000
N(1)	0.4771 ± 4	0.3475 ± 4	0.0412 ± 7
N(2)	0.6163 ± 5	0.4777 ± 5	0.1247 ± 9
N(3)	0.6290 ± 4	0.3011 ± 5	0.0413 ± 8
C(1)	0.5420 ± 6	0.2759 ± 5	0.0348 ± 9
C(2)	0.5213 ± 8	0.1693 ± 8	0.0281 ± 12
C(3)	0.4357 ± 7	0.1379 ± 7	0.0477 ± 12
C(4)	0.3709 ± 6	0.2119 ± 7	0.0655 ± 11
C(5)	0.3948 ± 5	0.3158 ± 6	0.0613 ± 10
C(6)	0.6631 ± 5	0.3882 ± 5	0.1035 ± 8
C(7)	0.7529 ± 5	0.3854 ± 7	0.1551 ± 9
C(8)	0.7915 ± 7	0.4686 ± 9	0.2393 ± 11
C(9)	0.7387 ± 6	0.5540 ± 8	0.2736 ± 10
C(10)	0.6528 ± 5	0.5555 ± 7	0.2149 ± 8

Anisotropic temperature factors are given also.

Structure

The molecule is centrosymmetrical and has approximate $2/m$ symmetry. The
configuration (idealized to $2/m$ symmetry) is shown in Fig. 214. Bond lengths
and angles are given in full. Mean values are:

a = 2.022 ± 7 Å	aa' = 85.6 ± 6°		
b = 1.369 ± 11	ab = 120.1 ± 9		
c = 1.403 ± 11	ag = 120.1 ± 9		
d = 1.404 ± 14	bc = 119.2		
e = 1.407 ± 15	cd = 119.5		
f = 1.373 ± 13	de = 120.0		
g = 1.364 ± 13	ef = 117.8		
h = 1.350 ± 10	fg = 122.7	± 12	
	gb = 120.3		
	bh = 124.3		
	hc = 116.5		
	hh' = 123.4		

The angle between adjacent pyridyl rings is 38.2°. Intermolecular distances
are consistent with van der Waals interaction.

Details of analysis

Intensity data were recorded on equi-inclination Weissenberg photographs
($hk0$ to $hk7$, $0kl$ to $2kl$, with CuK_α radiation; $hk8$, with CuK_β radiation) 1354
of a possible 1515 reflexions were observed and estimated visually. Refine-
ment was by full-matrix least squares to a final R index of 0.085, for all
reflexions.

1. H.C. FREEMAN, J.F. GELDARD, F. LIONS and M.R. SNOW, 1964. *Proc. Chem. Soc.*,
   58.

cis-(DIETHYLENETRIAMINE)MOLYBDENUM TRICARBONYL

$C_7H_{13}MoN_3O_3$ $H_2NCH_2CH_2NHCH_2CH_2NH_2 \cdot Mo \cdot (CO)_3$ F.W. = 283.1

I. The crystal and molecular structure of *cis-*(diethylenetriamine)-
   molybdenum tricarbonyl; the dependence of Mo–C bond length on bond order.
   F. A. COTTON and R. M. WING, 1965. *Inorg. Chem.*, **4**, 314-317.

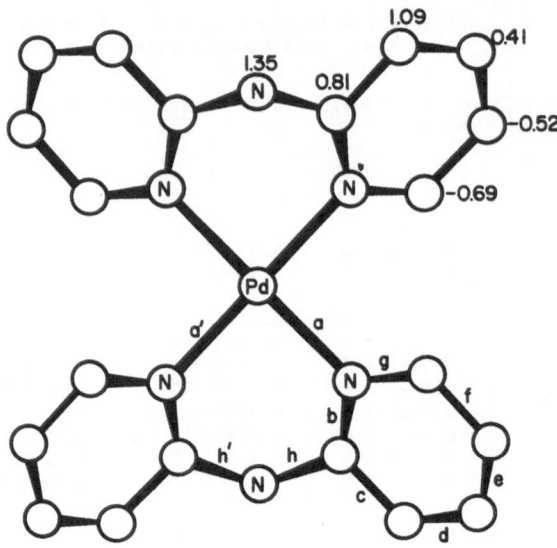

Fig. 214.   The $C_{20}H_{16}N_6Pd$ molecule (idealized to 2/m symmetry) viewed normal to
the plane containing Pd and the four adjacent atoms. Also given are
the distances (in Å) of some atoms from this plane.

Orthorhombic, a = 8.55 ± 1, b = 11.90 ± 1, c = 10.08 ± 1 Å, $[U = 1026$ Å$^3]$,
D_m = 1.83, Z = 4, D_x = 1.83.

Space group $P2_12_12_1$ (D_2^4)

Atomic positions

	x	y	z
Mo	0.3178 ± 2	0.3957 ± 1	0.8108 ± 2
C(1)	0.482 ± 2	0.428 ± 1	0.684 ± 3
C(2)	0.483 ± 2	0.408 ± 1	0.945 ± 2
C(3)	0.301 ± 2	0.558 ± 2	0.815 ± 2
O(1)	0.582 ± 2	0.454 ± 1	0.613 ± 2
O(2)	0.587 ± 2	0.417 ± 1	0.014 ± 2
O(3)	0.305 ± 2	0.656 ± 1	0.817 ± 2
N(1)	0.105 ± 2	0.360 ± 1	0.947 ± 2
N(2)	0.111 ± 2	0.358 ± 1	0.669 ± 2
N(3)	0.327 ± 2	0.202 ± 1	0.768 ± 2
C(4)	0.957 ± 3	0.343 ± 2	0.873 ± 3
C(5)	0.963 ± 2	0.392 ± 2	0.741 ± 3
C(6)	0.102 ± 3	0.238 ± 2	0.622 ± 3
C(7)	0.265 ± 3	0.180 ± 2	0.641 ± 2

Anisotropic temperature factors are given also.

Structure

The configuration of the molecule is illustrated in Fig. 215. The coordination of the molybdenum atom is distorted octahedral, with mean coordination
distances Mo-CO = 1.94 ± 1 Å and Mo-N = 2.32 ± 1 Å. Bond lengths and angles
are given in full. The intermolecular distances are consistent with van der
Waals interaction.

Details of analysis

The intensity data were measured with a four-circle diffractometer and
scintillation counter, using MoK_α radiation. 1500 reflexions were observed

in the range 2θ<59°. Refinement was by full-matrix least squares to a final *R*
index of 0.095.

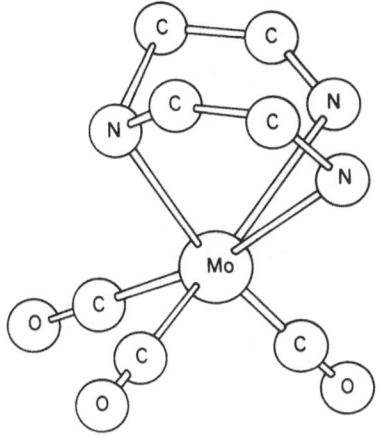

Fig. 215. The molecular configuration of *cis*-(diethylenetriamine)molybdenum
           tricarbonyl.

BENZENEDIAZONIUM COPPER(I) BROMIDE COMPLEX

$C_6H_5N_2Cu_2Br_3$ F.W. = 471.9

$(C_6H_5N_2)^+$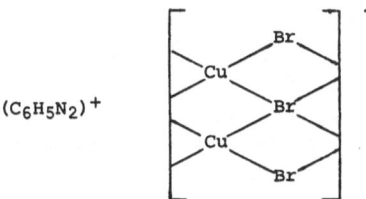

I. Crystal structure of the compound $C_6H_5N_2Cu_2Br_3$, an intermediate in the
    Sandmeyer reaction. C. RØMMING and K. WAERSTAD, 1965. *Chem. Communic.*,
    G. B., **14**, 299-300.

Orthorhombic, a = 17.69, b = 5.72, c = 11.01 Å, [U = 1114 Å³], z = 4,
[D_x = 2.81].

Space group $Pn2_1a$ (C_{2v}^9) or $Pnma$ (D_{2h}^{16}). *Pnma* confirmed by analysis. Molecular
symmetry, mirror plane.

Atomic positions are not given.

Structure

    The structure consists of normal benzenediazonium and $Cu_2Br_3^-$ ions. The
latter exist as endless chains extended along *b*, with pairs of copper atoms
alternating with triplets of bromine atoms.

Details of analysis

    Two-dimensional intensity data were recorded on integrated Weissenberg
photographs at -25°C. 245 reflexions of the *h0l* zone, and 70 of the *hk0*
zone were observed and measured photometrically. Refinement was by least
squares to final *R* indices of 0.09 (*h0l*) and 0.11 (*hk0*).

(PYRIDINE N-OXIDE)-COPPER(II) CHLORIDE

$C_5H_5NOCuCl$ F.W. = 229.6

I. Magnetic susceptibility and crystal structure of (pyridine N-oxide)-
    copper(II) chloride. H. L. SCHAFER, J. C. MORROW and H. M. SMITH, 1965.
    J. Chem. Phys., <u>42</u>, 504-508.

Monoclinic, a = 5.844 ± 5, b = 10.049 ± 5, c = 13.643 ± 5 Å, γ = 104°52 ± 10',
$[U$ = 774.5 Å], D_m = 1.9, z = 4, D_x = 1.97.

Space group $P2_1/b$ (C_{2h}^5)

Atomic positions

	x	y	z
Cu	0.266	0.094	0.016
Cl(1)	0.357	0.275	0.114
Cl(2)	0.595	0.138	0.924
O	0.975	0.007	0.091
N	0.940	0.034	0.177
C(1)	0.040	0.988	0.259
C(2)	0.943	0.012	0.352
C(3)	0.818	0.089	0.369
C(4)	0.745	0.144	0.291
C(5)	0.820	0.123	0.198

The standard deviations are estimated to be: Cu, 0.005 Å; Cl, 0.01 Å;
O, 0.03 Å; other, 0.05 Å. Isotropic temperature factors are given also.

Structure

As indicated in Fig. 216, the compound is an oxygen-bridged dimer. The
coordination of the copper atoms is intermediate between square-planar and
tetrahedral; the angle between the Cl-Cl and O-O vectors is 32°. Some mean
coordination distances and angles are: Cu-Cl, 2.20 Å; Cu-O, 2.05 Å; Cl-Cu-Cl,
99°; O-Cu-O, 76°. The intradimer Cu-Cu distance is 3.23 Å.

Details of analysis

The intensity data for the $0kl$ and $h0l$ zones were recorded on Weissenberg
photographs, using CuK_α radiation. 244 reflexions were observed and measured
visually. Refinement was by Fourier methods to a final R indices of 0.15 and
0.17 for the two zones.

IRON CUPFERRON

$C_{18}H_{15}FeN_6O_6$ F.W. = 467.2

I. The crystal structure of iron cupferron $Fe(O_2N_2C_6H_5)_3$. D. VAN DER HELM,
    L. L. MERRITT, JR., R. DEGEILH, and C. H. MacGILLAVRY, 1965. *Acta Cryst.*,
    <u>18</u>, 355-362.

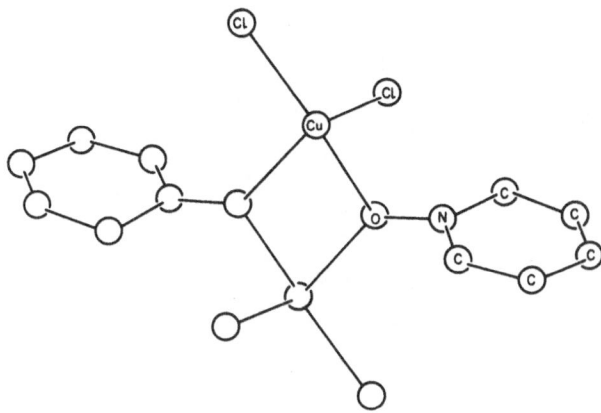

Fig. 216. The configuration of the (pyridine N-oxide)-copper(II) chloride dimer.

Monoclinic, a = 12.50 ± 1, b = 17.45 ± 5, c = 11.15 ± 4 Å, β = 122°19 ± 5', [U = 2055 Å], D_m = 1.50, Z = 4, [D_x = 1.51], (calibration with sodium chloride, a = 5.639 Å).

Space group $P2_1/a$ (C_{2h}^5)

Atomic positions

Atomic positions and isotropic temperature factors are given for the 31 non-hydrogen atoms.

Structure

Average bond lengths and angles for the cupferron group are given in Fig. 217. (Standard deviations for Fe-O bonds are about 0.01 Å; for the other bonds they vary considerably, ranging from 0.015 to 0.026 Å). The phenyl ring and the nitrosohydroxylamino groups are essentially planar, but not co-planar. The angle between the planes in any group is about 10°. A view of the molecule along [101] is given in Fig. 218. It is to be noted that because of the orientation of the cupferron group designated B-E, the molecule lacks the three-fold axis which it presumably could have had. Intermolecular distances appear to be normal van der Waals contacts.

Details of analysis

Three-dimensional intensity data were recorded on Weissenberg photographs ($h0l$ to $h7l$, and $hk0$ to $hk2$) with FeK_α radiation. 1710 reflexions were observed. No corrections were made for absorption, although the presence of absorption errors was noted. Refinement was by least squares and difference syntheses to a final R index of 0.15.

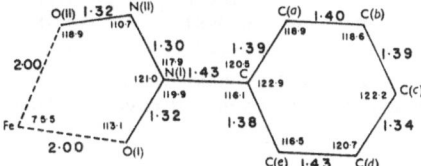

Fig. 217. Average bond lengths and angles for the cupferron group.

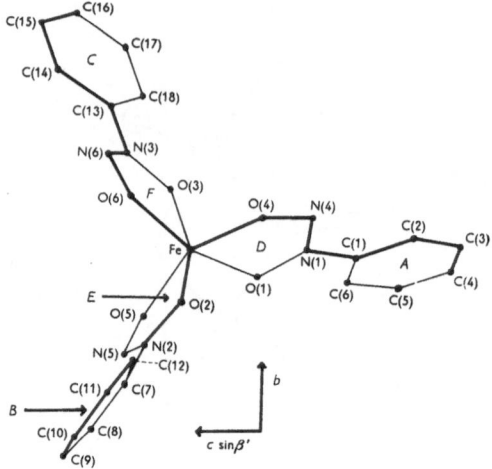

Fig. 218. The iron cupferron molecule viewed along [101].

BIS-SEMICARBAZIDE-COPPER(II) CHLORIDE

BIS-SEMICARBAZIDE-ZINC CHLORIDE

$C_2H_{10}Cl_2CuN_6O_2$	$M[OC(NH_2)NHNH_2]_2Cl_2$	F.W. = 284.6
$C_2H_{10}Cl_2N_6O_2Zn$	(M = Cu, Zn)	F.W. = 286.4

I. The crystal structures of semicarbazide complexes of copper(II) and zinc
chlorides. M. NARDELLI, G. F. GASPARRI, P. BOLDRINI and G. G. BATTISTINI,
1965. *Acta Cryst.*, <u>19</u>, 491-500.

COPPER COMPLEX

Monoclinic, a = 7.56 ± 1, b = 9.26 ± 1, c = 6.88 ± 1 Å, β = 102.2 ± 2°,
U = 471 Å^3, D_m = 2.16, Z = 2, D_x = 2.01. (For crystal data of the corresponding monosemicarbazide, see <u>1</u>).

Space group $P2_1/c$ (C_{2h}^5) Molecular symmetry, centre

Atomic positions

	x	y	z
Cu	0	0	0
Cl	0.2721	0.0710	0.3397
O	−0.1408	0.1678	0.0557
N(1)	−0.3933	0.2220	0.1604
N(2)	−0.2811	−0.0050	0.2044
N(3)	−0.1278	−0.0981	0.1873
C	−0.2643	0.1332	0.1378

Standard deviations are given in full. Mean values are 0.004 Å for chlorine
and 0.010 to 0.015 Å for carbon, nitrogen and oxygen. Anisotropic temperature
factors are given also.

ZINC COMPLEX

Monoclinic, a = 5.13 ± 1, b = 7.13 ± 1, c = 13.26 ± 1 Å, β = 109.7 ± 1°,
U = 457 Å^3, D_m = 2.10, Z = 2, D_x = 2.08.

(For crystal data of the corresponding monosemicarbazide, see 1.)

Space group $P2_1/c$ (C_{2h}^5) Molecular symmetry, centre.

Atomic positions

	x	y	z
Zn	0	0	0
Cl	0.3554	0.1995	0.1464
O	-0.0980	-0.1591	0.1123
N(1)	-0.0491	-0.4497	0.1818
N(2)	0.2140	-0.3544	0.0838
N(3)	0.2860	-0.2142	0.0224
C	0.0163	-0.3133	0.1265

Standard deviations are given in full. Mean values are 0.004 Å for chlorine, and 0.010 to 0.15 Å for carbon, nitrogen, and oxygen. Anisotropic temperature factors are given also.

Structure

In both complexes the semicarbazide ion is planar, with essentially the dimensions reported in 2. Each metal ion is coordinated to two nitrogen and two oxygen atoms at the corners of a slightly distorted rectangle. The semicarbazide ions are slightly inclined (15.5° for the copper complex, 12.0° for the zinc complex) to the plane of the rectangle. Octahedral coordination is completed by two chlorine atoms. Details are given in Figs. 219 and 220.

Details of analysis

The intensity data for both complexes were recorded on integrated and non-integrated Weissenberg photographs (Cu: $hk0$ to $hk5$. Zn: $h0l$ to $h6l$; $0kl$ to $4kl$) using CuK_α radiation. For the copper complex, 662 of 831 accessible reflexions were observed, and for the zinc complex, 892 of 987. The intensities were measured visually. Refinement was by differential syntheses to final R indices of 0.132 (copper complex) and 0.152 (zinc complex).

1. This volume, p. 407.

2. *Ibid.*, p. 46.

ACETYLACETONE-MONO-(o-HYDROXYANIL)COPPER(II)

$C_{11}H_{11}CuNO_2$ F.W. = 252.8

I. The crystal structure of acetylacetone-mono-(o-hydroxyanil)copper(II). G. A. BARCLAY and B. F. HOSKINS, 1965. *J. Chem. Soc.* 1979-1991.

Triclinic, a = 8.914 ± 10, b = 10.485 ± 10, c = 11.733 ± 10 Å, α = 101.6 ± 3, β = 110.6 ± 3, γ = 92.5 ± 3°, U = 997.8 Å³, D_m = 1.68, z = 4, D_x = 1.683.
Space group $P1$ (C_1^1) or $P\bar{1}$ (C_i^1). $P\bar{1}$ is consistent with structure analysis.

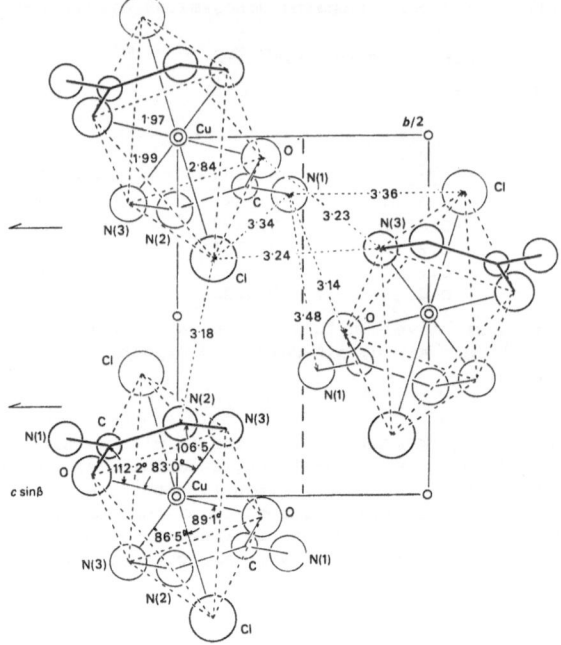

Fig. 219. The structure of the copper complex, viewed along *a*.

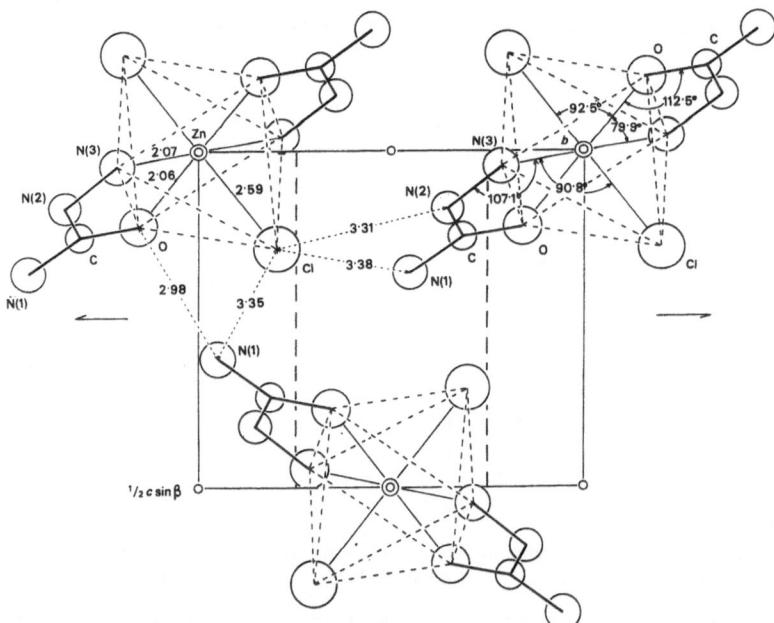

Fig. 220. The structure of the zinc complex, viewed along *a*.

Atomic positions

Atomic positions and isotropic temperature factors of the 30 non-hydrogen atoms are given. The standard deviation is 0.002 Å for copper, and ranges from 0.01 to 0.03 Å for the other atoms.

Structure

The compound exists as the dimer shown in Fig. 221. The dimer has an approximate centre of symmetry; the distances and angles shown are mean values. (Bond lengths and angles are given in full in the original.) The aromatic portion of each ligand is planar, as is the acetylacetone-imine portion. The angle between their mean planes in a given ligand is 16°. The coordination of Cu(1) is distorted square planar, while that of Cu(2) is, by reason of the close approach of an oxygen atom of a neighbouring dimer, distorted square pyramidal. The inter-dimer Cu-O distance is 2.64 Å.

Details of analysis

Intensity data were recorded on equi-inclination Weissenberg photographs (0*kl* to 5*kl*; *h*0*l* to *h*4*l*; *hk*0 to *hk*6) using CuK_α radiation. 2280 of a possible 2690 reflexions were observed and estimated visually. Refinement was by successive differential syntheses to a final R index of 0.15.

TRIBENZO[B,F,J][1.5.9]TRIAZACYCLODUODECINE NICKEL

NITRATE DIHYDRATE

$C_{21}H_{11}N_5O_6Ni.2H_2O$ F.W. = 524.1

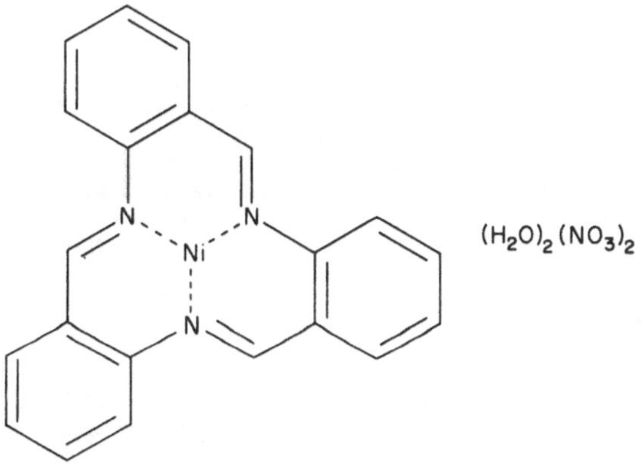

$(H_2O)_2(NO_3)_2$

I. The structure of a self-condensation product of *o*-aminobenzaldehyde in the presence of nickel ions. E. B. FLEISCHER and E. KLEM, 1965. *Inorg. Chem.*, <u>4</u>, 637-642.

Monoclinic, *a* = 20.0 ± 1, *b* = 16.70 ± 8, *c* = 7.24 ± 4 Å, β = 64°35 ± 30' [115°25'], *U* = 2185 Å³, D_m = 1.59, Z = 4, D_x = 1.60.

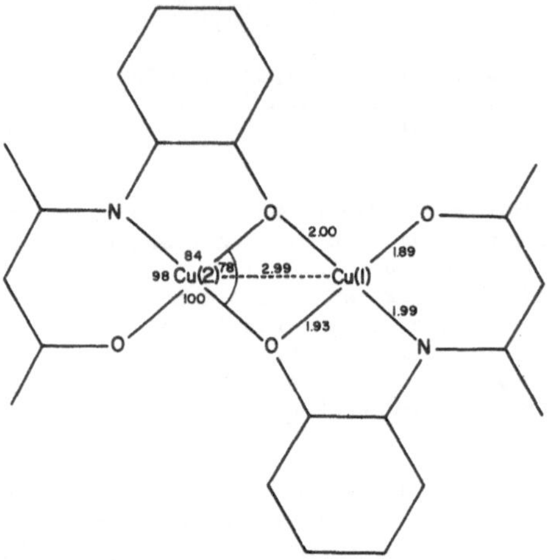

Fig. 221. Some bond lengths and angles in the $C_{11}H_{11}CuNO_2$ dimer. Standard
deviations are: Cu–Cu, 0.003; Cu–O, 0.01 Å; angles, 0.5°.

Space group $P2_1/a$ (C_{2h}^5)

Atomic positions

Atomic positions are given for the 35 non-hydrogen atoms. Standard
deviations are about 0.002 Å for nickel, and 0.01 to 0.02 Å for the other
atoms. Temperature factors (some anisotropic) are given also.

Structure

The tribenzo[b,f,j][1.5.9]triazacycloduodecine ligand (TRI) is not planar.
The nitrogen atoms are displaced about 0.35 Å to one side of the mean plane,
and the benzene rings are consequently twisted by about 13° out of the plane,
giving the ligand a propeller shape. The coordination of the nickel atom is
octahedral, with the three projecting nitrogen atoms of the TRI ligand constituting one face of the octahedron. The remaining face consists of two water molecules and one oxygen atom of a nitrate group. The nitrate groups form hydrogen
bonds (one for the coordinated group, and two for the uncoordinated group)
with water molecules. Bond lengths and angles are given in full. Mean coordination distances are: Ni–N, 2.03 ± 1 Å; Ni–O, 2.09 ± 1 Å.

Details of analysis

The intensity data were measured with a four-circle diffractometer and
scintillation counter, using CuK_α radiation. 1487 of 1750 accessible reflexions
were observed in the range 2θ<90°, and were measured by peak-counting. Refinement was by least squares to a final R index of 0.111.

TRIMETHYL-(8-QUINOLINATO)-PLATINUM(IV)

$C_{12}H_{15}NOPt$ $(CH_3)_3(C_9H_6NO)P$ F.W. = 384.4

I. The crystal structure of trimethyl-(8-quinolinato)-platinum(IV). J. E.
LYDON and M.R. TRUTER, 1965. *J. Chem. Soc.*, 6899–6910.

(For a preliminary report, see 1.)

Orthorhombic, a = 15.23 ± 2, b = 16.09 ± 2, c = 9.48 ± 2 Å, U = 2323 Å^3,
D_m = 2.17, Z = 8, D_x = 2.20. Cell and intensity data at 120 ± 5°K:
Space group $P2_12_12_1$ (D_2^4)

Structure

The analysis demonstrates that the two independent formula units associate as a dimer. As illustrated in Fig. 222, dimerization is effected by a symmetrical pair of oxygen bridges between the platinum atoms, which are separated by 3.384 ± 3 Å. Each platinum atom is six-coordinated, with two oxygen atoms at 2.24 ± 3 Å, one nitrogen atom at 2.13 ± 5 Å, and three methyl carbon atoms at 2.06 ± 5 A.

Details of analysis

Intensity data were recorded, at 120 ± 5°K, on equi-inclination Weissenberg photographs (0kl to 3kl; $h0l$ to $h5l$; $hk0$ to $hk6$.) using Cu$K\alpha$ radiation. 2152 of a possible 2624 reflexions were observed and measured visually. Absorption corrections were applied. Refinement was by least squares to a final R index of 0.10.

1. J.E. LYDON and M.R. TRUTER, 1964. *Proc. Chem. Soc.*, 193.

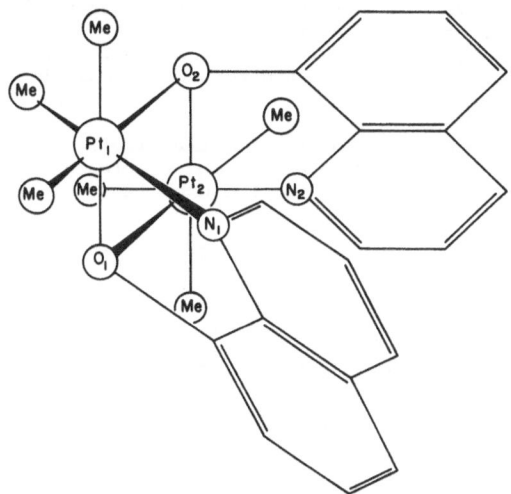

Fig. 222. A perspective view of the $C_{12}H_{15}NOPt$ dimer.

CATENA-DI-μ-HYDRAZINE-ZINC DIACETATE

$C_4H_{14}N_4O_4Zn$ $(NH_2NH_2)_2 \cdot Zn \cdot (CH_3COO)_2$ F.W. = 247.6

I. The crystal structure of catena-di-μ-hydrazine-zinc diacetate. A. FERRARI, A. BRAIBANTI, G. BIGLIARDI and A. M. LANFREDI, 1965. *Acta Cryst.*, <u>19</u>, 548-555.

Triclinic, a = 6.58 ± 2, b = 8.52 ± 1, c = 4.14 ± 1 Å, α = 90°, β = 90°25 ± 5, γ = 96°52 ± 12', [U = 230.4 Å³], D_m = 1.798, Z = 1, D_x = 1.783.

Space group P1 (C_1^1) or P$\bar{1}$ (C_i^1). P$\bar{1}$ confirmed by analysis. Molecular symmetry, centre.

Atomic positions

	x	y	z
Zn	0.0000	0.0000	0.0000
C(1)	0.7486 ± 8	0.2820 ± 8	0.8862 ± 24
C(2)	0.7434 ± 12	0.4509 ± 11	0.7637 ± 34
N(1)	0.2428 ± 7	0.0440 ± 8	0.6445 ± 21

	x	y	z
N(2)	0.1955 ± 7	0.1360 ± 7	0.3626 ± 21
O(1)	0.9138 ± 7	0.2232 ± 6	0.8438 ± 18
O(2)	0.5915 ± 7	0.2148 ± 7	0.0183 ± 24

Also given are anisotropic temperature factors for the above atoms, and the positions of the hydrogen atoms.

Structure

The octahedral coordination of the zinc atom is shown in Fig. 223. This atom is coordinated to two oxygen atoms of different acetate groups, and to four nitrogen atoms (in an accurately square array) of different hydrazine molecules. Bond lengths and angles are given in full. The N-N distance in the octahedron is 3.10 ± 1 Å, and the N-O distances are 2.92, 2.98, 3.18, and 3.19 Å (all ± 0.01 Å). Adjacent coordination octahedra are linked together by the covalent N-N bonds of the hydrazine molecules, as shown in Fig. 224, to form endless chains along c. Adjacent chains are held together by N-H...O bonds between hydrazine molecules and acetate groups to form sheets in the (010) planes. The acetate groups are accurately planar. The material is isostructural with the corresponding cadmium and manganese compounds.

Details of analysis

Intensity data were recorded on Weissenberg photographs ($hk0$ to $hk3$) using CuK$_\alpha$ radiation. 788 were observed and measured photometrically. Refinement was by differential syntheses to a final R index of 0.093.

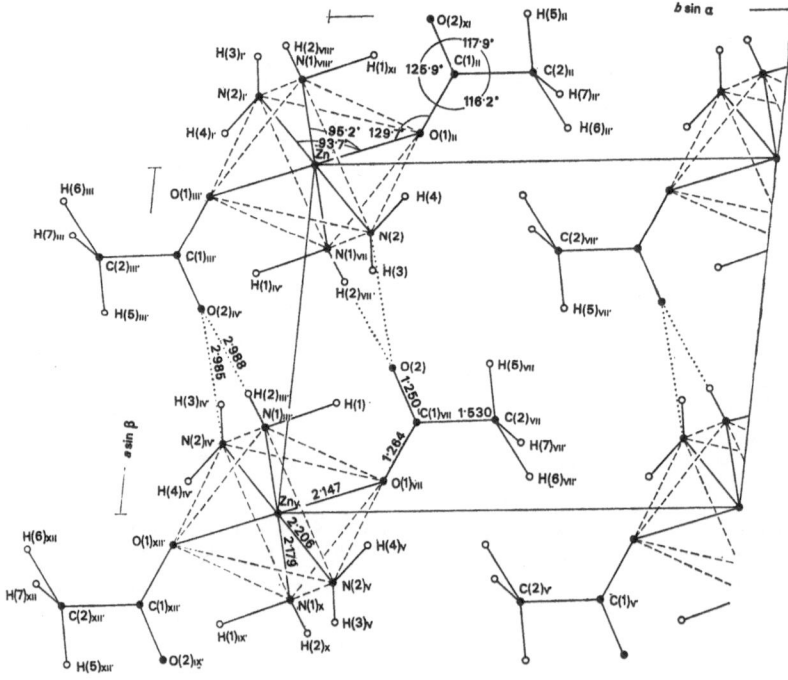

Fig. 223. One layer of the structure of $C_4H_{14}N_4O_4Zn$ viewed along c. Standard deviations are 0.007 to 0.014 Å, and 0.2 to 0.7°.

ZINC HYDRAZINECARBOXYLATE HYDRAZINE COMPLEX

$C_2H_{14}N_8O_4Zn$ $Zn(NH_2.NH_2)_2(NH_2.NH.COO)_2$ F.W. = 279.6

I. The crystal and molecular structure of *bis*(hydrazine)zinc *bis*(hydrazincarboxylate-N',O). A. FERRARI, A. BRAIBANTI, G. BIGLIARDI and A. M. LANFREDI, 1965. Z. *Kristallogr.*, **122**, 259–271.

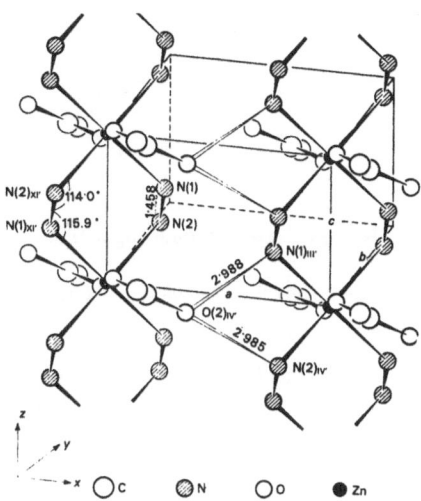

Fig. 224. Part of the structure of $C_4H_{14}N_4O_4Zn$, showing the chains of complexes.

Monoclinic, a = 8.412 ± 4, b = 7.384 ± 5, c = 9.582 ± 10 Å, β = 125°45 ± 27', U = 483 Å^3, D_m = 1.926, Z = 2, D_x = 1.936.

Space group $P2_1/c$ (C_{2h}^5) Molecular symmetry, centre.

Atomic positions

	x	y	z
Zn	0	½	½
O(1)	0.1055 ± 21	0.7271 ± 12	0.4579 ± 25
C(2)	0.2447 ± 21	0.8126 ± 15	0.5949 ± 28
N(3)	0.3260 ± 21	0.7283 ± 14	0.7532 ± 27
N(4)	0.2830 ± 21	0.5486 ± 15	0.7602 ± 30
O(5)	0.3114 ± 22	0.9646 ± 16	0.5977 ± 27
N(6)	0.1307 ± 22	0.3168 ± 15	0.4159,± 25
N(7)	0.2967 ± 27	0.3835 ± 20	0.4206 ± 30

Isotropic temperature factors are given also.

Structure

The zinc atom is coordinated to one nitrogen atom of each of two hydrazine groups, and to the terminal nitrogen atom and one oxygen atom of each of two hydrazinecarboxylate groups. The coordination is approximately octahedral. The molecular configuration is illustrated in Fig. 225. The structure is held together by N-H...O (2.91 ± 3 Å) and N-H...N (3.09 ± 5 Å) bonds.

Details of analysis

The intensity data were recorded on integrated Weissenberg photographs ($hk0$ to $hk5$) using CuK_α radiation. 548 reflexions were observed and measured photometrically. Refinement was by differential syntheses to a final R index of 0.17.

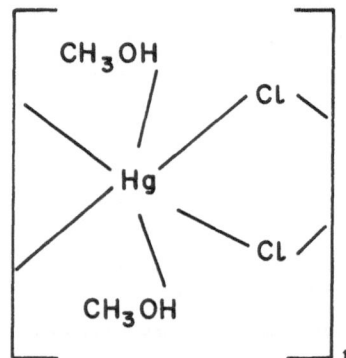

Fig. 225. The molecular configuration of $C_2H_{14}N_8O_4Zn$.

BIS-METHANOL, MERCURIC CHLORIDE

$C_2H_8HgCl_2O_2$ F.W. = 335.6

$$\left[\begin{array}{c} CH_3OH \quad \quad Cl \\ Hg \\ \quad Cl \\ CH_3OH \end{array} \right]_n$$

I.   Contribution a l'étude du système binaire chlorure mercurique-méthanol.
     F. MADAULE-AUBRY, 1965. *Ann. Chim. Fr.*, <u>10</u>, 367-394.

II. Structure cristalline du chlorure mercurique bis-méthanol. F. MADAULE-
     AUBRY and H. GILLIER-PANRAUD, 1965. *C.R. Acad. Sci. Paris*, <u>260</u>,
     6613-6615.

Triclinic, a = 3.94 ± 1, b = 6.87 ± 7, c = 11.52 ± 11 Å, α = 141.8 ± 3,
β = 92 ± 1, γ = 92 ± 1°, [U = 194.7 Å³], D_m = 2.7, Z = 1, D_x = 2.86.

Space group P1 (C_1^1) or P$\bar{1}$ (C_i^1). P$\bar{1}$ confirmed by analysis. Molecular symmetry,
centre for $HgCl_2$.

Atomic positions

	x	y	z
Hg	0	0	0
Cl	0.415	0.409	0.234
O	0.139	0.220	-0.123
C	0.063	-0.014	-0.332

Isotropic temperature factors are given also.

Structure

Adjacent $HgCl_2$ groups associate as planar ribbons extended along *a*, with Hg–Cl distances of 2.31 ± 3 Å and 3.07 ± 3 Å. Adjacent ribbons are separated by methanol molecules; octahedral coordination of the mercury atoms is completed by OH...Hg contacts of 2.82 ± 5 Å.

Details of analysis

The intensity data were recorded on Weissenberg photographs (0*kl* to 3*kl*) using CuK_α radiation. Refinement was by full-matrix least squares to a final *R* index of 0.129.

BIS-SALICYLALDEHYDATOCOPPER(II)

$C_{14}H_{10}O_2Cu$ $[O.C_6H_4.CHO]_2Cu$ F.W. = 305.7

I. The colour isomerism and structure of copper co-ordination compounds. Part VIII. The crystal structure of a second crystalline form of bis-salicylaldehydatocopper(II). D. HALL, A. J. McKINNON and T. N. WATERS, 1965. *J. Chem. Soc.*, 425–430.

   (For a report of another crystalline modification of this chelate, see 1.)

Monoclinic, a = 11.75 ± 3, b = 4.00 ± 2, c = 12.43 ± 3 Å, β = 90.3°, U = 584 Å³, D_m = 1.71, Z = 2, D_x = 1.74.

Space group $P2_1/c$ (C_{2h}^5) Molecular symmetry, centre.

Atomic positions

	x	y	z
Cu	0	0	0
O(1)	0.1234	0.3024	−0.0191
O(2)	0.0247	−0.0505	0.1553
C(1)	0.1078	0.0829	0.2046
C(2)	0.1967	0.2722	0.1620
C(3)	0.2020	0.3673	0.0547
C(4)	0.2972	0.5681	0.0209
C(5)	0.3831	0.6474	0.0915
C(6)	0.3767	0.5668	0.2010
C(7)	0.2843	0.3695	0.2356

The mean standard deviation for the light atoms is 0.021 Å.

Structure

The bond lengths and angles are given in Fig. 226. Each ligand is essentially planar, but the two ligands in a molecule are not coplanar; their mean planes are separated by 0.37 Å. The copper atom is coordinated to four oxygen atoms within the molecule, and to two hydroxy-oxygen atoms (distant 3.15 Å) in adjacent molecules. The coordination is thus distorted octahedral.

Details of analysis

Intensity data were recorded on Weissenberg photographs (*h*0*l* to *h*2*l*) using CuK_α radiation. 689 reflexions were observed and estimated visually. Refinement was by least squares to a final *R* index of 0.136.

1. *Structure Reports*, 29, 599.

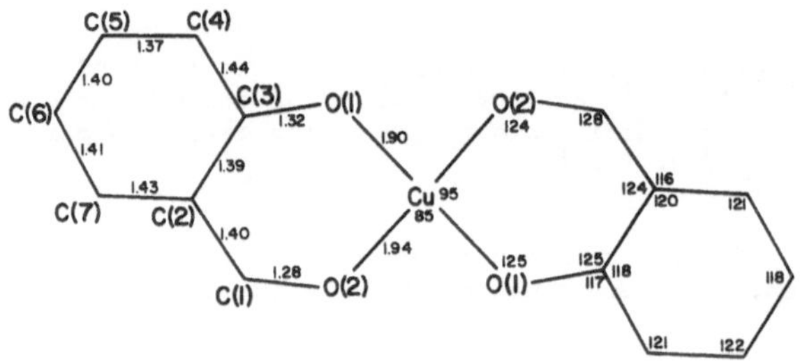

Fig. 226. Bond lengths and angles in bis-salicylaldehydatocopper(II).

CUPRIC TROPOLONE

$C_{14}H_{10}CuO_4$ F.W. = 305.8

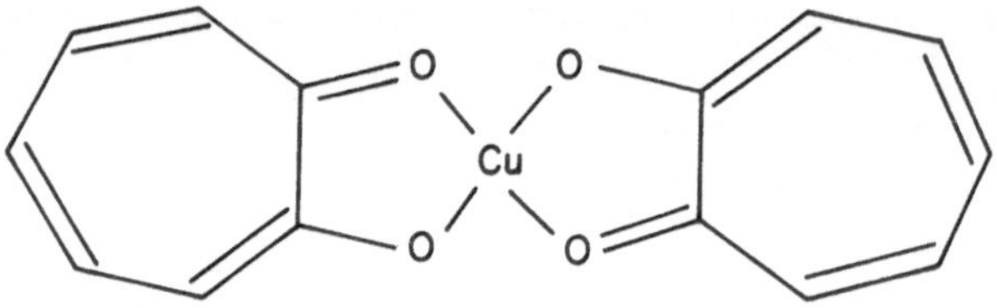

I. The crystal structure of cupric tropolone: a refinement. W. M. MACINTYRE, J. M. ROBERTSON and R.F. ZAHROBSKY, 1965. *Proc. Roy. Soc.*, A, <u>289</u>, 161–170.

(For earlier work, see <u>1</u>.)

Monoclinic, a = 3.800 ± 5, b = 13.82 ± 2, c = 11.60 ± 2 Å, β = 93.4 ± 2°, [U = 608.1 Å^3], D_m (from <u>1</u>) = 1.696, Z = 2, D_x = 1.67. (a and c have been reversed, compared to <u>1</u>.)

Space group $P2_1/c$ (C_{2h}^5) Molecular symmetry, centre.

Atomic positions

	x	y	z
Cu	0	0	0
O(1)	0.2848 ± 14	−0.0255 ± 5	0.1356 ± 5
O(2)	−0.0133 ± 12	0.1306 ± 3	0.0583 ± 4
C(1)	0.3111 ± 14	0.0506 ± 4	0.2034 ± 5
C(2)	0.1521 ± 14	0.1380 ± 4	0.1575 ± 5
C(3)	0.1514 ± 18	0.2323 ± 6	0.2090 ± 7
C(4)	0.2929 ± 20	0.2599 ± 6	0.3158 ± 7
C(5)	0.4808 ± 17	0.2038 ± 8	0.4023 ± 7
C(6)	0.5587 ± 21	0.1074 ± 6	0.3961 ± 6
C(7)	0.4914 ± 17	0.0360 ± 9	0.3088 ± 7

Anisotropic temperature factors are given also.

Structure

The copper and oxygen atoms are coplanar, as are the seven ring carbon atoms, but the two planes so defined are inclined to each other at about 4.9°. The bond lengths and angles are given in Fig. 227. The alternation of bond lengths around the ring is to be noted.

Details of analysis

The intensity data were measured with a four-circle diffractometer, using CuK_α radiation. Within the range 2θ<55°, 1155 reflexions (about 86% of those accessible) were observed. Refinement was by block-diagonal least squares to a final R index of 0.10.

1. *Structure Reports,* <u>15</u>, 473.

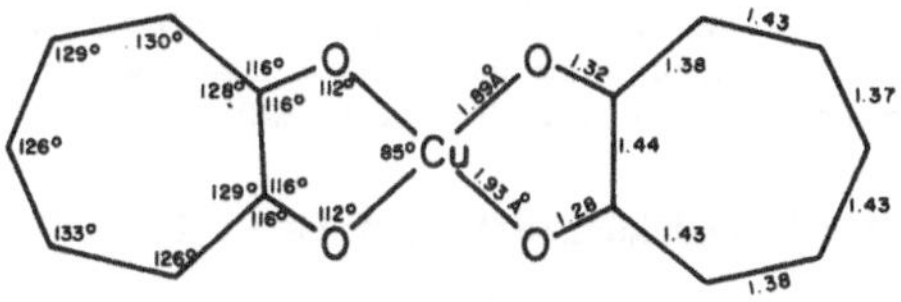

Fig. 227. Bond lengths and angles, in Å and degrees in cupric tropolone.
Standard deviations are 0.01 to 0.02 Å, and 0.5 to 1.4°.

CADMIUM n-BUTYL XANTHATE

$C_{10}H_{18}O_2S_4Cd$ F.W. = 410.9

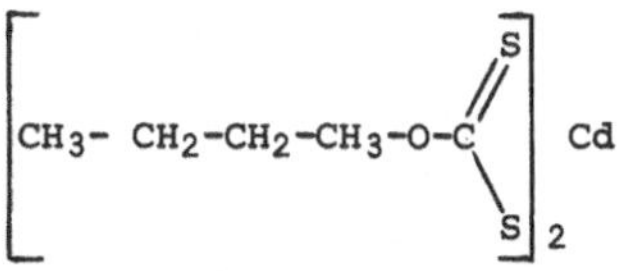

$CH_3- CH_2-CH_2-CH_3-O-C$ ... Cd ...

I. The crystal structure of cadmium n-butyl xanthate. H. M. RIETVELD and
E. N. MASLEN, 1965. *Acta Cryst.,* <u>18</u>, 429–436.

Monoclinic, a = 11.59 ± 17, b = 5.84 ± 9, c = 25.7 ± 4 Å, β = 101°44 ± 12',
U = 1705 Å³, D_m = 1.66, z = 4, D_x = 1.61.

Space group $P2_1/a$ (C_{2h}^5)

Atomic positions

	x	y	z
Cd	0.6231 ± 2	0.2940 ± 4	0.2436 ± 1
S(11)	0.3501 ± 7	0.1181 ± 15	0.2639 ± 4
S(12)	0.5377 ± 8	0.4377 ± 18	0.3222 ± 4
O(1)	0.3174 ± 19	0.4029 ± 38	0.3343 ± 09
C(11)	0.3929 ± 30	0.3239 ± 54	0.3062 ± 13
C(12)	0.3421 ± 28	0.6035 ± 55	0.3707 ± 16
C(13)	0.2356 ± 50	0.6184 ± 74	0.3940 ± 17
C(14)	0.2231 ± 36	0.4255 ± 92	0.4330 ± 16
C(15)	0.1053 ± 57	0.4376 ± 29	0.4529 ± 23
S(21)	0.6471 ± 7	0.8654 ± 14	0.2253 ± 4
S(22)	0.4925 ± 8	0.4999 ± 14	0.1663 ± 4
O(2)	0.4550 ± 17	0.9442 ± 30	0.1534 ± 8
C(21)	0.5275 ± 23	0.7757 ± 38	0.1811 ± 12
C(22)	0.3465 ± 27	0.8702 ± 49	0.1165 ± 13
C(23)	0.2852 ± 32	0.1097 ± 53	0.0958 ± 14
C(24)	0.1704 ± 33	0.0540 ± 74	0.0581 ± 14
C(25)	0.1026 ± 33	0.2784 ± 75	0.0428 ± 18

Anisotropic temperature factors are given also.

Structure

Bond lengths and angles are given in full. The values for the n-butyl
xanthate ions, although not very accurate, (standard deviations range from 0.03
to 0.08 Å) are consistent with the assumed formulation. The shapes of the two
butyl chains are quite different; one is extended, the other curled. Mean
values for the carbon-sulphur bonds are: C–S, 1.75 ± 3; C=S, 1.63 ± 3 Å. The
cadmium atom is coordinated to four sulphur atoms in an approximately tetrahedral array. Each of the four sulphur atoms is contributed by a different
xanthic radical. Cd–S distances range from 2.560 to 2.619 Å (± 0.010 Å), with
a mean value of 2.594 Å.

Details of analysis

Intensity data were recorded on equi-inclination Weissenberg photographs (h0l to h5l; 0kl to 2kl) with CuK$_\alpha$ radiation. 1296 of a possible 4330 reflexions were observed and measured visually. Refinement was by block-diagonal least squares to a final R index of 0.122.

BIS(*cis*-1,2-BIS(TRIFLUOROMETHYL)ETHYLENE-1,2-DITHIOLATE)COBALT
DIMER

$C_{16}F_{24}Co_2S_8$ F.W. = 1022.6

I. Molecular structure of the dimer of bis(*cis*-1,2-bis(trifluoromethyl)-ethylene-1,2-dithiolate)cobalt. J. H. ENEMARK and W. N. LIPSCOMB, 1965. *Inorg. Chem.*, **4**, 1729-1734.

Triclinic, a = 7.98 ± 2, b = 9.89 ± 1, c = 10.12 ± 2 Å, α = 103.0 ± 2, β = 98.5 ± 2, γ = 100.8 ± 2°, [U = 749 Å^3], D_m = 2.27, Z = 1, D_x = 2.28. *Space group* P1 (C_1^1) or P$\bar{1}$ (C_i^1) P$\bar{1}$ confirmed by analysis. Molecular symmetry, centre.

Atomic positions

	x	y	z
Co	0.6313	0.5976	0.4726
S(1)	0.9037	0.6953	0.5344
S(2)	0.5738	0.7958	0.5750
S(3)	0.3651	0.5402	0.3622
S(4)	0.6994	0.4460	0.3083
C(1)	0.9164	0.8611	0.6325
C(2)	0.7669	0.9073	0.6514
C(3)	1.0995	0.9486	0.6986
C(4)	0.7606	1.0534	0.7312
C(5)	0.3638	0.4308	0.2040
C(6)	0.5173	0.3859	0.1832
C(7)	0.1937	0.3805	0.1040
C(8)	0.5295	0.2796	0.0539
F(1)	1.1305	0.9761	0.8296
F(2)	1.1255	1.0737	0.6713
F(3)	1.2152	0.8855	0.6518
F(4)	0.8436	1.0875	0.8596
F(5)	0.8217	1.1541	0.6745
F(6)	0.6020	1.0680	0.7398
F(7)	0.0816	0.4487	0.1496
F(8)	0.2011	0.3994	-0.0149
F(9)	0.1272	0.2452	0.0835
F(10)	0.4839	0.3157	-0.0577
F(11)	0.4371	0.1548	0.0388
F(12)	0.6898	0.2623	0.0577

Anisotropic temperature factors are given also.

Structure

Pairs of the structural units shown above associate as centrosymmetrical dimers, with the cobalt atom of one unit bonded to a sulphur atom of the adjacent unit. The molecular structure is illustrated in Fig. 228. The coordination of the cobalt atom is square pyramidal, with the basal plane made up of sulphur atoms from the same unit, and the apical sulphur atom supplied by the adjacent unit. The cobalt atom is displaced 0.37 Å from the basal plane towards the apex. The cobalt atom is 2.38 ± 2 Å from the apical sulphur atom, and 2.16 ± 2 Å from the basal sulphur atoms. Other bond lengths and angles are given in full, and are consistent with expectation.

Details of analysis

The intensity data were recorded on equi-inclination Weissenberg photographs ($0kl$ to $7kl$) and precession photographs ($h0l$ to $h3l$; $hk0$ to $hk2$; hkk) using MoK$_\alpha$ radiation. 2031 of 2878 accessible reflexions were observed and measured visually. Refinement was by least squares to a final R index of 0.080.

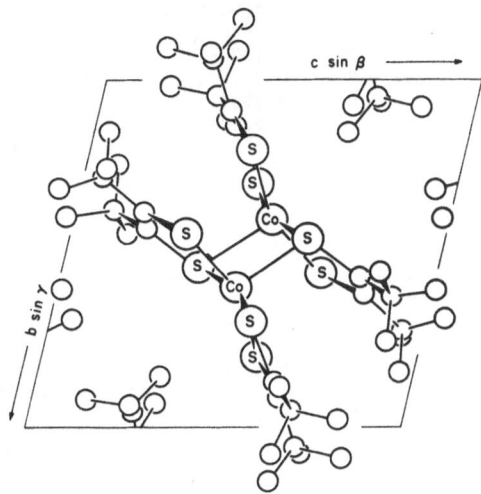

Fig. 228. Projection of the crystal structure of $C_{16}F_{24}Co_2S_8$ along a.

COPPER DIETHYLDITHIOCARBAMATE

$C_{10}H_{20}CuN_2S_4$ F.W. = 360.1

I. Structural studies of metal dithiocarbamates. II. The crystal and molecular structure of copper diethyldithiocarbamate. M. BONAMICO, G. DESSY, A. MUGNOLI, A. VACIAGO and L. ZAMBONELLI, 1965. *Acta Cryst.*, <u>19</u>, 886–897.

(For a preliminary account of this work, see 1.)

Monoclinic, a = 9.907 ± 10, b = 10.627 ± 5, c = 16.591 ± 10 Å, β = 113°52 ± 5', U = 1597.4 Å^3, D_m = 1.489, Z = 4, D_x = 1.498.

Space group $P2_1/c$ (C_{2h}^5)

Atomic positions

	x	y	z
Cu	0.1914 ± 1	0.0317 ± 1	0.0651 ± 1
S(1)	0.3344 ± 2	0.2116 ± 2	0.1106 ± 1
S(2)	0.2039 ± 2	0.0617 ± 2	0.2051 ± 1
S(3)	0.2462 ± 2	−0.0331 ± 2	−0.0506 ± 1
S(4)	0.0768 ± 2	−0.1649 ± 2	0.0279 ± 1
N(1)	0.3743 ± 6	0.2639 ± 7	0.2771 ± 4
N(2)	0.1374 ± 5	−0.2654 ± 5	−0.1032 ± 3
C(1)	0.3122 ± 8	0.1881 ± 8	0.2066 ± 4
C(2)	0.4620 ± 9	0.3736 ± 12	0.2741 ± 6
C(3)	0.3542 ± 9	0.2420 ± 12	0.3589 ± 5
C(4)	0.3632 ± 11	0.4880 ± 12	0.2392 ± 8
C(5)	0.4792 ± 14	0.1630 ± 17	0.4240 ± 7
C(6)	0.1520 ± 7	−0.1675 ± 7	−0.0502 ± 4
C(7)	0.2107 ± 8	−0.2675 ± 10	−0.1640 ± 5
C(8)	0.0453 ± 8	−0.3754 ± 9	−0.1046 ± 5
C(9)	0.1065 ± 12	−0.2292 ± 14	−0.2571 ± 5
C(10)	0.1369 ± 14	−0.4823 ± 13	−0.0489 ± 10

Also given are anisotropic temperature factors for the above atoms, and assumed positions for the hydrogen atoms.

Structure

Bond lengths and angles are given in Fig. 229. The ligand molecules are planar except for the terminal methyl groups. The molecule as a whole is not planar, however; details are given. A view of part of the structure is given in Fig. 230, showing the intermolecular Cu–S bonds which join pairs of molecules. The copper atom is seen to be coordinated to five sulphur atoms and one hydrogen atom which form a distorted octahedron. The shorter coordination distances (2.30 to 2.34 Å) are intramolecular.

Details of analysis

Intensity data were recorded on equi-inclination Weissenberg photographs ($0kl$ to $5kl$; $h0l$ to $h6l$) using CuK_α radiation. 2570 independent reflexions (about 75% of those accessible with the radiation used) were observed and estimated visually. Refinement was by successive differential syntheses to a final R index of 0.081.

1. M. BONAMICO *et al.,* 1963. *Rend. Accad. Lincei,* VIII, 35, 338.

MERCURY ETHYLMERCAPTIDE

$C_4H_{10}HgS_2$ CH$_2$–CH$_3$ F.W. = 333

|
S–Hg–S
|
CH$_3$–CH$_2$

I. Structures of mercury mercaptides. Part II. X–ray structural analysis of mercury ethylmercaptide. D. C. BRADLEY and N. R. KUNCHUR, 1965. *Can. J. Chem.,* 43, 2786–2792.

Monoclinic, $a = 7.54 \pm 2$, $b = 4.87 \pm 1$, $c = 23.80 \pm 4$ Å, $\beta = 95°$, [Given as
85° in original], [$U = 871$ Å^3], $D_m = 2.47$, $z = 4$, $D_x = 2.45$.

Space group Cc (C_s^4) or $C2/c$ (C_{2h}^6). [$C2/c$ was rejected for no apparent reason.
However, the deduced structure has (within the limited accuracy of the analysis)
a two-fold axis parallel to b.. It seems likely, therefore, that the space
group is in fact $C2/c$.]

Atomic positions

Atomic positions are given for the [probably incorrect] space group Cc.

Structure

The structure is essentially molecular, with no intermolecular attraction
stronger than van der Waals forces. The molecule is approximately planar,
with the conformation indicated above. The S–Hg–S group is linear, with Hg–S =
2.45 ± 5 Å.

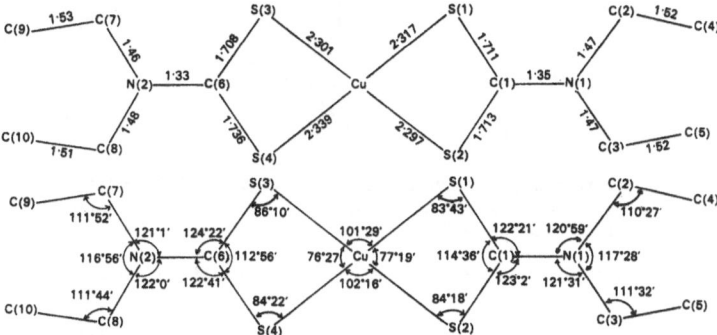

Fig. 229. Bond lengths and angles within the formula unit of copper diethyldithiocarbamate. Standard deviations of distances are: Cu–S, 0.002;
S–C, 0.008; C–N, 0.010; C–C, 0.016 Å.

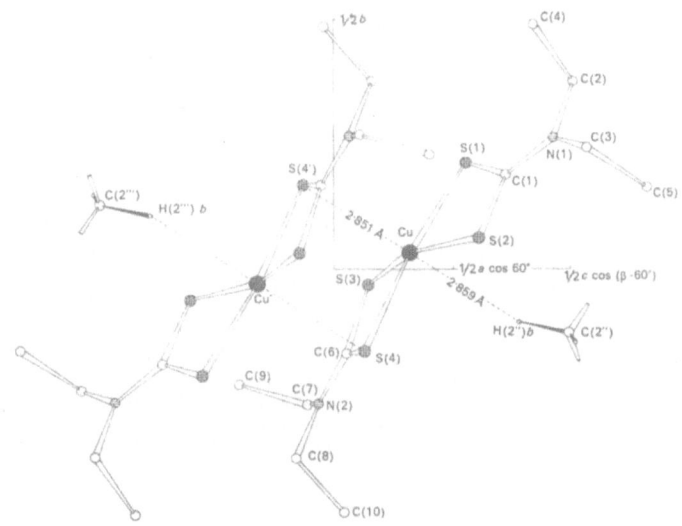

Fig. 230. A view of part of the structure of copper diethyldithiocarbamate,
showing short intermolecular distances.

Details of analysis

    h0l and *0kl* intensity data were recorded on Weissenberg photographs, using CuK_α radiation. 133 of a possible 153 reflexions were observed and measured visually. Refinement was by Fourier methods to a final *R* index of about 0.15.

1,6-DITHIACYCLODECA-*cis*-3,*cis*-8-DIENEBIS(MERCURIC CHLORIDE)

$C_8H_{12}Cl_4Hg_2S_2$ F.W. = 714.5

    I. Complexes of mercury. Part I. X-ray analysis of 1,6-dithiacyclodeca-*cis*-3,*cis*-8-dienebis(mercuric chloride). K. K. CHEUNG and G. A. SIM, 1965. *J. Chem. Soc.*, 5988-6004.

Monoclinic, a = 7.29, b = 17.01, c = 6.20 Å, β = 92°34', U = 768 Å^3, Z = 2, D_x = 3.08.

Space group $P2_1$ (C_2^2) or $P2_1/m$ (C_{2h}^2). $P2_1/m$ confirmed by analysis. Molecular symmetry, mirror plane.

Atomic positions

	x	y	z
Hg(1)	0.4874	0.75	0.1002
Hg(2)	0.1296	0.25	0.4068
Cl(1)	0.4538	0.6155	0.0959
Cl(2)	0.4704	0.25	0.3813
Cl(3)	0.1140	0.25	0.8108
S	-0.0248	0.3714	0.2462
C(1)	0.1642	0.5963	0.5308
C(2)	0.2548	0.5182	0.5858
C(3)	0.2510	0.4544	0.4704
C(4)	0.1554	0.4496	0.2340

    Standard deviations are given in full. Mean values are: Hg, 0.003 Å; Cl, 0.015 Å; S, 0.010 Å; C, 0.04 Å. Anisotropic temperature factors are given also.

Structure

    The polymeric nature of the structure is illustrated in Fig. 231. The co-ordination of Hg(2) is approximately tetrahedral, with two sulphur atoms at 2.53 Å and two chlorine atoms at 2.51 Å. Hg(1) exists in an almost undistorted $HgCl_2$ molecule, with Hg-Cl = 2.30 Å, and Cl-Hg-Cl = 168°. Bond lengths and angles are given in full.

Details of analysis

    The intensity data were recorded on equi-inclination Weissenberg photographs using CuK_α radiation. 1158 independent reflexions were observed and measured visually. Refinement was by least squares to a final R index of 0.136.

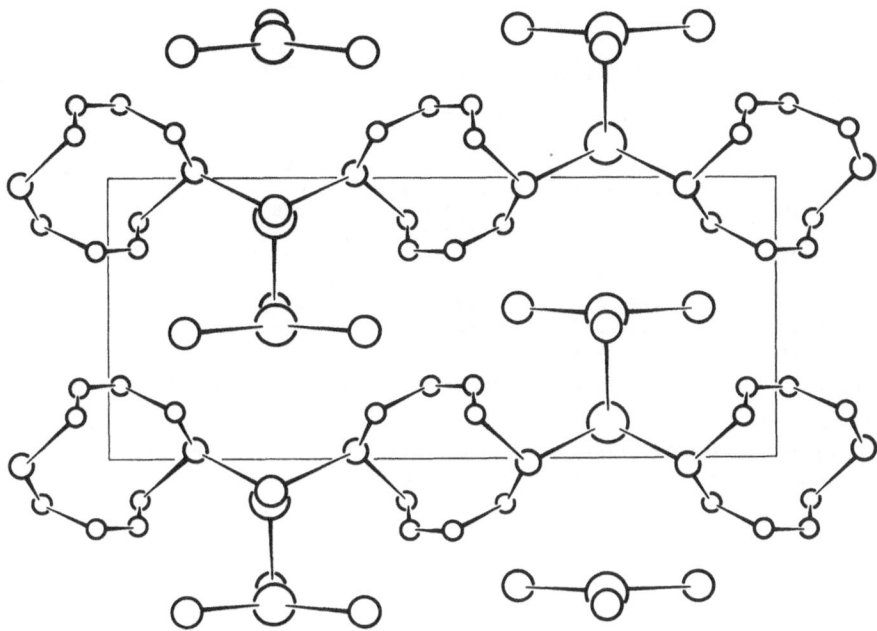

Fig. 231. The structure of $C_8H_{12}Cl_4Hg_2S_2$ viewed along c.

DI(TETRAMETHYLAMMONIUM)BIS(MALEONITRILE DITHIOLATE)NICKELATE(II)

$C_{16}H_{24}N_6S_4Ni$ F.W. = 487.2

I. Structure of di(tetramethylammonium)bis(maleonitrile dithiolate)nickelate (II). R. EISENBERG and J. A. IBERS, 1965. *Inorg. Chem.*, **4**, 605–608.

(For a preliminary report, see **1**.)

Triclinic, a = 10.18 ± 3, b = 15.79 ± 3, c = 8.04 ± 2 Å, α = 87.1 ± 1, β = 113.4 ± 1, γ = 91.9 ± 2°, U = 1185 Å³, D_m = 1.35, Z = 2, D_x = 1.37.
Space group $P1$ (C_1^1) *or* $P\bar{1}$ (C_i^1). $P\bar{1}$ confirmed by analysis. Molecular symmetry,

centre. (However, the parameters quoted above are for a centred cell, $C\bar{1}$. The structure is described in terms of this centred cell.)

Atomic positions

	x	y	z
Ni	0	0	0
S(1)	0.1851 ± 4	−0.0715 ± 3	0.1794 ± 5
S(2)	0.1349 ± 4	0.1000 ± 3	−0.0410 ± 5
C(1)	0.329 ± 1	−0.008 ± 1	0.182 ± 2
C(2)	0.470 ± 1	−0.040 ± 1	0.288 ± 2
N(1)	0.581 ± 2	−0.065 ± 1	0.369 ± 2
C(3)	0.310 ± 1	0.065 ± 1	0.087 ± 2
C(4)	0.423 ± 2	0.117 ± 1	0.088 ± 2
N(2)	0.515 ± 2	0.160 ± 1	0.095 ± 2
N(3)	0.034 ± 1	−0.178 ± 1	−0.389 ± 1
C(5)	0.064 ± 2	−0.226 ± 2	−0.208 ± 3
C(6)	0.136 ± 3	−0.106 ± 2	−0.374 ± 4
C(7)	−0.107 ± 2	−0.139 ± 2	−0.458 ± 3
C(8)	0.038 ± 2	−0.240 ± 2	−0.530 ± 3

Temperature factors (anisotropic for nickel and sulphur) are given also.

Structure

The bis(maleonitrile dithiolate)nickelate(II) anion is essentially planar, with approximate mmm (D_{2h}) symmetry. Bond lengths and angles are:

a	=	2.166 ± 6 Å	aa' =	91.5 ± 2°
b	=	1.77 ± 1	ab =	103.8 ± 5
c	=	1.33 ± 2	bc =	120 ± 1
d	=	1.41 ± 2	bd =	117 ± 1
e	=	1.13 ± 2	cd =	122 ± 1
f	=	1.51 ± 2	de =	178 ± 1

The tetramethylammonium cation is tetrahedral. Interionic distances are normal, and give no indication of any unusual interaction.

Details of analysis

The intensity data were recorded on equi-inclination Weissenberg photographs ($h0l$ to $h,10,l$) using MoK_α radiation. 996 reflexions were observed in the range $\theta < 22.2°$, and were measured visually. Corrections for absorption were applied. Refinement was by full-matrix least squares to a final R index of 0.104.

1. R. EISENBERG, J.A. IBERS, R.H.J. CLARK and H.B. GRAY, 1964. *J. Amer. Chem. Soc.*, **86**, 113.

NICKEL DIETHYLDITHIOCARBAMATE

$C_{10}H_{20}N_2NiS_4$ F.W. = 355.25

$$H_5C_2 \diagdown \atop H_5C_2 \diagup N-C \diagup{S}\diagdown{S} Ni \diagdown{S}\diagup{S} C-N \diagup{C_2H_5}\diagdown{C_2H_5}$$

I.   Structural studies of metal dithiocarbamates. I. The crystal and molecular structure of the α form of nickel diethyldithiocarbamate. M. BONAMICO, G. DESSY, C. MARIANI, A. VACIAGO, and L. ZAMBONELLI, 1965. *Acta Cryst.*, **19**, 619–626.

(For an account of earlier work, with which the present account is in substantial disagreement, see **1**.)

Monoclinic, a = 6.189 ± 10, b = 11.537 ± 5, c = 11.603 ± 10 Å, β = 95°51 ± 5', U = 824.2 Å^3, D_m = 1.437, Z = 2, D_x = 1.431.

Space group $P2_1/c$ (C_{2h}^5) Molecular symmetry, centre.

Atomic positions

	x	y	z
Ni	0	0	0
S(1)	0.3030 ± 3	0.0467 ± 2	0.1069 ± 2
S(2)	−0.0929 ± 3	0.1664 ± 2	0.0723 ± 2
N	0.2394 ± 13	0.2450 ± 6	0.2206 ± 6
C(1)	0.1604 ± 12	0.1658 ± 6	0.1444 ± 6
C(2)	0.1048 ± 18	0.3465 ± 9	0.2472 ± 9
C(3)	0.4540 ± 17	0.2338 ± 10	0.2879 ± 10
C(4)	0.1408 ± 27	0.4458 ± 12	0.1680 ± 14
C(5)	0.4351 ± 24	0.1722 ± 15	0.4028 ± 12

Also given are anisotropic temperature factors for the above atoms, and assumed positions for the hydrogen atoms.

Structure

The bond lengths and angles are given in Fig. 232, and a view of part of the structure in Fig. 233. The molecule is approximately planar except for the terminal methyl groups. The coordination of the nickel atom is distorted square planar.

Details of analysis

Intensity data were recorded on equi-inclination Weissenberg photographs ($0kl$ to $5kl$; $h0l$ to $h3l$) using CuK_α radiation. 1376 independent reflexions (about 73% of those accessible with the radiation used) were observed and esti-mated visually. Refinement was by successive differential syntheses to a final R index of 0.10.

1. *Structure Reports*, <u>24</u>, 604.

Fig. 232. Bond lengths and angles of nickel diethyldithiocarbamate. Mean standard deviations are: Ni–S, 0.002; S–C, 0.007; C(1)–N, 0.010; N–C, 0.015; C–C, 0.019 Å.

ORGANIC COMPOUNDS

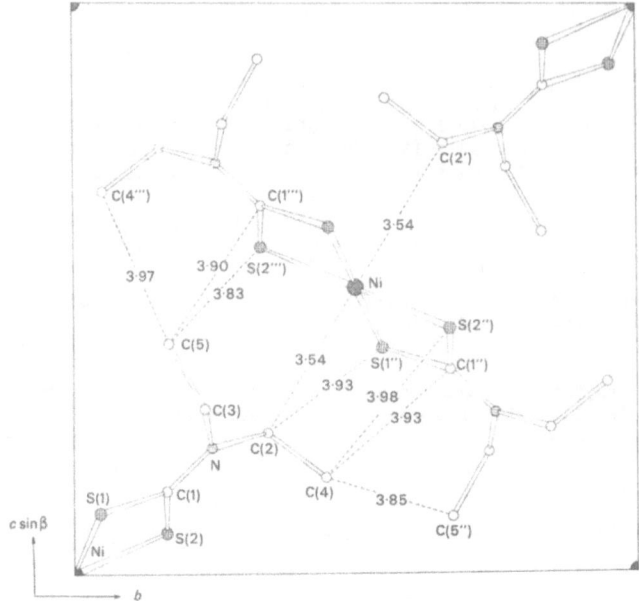

Fig. 233. Projection of nickel diethyldithiocarbamate structure along [100].
The dotted lines show contacts of less than 4.0 Å.

BIACETYLBIS(MERCAPTOETHYLIMINE)NICKEL(II)

$C_8H_{14}N_2S_2Ni$ F.W. = 261.1

I. The crystal and molecular structure of biacetylbis(mercaptoethylimine)-
nickel(II). Q. FERNANDO and P. J. WHEATLEY, 1965. *Inorg. Chem.*, **4**,
1726-1729.

Orthorhombic, a = 16.50 ± 2, b = 8.93 ± 2, c = 7.29 ± 2 Å, [U = 1074 Å^3], D_m = 1.64, Z = 4, D_x = 1.61.

Space group Pna2$_1$ (C_{2v}^9) or *Pnam* (D_{2h}^{16}). Pna2$_1$ is confirmed by analysis.

Atomic positions

	x	y	z
Ni	0.0725	0.0630	0.2384
S(1)	0.1839	0.1913	0.2568
S(2)	-0.0094	0.2494	0.2265
N(1)	-0.0148	-0.0676	0.2309
N(2)	0.1318	-0.1131	0.2421
C(1)	0.2493	0.0324	0.3121
C(2)	0.2225	-0.1067	0.2717
C(3)	-0.0973	-0.0176	0.2432
C(4)	-0.1053	0.1501	0.2132
C(5)	0.0878	-0.2392	0.2421
C(6)	0.0017	-0.2128	0.2361
C(7)	0.1244	-0.3928	0.2698
C(8)	-0.0619	-0.3269	0.2527

Coordinates in the original are given in Å. Standard deviations are given in full; mean values are: Ni, 0.006 Å; S, 0.01 Å; C and N, 0.02 to 0.08 Å. Anisotropic temperature factors are given also.

Structure

The configuration of the molecule is as represented above. The coordination of the nickel atom is approximately square planar; the mean coordination distances and angles are: Ni–S, 2.16 ± 1 Å; Ni–N, 1.85 ± 2 Å; N–Ni–N 83 ±1°; S–Ni–S, 97.3 ± 4°; N–Ni–S, (adjacent bonds) 90 ± 1°. Other bond lengths and angles are given in full. [It is erroneously reported that the ligands show a highly significant distortion from planarity, tending towards tetrahedral coordination. In fact the nickel, nitrogen, and sulphur atoms are essentially coplanar; their deviation from coplanarity is significant only at the 5% confidence level.] The carbon atoms adjacent to double bonds lie in the plane of the ligand atoms, but the remaining carbon atoms lie out of this plane by as much as 0.24 Å. The intermolecular distances are consistent with van der Waals interaction.

Details of analysis

The intensity data were recorded on equi-inclination Weissenberg photographs (h0l to h4l; hk0 to hk4) using CuK$_\alpha$ radiation. 895 independent reflexions were observed. Refinement was by differential Fourier syntheses to a final R index of 0.18.

ZINC DIETHYLDITHIOCARBAMATE

C$_{10}$H$_{20}$N$_2$S$_4$Zn   F.W. = 361.9

I. Structural studies of metal dithiocarbamates. III. The crystal and molecular structure of zinc diethyldithiocarbamate. M. BONAMICO, G. MAZZONE, A. VACIAGO and L. ZAMBONELLI, 1965. *Acta Cryst.*, 19, 898–909.

(For a preliminary report see 1.)

Monoclinic, a = 10.015 ± 10, b = 10.661 ± 5, c = 16.357 ± 10 Å, β = 110°58 ± 5', U = 1619.7 Å^3, D_m = 1.480, Z = 4, D_x = 1.485.

Space group P2$_1$/c (C_{2h}^5)

Atomic positions

	x	y	z
Zn	0.1697 ± 1	0.0725 ± 1	0.0540 ± 1
S(1)	0.3408 ± 3	0.2441 ± 2	0.1121 ± 2
S(2)	0.2169 ± 3	0.0725 ± 2	0.2063 ± 2
S(3)	0.2439 ± 3	−0.0386 ± 3	−0.0443 ± 2
S(4)	0.0535 ± 2	−0.1714 ± 2	0.0288 ± 1
N(1)	0.3872 ± 9	0.2672 ± 8	0.2834 ± 5
N(2)	0.1352 ± 9	−0.2624 ± 7	−0.0982 ± 5
C(1)	0.3206 ± 10	0.2027 ± 10	0.2086 ± 5
C(2)	0.4713 ± 11	0.3776 ± 10	0.2854 ± 8
C(3)	0.3690 ± 13	0.2342 ± 12	0.3670 ± 7
C(4)	0.3762 ± 16	0.4937 ± 12	0.2580 ± 10
C(5)	0.4865 ± 20	0.1536 ± 23	0.4258 ± 9
C(6)	0.1416 ± 11	−0.1687 ± 8	−0.0447 ± 5
C(7)	0.2123 ± 13	−0.2636 ± 12	−0.1595 ± 9
C(8)	0.0451 ± 12	−0.3740 ± 10	−0.1027 ± 8
C(9)	0.1191 ± 18	−0.2181 ± 19	−0.2506 ± 8
C(10)	0.1347 ± 17	−0.4817 ± 12	−0.0476 ± 13

Also given are anisotropic temperature factors for the above atoms, and
assumed positions for the hydrogen atoms.

Structure

The intramolecular bond lengths and angles are given in Fig. 234. The
ligand molecules are planar except for the terminal methyl groups. The molecule
as a whole is not planar, however; details are given. A view of part of the
structure is given in Fig. 235, showing the intermolecular Zn–S bonds which
join pairs of molecules as dimers. The zinc atom is thus somewhat irregularly
coordinated to five sulphur atoms. One Zn–S distance is much longer than the
others; surprisingly, this distance is intramolecular.

Details of analysis

Intensity data were recorded on equi-inclination Weissenberg photographs
($0kl$ to $5kl$; $h0l$ to $h6l$) using CuK$_\alpha$ radiation. 2652 independent reflexions
(about 75% of those accessible with the radiation used) were observed and estimated visually. Refinement was by successive differential syntheses to a final
R index of 0.106.

1. M. BONAMICO *et al.*, 1963. *Rend. Accad. Lincei*, VIII, <u>35</u>, 338.

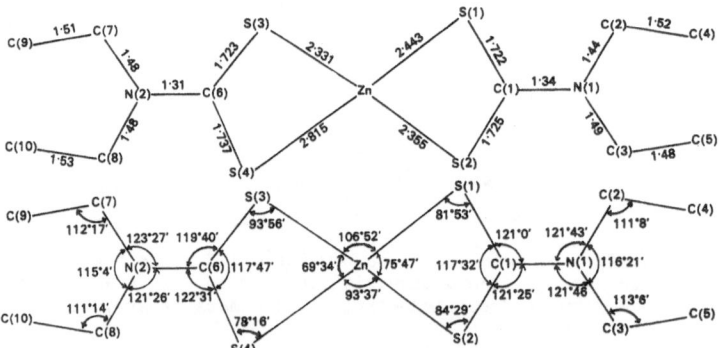

Fig. 234. Bond lengths and angles within the formula unit of zinc diethyldithiocarbamate. Standard deviations of distances are: Zn–S, 0.003; S–C,
0.010; C–N, 0.014; C–C, 0.020 Å.

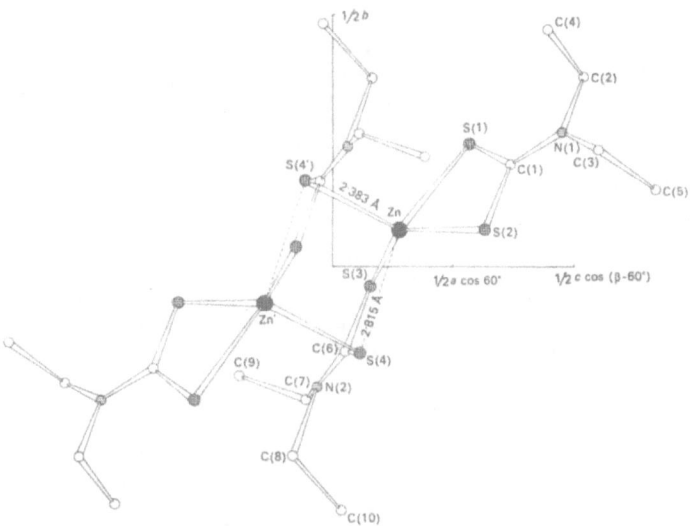

Fig. 235. A view of part of the molecule of zinc diethyldithiocarbamate, showing the short intermolecular distance.

DITHIOXOHEXACARBONYLDIIRON

$C_6Fe_2O_6S_2$ $[SFe(CO)_3]_2$ F.W. = 343.8

I. The molecular structure of a tricyclic complex, $[SFe(CO)_3]_2$. CHIN HSUAN WEI and L. F. DAHL, 1965. *Inorg. Chem.*, <u>4</u>, 1–11.

Triclinic, a = 6.63 ± 1, b = 7.85 ± 1, c = 11.46 ± 2 Å, α = 83°20 ± 10, β = 76°9 ± 10, γ = 78°25 ± 10', U = 566 Å^3, D_m = 1.92, Z = 2, D_x = 2.02.

Space group P1 (C_1^1) or P$\bar{1}$ (C_i^1). P$\bar{1}$ confirmed by analysis.

Atomic positions

	x	y	z
Fe(1)	0.3614 ± 3	0.4909 ± 2	0.2813 ± 2
Fe(2)	0.2113 ± 3	0.8117 ± 2	0.2380 ± 2
S(1)	0.0332 ± 5	0.6215 ± 4	0.3566 ± 3
S(2)	0.2622 ± 6	0.6919 ± 5	0.4172 ± 3
C(1)	0.3897 ± 26	0.2909 ± 18	0.3695 ± 14
C(2)	0.3371 ± 18	0.4106 ± 17	0.1496 ± 12
C(3)	0.6270 ± 19	0.5016 ± 16	0.2261 ± 14
C(4)	0.0472 ± 25	1.0192 ± 18	0.2749 ± 14
C(5)	0.1548 ± 20	0.7981 ± 16	0.0978 ± 12
C(6)	0.4523 ± 21	0.8950 ± 17	0.1770 ± 14
O(1)	0.4114 ± 20	0.1589 ± 16	0.4262 ± 12
O(2)	0.3146 ± 15	0.3478 ± 14	0.0667 ± 10
O(3)	0.7997 ± 17	0.5178 ± 14	0.1860 ± 10
O(4)	−0.0704 ± 21	1.1451 ± 18	0.2937 ± 12
O(5)	0.1215 ± 17	0.7823 ± 14	0.0059 ± 9
O(6)	0.6012 ± 20	0.9460 ± 15	0.1407 ± 12

Anisotropic temperature factors are given also.

Structure

The structure consists of discrete dimeric molecules of the configuration shown in Fig. 236. The molecule can be considered as resulting from the sharing of the sulphur atoms by two Fe(CO)$_3$S$_2$ molecules, each having the form of a distorted tetragonal pyramid. The dimer has non-crystallographic but fairly exact *mm*2 symmetry. The dihedral angle between FeS$_2$ groups is 59.7°. Bond angles are given in full. Intermolecular distances are consistent with van der Waals interaction.

Details of analysis

The intensity data were recorded on equi-inclination Weissenberg photographs, (0*kl* to 7*kl*) and precession photographs (*h0l* and *hk*0) using MoK$_\alpha$ radiation. 1536 independent reflexions were observed. Refinement was by full-matrix least squares to a final *R* index of 0.088. The thermal motion was analysed, and bond-length corrections were calculated, but not applied to the bond lengths reported here.

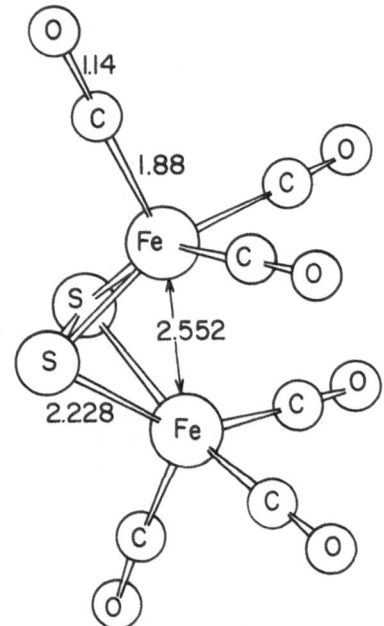

Fig. 236. Molecular structure of [SFe(CO)$_3$]$_2$. The distances shown are mean values for the bond type.

DITHIOXOHEXACARBONYLDIIRON, DITHIOXONONACARBONYLTRIIRON

(1:1 COMPLEX)

C$_{15}$Fe$_5$O$_{15}$S$_4$ S$_2$Fe$_3$(CO)$_9$S$_2$Fe$_2$(CO)$_6$ F.W. = 827.7

I. Crystal structure of a 1:1 mixture of two iron carbonyl sulphur complexes, S$_2$Fe$_3$(CO)$_9$ and S$_2$Fe$_2$(CO)$_6$. CHIN HSUAN WEI and L.F. DAHL, 1965. *Inorg. Chem.*, **4**, 493–499.

Orthorhombic, *a* = 13.23 ± 1, *b* = 11.08 ± 1, *c* = 17.95 ± 1 Å, *U* = 2631 Å^3, *D*$_m$ = 2.08, *Z* = 4, *D*$_x$ = 2.09.

Space group $Pna2_1$ (C_{2v}^9) or $Pnma$ (D_{2h}^{16}). $Pnma$ confirmed by analysis. Molecular symmetry, mirror plane.

Atomic positions

	x	y	z
Fe(1)	0.2176 ± 5	¼	0.0215 ± 4
Fe(2)	0.3248 ± 5	¼	0.1919 ± 4
Fe(3)	0.4059 ± 5	¼	0.0594 ± 4
S(1)	0.2891 ± 6	0.1208 ± 7	0.1007 ± 4
C(1)	0.0904 ± 43	¼	0.0594 ± 26
O(1)	0.0175 ± 25	¼	0.0925 ± 16
C(2)	0.1996 ± 24	0.1364 ± 36	−0.0427 ± 18
O(2)	0.1911 ± 16	0.0567 ± 20	−0.0845 ± 12
C(3)	0.4075 ± 28	¼	−0.0396 ± 21
O(3)	0.4136 ± 22	¼	−0.1039 ± 17
C(4)	0.4959 ± 23	0.1364 ± 28	0.0679 ± 15
O(4)	0.5545 ± 19	0.0568 ± 21	0.0803 ± 12
C(5)	0.2114 ± 41	¼	0.2384 ± 24
O(5)	0.1329 ± 24	¼	0.2726 ± 16
C(6)	0.3910 ± 25	0.1370 ± 32	0.2460 ± 16
O(6)	0.4272 ± 18	0.0571 ± 25	0.2752 ± 12
Fe(4)	0.2988 ± 5	¼	0.5640 ± 4
Fe(5)	0.2242 ± 6	¼	0.6947 ± 4
S(2)	0.1574 ± 6	0.1568 ± 7	0.5975 ± 4
C(7)	0.2942 ± 32	¼	0.4842 ± 31
O(7)	0.2842 ± 26	¼	0.4075 ± 18
C(8)	0.3898 ± 29	0.1434 ± 34	0.5696 ± 19
O(8)	0.4456 ± 16	0.0569 ± 19	0.5790 ± 11
C(9)	0.1265 ± 59	¼	0.7601 ± 37
O(9)	0.0613 ± 27	¼	0.8052 ± 20
C(10)	0.2986 ± 26	0.1389 ± 33	0.7346 ± 17
C(11)	0.3470 ± 17	0.0505 ± 22	0.7567 ± 13

Isotropic temperature factors are given also.

Structure

The structure consists of an ordered array of the two molecular species $S_2Fe_2(CO)_6$ and $S_2Fe_3(CO)_9$. The configuration of the first of these is essentially identical to that determined (much more accurately) in $\underline{1}$. The configuration of the second is given in Fig. 237. The crystallographic mirror plane contains all the iron atoms as well as C(1), O(1), C(3), O(3), C(5), and O(5). Details of the molecular packing are given. The intermolecular distances are consistent with van der Waals interaction.

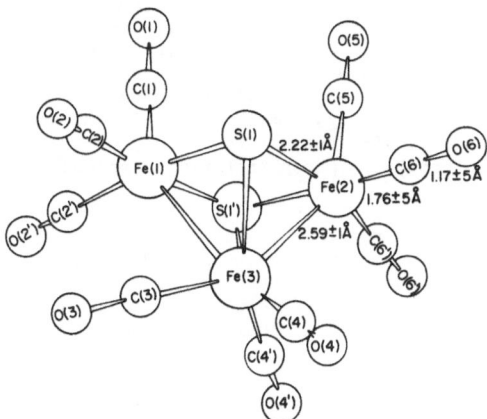

Fig. 237. Molecular configuration of $S_2Fe_3(CO)_9$. The distances shown are mean values for the bond type.

Details of analysis

The intensity data were recorded on equi-inclination Weissenberg photo-graphs ($h0l$ to $h10l$) and precession photographs ($h0l$ to $h2l$; $hk0$), using MoK_α radiation. 635 independent reflexions were observed and measured visually. Refinement was by full-matrix least squares to a final R index of 0.089.

1. This volume, p. 383.

CHLOROCARBONYL BIS(TRIPHENYL PHOSPHINE)IRIDIUM,

OXYGEN (1:1 COMPLEX)

$C_{37}H_{30}ClIrO_3P_2$ F.W. = 812.3

I. Structure of $IrO_2Cl(CO)(P(C_6H_5)_3)_2$, the oxygen adduct of a synthetic reversible oxygen carrier. S. J. LA PLACA and J. A. IBERS, 1965. *J. Amer. Chem. Soc.*, *87*, 2581-2586.
   (For a preliminary report, see 1.)

Triclinic, a = 19.02 ± 3, b = 9.83 ± 2, c = 9.93 ± 2 Å, α = 94.0 ± 1, β = 64.9 ± 1, γ = 93.2 ± 1°, U = 1676 Å , Z = 2, D_x = 1.61.

Space group P1 (C_1^1) or P$\bar{1}$ (C_i^1). P$\bar{1}$ confirmed by analysis.

Atomic positions

	x	y	z
Ir	0.2342 ± 1	0.2100 ± 2	0.0068 ± 2
P(1)	0.1335 ± 6	0.3207 ± 13	0.2136 ± 11
P(2)	0.3430 ± 7	0.1266 ± 15	−0.2024 ± 13
X(1)	0.2873 ± 9	0.1153 ± 18	0.1618 ± 16
X(2)	0.1467 ± 10	0.0293 ± 21	0.0019 ± 19
O(1).	0.224 ± 1	0.355 ± 3	−0.117 ± 2
O(2)	0.279 ± 2	0.397 ± 3	−0.073 ± 3

The carbonyl and chlorine positions are disordered, and indistinguishable; X(1) and X(2) denote the overlapped CO, Cl groups. Temperature factors (aniso-tropic for the iridium atom) are given also. Parameters for the phenyl rings (which were refined as rigid groups) are given, as well as the derived positions of the individual ring carbon atoms.

Structure

A general view of the complex is given in Fig. 238. (X(1) and X(2) are the positions of the disordered CO, Cl groups.) The oxygen atoms of the O_2 mole-cule, the two X groups, and the iridium atom are coplanar. The coordination polyhedron is completed by phosphorus atoms above and below this plane. (P(1)-Ir = 2.38 ± 1 Å, P(2)-Ir = 2.36 ± 1 Å, and P(1)-Ir-P(2) = 172.8 ± 5°.) Details of the basal plane of the coordination polyhedron are given in Fig. 239. The iridium atom can be described as six- or five-coordinated; in the latter case the O_2 molecule is counted as a single π-bonded ligand. The O–O distance (1.30 ± 3 Å) suggests that the oxygen molecule is in the state O_2^-; the significance of this result with respect to the oxygen-transport properties of the chlorocarbonyl bis(triphenyl phosphine)iridium molecule is discussed.

Details of analysis

The intensity data were recorded on equi-inclination Weissenberg photo-graphs ($h0l$ to $h6l$) using MoK_α radiation. 1128 of a possible 1880 reflexions in the sphere 2θ<18 were observed and measured visually. The observed intensities were corrected for absorption. Refinement was by full-matrix least squares, with the phenyl groups constrained to their known geometry, and the disordered CO, Cl positions approximated by a chlorine scattering-factor curve. The final R index was 0.071.

1. J.A. IBERS and S.J. LaPLACA, 1964. *Science*, *145*, 920.

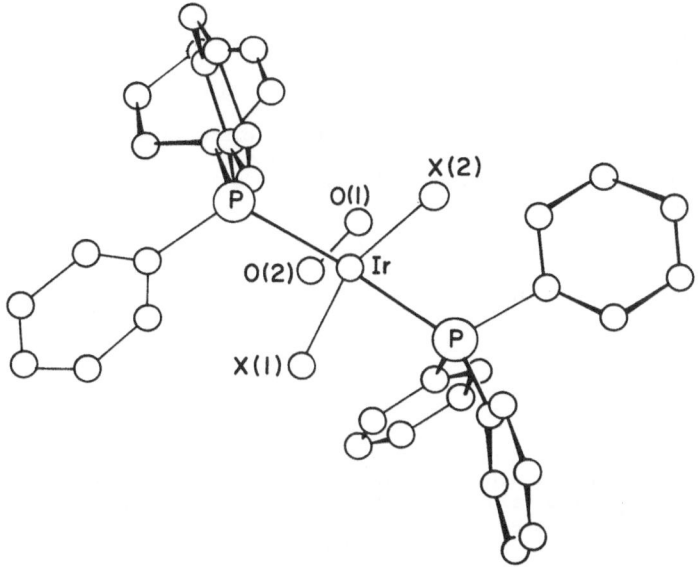

Fig. 238. A view of the molecular complex of $C_{37}H_{30}ClIrO_3P_2$.

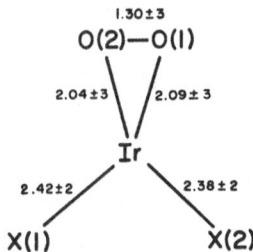

Fig. 239. The basal plane of the coordination polyhedron in $C_{37}H_{30}ClIrO_3P_2$.

BIS(DIPHENYLPHOSPHINO)ETHYLAMINE-MOLYBDENUM TETRACARBONYL

DICHLORO-BIS(DIPHENYLPHOSPHINO)ETHYLAMINE-PALLADIUM(II)

$C_{30}H_{25}NO_4P_2Mo$ $Mo(CO)_4[(C_6H_5)_2P.N(C_2H_5).P(C_6H_5)_2]$ F.W. = 621.4

$C_{26}H_{25}Cl_2NP_2Pd$ $Pd[(C_6H_5)_2P.N(C_2H_5).P(C_6H_5)_2]Cl_2$ F.W. = 590.8

I. X-ray studies of aminophosphine complexes of molybdenum and palladium, and of an aminophosphonium iodide. D. S. PAYNE, J. A. A. MOKUOLU and J. C. SPEAKMAN, 1965. *Chem. Communic.*, GB, 599.

MOLYBDENUM COMPOUND

Orthorhombic, a = 16.69, b = 17.21, c = 20.16 Å, [U = 5791 Å^3], z = 8, [D_x = 1.42].

Space group Pbcn (D_{2h}^{14})

Atomic positions are not given.

Structure

The coordination of the molybdenum atom is distorted octahedral, with the aminophosphine ligands occupying two adjacent sites. Mean bond lengths and angles are given; some values are: Mo–C, 1.990 ± 11 Å; Mo–P, 2.505 ± 5 Å; P–Mo–P, 64.8 ± 2°.

PALLADIUM COMPOUND

Orthorhombic, a = 20.90, b = 17.60, c = 13.87 Å, [U = 5102 Å³], Z = 8, [D_x = 1.54].

Space group Pbca (D_{2h}^{15})

Atomic positions are not given.

Structure

The coordination of the palladium atom is approximately square planar, with the chlorine (or phosphorus) atoms *cis* to each other. Mean bond lengths and angles are given; some values are: Pd–Cl, 2.367 ± 3 Å; Cl–Pd–Cl, 94.8 ± 2°; Pd–P, 2.224 ± 3 Å; P–Pd–P, 71.4 ± 2°.

Details of analysis

The intensity data were recorded photographically, using CuK_α radiation, and estimated visually. Refinement was by three-dimensional least squares. Final R indices were: molybdenum compound, 0.095 (1700 reflexions); palladium compound, 0.109 (2735 reflexions).

HYDRIDOCHLOROBIS(DIPHENYLETHYLPHOSPHINE)PLATINUM

$C_{28}H_{31}ClP_2Pt$ $PtHCl[P(C_6H_5)C_2H_5]_2$ F.W. = 660.0

I.   Structure of hydridochlorobis(diphenylethylphosphine)platinum. R. EISEN-
     BERG and J. A. IBERS, 1965. *Inorg. Chem.*, <u>4</u>, 773–778.

Monoclinic, a = 11.80 ± 2, b = 16.93 ± 3, c = 14.31 ± 2 Å, β = 108.4 ± 3°, U = 2713 Å³, D_m = 1.60, Z = 4, D_x = 1.616.

Space group P2₁/c (C_{2h}^5)

Atomic positions

Atomic positions are given for the 32 non-hydrogen atoms. Average standard deviations are: Pt, 0.001 Å; P and Cl, 0.008 Å; C, 0.02 Å. Temperature factors (anisotropic for Pt, Cl and P) are given also.

Structure

The molecular configuration is illustrated in Fig. 240. The coordination of the platinum atom is distorted square planar, with the phosphorus atoms occupying *trans* positions, and the chlorine atom occupying one of the two remaining positions. The fourth position is believed to be occupied by the hydrogen atom, which was, however, not located. Coordination distances and angles are: Pt–P, 2.268 ± 8 Å, Pt–Cl, 2.422 ± 9 Å; P–Pt–P, 188.8 ± 3°; P–Pt–Cl, 92.6 ± 4° and 94.5 ± 4°. Other selected distances and angles are given. The Pt, Cl, P portion of the molecule deviates moderately from planarity.

Details of analysis

The intensity data were recorded on equi-inclination Weissenberg photographs (h0l to h, 14,l) using MoK_α radiation. 1711 reflexions were observed in the range θ<20°, and were measured visually. Refinement was by full-matrix least squares, with the phenyl groups constrained to their known geometry, to a final R index of 0.076.

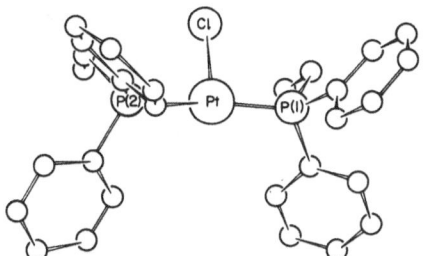

Fig. 240. A perspective drawing of the $C_{28}H_{31}ClP_2Pt$ molecule.

TRISTRIPHENYLPHOSPHINE RHODIUM CARBONYL HYDRIDE

$C_{55}H_{46}OP_3Rh$ $RhH(CO)(P(C_6H_5)_3)_3$ F.W. = 918.8

I. Crystal and molecular structure of tristriphenylphosphine rhodium car-
    bonyl hydride. S. J. LA PLACA and J. A. IBERS, 1965. *Acta Cryst.*, <u>18</u>,
    511–519.

    (For a preliminary account of this work see <u>1</u>.)

Monoclinic, a = 10.11 ± 5, b = 33.31 ± 15, c = 13.33 ± 7 Å, β = 90.0 ± 1°,
[U = 4489 Å³], D_m = 1.33, Z = 4, D_x = 1.36.

Space group $P2_1/n$ (C_{2h}^5)

Atomic positions

	x	y	z
Rh	0.22696 ± 17	0.11330 ± 56	0.22315 ± 14
P(1)	0.28161 ± 55	0.11137 ± 21	0.05283 ± 48
P(2)	0.18089 ± 58	0.17624 ± 18	0.28816 ± 50
P(3)	0.31340 ± 56	0.06639 ± 18	0.33328 ± 50
C	0.0640 ± 26	0.09039 ± 75	0.2064 ± 19
O	-0.0415 ± 18	0.07631 ± 52	0.1966 ± 13
H	0.377 ± 12	0.1290 ± 36	0.2168 ± 90

    Temperature factors (anisotropic for rhodium and phosphorus) are given
for the non-hydrogen atoms listed. The hydrogen atom specified above is coordinated to the rhodium atom. The positions of the 54 ring carbon atoms are given
also.

Structure

    A perspective drawing of the molecule is given in Fig. 241. The coordin-
ation of the rhodium atom is trigonal bipyramidal, with three phosphorus atoms
in the basal plane, the hydrogen atom at one apex and the carbonyl carbon at
the other. The rhodium is displaced 0.355 Å from the basal plane, towards the
carbonyl carbon. Distances and angles are given for the coordination polyhedron.
Some distances (averaged where appropriate) are: Rh-H, 1.60 ± 12; Rh-P, 2.322
± 8; Rh-C, 1.829 ± 28; Rh-O, 3.002 ± 20; P-P, 3.975 ± 12 Å. The mean P-C_6H_5
distance is 1.83 ± 1 Å.   There are no unusual intermolecular distances.

Details of analysis

    Intensity data were recorded on integrated equi-inclination Weissenberg
photographs (0kl to 7kl) using CuK_α radiation. Within the range θ<40°, 1480
of a possible 2650 reflexions were observed and measured visually. Refinement
was by full-matrix least squares, with the nine phenyl rings constrained to
their known geometry. The unique hydrogen atom was located from a difference map.
The final R index was 0.072.

1. S. J. LA PLACA and J. A. IBERS, 1963. *J. Amer. Chem. Soc.*, **85**, 3501.

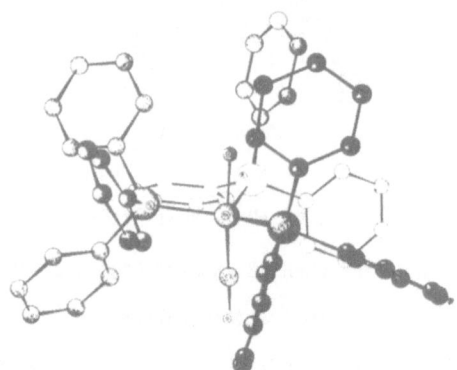

Fig. 241. A perspective drawing of one $C_{55}H_{46}OP_3Rh$ molecule. Ring hydrogens are not shown.

DICHLOROTRIS(TRIPHENYLPHOSPHINE)RUTHENIUM(II)

$C_{54}H_{45}Cl_2P_3Ru$ $RuCl_2[P(C_6H_5)_3]_3$ F.W. = 959.4

I. A five-coordinated d^6 complex: structure of dichlorotris(triphenyl-phosphine)ruthenium(II). S. J. LAPLACA and J. A. IBERS, 1965. *Inorg. Chem.*, **4**, 773–778.

Monoclinic, a = 18.01 ± 4, b = 20.22 ± 4, c = 12.36 ± 2 Å, β = 90.5 ± 3°, [U = 4501 Å³], D_m = 1.43, Z = 4, D_x = 1.415.

Space group $P2_1/c$ (C_{2h}^5)

Atomic positions

    Atomic positions are given for the 60 non-hydrogen atoms. Average standard deviations are: Ru, 0.0012 Å; P and Cl, 0.0036 Å; C, 0.011 Å. Temperature factors (anisotropic for Ru, P, and Cl) are given also.

Structure

    The molecular configuration is illustrated in Fig. 242. The coordination of the ruthenium atom is distorted square pyramidal; two chlorine atoms occupy *trans* basal positions, and three phosphorus atoms occupy the remaining positions. The ruthenium atom lies 0.46 Å from the basal plane, towards the apex. Coordination distances and angles are given in full. Some mean distances are: Ru-Cl, 2.388 ± 7 Å; Ru-P (basal), 2.393 ± 6 Å; Ru-P (apical), 2.230 ± 8 Å. It is suggested that the typical octahedral coordination of the ruthenium atom is prevented by the close approach of a phenyl hydrogen atom, which blocks the unused octahedral site.

Details of analysis

    The intensity data were recorded on equi-inclination Weissenberg photographs ($hk0$ to $hk8$) using CuK_α radiation. 1778 reflexions were measured within the range $\theta < 42°$, and were measured visually. Refinement was by full-matrix least squares, with the phenyl groups constrained to their known geometry, to a final R index of 0.063.

TRIS-(*o*-DIPHENYLARSINOPHENYL)-ARSINERUTHENIUM DIBROMIDE

$C_{54}H_{42}As_4Br_2Ru$ F.W. = 1251.5

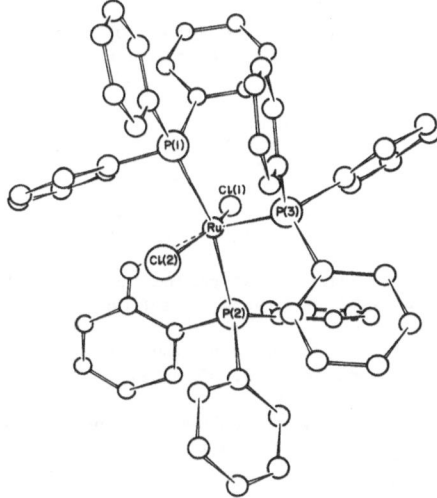

$Ru\,Br_2$

I.   The crystal structure·of tris-(o-diphenylarsinophenyl)-arsineruthenium
     dibromide. R. H. B. MAIS and H. M. POWELL, 1965. *J. Chem. Soc.*, 7471–
     7481.

Orthorhombic, a = 31.53 ± 8, b = 11.40 ± 10, c = 13.03 ± 2 Å, U = 4684.2 Å³,
D_m = 1.79, Z = 4, D_x = 1.76.

Space group *Pna*2_1 (C_{2v}^9) or *Pnam* (D_{2h}^{16}). *Pna*2_1 confirmed by analysis.

Atomic positions

     Atomic positions and isotropic temperature factors are given for the 61
non-hydrogen atoms.

Fig. 242. A perspective view of the $C_{54}H_{45}Cl_2P_3Ru$ molecule. The only hydrogen
           atom shown is that which blocks the unused octahedral site.

Structure

     Bond lengths and angles are given in full. The form of the molecule is
shown in Fig. 243. In spite of stereochemical difficulties the coordination of

the ruthenium atom does not depart greatly from octahedral. Ru-As distances range from 2.308 to 2.472 Å, and Ru-Br distances from 2.610 to 2.615 Å (all ± 0.005 Å). Bond angles subtended by ruthenium which would ideally be 90° do not depart from this value by more than 7°.

Details of analysis

Intensity data were recorded on equi-inclination Weissenberg photographs about b, using CuK_α radiation. 2500 reflexions were observed and estimated visually. Refinement was by least squares to a final R index of 0.087.

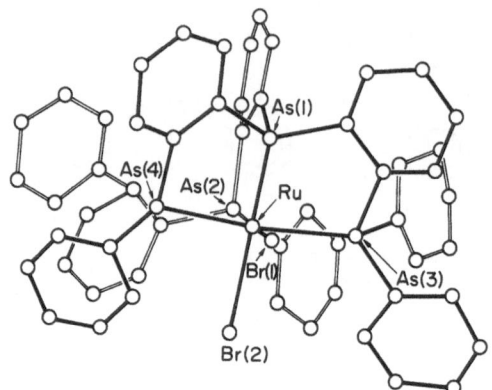

Fig. 243. The molecular configuration of $C_{54}H_{42}As_4Br_2Ru$.

MISCELLANEOUS

The following papers are deferred to later volumes because of the appearance of more complete accounts, or the reasonable expectation that such will appear.

[The crystal structure of tetraaquo-dimonoethanolamine-cuprosulfate. $Cu(OHCH_2CH_2NH_2)_2(H_2O)_4SO_4$.] G. K. ABDULLAEV and KH. S. MAMEDOV, 1965. *Ž. Strukt. Khim. SSSR*, 6, 171-172   [*J. Struct. Chem.*, 6, 159-160.]

Molecular structure of xylerythrin, a fungus pigment. S. ABRAHAMSSON and M. INNES, 1965. *Acta Chem. Scand.*, 19, 2246   [See *Acta Cryst.*, 21, (1966).]

The crystal structure of the etherate of sodium hydridodiethylberyllate. G. W. ADAMSON and H. M. M. SHEARER, 1965. *Chem. Communic.*, GB, 240.

The molecular structure of cadrela odorata substance B. S. A. ADEOYE and D. A. BEKOE, 1965. *Chem. Communic.*, GB, 301-302.

The molecular structure of dimethyl-π-cyclopentadienyl-methylcyclopentadiene-rhenium. N. W. ALCOCK, 1965. *Chem. Communic.*, GB, 177-178 [See *J. Chem. Soc. A*, 2001 (1967).]

Structure du *p*-aminobenzamide. M. ALLEAUME and J. DECAP, 1965. *C. R. Acad. Sci. Paris*, 260, 5790-5792.

Structure de la diaminodiphénylsulfone. *Idem.*, 1965. *Ibid.*, 261, 1693-1694.

Structure du sulfanilamide monohydrate. *Idem.*, 1965. *Ibid.*, 261, 4111-4114 [See *Acta Cryst.*, B24, 214 (1968).]

A new type of bis arene-metal complex. G. ALLEGRA, A. IMMIRZI and L. PORRI, 1965. *J. Amer. Chem. Soc.*, 87, 1394-1395.

The structure of 2,4,6-trimethoxy-1,3,5-trimethyl-2,4,6-trioxocyclotriphos-phazane, $N_3Me_3P_3O_3(OMe)_3$. G. B. ANSELL and G. J. BULLEN, 1965. *Chem. Communic.*, *GB*, 493-494 [See *J. Chem. Soc. A*, 3026 (1968).]

Crystal structure of L(+) cysteine hydrochloride monohydrate. R. R. AYYAR and R. SRINIVASAN, 1965. *Curr. Sci.*, 34, 449-450.

The stereochemistry of tricarbonylcycloheptatriene chromium derivatives. P. E. BAIKIE, O. S. MILLS, P. L. PAUSON, G. H. SMITH and J. VALENTINE, 1965. *Chem. Communic.*, *GB*, 425 [See *J. Chem. Soc. A*, 2704 (1968).]

Stereochemistry of 1:1 adducts of aluminium compounds: crystal structure of aluminium borohydride-trimethylamine. N. A. BAILEY, P. H. BIRD and M. G. H. WALLBRIDGE, 1965. *Chem. Communic.*, *GB*, 438-439 [See *Chem. Communic.*, *GB*, 286 (1966).]

Unusual coordination of the palladous ion in the structure of *trans*-di-iodobis-(dimethylphenylphosphine)palladium(II). N. A. BAILEY, J. M. JENKINS, R. MASON and B. L. SHAW, 1965. *Chem. Communic.*, *GB*, 237-238 [See *J. Chem. Soc. A*, 2594 (1968).]

Structure du bis-diéthyldithiocarbamate de phénylarsine. R. BALLY, 1965. *C. R. Acad. Sci. Paris*, 261, 3617-3618 [See *Acta Cryst.*, 23, 295 (1967).]

The structure of neothiobinupharidine. G. I. BIRNBAUM, 1965. *Tetrahedron Letters*, 4149-4152 [See *Acta Cryst.*, 23, 526 (1967).]

The molecular structure of dichloro-1,4,8,11-tetra-azacyclotetradecanenickel-(II). B. BOSNICH, R. MASON, P. J. PAULING, G. B. ROBERTSON and M. L. TOBE, 1965. *Chem Communic.*, *GB*, 97-98.

Zur Kristallstruktur des 4,4'-Dimethoxy-α,β-diäthylstilbens und des 4,4'-Dichlor-α,β-diathylstilbens. H. BOETTICHER, K. PLIETH and H. REPMANN, 1965. *Naturwissenschaften*, 52, 390.

Structure cristalline de la bromo-2-méthyl-3-naphtoquinone-1,4. M. BRETON, 1965. *C. R. Acad. Sci. Paris*, 260, 5275-5277.

The crystal structure and stereochemistry of laurencin. A. F. CAMERON, K. K. CHEUNG, G. FERGUSON and J. M. ROBERTSON, 1965. *Chem. Communic.*, *GB*, 638 [See *J. Chem. Soc. B*, 559,(1969).]

Structure cristalline de l'acide brassylique. G. CAMY, J. HOUSTY and M. HOSPITAL, 1965. *C. R. Acad. Sci. Paris*, 260,6383-6384 [See *Acta Cryst.*, B24, 486 (1968).]

[Crystal structure of 3-hydroxy-5-phenylisooxazole.] M. CANNAS and G. MOCCI, 1965. *Ric. Sci., Rend., Sez. A*, 8, 467-468.

Stereochemistry of copper in bis(N-t-butylsalicylaldiminato)copper(II). T.P. CHEESEMAN, D. HALL and D. N. WATERS, 1965. *Nature, Lond.*, 205, 494-495 [See *J. Chem. Soc. A*, 685 (1966).]

Structures of some mercury(II) complexes. K. K. CHEUNG, R. S. McEWEN and G. A. SIM, 1965. *Nature, Lond.*, 205, 383-384.

The molecular structure of acetylacetonato(cyclo-octa-2,4-dienyl)palladium. M. R. CHURCHILL, 1965. *Chem. Communic.*, *GB*, 625-626 [See *Inorg. Chem.*, 5, 1608 (1966).]

The structure and bonding of a cobaltcyclopentadiene complex. M. R. CHURCHILL, 1965. *J. Organometal. Chem. Netherl.*, *4*, 258-260.

Structure cristalline du pyrocatechol. J. CLASTRE and A. LAMARQUE, 1965. *C. R. Acad. Sci. Paris*, *260*, 2518-2520. [See C. J. BROWN, 1966. *Acta Cryst.*, *21*, 170.]

[Absolute configuration of olefinic distereoisomers of Pt(II).] P. CORRADINI and C. PEDONE, 1965. *Chim. e Industr.*, *47*, 664.

An authenticated perchlorate complex. F. A. COTTON and D. L. WEAVER, 1965. *J. Amer. Chem. Soc.*, *87*, 4189-4190.

New rhenium(V) oxyhalide complexes. F. A. COTTON and S. J. LIPPARD, 1965. *Chem. Communic.*, *GB*, 245-246. [See *Inorg. Chem.*, *5*, 416 (1966).]

[Preliminary results on the crystal structure of 1,4,5,8-tetrabromonaphthalene.] M. A. DAVYDOVA and JU. T. STRUČKOV, 1965. *Ž. Strukt. Khim. SSSR*, *6*, 922-923. [*J. Struct. Chem.*, *6*, 887.]

Zur Kristallstruktur von all-trans-cyclododecatrien-1,5,9-nickel(0), ($C_{12}H_{18}Ni$). H. DIETRICH and H. SCHMIDT, 1965. *Naturwissenschaften*, *52*, 301.

Five-coordinated structure of the cobalt(II) chloride complex with N-methylated diethylenetriamine. M. DI VAIRA and P. L. ORIOLI, 1965. *Chem. Communic.*, *GB*, 590.

The structure of dinitro-(NNN'N'-tetramethylethylenediamine)nickel(II). M. G. B. DREW and D. ROGERS, 1965. *Chem. Communic.*, *GB*, 476-477.

Étude physicochemique des phénols. III. Caracterisation propriétés physiques et étude structurale préliminaire du méthyl-3-isopropyl-2-phénol (thymol vicinal). P. DUCROS, R. PERRIN, A. THOZET and M. PERRIN, 1965. *Bull. Soc. Chim. Fr.*, 1631-1633.

Structure analysis of nickel(II)-1,8,8,13,13-pentamethyl-5-cyano-trans-corrin chloride. J. D. DUNITZ and E. F. MEYER, 1965. *Proc. Roy. Soc., Ser. A*, *288*, 324-330.

Die Kristallstruktur von 1,4-trans-Diaminocyclohexandihydrochlorid. J. D. DUNITZ and E. F. MEYER, 1965. *Helv. Chem. Acta*, *48*, 1450-1456. [See *Ibid*, *49*, 2502.]

Trigonal prismatic coordination. The molecular structure of tris(cis-1,2-diphenylethene-1,2-dithiolato)rhenium. R. EISENBERG and J. A. IBERS, 1965. *J. Amer. Chem. Soc.*, *87*, 3776-3778. [See *Inorg. Chem.*, *5*, 411 (1966).]

Die Struktur des Alkaloids C-Calebassin. M. FEHLMANN, H. KOYAMA and A. NIGGLI, 1965. *Helv. Chem. Acta*, *48*, 303-304.

C-H...O hydrogen bonding. G. FERGUSON and J. TYRRELL, 1965. *Chem. Communic.*, *GB*, 196-197.

The bacterial pigment from pseudomonas lemmonieri; structure of a degradation product. G. FERGUSON, D. R. POLLARD, J. M. ROBERTSON, D. M. HAWLEY, G. O. P. DOHERTY, N. B. HAYNES, D. W. MATHIESON, W. B. WHALLEY and T. H. SIMPSON, 1965. *Chem. Communic. GB*, 640-642.

Crystal-molecular structure and magnetic properties of $Cr_3(CH_3COO)_6OCl.5H_2O$. B. N. FIGGIS and G. B. ROBERTSON, 1965.. *Nature, Lond.*, *205* 694-695.

The structures of bishistidino-nickel(II), -cobalt(II) and -cadmium(II). K. A. FRASER, H. A. LONG, R. CANDLIN and M. M. HARDING, 1965. *Chem. Communic.*, *GB*, 344-345. [See *J. Chem. Soc. A*, 415 (1967).]

Model compounds for metal-protein interaction: the crystal structure of the copper(II) complex of β-alanyl-L-histidine (carnosine). H. C. FREEMAN and J. T. SZYMANSKI, 1965. *Chem. Communic.*, *GB*, 598-599. [See *Acta Cryst.*, 22, 406 (1967).]

Neutron diffraction study of deuterated oxalic acid dihydrate, $(COOD)_2.2D_2O$. F. FUKUSHIMA, H. IWASAKI, Y. SAITO, S. SATO and S. HOSHINO, 1965. *Bull. Chem. Soc. Japan*, 38, 151-152. [See *Acta Cryst.*, 23, 64, (1967).]

The crystal structure of N-(α-glutarimido)-4-bromophthalimide. S. FURBERG and C. S. PETERSEN, 1965. *Acta Chem. Scand.*, 19, 253-254.

A preliminary crystal structure analysis of the *trans*-dichloro-(*cis*-2-butene) [(S)-α-phenethylamine]platinum(II) complex. P. GANIS and C. PEDONE, 1965. *Ric. Sci.*, *Rend.*, *Sez. A*. 8, 1462-1468.

Structure atomique de la bromo-2-amino-3-naphtoquinone-1,4. J. GAULTIER and C. HAUW, 1965. *C. R. Acad. Sci. Paris*, 260, 2831-2832. [See *Acta Cryst.*, 20, 620 (1966).]

Structure cristalline de l'hydroxy-4 coumarine monohydrate. J. GAULTIER and C. HAUW, 1965. *C. R. Acad. Sci. Paris*, 260, 5787-5789. [*See Acta Cryst.*, 20, 646 (1966).]

Structure cristalline de l'α-naphtohydroquinone. J. GAULTIER and C. HAUW, 1965. *C. R. Acad. Sci. Paris*, 261, 3818-3819. [See *Acta Cryst.*, 23, 1016 (1967).]

Structure cristalline de la méthyl-2-α-naphtohydroquinone. J. GAULTIER and C. HAUW, 1965. *C. R. Acad. Sci. Paris*, 261, 4109-4110. [See *Acta Cryst.*, B25, 51 (1969).]

The conformation of a constrained eight-membered ring. K.-T. GO and I. C. PAUL, 1965. *Tetrahedron Letters*, 4265-4271. See *J. Chem. Soc. B*, 33 (1969).

[Structure of zinc and copper glutamate.] C. GRAMACCIOLI, 1965. *Chim. e Industr.*, *Ital.*, 47, 772. [See *Acta Cryst.*, 21, 594, 600 (1966).

Structures involving unshared electron pairs: coordination of antimony in racemic potassium antimony tartrate. D. GRDENIĆ and B. KAMENAR, 1965. *Acta Cryst.*, 19, 197-199. [See *Acta Cryst.*, B26, 181 (1970).]

Die Strukturen zweier 1:2-Chromcomplexe von *o,o'*-Dihydroxydiaryl-*trans*-azo-farbstoffmolekeln. R. GRIEB and A. NIGGLI, 1965. *Helv. Chim. Acta*, 48, 317-320.

Crystal structure of 3,6-spirodicyclooctylidene-1,2,4,5-tetra-oxa-cyclohexane ("dimeric cyclooctanone peroxide"). P. GROTH, 1965. *Acta Chem. Scand.*, 19, 1497. [See *Ibid.*, 21, 2695 (1967).]

Structure of the pyridine adduct of bis-(N-phenylsalicylaldiminato)copper(II). D. HALL, S. V. SHEAT and T. N. WATERS, 1965. *Chem. and Industr.*, 1428. [See *J. Chem. Soc. A*, 2721 (1969).]

The structure of a pentaoxyphosphorane by x-ray analysis. W. C. HAMILTON, S. J. LAPLACA and F. RAMIREZ, 1965. *J. Amer. Chem. Soc.*, 87, 127-128. [See *Ibid.*, 89, 2268, (1967).]

Beitrag zur Struktur der substituierten Triarylphosphine, -phosphinoxyde und -phosphinsulfide. C. HARTLEY, H. J. KUHN, K. PLIETH and P. ZAESKE, 1965. *Naturwissenschaften*, 52, 12-13.

The conformation of non-aromatic ring compounds. XXVI. The absolute configuration of 2α,3β-di-bromo-5α-cholestane. E. VAN HEIJKOOP, H. J. GEISE and C. ROMERS, 1965. *Rev. Trav. chim. Pays-Bas*, 84, 1626-1628. [See *Acta Cryst.*, 20, 257 (1966).]

On the structure-dependent behaviour of ethylenediaminetetraacetato complexes of the rare earth Ln^{3+} ions. J. L. HOARD, B. LEE and M. D. LIND, 1965. *J. Amer. Chem. Soc.*, 87, 1612-1613.

Stereochemistry of the eight-coordinate bisnitrilotriacetatozirconate(IV) ion. J. L. HOARD, E. WILLSTADTER and J. V. SILVERTON, 1965. *J. Amer. Chem. Soc.*, 87, 1610-1611. [See *ibid.*, 90, 2300 (1968).]

The structure of 3,5-diacetamido-1,2-dithiolium iodide. A. HORDVIK and H. M. KJOGE, 1965. *Acta Chem. Scand.*, 19, 523-524. [See *Ibid.*, 20, 1923 (1966).]

Structure of thiuret hydrochloride hemihydrate. A. HORDVIK and J. SUNDSFJORD, 1965. *Acta Chem. Scand.*, 19, 753-754. [See *Ibid.*, 20, 1907 (1966).]

Structure cristalline de l'adipamide. M. HOSPITAL and J. HOUSTY, 1965. *C. R. Acad. Sci. Paris*, 260, 5793-5794. [See *Acta Cryst.*, 20, 626 (1966).]

Structure crystalline du pimélamide. M. HOSPITAL and J. HOUSTY, 1965. *C. R. Acad. Sci. Paris*, 261, 3820-3821.

Structure cristalline du glutaramide. M. HOSPITAL and J. HOUSTY, *C. R. Acad. Sci. Paris*, 260, 5041-5042. [See *Acta Cryst.*, 21, 413 (1966).]

Determination by neutron and x-ray diffraction of the absolute configuration of an enzymatically formed α-monodeuterioglycolate. C. K. JOHNSON, E. J. GABE, M. R. TAYLOR and I. A. ROSE, 1965. *J. Amer. Chem. Soc.*, 87, 1802-1804.

[X-ray study of terferrocenyl.] Z. L. KALUSKI and JU. T. STRUČKOV, 1965. *Ž. Strukt. Khim. SSSR*, 6, 316-318. [*J. Struct. Chem.*, 6, 296-298.] [See *Bull. Acad. Polon., Sci., Ser. Sci. Chim.*, 14, 607 (1966).]

[X-ray study of bis-chloroferrocenyl.] Z. L. KALUSKI and JU. T. STRUČKOV, 1965. *Ž. Strukt. Khim. SSSR*, 6, 475-476. [*J. Struct. Chem.*, 6, 456-457.] [See *Bull. Acad. Polon. Sci., Ser. Sci. Chim.*, 14, 719 (1966).]

[Structure of bis-acetylferrocenyl.] Z. L. KALUSKI and JU. T. STRUČKOV, 1965. *Ž. Strukt. Khim. SSSR*, 6, 921-922. [*J. Struct. Chem.*, 6, 885-886.]

The molecular structure of Eschenmoser's "pseudocorrin". B. KAMENAR, B. F. HOSKINS and J. D. WRIGHT, 1965. *Proc. Roy. Soc., Ser. A*, 288, 331-333.

The molecular structure of the crystalline complex ethyladenine:methylbromouracil. L. KATZ, K. TOMITA and A. RICH, 1965. *J. Molec. Biol.*, 13, 340-350. [See *Acta Cryst.*, 21, 754 (1966).]

The complete structure of the triterpene arborinol. O. KENNARD, L. RIVA DI SANSEVERINO, H. VORBRUGGEN and C. DJERASSI, 1965. *Tetrahedron Letters*, 3433-3438. [See *Tetrahedron*, 23, 131 (1967).]

The molecular structures of bis-(o-dimethylarsinophenyl)methylarsinecopper-manganese pentacarbonyl, triphenylgermanium-manganese pentacarbonyl, and triphenylphosphinegold-cobalt tetracarbonyl. B. T. KILBOURN, T. L. BLUNDELL and H. M. POWELL, 1965. *Chem. Communic.*, GB, 444-445.

Molekül, Koordination und Bindungscharakter im kristallisierten HgCl₂-Kollidin. S. KULPE, 1965. *Z. Chem.*, 5, 306-308. [See *Z. anorg. Chem.*, 349, 314-323 (1967).]

Étude de la structure cristalline du chlorhydrate d'acétyl-2-pyridine. A. LAURENT, 1965. *C. R. Acad. Sci. Paris*, 261, 446-447. [See *Acta Cryst.*, 21, 710 (1966).]

The molecular structure of molybdenum(II) acetate. D. LAWTON and R. MASON, 1965. *J. Amer. Chem. Soc.*, 87, 921-922.

Structure and bonding in a ten-coordinate lanthanum(III) chelate of ethylenediaminetetraacetic acid. M. D. LIND, B. LEE and J. L. HOARD, 1965. *J. Amer. Chem. Soc.*, 87, 1611-1612.

Crystal structure of benzo-1,2:4,5-dicyclobutene. S. G. G. MACDONALD, J. LAWRENCE and M. P. CAVA, 1965. *Chem. and Industry*, 86.

The stereochemistry of acorone. C. E. McEACHAN, A. T. McPHAIL and G. A. SIM, 1965. *Chem. Communic.*, GB, 276-277. [See *J. Chem. Soc. A*, 579 (1966).]

Hydroxyl-benzene hydrogen bonding: an x-ray study. A. T. McPHAIL and G. A. SIM, 1965. *Chem. Communic.*, GB, 124-126.

The constitution and absolute steroechemistry of verrucarin A. A. T. McPHAIL and G. A. SIM, 1965. *Chem. Communic.*, GB, 350-352. [See *J. Chem. Soc.*, A, 1394 (1966).]

The molecular structure of bicyclo[3.2.1]-octadienyliron tricarbonyl tetrafluoroborate. T. N. MARGULIS, L. SCHIFF and M. ROSENBLUM, 1965. *J. Amer. Chem. Soc.*, 87, 3269-3270.

Crystal structure of 1,1-di-*p*-tolyl-2-bromoethylene. C. MARIANI, A. MUGNOLI and G. L. CASALONE, 1965. *Atti Accad. Maz. Nazl. Lincei, Rend.*, Cl. Sci. Fis. Mat. Nat., 38, 880-885. [See *Acta Cryst.*, 22, 228 (1967).]

The crystal structure of tetramethylthiuram disulphide. K. MAROY, 1965. *Acta Chem. Scand.*, 19, 1509.

The crystal structure of ethyl stearate. A. McL. MATHIESON and H. K. WELSH, 1965. *Acta Cryst.*, 18, 953-955. [See S. ALEBY, 1968. *Acta Chem. Scand.*, 22, 811.]

Structures of leurocristine (vincristine) and vincaleukoblastine. X-ray analysis of leurocristine methiodide. J. W. MONCRIEF and W. N. LIPSCOMB, 1965. *J. Amer. Chem. Soc.*, 87, 4963-4964. [see *Acta Cryst.*, 21, 322-331 (1966).]

The structure of anemonin. R. M. MORIARTY, C. R. ROMAIN, I. L. KARLE and J. KARLE, 1965. *J. Amer. Chem. Soc.*, 87, 3251-3252. [See *Acta Cryst.*, 20, 555 (1966).]

Configuration of thiosemicarbazide molecule in monochloromonothiosemicarbazide-silver. M. NARDELLI, G. F. GASPARRI, G. G. BATTISTINI and A. MUSATTI, 1965. *Chem. Communic.*, GB, 187-188.

The x-ray investigation of eupteleogenin iodoacetate. M. NISHIKAWA, K. KAMIYA, T. MURATA, Y. TOMIIE and I. NITTA, 1965. *Tetrahedron Letters*, 3223-3225.

Structure cristalline de l'iodométhylate de phyllochrysine. C. PASCARD-BILLY, 1965. *C. R. Acad. Sci. Paris*, 260, 555-556. [See *Bull. Soc. Chim. Fr.* 369 (1966).]

The crystal structure of the monorubidium salt of furantetracarboxylic acid.
I. C. PAUL, 1965. *Chem. Communic., GB*, 461-464. [See *Acta Cryst.*, 22, 559
(1967).]

Coordination of transition metals by the perchlorate ion: the crystal structure
of Co(PhMeAsO)$_4$(ClO$_4$)$_2$. P. PAULING, G. B. ROBERTSON and G. A. RODLEY, 1965.
Nature, Lond., 207, 73-74.

The stable conformation of N-methylacetanilide. B. F. PEDERSEN and B. PEDERSEN,
1965. *Tetrahedron Letters*, 2995-3001. [See *Acta Chem. Scand.*, 21, 1415
(1967).]

Die absolute Konfiguration des rechtsdrehenden Methyl-*n*-propyl-phenyl-benzyl-
phosphoniumbromids. A. F. PEERDEMAN, J. P. C. HOLST, L. HORNDER and H. WINKLER,
1965. *Tetrahedron Letters*, 811-845.

Bestimmung der Struktur von Diphenantridylmethan mit der Methode der diffusen
Streuung. H. POPPE and W. HOPPE, 1965. *Z. Kristallogr.*, 122, 298-306.

Formation of dichloro(2,7-dimethylocta-2,6-diene-1,8-diyl)ruthenium(IV) from
RuCl$_3$ and isoprene. L. PORRI, M. C. GALLAZZI, A. COLOMBO and G. ALLEGRA, 1965.
Tetrahedron Letters, 4187-4189.

Dibutadienerhodium(I) chloride. L. PORRI, A. LIONETTI, G. ALLEGRA and A.
IMMIRZI, 1965. *Chem. Communic., GB*, 336-337.

X-ray analysis of the crystal structure of L-valine hydrochloride monohydrate.
S. T. RAO and R. PARTHASARATHY, 1965. *Curr. Sci.*, 34, 628-629. See *Z. Kristallogr.*, 128, 339 (1969).

Stereochemistry of a Friedel-Crafts intermediate: crystal structure of benzoyl
chloride-aluminium chloride. S. E. RASMUSSEN and N. C. BROCH, 1965. *Chem.
Communic., GB*, 289-290. [See *Acta Chem. Scand.*, 20, 1351 (1966).]

Étude radiocristallographique de la dichloro-2,3-*p*-bénzoquinone. B. REES,
R. HASER and R. WEISS, 1965. *C. R. Acad. Sci. Paris*, 261, 450-451. [See
Bull. Soc. Chim. Fr., 2666 (1966).]

Structure du N-(bromo-1-diméthyl-3,5-phényl)-benzènesulfonamide. B. RÉRAT,
G. DAUPHIN, A. KERGOMARD and C. RÉRAT, 1965. *C. R. Acad. Sci. Paris*, 261,
139-141. [See *Acta Cryst.*, B25, 1392 (1969).]

Sur la structure du violurate de baryum. C. RICHE, 1965. *C. R. Acad. Sci.
Paris*, 260, 2516-2517.

The conformation of non-aromatic ring compounds. XV. The crystal structure
of 4-bromo-9β,10α-pregn-4,6-dione-3,20-dione, a preliminary communication.
C. ROMERS, E. VAN HEIJKOOP, B. HESPER and H. J. V. H. GEISE, 1965. *Rec. trav.
Chim., Pays-Bas*, 84, 885-888. [See *Acta Cryst.*, 20, 363 (1966).]

Crystal structure of thiodiglycollic acid. S. ROY, 1965. *Indian J. Phys.*
39, 49-50.

The structure of five-coordinated high-spin complexes of nickel(II) and cobalt
(II) with N-β-diethylamineethyl-5-chlorosalicylaldimine. L. SACCONI, P. L.
ORIOLI and M. DI VAIRA, 1965. *J. Amer. Chem. Soc.*, 87, 2059. [See *Ibid.*,
88, 4383 (1966).]

X-ray structure determination of bromonoranisatinone, a derivative of anisatin.
N. SAKABE, Y. HIRATA, A. FURUSAKI, Y. TOMIIE and I. NITTA, 1965. *Tetrahedron
Letters*, 4795-4796.

Preliminary crystal analysis of p-nitrobenzoic acid. T. D. SAKORE and L. M. PANT, 1965. *Indian J. Pure Appl. Phys.*, 3, 143-147. [See *Acta Cryst.*, 21, 715 (1966).]

Structure of the complex $CaBr_2.10H_2O(CH_2)_6N_4$. P. DE SANTIS, A. L. KOVACS, A. M. LIQUORI and L. MAZZARELLA, 1965. *J. Amer. Chem. Soc.*, 87, 4965-4966. [See *Acta Cryst.*, 22, 65-74 (1967).]

The crystal structure of 8,8-dicyanoheptafulvene. H. SHIMANOUCHI, T. ASHIDA, Y. SASADA, M. KAKUDO, I. MURATA and Y. KITAHARA, 1965. *Bull. Chem. Soc. Japan*, 38, 1230. [See *Ibid.*, 39, 2322 (1966).]

Structure cristalline de l'acide undécandeioique (α). A. SINTES, J. HOUSTY and M. HOSPITAL, 1965. *C. R. Acad. Sci. Paris*, 260, 6105-6106. [See *Acta Cryst.*, 21, 965 (1966).]

The crystal and molecular structure of $MoS_6C_6H_6$. A. E. SMITH, G. N. SCHRAUZER, V. P. MAYWEG and W. HEINRICH, 1965. *J. Amer. Chem. Soc.*, 87, 5798-5799.

[The structure of barene.] V. I. STANKO and JU. T. STRUČKOV, 1965. *Ž. Obšč. Khim.*, SSSR, 35, 930-931 [*J. Gen. Chem.*, USSR, 35, 935].

[Two modifications of diacidodiethylenediaminonickel, $Nien_2NO_2NCS$.] A. E. SVELASVILI, M. A. PORAJ-KOŠIC and A. S. ANCYSKINA, 1965. *Ž. Strukt. Khim. SSSR*, 6, 168-170. [*J. Struct. Chem.*, 6, 155-156.]

[The *cis*-octahedral structure of the diacidobis(ethylenediamine)nickel compounds $Nien_2NCSCl$ and $Nien_2NCSBr$.] A. E. SVELASVILI, M. A. PORAJ-KOŠIC and A. S. ANCYSKINA, 1965. *Ž. Strukt. Khim. SSSR*, 6, 170-171. [*J. Struct. Chem.*, 6, 157-158.]

The crystal structure of hexachlorocyclopropane C_3Cl_6. T. TAKANO, T. CHIBA, Y. SASADA, M. KAKUDO, S. NOZAKURA and S. MURAHAHI, 1965. *Bull. Chem. Soc. Japan*, 38, 157-158.

The crystal and molecular structures of 7-chloro and 7-bromo-4-hydroxytetra-cycloxide. J. H. VAN DEN HENDE, 1965. *J. Amer. Chem. Soc.*, 87, 929-931.

The molecular structure of a Dies-Alder adduct of an azepine. J. H. VAN DEN HENDE and A. S. KENDE, 1965. *Chem. Communic.*, GB, 384-385.

$Na_3Nd(OCOCH_2OCH_2OCO)_3$ - a case of 9-coordination. N.-G. VANNERBERG and J. ALBERTSSON, 1965. *Acta Chem. Scand.*, 19, 1760-1761.

The crystal structure of magnesium hexaantipyrine perchlorate. M. VIJAYAN and M. A. VISWAMITRA, 1965. *Indian J. Pure Appl. Phys.*, 3, 357-358. [See *Acta Cryst.*, 23, 1000 (1967).]

Structure of a cyclic hexameric phosphonitrilic molecule. A. J. WAGNER and A. VOS, 1965. *Rec. trav. chim. Pays-Bas*, 84, 603-605. [See *Acta Cryst.*, B24, 1423 (1968).]

Crystal structure of monomethyltriethyl titanate. R. D. WITTERS and C. N. CAUGHLAN, 1965. *Nature, Lond.*, 205, 1312-1313.

A new type of cyclic transition metal complex $[Ni(SC_2H_5)_2]_6$. P. WOODWARD, L. F. DAHL, E. W. ABEL and B. C. CROSSE, 1965. *J. Amer. Chem. Soc.*, 87, 5251-5253.

Structure of 9,9,10,10-tetrachloroanthracene. N. F. YANNONI, A. P. KRUKONIS and J. SILVERMAN, 1965. *Science*, 148, 231-232. [See *Acta Cryst.*, 21, 390 (1966).]

RESULTS OBTAINED BY GAS ELECTRON DIFFRACTION (ROTATING SECTOR).

Substance	Molecular symmetry	Bond lengths and angles in Å and degrees	Remarks	Reference
$B(CH_3)_3$ Trimethylborine	$6m2$ (D_{3h})	B–C = 1.5783 ± 11 Å C–H = 1.1138 ± 15 C–B–C = 119.4 ± 0.3° B–C–H = 111.9 ± 0.2	Least-squares fitting of radial distribution curves, and of total intensity.	3
CCl_4 Carbon tetrachloride	$\bar{4}3m$ (T_d)	C–Cl = 1.769 ± 2	Least-squares fitting of radial distribution curves.	13
CH_3NH_2 Methylamine		C–N = 1.4667 ± 21	Least-squares fitting of radial distribution curves, and of total intensity.	14
CD_3ND_2 Deuteromethylamine		C–N = 1.4679 ± 21	As above.	14
CF_3NO Trifluoronitrosomethane	m (C_s) Eclipsed conformation. (N–O cis to one C–F bond)	C–F = 1.321 ± 4 C–N = 1.555 ± 15 N–O = 1.171 ± 8 F–C–F = 111.9 ± 0.4 C–N–O = 121.0 ± 1.6	Comparison of theoretical and observed radial distribution and other molecular scattering curves.	10
CH_3PF_4 Tetrafluoromethylphosphorane	trigonal bipyramid, with one CH_3 group on an equatorial site.	$P-F^{mean}$ = 1.577 ± 1 $P-F^{eq.}$ = 1.5430 ± 37 $P-F^{eq.}$ = 1.6120 ± 35 $P-C^{ax.}$ = 1.7800 ± 46 C–H = 1.099 ± 31 $F^{ax.}-P-C$ = 91.9 ± 0.4 $F^{eq.}-P-C$ = 122.2 ± 0.9	Least squares fitting of radial distribution curves, and of total intensity.	6
$(CH_3)_2PF_3$ Trifluorodimethylphosphorane	trigonal bipyramid, with CH_3 groups on equatorial sites.	$P-F^{mean}$ = 1.6140 ± 13 $P-F^{eq.}$ = 1.5530 ± 58 $P-F^{eq.}$ = 1.6430 ± 29 $P-C^{ax.}$ = 1.7980 ± 41	As above	6

Compound	Symmetry / Structure	Parameters	Method	Ref
		$C-H = 1.107 \pm 12$ $F ax.-p-F = 89.9 \pm 0.3$ $F ax.-p-C eq. = 118.0 \pm 0.8$		
$(CH_3)_3PO$ Trimethylphosphine oxide	$3m$ (C_{3v}) H atom in each methyl group is trans to O.	$P-O = 1.48$ $C-H = 1.10$ $P-C = 1.813$ $C-P-O = 112.3$ $C-P-C = 106.0$	Empirical fitting of radial distribution curves.	21
$P(OC_2H_5)_3$ Triethyl phosphite	$3m$ (C_{3v}) pyramidal	$P-O = 1.58 \pm 2$ $C-H = 1.10 \pm 3$ $P-O-C = 113 \pm 3$ $O-P-O = 100 \pm 4$	Comparison of theoretical and observed radial distribution curves.	17
$P(OC_2H_3)_3$ Trivinyl phosphite	$3m$ (C_{3v}) pyramidal	$P-O = 1.60 \pm 1$ $C-O = 1.33 \pm 2$ $P-O-C = 118 \pm 3$ $O-P-O = 104 \pm 4$	As above.	17
C_2H_6 Ethane	$3m$ (C_{3v})	$C-C = 1.5340 \pm 11$ $C-H = 1.1122 \pm 12$ $C-C-H = 111.0 \pm 0.2$	Least-squares fitting of radial distribution curves, and of total intensity.	8
C_2D_6 Deuteroethane	As above.	$C-C = 1.5323 \pm 11$ $C-D = 1.1071 \pm 12$ $C-C-D = 111.0 \pm 0.2$	As above.	8
C_2H_L	mmm (D_{2h})	$C-C = 1.3369 \pm 16$ $C-H = 1.1030 \pm 18$ $C-C-H = 121.4 \pm 0.6$	As above.	9

Substance	Molecular symmetry	Bond lengths and angles	Remarks	Reference
C_2D_4 Deuteroethylene	As above.	$C-C = 1.338 \pm 3$ $C-D = 1.099 \pm 3$ $C-C-D = 121.4 \pm 0.8$	As above.	9
H_2CCCl_2 cis-dichloroethylene	mm2 (C_{2v})	$C-C = 1.354 \pm 5$ $C-Cl = 1.718 \pm 7$ $C-H = 1.075 \pm 15$ $C-C-Cl = 123.8 \pm 0.5$ $C-C-H = 132 \pm 3$	Comparison of theoretical and observed radial distribution and other molecular scattering curves.	11
H_3COCH_2Cl Monochloro-dimethyl ether	1 (C_1) Chlorine atom gauche to O-CH$_3$ bond	$H_3ClC-O = 1.368$ $H_3C-O = 1.414$ $C-Cl = 1.813$ $C-H = 1.120$ $C-O-C = 113.2$ $O-C-Cl = 112.3$	Fitting of radial distribution curves; refinement by steepest ascents.	15
$(CF_3)_2NN(CF_3)_2$ Tetrakis(trifluoromethyl)hydrazine	2 (C_2)	$N-N = 1.402 \pm 20$ $C-N = 1.433 \pm 10$ $C-F = 1.325 \pm 5$ $C-N-C = 121.2 \pm 1.5$ $N-N-C = 119.0 \pm 1.5$	Least-squares fitting of radial distribution curves, and of total intensity.	7
C_3H_7CHO Isopropyl carboxaldehyde	None; predominantly gauche	$C-C^{av}, = 1.528 \pm 3$ $C-O = 1.206 \pm 12$ $C-H = 1.127 \pm 35$ $C-C-O = 123.3 \pm 1.9$ $C-C-H = 110.8 \pm 3.2$	As above	12
C_3H_5CHO Cyclopropyl carboxaldehyde	m (C_2) s-trans and s-cis isomers present in nearly-equal concentrations	$C-C = 1.507 \pm 2$ $C-O = 1.216 \pm 2$ $C-H = 1.115 \pm 37$	As above	4

Compound	Structure	Parameters	Method	Ref.
$C_3H_5COCH_3$ Cyclopropyl methyl ketone	m (C_s) Predominant isomer has C–O group *cis* to ring.	C–C = 1.510 ± 3 C–O = 1.225 ± 2 C–H = 1.126 ± 50 C–C–O = 121.8 ± 2.0 C–C–H = 117.2 ± 3.0	As above.	5
C_3H_5COCl Cyclopropane-carboxylic acid chloride	m (C_s) As above.	C–C = 1.506 ± 2 C–O = 1.197 ± 25 C–Cl = 1.797 ± 9 C–H = 1.105 ± 60 C–C–O = 127.6 ± 3.0 C–C–Cl = 111.0 ± 2.5 O–C–Cl = 122.4 ± 2.0 C–C–H = 120.7 ± 4.0	As above.	5
$HC(C_6H_5)_3$ Triphenyl-methane	3 (C_3) Pyramidal; rings inclined at 45° to plane of 3-fold axis and central C–C bond.	$C–C_{arom.}$ = 1.40_3 $C–C_{sing.}$ = 1.53 $C–H_{sing.}$ = 1.08_4 H_5C_6–C–C_6H_5 = 112	Empirical fitting of radial distribution curves.	1
$C(C_6H_5)_3^-$ Triphenylmethyl (free radical).	As above; inclination of rings 40–45°.	$C–C_{arom.}$ = 1.395 $C–C_{sing.}$ = 1.48 C–H = 1.11 H_5C_6–C–C_6H_5 = 117 ± 1	As above.	2
$C_6H_5N(CH_3)_2$ N-Dimethylaniline	m (C_s) Both methyl groups on same side of ring.	C–C = 1.40 $N–CH_3$ = 1.46 $C–N–C_{av.}$ = 116	As above.	20

Substance	Molecular symmetry	Bond lengths and angles	Remarks	Reference
$C_6H_5SiH_2Cl$ Phenylmono-chlorosilane	mm2 (C_{2v}) (Free rotation about Si–C)	C–C = 1.40 ± 1 Si–Cl = 2.076 ± 10 Si–C = 1.81 ± 3 Cl–Si–C = 110 ± 3	As above	18
$C_6H_5C_3H_5$ Phenylcyclopropane	m (C_S) 6-ring plane bisects 3-ring. (predominant isomer)	C–C (3-ring) = 1.50 ± 2 C–H (" ") = 1.08 ± 2 The angle between the bridging bond and the 3-ring is 128.5 ± 0.3°.	As above.	19
$C_{20}H_{12}$ Perylene	mmm (D_{2h})	Bridging, or *peri* bond length = 1.493 ± 15	As above.	16

ORGANIC COMPOUNDS   **405**

REFERENCES FOR PRECEEDING TABLE

1. An electron diffraction investigation of triphenylmethane in the gas phase.
   P. ANDERSEN, 1965. *Acta Chem. Scand.*, 19, 622-628.

2. An electron diffraction investigation of the free radical triphenylmethyl
   in the gas phase. P. ANDERSEN, 1965. *Acta Chem. Scand.*, 19, 629-637.

3. Electron diffraction study of the structure of $B(CH_3)_3$. L. S. BARTELL and
   B. L. CARROLL, 1965. *J. Chem. Phys.*, 42, 3076-3078.

4. Electron diffraction study of the structure and internal rotation of cyclopropyl carboxaldehyde. L. S. BARTELL and J. P. GUILLORY, 1965. *J. Chem.
   Phys.*, 43, 647-653.

5. Electron diffraction study of the structure and conformational behavior of
   cyclopropyl methyl ketone and cyclopropane carboxylic acid chloride. L. S.
   BARTELL, J. P. GUILLORY and A. T. PARKS, 1965. *J. Phys. Chem.*, 69, 3043-
   3048.

6. Structure and bonding in CH_3PF_4 and $(CH_3)_2PF_3$. An electron diffraction
   study. L. S. BARTELL and K. W. HANSON, 1965. *Inorg. Chem.*, 4, 1777-1782.

7. Molecular structure and bonding of $N_2(CF_3)_4$. An electron diffraction study.
   L. S. BARTELL and H. K. HIGGINBOTHAM, 1965. *Inorg. Chem.*, 4, 1346-1350.

8. Electron diffraction study of the structure of ethane and deuteroethane.
   L. S. BARTELL and H. K. HIGGINBOTHAM, 1965. *J. Chem. Phys.*, 42, 851-856.

9. Electron diffraction study of the structure of C_2H_4 and C_2D_4. L. S. BARTELL,
   E. A. ROTH, C. D. HOLLOWELL, K. KUCHITSU and J. E. YOUNG, 1965. *J. Chem.
   Phys.*, 42, 2683-2686.

10. An electron diffraction study of trifluoronitrosomethane. M. I. DAVIS,
    J. E. BOGGS, D. COFFEY and H. P. HANSON, 1965. *J. Phys. Chem.*, 69, 3727-
    3730.

11. A gas phase electron diffraction study of *cis*-dichloroethylene. M. I. DAVIS
    and H. P. HANSON, 1965. *J. Phys. Chem.*, 69, 4091-4097.

12. Electron diffraction study of the structure and internal rotation of iso-
    propyl carboxaldehyde. J. P. GUILLORY and L. S. BARTELL, 1965. *J. Chem.
    Phys.*, 43, 654-657.

13. Neubestimmung der Struktur von Tetrachlorokohlenstoff durch Elektronen-
    beugung an Gasen bei zwei verschiedenen Wellenlangen. J. HAASE and W. ZEIL,
    1965. *Zeit. Phys. Chem. Frankf.*, 45, 202-208.

14. Electron diffraction study of CH_3NH_2 and CD_3ND_2. H. K. HIGGINBOTHAM and
    L. S. BARTELL, 1965. *J. Chem. Phys.*, 42, 1131-1132.

15. Electron diffraction by gases. The molecular structure of monochlorodimethyl
    ether. M. C. PLANJE, L. H. TONEMAN and G. DALLINGA, 1965. *Rec. trav. chim.
    Pays-Bas*, 84, 232-240.

16. An electron diffraction investigation of the *peri*-bonds in perylene. M.
    TRAETTEBERG, 1965. *Proc. Roy. Soc., Ser. A*, 283, 557-575.

17. [An electron diffraction study of the molecules of triethyl and trivinyl
    phosphites in the vapour phase.] L. V. VILKOV, P. A. AKIŠIN and G. E.
    SALOVA, 1965. *Ž. Strukt. Khim.*, 6,,355-360. [*J. Struct. Chem.*, 6, 339-343.]

18. [An electron diffraction study of the structure of the molecule of phenyl-
     monochlorosilane.]   L. V. VILKOV and V. S. MASTRJUKOV, 1965. *Dokl. Akad.*
     Nauk SSSR, <u>162</u>, 1306-1309.

19. [Investigation of the structure of phenylcyclopropane molecules by electron
     diffraction.] L. V. VILKOV and N. I. SADOVA, 1965. *Dokl. Akad. Nauk SSSR,*
     <u>162</u>, 565-568.

20. [Investigation of the structure of molecules of trivalent nitrogen compounds
     by electron diffraction.]   N-Dimethylaniline. L. V. VILKOV and T. P.
     TIMASEVA, 1965. *Dokl. Akad. Nauk SSSr,* <u>161</u>, 351-354.

21. The molecular structure of trimethylphosphine oxide. H. K. WANG, 1965.
     Acta Chem. Scand., <u>19</u>, 879-882.

TABLE OF CRYSTAL DATA

Compounds are listed in order of carbon and hydrogen content, but the arrangement is modified where necessary to permit the grouping of related molecules. It is to be noted that in many cases the tabulated space group has not been uniquely determined, and is only one of two or more possibilities.

Compound	Formula	Class	a(Å) / α(Deg.)	b(Å) / β(Deg.)	c(Å) / γ(Deg.)	D_m	Z	D_x	S.G.	Ref.
Semicarbazide-copper(II) chloride	$CH_5Cl_2CuN_3O$ / $Cu[OC(NH_2)NHNH_2]Cl_2$	Ortho.	6.9$_0$	10.2$_2$	8.2$_6$	2.39	4	2.39	$Pnma$ (C_{2h}^5)	79
Semicarbazide-zinc(II) chloride	$Zn[OC(NH_2)NHNH_2]Cl_2$	Mono.	12.7$_1$ / –	7.4$_7$ / 110.3	14.95 / –	2.10	8	2.10	$P2_1/c$ (C_{2h}^5)	79
Guanidinium perchlorate	$CH_6ClN_3O_4$ / $CN_3H_5.HClO_4$	Cub.	5.34±5 (Original in kX.)	–	–	1.743	1	1.772		105
Caesium oxalate monohydrate	$Cs_2C_2O_4.H_2O$	Mono.	6.17 / –	11.04 / 114.0	6.19 / –	3.23	2	3.20	$P2_1/c$ (C_{2h}^5)	88
Ammonium dioxovanadium (V) bisoxalate dihydrate	$(NH_4)_3\,VO_2(C_2O_4)_2\,.2H_2O$	Ortho.	15.75 / –	11.09 / –	8.015 / –	1.665	4	1.66	–	99
Ammonium dioxovanadium (IV) trisoxalate hexahydrate	$(NH_4)_2\,(VO)_2(C_2O_4)_3\,.6H_2O$	Ortho.	11.49	19.91	8.62	1.802	4	1.824	–	100
Ammonium oxovanadium (IV) bisoxalate dihydrate	$(NH_4)_2\,VO(C_2O_4)_2\,.2H_2O$	Mono.	13.26	9.27	8.87	1.926	4	1.943	–	100
Cadmium oxalate tri-hydrate	$CdC_2O_4.3H_2O$	Tric.	7.36±4 / 135.9±5	9.39±4 / 132.5±5	9.06±4 / 68.4±5	2.79	2	2.73	$P1$ (C_1^1)	22

Compound	Formula	Class	a α	b β	c γ	D_m	Z	D_x	S.G.	Ref.
Calcium oxalate mono-hydrate	$CaC_2O_4 \cdot H_2O$	Mono.	6.28 —	14.46 109.4	11.10 —	—	8	2.254	$P2_1/c$ (C_{2h}^5)	5
Plutonium oxalate decahydrate	$Pu_2(C_2O_4)_3 \cdot 10H_2O$	Mono.	11.595±3 —	9.599±5 118.94±1	10.171±2 —					58
Lanthanum oxalate decahydrate	La_2 "	"	11.830±4 —	9.658±6 119.14±1	10.492±3 —					58
Neodymium oxalate decahydrate	Nd_2 "	"	11.662±3 —	9.669±2 118.62±1	10.227±3 —					58
Gadolinium oxalate decahydrate	Gd_2 "	"	11.516±3 —	9.640±1 118.81±1	10.097±3 —					58
Plutonyl oxalate tri-hydrate	$PuO_2C_2O_4 \cdot 3H_2O$	Mono.	5.619±3 —	16.865±6 98.43±1	9.420±4 —					58
Uranyl oxalate tri-hydrate	UO_2 " "	"	5.623±5 —	17.065±5 98.74±1	9.451±3 —					58
Plutonium oxalate hexa-hydrate	$Pu(C_2O_4)_2 \cdot 6H_2O$	Tric. [Mono.?]	6.377±4 91.74±1	6.377±4 91.74±1	7.931±5 89.26±1	—	—	—	$P1$ (C_1^1)	58
Uranium oxalate hexa-hydrate	$U(C_2O_4)_2 \cdot 6H_2O$	Tric. [Mono.?]	6.388±3 91.64±1	6.388±3 91.74±1	7.881±4 89.45±1	—	—	—	$P1$ (C_1^1)	58

Name	Formula	System	a	b	c	β	D	Z	D	Space group	Ref
Plutonium oxalate dihydrate	$Pu(C_2O_4)_2 \cdot 2H_2O$	Tetr.	10.527±5	-	8.861±5	-	-	-	-	-	58
Thorium oxalate dihydrate	$Th(C_2O_4)_2 \cdot 2H_2O$	Ortho.	10.504±4	9.735±4	8.506±4	-	-	-	-	-	58
Uranium oxalate dihydrate	$U(C_2O_4)_2 \cdot 2H_2O$	Ortho.	10.479±3	9.443±3	8.572±3	-	-	-	-	-	58
Plutonium oxalate dihydrate (acetone solvate)	$Pu(C_2O_4)_2 \cdot 2H_2O \cdot xC_3H_6O$	Tetr.	8.642±2	-	8.088±5	-	-	-	-	-	58
Sodium acetate trihydrate	$C_2H_3NaO_2 \cdot 3H_2O$ / $NaCH_2COOH \cdot 3H_2O$	Mono.	12.32±1	10.42±1	10.38±1	111.71±15	1.456	8	1.458	$C2/c$ (C_{2h}^6)	61
Nickel biuret	$C_2H_5NiN_3O_2$ / $Ni(H_2NCONHCONH_2)$	Mono.	10.68	9.82	4.31	101.3	2.653	2	2.54	-	42
Cobalt biuret	$C_2H_5CoN_3O_2$ / $Co(H_2NCONHCONH_2)$	Mono.	6.06	11.05	11.23	110.9	2.542	2	2.436	-	42
Cadmium hydrazinecarboxylate	$C_2H_6CdN_4O_4$ / $Cd(N_2H_3CO_2)_2$	Ortho.	5.372	9.116	6.444	-	2.745	2	2.762	$P2_12_12$ (D_2^3)	21
Manganese hydrazine-carboxylate dihydrate	$C_2H_6MnN_4O_4$ / $Mn(N_2H_3CO_2)_2$	Ortho.	9.862	11.052	7.847	-	1.893	4	1.872	$Pbam$ (C_{2v}^8)	21

Compound	Formula	Class	a α	b β	c γ	D_m	Z	D_x	S.G.	Ref.
Zirconyl perchlorate, dimethyl sulphoxide (1:8)	$(C_2H_6SO)_8ZrO(ClO_4)_2$ $[ZrO.6(CH_3)_2SO](ClO_4)_2-$ $.2(CH_3)_2SO$	Hexa.	12.45 -	- -	15.88 -	1.428	2	1.448	-	64
Thorium perchlorate, dimethyl sulphoxide (1:12)	$(C_2H_6SO)_{12}Th(ClO_4)_2$ $[Th.6(CH_3)_2SO](ClO_4)_2-$ $.6(CH_3)_2SO$	Hexa.	13.10 -	- -	14.40 -	2.124	2	2.104	-	64
B-Dichlorobarene	$C_2H_{10}B_{10}Cl_2$ HC⟨$B_{10}H_8Cl_2$⟩HC	Ortho.	12.13±7 -	6.95±3 -	11.61±5 -	-	4	1.45	$Pnam$ (D_{2h}^{16})	104
B-Bromobarene	$C_2H_{11}B_{10}Br$ HC⟨$B_{10}H_9Br$⟩HC	Ortho.	12.79±5 -	14.73±4 -	11.59±2 -	-	4	1.37	$Pcam$ (D_{2h}^{11})	104
B-Iodobarene	$C_2H_{11}B_{10}I$ HC⟨$B_{10}H_9I$⟩HC	Ortho.	13.24±4 -	10.93±4 -	7.07±4 -	1.7	4	1.76	$Cmca$ (D_{2h}^{18})	104
B-Diiodobarene	$C_2H_{10}B_{10}I_2$ HC⟨$B_{10}H_8I_2$⟩HC	Mono.	21.24±14 -	7.77±9 91±1	14.59±13 -	2.18	8	2.20	$C2/c$ (C_{2h}^6)	104

Compound	Formula	Crystal system					Z	Density	Space group	Ref
B-Triiodobarene	$C_2H_9B_{10}I_3$	Mono.	7.84±8 —	24.51±4 112.7±0.5	8.19±7 —	–	4	2.40	$P2_1/a$ (C_{2h}^5)	104
B-Dichloro-C-methylbarene	$C_3H_{12}B_{10}Cl_2$	Hexa.	7.14±3 —	– —	38.10±23 —	1.3	6	1.35	$P6/mmm$ (D_{6h}^1)	104
B-Trichloro-C-methylbarene	$C_3H_{11}B_{10}Cl_3$	Mono.	7.60±5 —	13.64±10 125±1	15.38±12 —	1.37	4	1.34	$P2_1/a$ (C_{2h}^5)	104
B-Dibromo-C-methylbarene	$C_3H_{12}B_{10}Br_2$	Hexa.	7.38±1 —	– —	39.92±30 —	1.67	6	1.68	$P6/mmm$ (D_{6h}^1)	104
1-Bromo-2-barenyl-ethane	$C_4H_{15}B_{10}Br$	Ortho.	14.77±7 —	7.53±7 —	22.51±11 —	–	8	1.34	$Pcam$ (D_{2h}^{11})	104
C-p-Bromophenyl-barene	$C_8H_{15}B_{10}Br$	Mono.	7.55±3 —	8.64±3 94±1	21.89±10 —	–	4	1.40	$P2_1/c$ (C_{2h}^5)	104
Bis-(C-vinylbarenyl)mercury	$C_8H_{26}B_{20}Hg$	Ortho.	8.87±3 —	9.69±3 —	14.77±8 —	–	2	1.42	$Pcmm$ (D_{2h}^5)	104

Compound	Formula	Class	a α	b β	c γ	D_m	Z	D_x	S.G.	Ref.
C-Vinylbarenylmethyl-mercury	$C_5H_{14}B_{10}Hg$ H_2CCHC—$B_{10}H_{10}$ / $HCHgC$	Mono.	12.07±4 –	7.09±2 98.3±0.5	15.05±4 –	–	4	2.02	$P2_1/a$ (C_{2h}^5)	104
Neobarene	$C_2H_{12}B_{10}$ HC—$B_{10}H_{10}$—CH	Hexa.	15.79±8 –	– –	6.96±2 –	–	6	0.96	–	104
B-Iodoneobarene	$C_2H_{11}B_{10}I$ HC—$B_{10}H_9I$—CH	Mono.	13.15±4 –	7.12±2 105±1	11.25±4 –	1.7	4	1.77	$P2_1/a$ (C_{2h}^5)	104
B-Diiodoneobarene	$C_2H_{10}B_{10}I_2$ HC—$B_{10}H_8I_2$—CH	Ortho.	14.64±6 –	22.48±6 –	14.93±7 –	2.1	16	2.15	$Cmcm$ (D_{2h}^{17})	104

Compound	Formula	System	a	b / β	c	d_obs	Z	d_calc	Space group	Ref
B-Decachloroneobarene	$C_2H_2B_{10}Cl_{10}$	Mono.	14.46±8 / –	14.15±10 / 110.1±0.5	8.66±4 / –	–	4	1.96	$P2_1/a$ (C_{2h}^5)	104
Malonamide	$C_3H_6N_2O_2$ $CONH_2 \cdot CH_2 \cdot CONH_2$	Mono.	13.63±2 / –	9.47±1 / 113.2	8.06±1 / –	–	8	1.416	$P2_1/c$ (C_{2h}^5)	55
Adipamide	$C_6H_{12}N_2O_2$ $CONH_2 \cdot (CH_2)_4 \cdot CONH_2$	Mono.	6.89±1 / –	5.14±1 / 111	10.67±1 / –	–	2	1.367	$P2/c$ (C_{2h}^4)	55
Suberamide	$C_8H_{16}N_2O_2$ $CONH_2 \cdot (CH_2)_6 \cdot CONH_2$	Mono.	14.44±2 / –	5.13±1	14.17±2	–	4	1.225	$C2/c$ (C_{2h}^6)	55
Trimethyltin hydroxide	$C_3H_{10}OSn$ $(CH_3)_3SnOH$	Mono.	13.34 / –	33.20 / 90	22.42 / –	–	64	–	Pn (C_s^2)	63
Iron(II)hydrazine-carboxylate	$C_3H_{16}N_8O_7Fe$ $Fe(II)(N_2H_3COO)_2(N_2H_5)_2\text{-}CO_3$	Mono.	12.244±5 / –	11.055±6 / 121.6±0.2	10.211±5 / –	1.879	4	1.873	$C2/c$ (C_{2h}^6)	20
Cobalt(II)hydrazine-carboxylate	$Co(II)(N_2H_3COO)_2(N_2H_5)_2\text{-}CO_3$	Mono.	12.029±14 / –	10.89±2 / 120.2±0.2	10.24±1 / –	1.915	4	1.922	"	20

Compound	Formula	Class	a / α	b / β	c / γ	D_m	Z	D_x	S.G.	Ref.
Nickel(II)hydrazine-carboxylate	$Ni(II)(N_2H_3COO)_2(N_2H_5)_2$ CO_3	Mono.	12.12±2 / -	10.858±5 / 120.1±0.3	10.255±6 / -	1.868	4	1.889	$C2/c$ (C_{2h}^6)	20
Zinc(II)hydrazine-carboxylate	$Zn(II)(N_2H_3COO)_2(N_2H_5)_2$ CO_3	Mono.	11.970±5 / -	10.897±5 / 119.8±0.1	10.292±7 / -	1.947	4	1.948	"	20
Fumaric acid (Sublimed)	$C_4H_4O_4$ HOOCHCHCOOH	Tric.	4.52±2 / 136.7±3	7.51±2 / 110.6±3	5.40±4 / 72.8±3	1.60	1	1.63	$P\bar{1}$ (C_i^1)	98
Copper tartrate trihydrate	$C_4H_4CuO_6 \cdot 3H_2O$ $CuC_4H_4O_6 \cdot 3H_2O$	Tric.	8.42±7 / 97±1	12.33±10 / 80.7±0.7	8.82±7 / 115.7±0.5	2.05	4	2.17	$P1$ (C_1^1)	22
1,1,4,4-Tetrachloro-1,4-digermacyclohexa-2,5-diene	$C_4H_4Cl_4Ge_2$	Mono.	6.2_6 / -	7.4_4 / 107.5	12.1_6 / -	2.13	4	2.10	$P2_1/c$ (C_{2h}^5)	17
1,1,4,4-Tetraiodo-1,4-digermacyclohexa-2,5-diene	$C_4H_4I_4Ge_2$	Mono.	8.5_6 / -	7.3_4 / 92.5	10.8_1 / -	-	4	3.45	$P2_1/c$ (C_{2h}^5)	17
2,3-dibromo-1,4-dithiane	$C_4H_6Br_2S_2$	Ortho.	14.54 / -	7.52 / -	7.60 / -	2.20	4	2.19	$Pnam$ (D_{2h}^{16})	60
Mercury(II) acetate	$C_4H_6HgO_4$ $Hg(CH_3COO)_2$	Mono.	4.62 / -	20.14 / 107.9	7.15 / -	3.286	4	3.33	$P2_1$ (C_2^2)	94
Mercury(I) acetate	$C_4H_6Hg_2O_4$ $Hg_2(CH_3COO)_2$	Mono.	5.18 / -	5.96 / 100.05	12.17 / -	4.59	4	4.65	$C2$ (C_2^3)	93

Compound	Formula	Structural formula	System	a (/angle)	b (/angle)	c (/angle)	d_{calc}	Z	d_{obs}	Space group	Ref.
DL-Aspartic acid (Modification II)	$C_4H_7NO_4$	$COOHCH_2CH(NH_2)COOH$	Ortho.	15.00 / —	6.87 / —	10.26 / —	1.663	8	1.67	$Pmmn$ (D_{2h}^1)	78
Copper formate, hemidioxane	$C_4H_6CuO_5$	$Cu(HCOO)_2 \cdot \tfrac{1}{2}C_4H_8O_2$	Tric.	6.62 / 112.5	9.05 / 98.5	13.31 / 112.5	2.03	4	—	$P\bar{1}$ (C_i^1)	23
tert-Lithium butoxide	C_4H_9OLi		Mono.	29.63 / —	17.67 / 93	10.32 / —	0.897	36	0.886	$C2$ (C_2^3)	56
tert-Butyllithium	C_4H_9Li		Ortho.	13.99 / —	75.17 / —	17.50 / —	0.8	160	0.924	—	56
n-Butyllithium, tert-lithium butoxide (addition compound)	$C_8H_{18}Li_2O$	C_4H_9Li,C_4H_9LiO	Mono.	11.28 / —	23.64 / 117	18.37 / —	0.8	16	0.877	$P2_1/c$ (C_{2h}^5)	56
tert-Butyllithium, tert-lithium butoxide (addition compound)	$C_8H_{18}LiO$	C_4H_9Li,C_4H_9LiO	Mono.	20.34 / —	16.86 / 124	35.48 / —	—	36	0.854	$P2_1/c$ (C_{2h}^5)	56
Aluminium chloride, diethyl ether (1:1 complex)	$C_4H_{10}AlCl_3O$	$AlCl_3 \cdot (C_2H_5)_2O$	Ortho.	11.68±1 / —	6.96±1 / —	12.70±1 / —	1.30	4	1.33	$P2_12_12_1$ (D_2^4)	101
Tetramethylammonium chloride, hydrogen chloride	$C_4H_{13}Cl_2N$	$(CH_3)_4N^+ \cdot HCl_2^-$	Ortho.	9.27±1 / —	7.73±1 / —	11.59±1 / —	1.15	4	1.168	$Pnma$ (D_{2h}^{16})	110
Nickel biguanide chloride dihydrate	$C_4H_{14}Cl_2NiN_{10} \cdot 2H_2O$	$Ni(C_2H_7N_5)_2Cl_2 \cdot 2H_2O$	Tric.	6.78 / 113.6	9.56 / 98.4	12.70 / 100.5	1.711	2	1.696	$P1$ (C_1^1)	45
Copper biguanide chloride dihydrate	$C_4H_{14}Cl_2CuN_{10} \cdot 2H_2O$	$Cu(C_2H_7N_5)_2Cl_2 \cdot 2H_2O$	Tric.	6.78 / 113.4	9.58 / 99.0	12.71 / 99.7	1.72	2	1.704	$P1$ (C_1^1)	45

Compound	Formula	Class	a / α	b / β	c / γ	D_m	Z	D_x	S.G.	Ref.
Copper biguanide nitrate dihydrate	$C_4H_{14}CuN_{12}O_6 \cdot 2H_2O$, $Cu(C_2H_7N_5)_2(NO_3)_2 \cdot 2H_2O$	Tric.	3.62 / 106.1	10.47 / 104.3	12.21 / 92.5	1.826	1	1.656	$P1$ (C_1^1)	45
Platinium biguanide perchlorate	$C_4H_{14}Cl_2O_8N_{10}Pt$, $Pt(C_2H_7N_5)_2 \cdot (ClO_4)_2$	Tric.	9.19 / 116.2	10.96 / 123.1	14.31 / 90.7	2.569	2	2.670	$P1$ (C_1^1)	45
Ruthenium biguanide sulphate heptahydrate	$C_4H_{14}N_{10}O_{12}RuS_3 \cdot 7H_2O$, $Ru(C_2H_7N_5)_2(SO_4)_3 \cdot 7H_2O$	Mono.	20.26 / -	11.65 / 117.8	22.19 / -	1.755	4	1.754	$P2_1/c$ (C_{2h}^5)	45
Uric Acid	$C_5H_4N_4O_3$	Mono.	14.464±3 / -	7.403±2 / 65.10	6.208±1 / -	1.844	4	1.851	$P2_1/a$ (C_{2h}^5)	97
Uric acid dihydrate	$C_5H_4N_4O_3 \cdot 2H_2O$	Ortho.	7.40±1 / -	17.55±1 / -	6.350±5 / -	1.650	4	1.643	-	97
1-Methyl-5-bromour- acil, 9-methyladenine (1:1)	$C_5H_5N_2O_2Br$, $C_6H_7N_5$	Tric.	9.26±5 / 78.0±0.5	7.40±5 / 102.0±0.5	10.78±5 / 80.0±0.5	1.79	2	1.82	$P1$ (C_1^1)	9
1-Methyl-5- bromouracil	$C_5H_5N_2O_2Br$	Mono.	7.16±5 / -	12.35±5 / 95.0±0.5	7.65±5 / -	-	4	1.87	$P2_1/c$ (C_{2h}^5)	9
9-Methylguanine	$C_6H_7N_5O$	Mono.	10.50±5 / -	23.83±5 / 95.0±0.5	6.75±5 / -	-	8	1.31		9
Benzotrifuroxan, pyrene (1:1)	$C_6N_6O_6 \cdot C_{16}H_{10}$	Mono.	20.6 / -	29.6 / 111	6.7 / -	1.58	8	1.60	Cc (C_s^4)	15

			a	α	b	β	c	γ	d_m	Z	d_c	Space group	Ref.
Benzotrifuroxan, perylene (1:1)	$C_6N_6O_6,C_{20}H_{12}$	Mono.	19.2	-	6.95	98	15.6	-	1.56	4	1.56	$P2_1/a$ (C_{2h}^5)	15
Benzotrifuroxan, diphenyl (1:1)	$C_6N_6O_6,C_{12}H_{10}$	Ortho.	7.04	-	15.1	-	16.8	-	1.52	4	1.51	$Pbcn$ (D_{2h}^{14})	15
Hexafluorobenzene, anthracene (1:1)	$C_6F_6,C_{14}H_{10}$	Mono.	9.03	-	12.2	95	7.26	-	-	2	1.51	$C2$ (C_2^3)	15
Hexafluorobenzene, pyrene (1:1)	$C_6F_6,C_{16}H_{10}$	Mono.	9.88	-	13.5	113	6.98	-	-	2	1.52	$C2$ (C_2^3)	15
Hexafluorobenzene, perylene (1:1)	$C_6F_6,C_{20}H_{12}$	Mono.	17.6	-	7.73	106	7.31	-	-	2	1.53	$P2_1/a$ (C_{2h}^5)	15
Tetracyanoethylene, naphthalene (1:1)	$C_6N_4,C_{10}H_8$	Mono.	7.24	-	12.8	94	7.33	-	1.24	2	1.25	$C2$ (C_2^3)	15
Tetracyanoethylene, pyrene (1:1)	$C_6N_4,C_{16}H_{10}$	Mono.	14.2	-	7.2	93	7.9	-	1.33	2	1.30	$P2_1/a$ (C_{2h}^5)	15
Tetracyanoethylene, perylene (1:1)	$C_6N_4,C_{20}H_{12}$	Mono.	15.7	-	8.19	96	7.37	-	1.36	2	1.36	$P2_1/a$ (C_{2h}^5)	15
Benzotrifuroxan	$C_6N_6O_6$	Ortho.	6.96	-	19.7	-	6.58	-	1.86	4	1.86	$Pna2_1$ (C_{2v}^9)	15
Benzotrifuroxan, benzene (1:1)	$C_6N_6O_6,C_6H_6$	Mono.	15.41	-	7.36	117	13.75	-	-	4	1.59	$P2_1/a$ (C_{2h}^5)	15

Compound	Formula	Class	a / α	b / β	c / γ	D_m	Z	D_x	S.G.	Ref.
Benzotrifuroxan, naphthalene (1:1)	$C_6N_6O_6 \cdot C_{10}H_8$	Trig.	12.6 / –	– / –	6.92 / –	1.50	3	1.50	$R3$ (C_3^4)	15
Benzotrifuroxan, anthracene (1:1)	$C_6N_6O_6 \cdot C_{14}H_{10}$	Ortho.	15.8 / –	17.3 / –	6.8 / –	1.53	4	1.54	$P2_12_12_1$ (D_2^4)	15
1,2,4,5-tetrachlorobenzene (α Form: 150°K)	$C_6H_2Cl_4$	Tric.	9.60 / 95	10.59 / 102.5	3.76 / 92.5	–	2	1.93	$P1$ (C_1^1)	52
1,2,4,5-Tetrachlorobenzene (β Form: 300°K)	$C_6H_2Cl_4$	Mono.	9.73 / –	10.63 / 103.5	3.86 / –	–	2	1.85	$P2_1/a$ (C_{2h}^5)	52
2,5-Dichloronitrobenzene	$C_6H_3Cl_2NO_2$	Tric.	7.35±2 / 92.7±0.2	8.970±6 / 113.0±0.2	7.38±2 / 60.4±0.2	–	2	–	$P1$ (C_1^1)	3
Antimony pentachloride, o-chlorophenyl diazonium chloride (double salt)	$C_6H_4Cl_7N_2Sb$ (o-$ClC_6H_4N_2Cl$) ($SbCl_5$)	Mono.	13.93±3 / –	8.10±2 / 108.0±0.5	16.27±3 / –	–	4	–	Pc (C_s^2)	91
Sulphur derivative of phosphobenzene	$(C_6H_5PS)_n$ ($n = 3?$)	Mono.	8.90 / –	13.71 / 107.5	8.35 / –	1.438	2	1.437	$P2_1$ (C_2^2)	33
Allyltricarbonyl iron iodide	$C_6H_5FeIO_3$ $[C_3H_5Fe(CO)_3]I$	Ortho.	7.40±4 / –	11.02±4 / –	13.52±4 / –	2.01	4	2.13	$P2_1/c$ (C_{2h}^5)	16

Compound	Formula	System	a	b	c	Angle(s)	D	Z	D	Space group	Ref.
Allyltricarbonyl iron nitrate	$C_6H_5FeNO_6$ $[C_3H_5Fe(CO)_3]NO_3$	Tetr.	11.23±4	–	15.43±4	–	1.6	8	1.73	$P\bar{4}2_1c$ (D_{2d}^4)	16
γ-o-Nitroaniline	$C_6H_6N_2O_2$ $H_2N.C_6H_4.NO_2$	Mono.	15.28±3	10.00±5	8.54±3	105.5±1.0	1.442	8	1.46	$P2_1/a$ (C_{2h}^5)	53
β-o-Nitroaniline	$C_6H_6N_2O_2$ $H_2N.C_6H_4.NO_2$	Mono.	13.80±5	3.90±4	23.40±5	94.5±0.5	1.43	8	1.46	$P2_1/c$ (C_{2h}^5)	53
Copper propionate	$C_6H_{10}CuO_4$ $Cu(CH_3CH_2COO)_2$	Tric.	9.58	5.22	8.55	105.9, 92.5, 91.1	–	–	–	$P1$ (C_1^1)	89
Copper propionate monohydrate	$C_6H_{10}CuO_4.H_2O$ $Cu(CH_3CH_2COO)_2.H_2O$	Mono.	15.2	17.4	15.4	94.3	1.42	16	1.48	$P2_1/a$ (C_{2h}^5)	90
Dibromobis(1,1-dimethylurea)cobalt hexahydrate	$C_6H_{16}Br_2CoN_4O_2.6H_2O$ $CoBr_2.2OCNH_2N(CH_3)_2.6H_2O$	Ortho.	17.00±7	8.03±1	7.53±1	–	1.66	2	1.63	$C222$ (D_2^6)	13
Dibromobis(1,1-dimethylurea)nickel hexahydrate	$C_6H_{16}Br_2NiN_4O_2.6H_2O$ $NiBr_2.2OCNH_2N(CH_3)_2.6H_2O$	Ortho.	17.02±2	7.96±1	7.45±1	–	1.65	2	1.65	$C222$ (D_2^6)	13
Dichlorobis(1,1-dimethylurea)cobalt hexahydrate	$C_6H_{16}Cl_2CoN_4O_2.6H_2O$ $CoCl_2.2OCNH_2N(CH_3)_2.6H_2O$	Ortho.	16.29±3	7.80±1	7.34±1	–	–	2	1.40	$C222$ (D_2^6)	13
Dichlorobis(1,1-dimethylurea)nickel hexahydrate	$C_6H_{16}Cl_2NiN_4O_2.6H_2O$ $NiCl_2.2OCNH_2N(CH_3)_2.6H_2O$	Ortho.	16.34±1	7.77±2	7.29±3	–	–	2	1.45	$C222$ (D_2^6)	13
Dibromobis(1,1-dimethylurea)zinc	$C_6H_{16}Br_2N_4O_2Zn$ $ZnBr_2.2OCNH_2N(CH_3)_2$	Mono.	16.17±9	12.70±2	7.62±1	117.3	–	4	–	$P2_1/c$ (C_{2h}^5)	13

420

Compound	Formula	Class	a / α	b / β	c / γ	D_m	Z	D_x	S.G.	Ref.
Dichlorobis(1,1-di-methylurea)zinc	$C_6H_{16}Cl_2N_4O_2Zn$ $ZnCl_2 \cdot 2OCNH_2N(CH_3)_2$	Ortho.	13.74±2 —	18.77±5 —	10.50±1 —	1.59	8	1.53	Pbca (D_{2h}^{15})	13
Diiodobis(1,1-di-methylurea)zinc	$C_6H_{16}I_2N_4O_2Zn$ $ZnI_2 \cdot 2OCNH_2N(CH_3)_2$	Mono.	17.29±2 —	12.95±2 116.4	7.76±3 —	2.13	4	2.12	$P2_1/c$ (C_{2h}^5)	13
Cycloalliin hydro-chloride monohydrate	$C_6H_{11}ClNO_3S \cdot H_2O$	Mono.	5.28 —	12.35 107.14	8.42 —	1.46	2	1.47	$P2_1$ (C_2^2)	59
L-Arginine hydro-chloride monohydrate	$C_6H_{15}ClN_4O_2 \cdot H_2O$	Mono.	11.07 —	8.50 91	11.22 —	—	4	—	$P2_1$ (C_2^2)	72
L-Arginine hydro-bromide monohydrate	$C_6H_{15}BrN_4O_2 \cdot H_2O$	Mono.	11.26 —	8.65 91.5	11.25 —	—	4	—	$P2_1$ (C_2^2)	72
Bis(1,3-diamino-propane)copper(II) chloride monohydrate	$C_6H_{20}Cl_2Cu_2N_2 \cdot H_2O$ $(C_3H_{10}NCu)_2Cl \cdot H_2O$	Tetr.	9.303±2 —	— —	7.582±2 —	1.52	2	1.52	$P\bar{4}n2$ (D_{2d}^8)	80
3-Nitro-4-toluidine	$C_7H_8N_2O_2$ $C_6H_3NO_2CH_3NH_2$	Mono.	12.41±1 —	9.15±1 103.5±0.5	13.52±1 —	—	8	—	Aa (C_s^4)	81
Methyl-2-chloro-mercury-2-deoxy-α-D-talopyranoside	$C_7H_{13}O_5HgCl$	Ortho.	6.68±1 —	13.6±1 —	11.9±1 —	—	4	2.53	$P2_12_12_1$ (D_2^4)	7
p-Chlorophenoxy-acetic acid	$C_8H_7O_3Cl$ $ClC_6H_4OCH_2COOH$	Tric.	7.14 88.4	7.44 120.7	8.97 94.0	1.498	2	1.514	P1 (C_1^1)	44

Compound	Formula	System	a	α	b	β	c	γ	Density (obs)	Z	Density (calc)	Space group	Ref.
2,3-Dimethylphenol	$C_8H_{10}O$	Ortho.	4.81±2	–	5.92±3	–	24.65±4	–	1.11	4	1.15	$P2_12_12_1$ (D_2^4)	43
2,5-Dimethylphenol	$C_8H_{10}O$	Mono.	5.94±2	–	4.91±2	109.1±0.1	12.48±3	–	1.13	2	1.18	$P2_1$ (C_2^2)	43
2,6-Dimethylphenol	$C_8H_{10}O$	Mono.	10.16±3	–	4.49±2	92.0±0.1	15.53±3	–	1.13	4	1.14	$P2_1/c$ (C_{2h}^5)	43
3,4-Dimethylphenol	$C_8H_{10}O$	Tric.	10.48±3	110.1±1.0	10.85±3	124.1±1.0	14.27±3	100.0±1.0	1.11	6	1.13	$P1$ (C_1^1)	43
3,5-Dimethylphenol	$C_8H_{10}O$	Mono.	8.59±2	–	14.00±3	91.0±0.1	12.10±3	–	1.08	8	1.11	$P2_1/a$ (C_{2h}^5)	43
Allylcyclopentadienyl palladium	$C_8H_{10}Pd$ $(C_3H_5)Pd(C_5H_5)$	Ortho.	5.77±2	–	10.44±2	–	13.50±3	–	1.7	4	1.73	$Pnam$ (D_{2h}^{16})	16
Lithium dimedonesulphonate dihydrate (Dimedonesulphonic acid is 5,5-dimethyl-2-sulpho-1,3-cyclohexanedione.)	$C_8H_{11}LiO_5S.2H_2O$	Mono.	29.96	–	6.57	97.0	6.15	–		4			4
Caesium dimedone-sulphonate (Below 70°C.)	$C_8H_{11}CsO_5S$	Mono.	30.60	–	5.49	96.7	13.49	–		8			4
Caesium dimedone-sulphonate (Above 70°C.)	$C_8H_{11}CsO_5S$	Ortho.	27.05	–	8.36	–	10.07	–		8			4

Compound	Formula	Class	a α	b β	c γ	D_m	Z	D_x	S.G.	Ref.
1,1,4,4-Tetramethyl-1,4-digermacyclohexa-2,5-diene	$C_8H_{16}Ge_2$	Tric.	6.1_5 97.0	8.7_5 $112._5$	12.6_5 86.0	-	4	1.37	$P\bar{1}$ (C_i^1)	17
Bis(phosphorodi-bromidato)-bis(ethyl acetate)calcium	$C_8H_{16}Br_4CaO_6P_2$ $Ca(PO_2Br_2)_2 \cdot 2CH_3COOC_2H_5$	Mono.	14.7 ± 2 -	14.3 ± 4 96 ± 2	10.51 ± 1 -	1.99	4	[1.9]	$P2_1/a$ (C_{2h}^5)	48
trans-Tetrachlorobis-(tetrahydrothiophen)-tin(IV)	$C_8H_{16}Cl_4S_2Sn$ $SnCl_4(C_4H_8S)_2$	Mono.	10.1 -	9.3 98	7.9 -	1.9	2	1.98	$P2_1/n$ (C_{2h}^5)	12
Triketoindane (Orthorhombic form)	$C_9H_4O_3$	Ortho.	15.524 ± 8 -	14.160 ± 8 -	6.380 ± 5 -	1.52	8	1.526	$Pbca$ (D_{2h}^{15})	18
2-Chloroquinoline	C_9H_6ClN	Ortho.	10.38 2 -	18.42 4 -	4.000 6 -		4		$P2_12_12_1$ (D_2^4)	73
Bis(8-quinolinethio-lato)palladium	$C_{18}H_{12}N_2PdS_2$ $Pd(C_9H_6NS)_2$	Tric.	7.76 104.55	8.04 77.30	7.78 125.70	1.83	1	1.87	$P\bar{1}$ (C_i^1)	87
5-Formylvanillic acid (Triclinic form)	$C_9H_8O_5$	Tric.	6.91 125.4	7.93 125.0	9.36 58.7	2.0	2	1.965	$P\bar{1}$ (C_i^1)	77
Pyromellitic dian-hydride	$C_{10}H_2O_6$	Mono.	10.7 -	7.48 90	10.7 -	1.70	4	1.71	$P2_1/n$ (C_{2h}^5)	15

Name	Formula	System	a	b	β	c	d_{obs}	Z	d_{calc}	Space group	Ref.
Pyromellitic dianhydride, benzene (1:1)	$C_{10}H_2O_6,C_6H_6$	Mono.	12.9	6.7	96	7.68	-	2	1.37	$P2_1/a$ (C_{2h}^5)	15
Pyromellitic dianhydride, naphthalene (1:1)	$C_{10}H_2O_6,C_{10}H_8$	Mono.	9.19	13.0	104	6.81	1.45	2	1.47	$C2$ (C_2^3)	15
Pyromellitic dianhydride, pyrene (1:1)	$C_{10}H_2O_6,C_{16}H_{10}$	Mono.	13.9	9.25	94	7.24	1.50	2	1.49	$P2_1/a$ (C_{2h}^5)	15
β-Chloronaphthalene	$C_{10}H_7Cl$	Mono.	7.667±5	5.94±2	99.0±0.1	17.93±5	-	4		$P2_1/c$ (C_{2h}^5)	76
Dichloroquinaldine	$C_{10}H_7Cl_2N$ / $C_9H_6NCHCl_2$	Ortho.	9.59±6	7.03±3		14.41±3	1.449	4	1.451	$Ccca$ (D_{2h}^{22})	62
Trichloroquinaldine	$C_{10}H_6Cl_3N$ / $C_9H_6NCCl_3$	Mono.	9.57±3	9.95±1	99.0±0.3	10.91±3	1.573	4	1.594	$P2_1/c$ (C_{2h}^5)	62
7-Chloro-4-methyl coumarin	$C_{10}H_7O_2Cl$	Ortho.	11.09	4.08		9.61	1.45	2	1.48	$P222_1$ (D_2^2)	95
β-Naphthol (Metastable modification)	$C_{10}H_7O$ / $C_{10}H_6OH$	Mono.	8.29	5.95	123	9.01	-	2	-	$P2_1/a$ (C_{2h}^5)	30
Methyl cinnamate	$C_{10}H_{10}O_2$ / $C_6H_5CHCHCO.OCH_3$	Mono.	21.9	5.8	104	20.99	-	12	1.25	$P2_1/c$ (C_{2h}^5)	69
Methyl p-chloro-cinnamate	$C_{10}H_9ClO_2$ / $ClC_6H_4CHCHCO.OCH_3$	Mono.	8.77	5.84	95.6	18.75	-	4	1.37	$P2_1/c$ (C_{2h}^5)	69

Compound	Formula	Class	a α	b β	c γ	D_m	Z	D_x	S.G.	Ref.
Dicyclopentadienyl mercury	$C_{10}H_{10}Hg$ $Hg(C_5H_5)_2$	Tric.	13.1_1 99.8 (A sub-cell; measurements at 90°K.)	13.4_1 101.4	5.8_4 88.2	2.21	4	2.23	$P1$ (C_1^1)	16
1-Ephedrine copper(II) chelate, 1/3 cyclohexane	$(C_{10}H_{14}ON)_2Cu.1/3C_6H_{12}$	Trig.	11.73 –	– –	14.94 –	–	–	–	$P312$ (D_3^1)	2
1-Ephedrine copper(II) chelate, 2/3 mesitylene	$(C_{10}H_{14}ON)_2Cu.2/3C_9H_{12}$	Hex.	11.84 –	– –	95.11 –	–	–	–	$P6_122$ (D_6^2)	2
1-Ephedrine copper(II) chelate, 1/6 hexane	$(C_{10}H_{14}ON)_2Cu.1/6C_6H_{12}$	Ortho.	16.37	19.27	42.85	–	–	–	$P2_12_12_1$ (D_2^4)	2
Ferricynium ferrichloride	$C_{10}H_{10}Cl_4Fe_2$ $(C_5H_5)_2Fe\ ^+(FeCl_4)^-$	Ortho.	13.76±2 –	12.14±5 –	8.80±2 –	–	4	1.74	$Pnam$ (D_{2h}^{16})	16
1,2-Disilacenaphthene	$C_{10}H_{10}Si_2$	Tric.	13.56±3 113.4±0.3	8.51±2 79.2±0.3	9.78±3 107.8±0.3	1.16	4	1.26	$P1$ (C_1^1)	6
2,4,6-Trinitrophene-tole, potassium ethoxide (1:1)	$C_{10}H_{12}KN_3O_8$ $C_2H_5OC_6H_2(NO_2)_3,KOC_2H_5$	Tric.	14.7443±8 105.901±6	10.2848±4 104.05±1	9.9919±4 97.15±1	1.640	4	1.638	$P\bar{1}$ (C_i^1)	34
2,4,6-Trinitrophene-tole, caesium ethoxide (1:1)	$C_{10}H_{12}CsN_3O_8$ $C_2H_5OC_6H_2(NO_2)_3,CsOC_2H_5$	Mono.	15.564±3 –	10.54±2 110.31±1	19.919±1 –	1.878	8	1.890	$P2_1/c$ (C_{2h}^5)	34

Compound	Formula	System	a	b	c	angle	d_{obs}	Z	d_{calc}	Space group	Ref.
Bicyclo [3.3.1]nonane	$C_{10}H_{16}$	Cub.	9.57	–	–	–	–	–	–	$F\bar{4}3m$ (T_d^2)	67
Hexaantipyrine magnesium perchlorate	$(C_{11}H_{12}ON_2)_6Mg(ClO_4)_2$	Trig.	14.06	–	9.76	–	1.375	1	1.343	$P\bar{3}$ (C_{3i}^1)	108
Hexaantipyrine calcium perchlorate	$(C_{11}H_{12}ON_2)_6Ca(ClO_4)_2$	Trig.	14.33	–	9.78	–	1.330	1	1.310	$P\bar{3}$ (C_{3i}^1)	108
Hexaantipyrine zinc perchlorate	$(C_{11}H_{12}ON_2)_6Zn(ClO_4)_2$	Trig.	14.14	–	9.61	–	1.413	1	1.390	$P\bar{3}$ (C_{3i}^1)	108
Pentaantipyrine copper perchlorate	$(C_{11}H_{12}ON_2)_5Cu(ClO_4)_2$	Mono.	19.52	16.27	18.99	102.6	1.372	4	[1.36]	$P2_1/a$ (C_{2h}^5)	108
Aluminium mellitate hexahydrate	$C_{12}O_{12}Al_2 \cdot 6H_2O$ $Al_2 C_6(COO)_6 \cdot 6H_2O$	Tetr.	14.04±5	–	14.75±5	–	1.278	4	1.14	--	50
Hexacyanobenzene	$C_{12}N_6$	Cub.	10.82±1	–	–	–	1.20	4	1.196	$Pa3$ (T_h^6)	70
Bis-o-phenylene disulphide	$C_{12}H_8S_4$	Tric.	7.17	7.39	7.30	114.6, 95.7, 116.6	1.590	1	1.58	$P1$ (C_1^1)	75
o-Phenanthroline hydrate	$C_{12}H_8N_2 \cdot H_2O$	Trig.	17.67±5	–	8.55±3	–	1.25	9	1.28	$P31m$ (C_{3v}^2)	35
Aldrin	$C_{12}H_8Cl_6$	Mono.	10.76	14.70	8.98	93	1.70	4	1.705	$P2_1/n$ (C_{2h}^5)	109

Original in kX.

Compound	Formula	Class	a / α	b / β	c / γ	D_m	Z	D_x	S.G.	Ref.
Phenothiazine	$C_{12}H_9NS$	Ortho.	7.94±1 / -	21.02±1 / -	5.91±1 / -	1.355	4	1.342	$Pnma$ (D_{2h}^{16})	39
Tris(1,10-phenanthroline)-iron dibromide heptahydrate (monoclinic form)	$C_{36}H_{24}Br_2FeN_6 \cdot 7H_2O$ $Fe(C_{12}H_8N_2)_3Br_2 \cdot 7H_2O$	Mono.	35.4±1 / -	37.3±1 / 94	11.02±2 / -	-	16	-	-	1
Tris(1,10-phenanthroline)-iron dibromide heptahydrate (tetragonal form)	$C_{36}H_{24}Br_2FeN_6 \cdot 7H_2O$ $Fe(C_{12}H_8N_2)_3Br_2 \cdot 7H_2O$	Tetr.	36.04±8 / -	- / -	11.02±2 / -	-	16	-	-	1
Tris(1,10-phenanthroline)-cobalt dibromide heptahydrate	$C_{36}H_{24}Br_2CoN_6 \cdot 7H_2O$ $Co(C_{12}H_8N_2)_3Br_2 \cdot 7H_2O$	Ortho.	21.70±5 / -	22.60±5 / -	15.64±6 / -	-	8	-	-	1
Tris(1,10-phenanthroline)-nickel dibromide heptahydrate	$C_{36}H_{24}Br_2NiN_6 \cdot 7H_2O$ $Ni(C_{12}H_8N_2)_3Br_2 \cdot 7H_2O$	Ortho.	21.71±5 / -	22.94±4 / -	15.76±6 / -	-	8	-	-	1
2-Silaphenalane	$C_{12}H_{12}Si_2$	Mono.	12.42±3 / -	5.09±2 / 96.2±0.3	15.54±4 / -	1.21	4	1.26	$P2_1/c$ (C_{2h}^5)	6
5,10-Dihydrosilanthrene	$C_{12}H_{12}Si_2$	Ortho.	12.39±3 / -	12.23±3 / -	8.25±3 / -	1.12	4	1.13	$Cmc2$ (C_{2v}^{12})	6
1,6-Disilapyracene	$C_{12}H_{12}Si_2$	Mono.	7.66±2 / -	5.48±2 / 120.0±0.3	15.57±5 / -	1.23	2	1.25	Pc (C_s^2)	6

Compound	Formula	System	a	b	c	D_m	Z	D_c	Space group	Ref.
1,5-Disilapyracene	$C_{12}H_{12}Si_2$	Mono.	7.69±3	5.37±2 120.0±0.3	15.38±4	1.26	2	1.29	Pc (C_s^2)	6
Calcium 2-keto-D-gluconate trihydrate	$C_{12}H_{18}CaO_{14}\cdot3H_2O$ $Ca(C_6H_9O_7)_2\cdot3H_2O$	Ortho.	10.43±9	18.33±6	9.50±5	–	4	–	$P2_12_12_1$ (D_2^4)	10
Benzaldehyde phenylhydrazone	$C_{13}H_{12}N_2$ $C_6H_5CH{:}N.NH.C_6H_5$	Mono.	5.95±5	17.76±5 92.5	15.15±5	1.16	6	1.21	$P2_1/c$ (C_{2h}^5)	107
Securinine hydrobromide	$C_{13}H_{15}O_2N.HBr$	Ortho.	12.31	14.17	7.23	1.572	4	1.570	$P2_12_12_1$ (D_2^4)	57
Securinine hydrobromide monohydrate	$C_{13}H_{15}O_2N.HBr.H_2O$	Ortho.	9.64	21.80	6.55	–	4	1.510	$P2_12_12_1$ (D_2^4)	57
2'-Fluorobiphenyl-4-carboxylic acid	$C_{13}H_9FO_2$	Mono.	3.86	34.35 102.3	7.78	1.44	4	1.43	$P2_1/c$ (C_{2h}^5)	46
2'-Chlorobiphenyl-4-carboxylic acid	$C_{13}H_9ClO_2$	Mono.	3.94	35.81 101.1	7.61	1.46	4	1.47	$P2_1/c$ (C_{2h}^5)	46
2'-Bromobiphenyl-4-carboxylic acid	$C_{13}H_9BrO_2$	Mono.	4.01	35.81 not given	7.97	1.70	4	1.71	$P2_1$ (C_2^2)	46
2'-Iodobiphenyl-4-carboxylic acid	$C_{13}H_9IO_2$	Tric.	4.15 99.1	7.73 91.4	17.79 97.6	1.90	2	1.90	$P1$ (C_1^1)	46
4-Acetyl-2'-fluorobiphenyl	$C_{14}H_{11}FO$	Mono.	13.69	5.97 116.2	14.77	1.31	4	1.31	$P2_1/c$ (C_{2h}^5)	46

Compound	Formula	Class	a α	b β	c γ	D_m	Z	D_x	S.G.	Ref.
4-Acetyl-2'-chloro-biphenyl	$C_{14}H_{11}ClO$	Mono.	4.00 -	38.51 100.1	7.52 -	1.36	4	1.34	$P2_1/c$ (C_{2h}^5)	46
4-Acetyl-2'-bromo-biphenyl	$C_{14}H_{11}BrO$	Ortho.	23.43 -	12.70 -	8.07 -	1.53	8	1.52	$Pbca$ (D_{2h}^{15})	46
4-Acetyl-2'-iodo-biphenyl	$C_{14}H_{11}IO$	Ortho.	23.91 -	12.49 -	8.45 -	1.71	8	1.70	$Pbca$ (D_{2h}^{15})	46
Mercury dibenzyl $(C_6H_5CH_2)_2Hg$	$C_{14}H_{14}Hg$	Tetr.	12.91±2 -	- -	7.08±3 -	2.19	4	2.155	$P4_2/n$ (C_{4h}^4)	8
Diisothiocyanatobis-(aniline)zinc	$C_{14}H_{14}N_4S_2Zn$ $Zn(NCS)_2(C_6H_5NH_2)_2$	Ortho.	14.56±5 -	9.10±5 -	12.7±1 -	1.4	4	1.45	$Ima2$ (C_{2v}^{22})	102
(2-Butyne)dicyclopenta-dienylnickel	$C_{14}H_{16}Ni_2$	Mono.	6.15±3 -	12.37±4 100.6±0.1	8.75±3 -	1.52	2	1.53	$P2_1$ (C_2^2)	74
Tetracarbonyl[o-phenyl-ene-bis(dimethylarsine)]-chromium	$C_{14}H_{16}As_2CrO_4$ $Cr(CO)_4C_6H_4[As(CH_3)_2]_2$	Mono.	17.4 -	9.0 117.6	12.9 -	1.67	4	1.67	$P2_1/a$ (C_{2h}^5)	24
Tetracarbonyl[o-phenyl-ene-bis(dimethylarsine)]-molybdenum	$C_{14}H_{16}As_2MoO_4$ $Mo(CO)_4C_6H_4[As(CH_3)_2]_2$	Ortho.	17.1 -	16.3 -	12.8 -	1.80	8	1.84	$Pbca$ (D_{2h}^{15})	24
Tetracarbonyl[o-phenyl-ene-bis(dimethylarsine)]-tungsten	$C_{14}H_{16}As_2O_4W$ $W(CO)_4C_6H_4[As(CH_3)_2]_2$	Ortho.	17.1 -	16.3 -	12.8 -	-	8	-	$Pbca$ (D_{2h}^{15})	24

Compound	Formula	System	a	b	c	α	β	γ	D_o	Z	D_c	Space group	Ref.
Dichlorobis(2,6-di-methyl-γ-pyrone)-copper	$C_{14}H_{16}Cl_2CuO_4$ $CuCl_2.2C_7H_8O_2$	Tric.	7.80±1	15.21±1	7.74±2	101.0	102.6	107.6	1.57	2	1.59	$P1$ (C_1^1)	14
Dichlorobis(2,6-di-methyl-γ-pyrone)zinc	$C_{14}H_{16}Cl_2O_4Zn$ $ZnCl_2.2C_7H_8O_2$	Mono.	9.21±1	13.07±2	14.72±2	–	107.3		1.55	4	1.51	$C2/c$ (C_{2h}^6)	14
Cadmium chloride, 2,6-dimethyl-γ-pyrone (3:2 complex) dihydrate	$C_{14}H_{16}Cd_3Cl_6O_4.2H_2O$ $3CdCl_2.2C_7H_8O_2.2H_2O$	Tric.	10.58±5	15.22±2	3.87±1	89.2	90.0	82.3	2.26	1	2.26	$P1$ (C_1^1)	14
2-Mono-11-bromoun-decanoin	$C_{14}H_{27}BrO_4$	Tric.	5.90±4	4.94±3	32.2±2	91.5±0.6	91.5±0.6	118.5±0.8	–	–	–	$P1$ (C_1^1)	66
Methyl-4,6-benzal-α-D-idoside	$C_{14}H_{18}O_6$	Ortho.	16.72±1	10.210±5	7.920±5	–	–	–	1.389	4	1.387	$P2_12_12_1$ (D_2^4)	29
Methyl-4,6-benzal-α-D-idoside hemihydrate	$C_{14}H_{18}O_6.\frac{1}{2}H_2O$	Ortho.	22.22±1	11.310±5	11.050±5	–	–	–	1.386	8	1.389	$P2_12_12_1$ (D_2^4)	29
Methyl-4,6-benzal-3-methyl-β-idoside	$C_{15}H_{20}O_6$	Ortho.	22.38±1	13.000±7	5.000±3	–	–	–	1.359	4	1.353	$P2_12_12_1$ (D_2^4)	29
2,3-Dimethyl-4,6-benzal-methyl-β-idoside	$C_{16}H_{22}O_6$	Ortho.	20.76±1	9.100±5	8.400±4	–	–	–	1.295	4	1.299	$P2_12_12_1$ (D_2^4)	29
Potassium myristate	$C_{14}H_{27}KO_2$ $CH_3(CH_2)_{12}COOK$	Tric.	4.13±2 91.1±0.3	5.65±2 91.4±0.3	34.28±8 92.4±0.3				1.125	2	1.105	$P1$ (C_1^1)	36

Compound	Formula	Class	a / α	b / β	c / γ	D_m	Z	D_x	S.G.	Ref.
1:1 Acid potassium myristate	$C_{28}H_{55}KO$ [CH$_3$(CH$_2$)$_{12}$COOK]-[CH$_3$(CH$_2$)$_{12}$COOH]	Tric.	13.52±4 / 87.6±0.3	8.78±3 / 93.2±0.3	40.40±8 / 110.0±0.3	1.096	6	1.104	$P1$ (C_1^1)	36
Dibenzoylmethane	$C_{15}H_{12}O_2$	Ortho.	8.8 / –	10.8 / –	25.0 / –	1.20	8	1.185	$Pbca$ (D_{2h}^{15})	92
pp'-Dibromodibenzoyl-methane	$C_{15}H_{10}Br_2O_2$	Ortho.	31.20 / –	7.05 / –	6.08 / –	1.88	4	1.89	$Cmma$ (D_{2h}^{21})	92
Difluoro[1,3-bis(p-bromophenyl)-1,3-propanedionato]-boron	$C_{15}H_9BBr_2F_2O_2$	Mono.	6.87 / –	30.10 / 96	7.46 / –	1.89	4	1.83	$P2_1/c$ (C_{2h}^5)	92
mm'-Dichlorodibenzoyl-methane	$C_{15}H_{10}Cl_2O_2$	Ortho.	26.48 / –	4.05 / –	12.79 / –	1.80	4	1.85	$Pnca$ (D_{2h}^{14})	92
mm'-Dibromodibenzoyl-methane	$C_{15}H_{10}Br_2O_2$	Ortho.	30.08 / –	3.85 / –	11.12 / –	1.50	4	1.52	$Pca2_1$ (C_{2v}^5)	92
Retamine	$C_{15}H_{26}N_2O$	Mono.	11.873 / –	6.774 / 105.1	9.227 / –	–	–	–	$P2_1$ (C_2^2)	19
Retamine hydro-bromide	$C_{15}H_{26}N_2O.HBr$	Mono.	9.319 / –	11.298 / 100.3	7.298 / –	–	–	–	$P2_1$ (C_2^2)	19

431

Compound	Formula	System	a	b (angle)	c (angle)	D_m	Z	D_c	Space group	Ref
Dicyclopentadienyl-(1-hexyne)dinickel	$C_{16}H_{20}Ni_2$	Mono.	23.87±6	5.75±3 / 92.6±0.1	10.63±4 / –	1.50	4	1.50	$P2_1/n$ (C_{2h}^5)	75
Monobenzoylosmocene	$C_{17}H_{14}Os$ $C_5H_5.Os.C_5H_4.CO.C_6H_5$	Mono.	6.07	15.49 / 106.7	14.53 / –	2.18	4	2.15	$P2_1/c$ (C_{2h}^5)	71
Chlorpromazine base	$C_{17}H_{19}ClN_2S$	Ortho.	23.50±4	15.20±2 / –	9.23±1 / –	1.289	8	1.285	$Pbca$ (D_{2h}^{15})	39
Chlorpromazine hydrochloride	$C_{17}H_{20}Cl_2N_2S$	Mono.	11.99±2	32.33±6 / 99	9.89±2 / –	1.31	8	1.25	$P2_1/c$ (C_{2h}^5)	39
Tri-O-acetyl-D-xylopyranose-2,4-phenyl-hydrazone	$C_{17}H_{20}N_4O_{11}$	Mono.	13.15±1	6.00±1 / 98.6	13.27±3 / –	1.42	2	[1.47]	$P2_1$ (C_2^2)	32
2,2-Diphenyl-1-picryl-hydrazyl	$C_{18}H_{12}N_5O_6$	Tric.	13.58±1 / 92.2±0.1	18.91±1 / 101.6±0.1	7.555±5 / 95.0±0.1	–	4	1.385	$P1$ (C_1^1)	111
Triphenyltin chloride	$C_{18}H_{15}ClSn$ $(C_6H_5)_3SnCl$	Mono.	18.65	9.59 / 105.5	19.02 / –	1.4	8	1.57	Pa (C_s^2)	16
Dicyclopentadienyl-(ethynylbenzene)-dinickel	$C_{18}H_{16}Ni_2$	Tric.	5.73 / 101.7	10.71 / 108.9	13.08 / 90.0	1.57	2	1.57	$P1$ (C_1^1)	75
Monorden (Radicicol)	$C_{18}H_{17}O_6Cl$	Mono.	9.16±3	15.01±3 / 100.2±0.2	12.35±3 / –	1.42	4	1.45	–	31
Monorden, hemi-chloroform	$C_{18}H_{17}O_6Cl.\frac{1}{2}CHCl_3$	Mono.	9.12±2 / –	23.18±4 / 99.7±0.2	8.95±2 / –	1.49	4	1.51	$P2$ (C_2^1)	31

Compound	Formula	Class	a / α	b / β	c / γ	D_m	z	D_x	S.G.	Ref.
Cycloneosamandione hydroiodide	$C_{19}H_{30}INO_2$	Ortho.	10.87 / –	9.59 / –	19.51 / –	–	4	–	$P2_12_12_1$ (D_2^4)	49
Sodium warfarin, 2-propanol, water (clathrate)	$(C_{19}H_{15}NaO_4)$, (C_3H_8O), (H_2O)	Mono.	15.4±2 / –	11.4±2 / 107	22.7±2 / –	–	–	–	$P2_1/c$ (C_{2h}^5)	54
Tetra-O-acetyl-D-gluco-pyranose-2,4-dinitro-phenylhydrazone	$C_{20}H_{24}N_4O_{13}$	Ortho.	8.38±3 / –	15.43±3 / –	19.60±4 / –	1.41	4	1.39	$P2_12_12_1$ (D_2^4)	32
Bis(2,4-pentanedion-ato)bis(pyridine)-nickel(II)	$C_{20}H_{24}N_2NiO_2$ $(C_5H_7O_2)_2(C_5H_5N)_2Ni$	Mono.	8.28 / –	9.67 / 116.8	14.65 / –	1.3	2	1.32	$P2_1/c$ (C_{2h}^5)	51
Sodium thymidylyl-(5'—3')-thymidylate-(5') dodecahydrate	$C_{20}H_{25}N_4Na_3O_{15}P_2 \cdot 12H_2O$	Ortho.	16.06±4 / –	15.13±4 / –	15.65±4 / –	1.588	4	1.587	$P2_12_12$ (D_2^3)	26
'Stelazine'	$C_{21}H_{24}F_3N_3S$	Mono.	18.86±4 / –	9.27±3 / 111.	25.67±7 / –	1.27	8	1.29	Cc (C_s^4)	39
'Torecan'	$C_{22}H_{29}N_2S_2$	Ortho.	12.07±1 / –	19.99±2 / –	9.25±1 / –	1.21	4	1.19	$P2_12_12_1$ (D_2^4)	39
Pseudo-akuammigine iodomethylate	$C_{23}H_{29}IN_2O_3$	Ortho.	6.28 / –	11.88 / –	27.57 / –	1.54	4	1.51	$P2_12_12_1$ (D_2^4)	38
Robustic acid methyl ether	$C_{23}H_{22}O_6$	Tric.	11.83 / 94	9.09 / 96	9.36 / 100	1.35	2	1.33	$P1$ (C_1^1)	96

Name	Formula	System	a	b	c	d	Z	d	Space group	Ref
3,7-Dideoxy-7-iodo-heptulosonate methyl ester	$C_{23}H_{23}IO_8$	Mono.	20.99±6 —	6.01±2 108.1±0.1	19.56±6 —	1.571	4	1.57	$A2$ (C_2^3)	28
1,1'-Dibenzoylruthenocene	$C_{24}H_{18}O_2Ru$	Mono.	7.34±2 —	12.02±2 90	20.02±4 —	1.65	4	1.65	–	106
Copper N-phenyl-5-bromosalicylaliminate	$C_{24}H_{18}Br_2CuN_2O_2$ $Cu(C_{12}H_9BrNO)_2$	Mono.	9.69±1 —	9.94±1 108	13.35±4 —	–	2	1.55	$P2_1/c$ (C_{2h}^5)	103
Copper N-p-bromophenyl-salicylaliminate	$C_{24}H_{18}Br_2CuN_2O_2$ $Cu(C_{12}H_9BrNO)_2$	Mono.	1.400±4 —	10.55±2 98	8.14±1 —	–	2	1.71	$P2_1/c$ (C_{2h}^5)	103
Copper N-p-iodophenyl-salicylaliminate	$C_{24}H_{18}CuI_2N_2O_2$ $Cu(C_{12}H_9INO)_2$	Mono.	14.45±4 —	10.60±4 98	8.14±2 —	–	2	1.88	$P2_1/c$ (C_{2h}^5)	103
Copper N-phenylsali-cylaliminate	$C_{24}H_{20}CuN_2O_2$ $Cu(C_{12}H_{10}NO)_2$	Mono.	11.90±1 —	7.96±2 123	13.37±4 —	1.36	2	1.41	$P2_1/c$ (C_{2h}^5)	103
Copper N-phenyl-4-methylsalicyl-aliminate	$C_{24}H_{24}CuN_2O_2$ $Cu(C_{12}H_{12}NO)_2$	Mono.	12.72±5 —	10.36±2 102	8.92±3 —	1.36	2	1.37	$P2_1/c$ (C_{2h}^5)	103
Cobalt(II) N-p-methoxy-phenylsalicylaliminate	$C_{24}H_{24}CoN_2O_4$ $Co(C_{12}H_{12}NO_2)_2$	Tric.	11.53±2 90	9.16±2 95	12.10±3 70	–	2	1.32	$P1$ (C_1^1)	103
Dicyclopentadienyl-(ethynylbenzene)di-nickel	$C_{26}H_{20}Ni_2$	Tric.	10.90±4 118.8±0.1	11.57±4 91.1±0.1	18.99±4 90.5±0.1	1.41	2	1.42	$P1$ (C_1^1)	75

Compound	Formula	Class	a α	b β	c γ	D_m	Z	D_x	S.G.	Ref.
1,1,4,4-Tetrachloro-2,3,5,6-tetraphenyl-1,4-digermacyclohexa-2,5-diene	$C_{28}H_{20}Cl_4Ge_2$	Mono.	17.0_5 –	14.55 105.0	5.9_3 –	–	4	1.51	$C2$ (C_2^3)	17
1,1,4,4-Tetrabromo-2,3,5,6-tetraphenyl-1,4-digermacyclohexa-2,5-diene	$C_{28}H_{20}Br_4Ge_2$	Mono.	16.6_9 –	16.9_1 116.0	6.6_8 –	1.5	4	1.61	$C2$ (C_2^3)	17
1,1,4,4-Tetraphenyl-1,4-digermacyclohexa-2,5-diene	$C_{28}H_{24}Ge_2$	Tric.	10.88 105.0	11.7_9 90.0	10.2_3 105.5	1.36	4	1.38	$P1$ (C_1^1)	17
Tetramethyltetraphenylcyclotetrasiloxane (1,1,3,3 – 2,2,4,4 ?)	$C_{28}H_{32}O_4Si_4$	Tric.	11.05 ± 4 146.8 ± 0.8	8.33 ± 3 119.0 ± 0.8	17.47 ± 6 59.0 ± 0.8	1.18	1	1.20	$P1$ (C_1^1)	6
Tetramethyltetraphenylcyclotetrasiloxane (1,1,2,2 – 3,3,4,4 ?)	$C_{28}H_{32}O_4Si_4$	Mono.	12.64 ± 6 –	26.19 ± 5 96.8 ± 0.5	9.13 ± 2 –	1.15	4	1.17	$P2_1/n$ (C_{2h}^5)	6
Monobromoduclauxin	$C_{29}H_{21}O_{11}Br$	Ortho.	15.01 –	18.83 –	9.15 –	1.57	4	1.61	$P2_12_12_1$ (D_2^4)	83
π-Cyclopentadienyl-triphenylphosphine-monocarbonyl-σ-phenyl iron	$C_{30}H_{25}FeOP$	Mono.	11.25 ± 2 –	14.39 ± 3 90.0 ± 0.5	15.31 ± 4 –	1.3	4	1.32	$P2_1/n$ (C_{2h}^5)	16
Dipotassium dipyraphene	$C_{30}H_{20}N_4O_4K_2$	Mono.	16.39 ± 1 –	12.36 ± 3 116.4 ± 0.2	15.50 ± 1 –	–	4	1.37	$P2_1/a$ (C_{2h}^5)	40

Compound	Formula	System					Z		Space group	Ref.
Disodium dipyraphene	$C_{30}H_{20}N_4Na_2O_4$	Mono.	14.77±1 / –	14.28±1 / 104.6±0.2	15.65±1 / –	–	4	1.14	$P2_1/c$ (C_{2h}^5)	40
Three-pyraphene	$C_{45}H_{32}N_6O_6$	Mono.	14.92±1 / –	24.80±3 / 110.8±0.3	12.28±1 / –	–	4	1.18	$P2_1/c$ (C_{2h}^5)	40
1,1,4,4-Tetramethyl-2,3,5,6-tetraphenyl-1,4-digermacyclohexa-2,5-diene	$C_{32}H_{32}Ge_2$	Tric.	12.76 / 110.0	9.36 / 97.5	6.30 / 92.0	–	2	1.34	$P1$ (C_1^1)	17
1,1,4,4-Tetramethyl-2,3,5,6-tetraphenyl-1,4-disilacyclohexa-2,5-diene	$C_{32}H_{32}Si_2$	Tric.	12.61 / 110.0	9.24 / 95.5	6.32 / 92.0	1.14	2	1.14	$P1$ (C_1^1)	17
Copper phthalocyanine (high-density β-form)	$C_{32}H_{16}N_8Cu$	Mono.	19.61 / –	4.80 / 121.51	14.74 / –	1.625	2	1.62	$P2_1/a$ (C_{2h}^5)	41
Nickel phthalocyanine (high-density β-form)	$C_{32}H_{16}N_8Ni$	Mono.	19.48 / –	4.70 / 122.33	14.82 / –	1.61	2	1.65	$P2_1/a$ (C_{2h}^5)	41
Cobalt phthalocyanine (high-density β-form)	$C_{32}H_{16}N_8Co$	Mono.	19.39 / –	4.79 / 120.66	14.57 / –	1.632	2	1.63	$P2_1/a$ (C_{2h}^5)	41
Iron phthalocyanine (high-density β-form)	$C_{32}H_{16}N_8Fe$	Mono.	19.39 / –	4.78 / 120.74	14.55 / –	1.59– / 1.62	2	1.62	$P2_1/a$ (C_{2h}^5)	41
Manganese phthalocyanine (high-density β-form)	$C_{32}H_{16}N_8Mn$	Mono.	19.17 / –	4.77 / 119.4	14.43 / –	1.62	2	1.64	$P2_1/a$ (C_{2h}^5)	41

Compound	Formula	Class	a / α	b / β	c / γ	D_m	Z	D_x	S.G.	Ref.
Zinc phthalocyanine (high-density β-form)	$C_{32}H_{16}N_8Zn$	Mono.	19.13 / -	4.86 / 120.1	14.50 / -	1.630	2	1.63	$P2_1/a$ (C_{2h}^5)	41
Tetracyclopenta-dienyl(diphenylbuta-diyne)tetranickel	$C_{36}H_{30}Ni_4$	Ortho.	17.60±5 / -	20.17±5 / -	8.67±3 / -	1.50	4	1.50	$P2_12_12_1$ (D_2^4)	75
β-Tri-11-bromounde-canoin	$C_{36}H_{65}Br_3O_6$	Tric.	12.4±1 / 90.5±0.6	5.22±5 / 96±1	31.6±3 / 101±1	-	2	-	$P\bar{1}$ (C_i^1)	65
Basic zinc OO-di-isopropyl phosphoro-dithioate	$C_{36}H_{84}O_{13}P_6S_{12}Zn_4$ $Zn_4[S_2P(OCH(CH_3)_2)_2]_6O$	Mono.	24.93±8 / -	22.76±5 / 113.5±0.2	13.67±2 / -	1.47	4	1.45	$P2_1/a$ (C_{2h}^5)	25
Basic zinc OO-di-n-butyl phosphoro-dithioate	$C_{48}H_{108}O_{13}P_6S_{12}Zn_4$ $Zn_4[S_2P(O(CH_2)_3CH_3)_2]_6O$	Trig.	23.9±2 / -	- / -	12.84±8 / -	1.34	3	1.35	$R3$ (C_3^4)	25
Cycloveratryl, ben-zene (clathrate)	$C_{54}H_{60}$, $1.19C_6H_6$	Mono.	22.71 / -	9.61 / 91.5	24.49 / -	1.236	4	-	Cc (C_s^4)	11
Cycloveratryl, chloro-benzene (clathrate)	$C_{54}H_{60}$, $1.11C_6H_5Cl$	Mono.	22.82 / -	9.64 / 93	24.50 / -	1.258	4	-	Cc (C_s^4)	11
Cycloveratryl, toluene (clathrate)	$C_{54}H_{60}$, $0.20C_7H_8$	Mono.	21.75 / -	9.73 / 92	23.20 / -	1.250	4	-	Cc (C_s^4)	11
Cycloveratryl, chloro-form (clathrate)	$C_{54}H_{60}$, $2.92CHCl_3$	Mono.	20.59 / -	9.78 / 93	23.98 / -	-	4	-	Cc (C_s^4)	11

Compound	Formula	System	a	b	c	β	D_{obs}	Z	D_{calc}	Space group	Ref
Cyclveratryl, acetone (clathrate)	$C_{54}H_{60}, 0.55C_3H_6O$	Mono.	20.82	8.39	30.22	102	1.226	4	–	Cc (C_s^4)	11
Cycloveratryl, carbon disulphide (clathrate)	$C_{54}H_{60}, 0.96CS_2$	Mono.	20.82	8.28	30.29	103	1.225	4	–	Cc (C_s^4)	11
Cycloveratryl, butyric acid (clathrate)	$C_{54}H_{60}, 1.95C_4H_8O_2$	Mono.	21.09	8.07	34.66	92.8	1.232	4	–	Cc (C_s^4)	11
4-Androsten-3,17-dione	$C_{19}H_{26}O_2$	Ortho.	12.963±9	16.929±9	7.366±9	–	1.178	4	1.175	$P2_12_12_1$ (D_2^4)	84
1,4-Androstadien-17β-ol-3-one	$C_{19}H_{26}O_2$	Ortho.	12.302±9	18.644±9	7.065±9	–	1.164	4	1.174	$P2_12_12$ (D_2^3)	84
5α-Androstan-3,17-dione	$C_{19}H_{28}O_2$	Mono.	21.337±9	6.186±9	12.704±9	91.27	1.147	4	1.143	$A2$ (C_2^3)	84
4-Androsten-17β-ol-3-one	$C_{19}H_{28}O_2$	Mono.	14.691±9	11.093±9	10.872±9	113.23	1.186	4	1.177	$P2_1$ (C_2^2)	84
5α-Androstan-17β-ol-3-one	$C_{19}H_{30}O_2$	Mono.	11.614±9	8.096±9	9.422±9	99.23	1.132	2	1.103	$P2_1$ (C_2^2)	84
5-Androsten-3β,17β-diol	$C_{19}H_{30}O_2$	Ortho.	12.146±9	23.434±9	6.248±9	–	1.147	4	1.084	$P2_12_12_1$ (D_2^4)	84
4-Androsten-3,11,17-trione	$C_{19}H_{24}O_3$	Ortho.	9.263±9	26.531±9	6.477±9	–	1.264	4	1.253	$P2_12_12_1$ (D_2^4)	84

Compound	Formula	Class	a / α	b / β	c / γ	D_m	Z	D_x	S.G.	Ref.
4-Androsten-17β-ol-3-one 17-acetate	$C_{21}H_{30}O_3$	Ortho.	12.800±9 / -	18.169±9 / -	7.856±9 / -	1.162	4	1.201	$P2_12_12_1$ (D_2^4)	84
5α-Androstan-17β-ol-3-one 17-benzoate	$C_{26}H_{34}O_3$	Mono.	10.860±9 / -	16.241±9 / 99.17	6.236±9 / -	1.148	2	1.207	$P2_1$ (C_2^2)	84
5α-Androstan-3α-ol-17-one, p-bromophenol (1:1)	$C_{19}H_{30}O_2,C_6H_5OBr$	Tric.	9.584±4 / 62.06	11.272±4 / 78.09	6.525±4 / 71.54	1.315	1	1.306	$P1$ (C_1^1)	37
Δ^5-Androsten-3β-ol-17-one, p-bromophenol (3:1)	$(C_{19}H_{28}O_2)_3,C_6H_5OBr$	Tric.	11.599±4 / 87.64	12.412±4 / 84.19	11.041±4 / 88.77	1.263	1	1.273	$P1$ (C_1^1)	37
5β-Androstan-3α-ol-17-one, p-bromophenol (1:1)	$(C_{19}H_{30}O_2),C_6H_5OBr$	Tric.	10.763±4 / 93.90	13.645±4 / 96.34	7.073±4 / 105.80	1.222	1	1.260	$P1$ (C_1^1)	37
Δ^4-Androsten-17β-ol-3-one, p-bromophenol (1:1)	$C_{19}H_{28}O_2,C_6H_5OBr$	Ortho.	13.113±4 / -	22.753±4 / -	7.667±4 / -	1.325	4	1.340	$P2_12_12_1$ (D_2^4)	37
5α-Androstan-17β-ol-3-one, p-bromophenol (1:1)	$C_{19}H_{30}O_2,C_6H_5OBr$	Tric.	13.318±4 / 91.69	13.488±4 / 90.83	7.533±4 / 120.41	1.274	[2]	1.319	$P1$ (C_1^1)	37
1,4-Androstandien-17β-ol-3-one, p-bromophenol (1:1)	$C_{19}H_{26}O_2,C_6H_5OBr$	Tric.	10.342±4 / 113.53	10.566±4 / 103.54	6.151±4 / 116.36	1.366	1	1.385	$P1$ (C_1^1)	37
5β-Androstan-17β-ol-3-one, p-bromophenol (1:1)	$C_{19}H_{30}O_2,C_6H_5OBr$	Tric.	11.794±4 / 85.64	13.739±4 / 89.19	7.255±4 / 84.62	1.322	2	1.319	$P1$ (C_1^1)	37

439

Compound	Formula	System	a	b	c	β	ρ_{obs}	Z	ρ_{calc}	Space group	Ref
Testosterone formate	$C_{20}H_{28}O_3$	Mono.	21.7±2	12.5±1	12.9±1	96±1	1.22	8	1.21	$P2_1$ (C_2^2)	47
Testosterone acetate	$C_{21}H_{30}O_3$	Ortho.	12.6±1	18.1±2	7.8±1	-	1.22	4	1.22	$P2_12_12_1$ (D_2^4)	47
Testosterone propionate	$C_{22}H_{32}O_3$	Ortho.	12.6±1	20.3±2	7.6±1	-	1.17	4	1.17	$P2_12_12_1$ (D_2^4)	47
Testosterone butyrate	$C_{23}H_{34}O_3$	Ortho.	12.3±1	16.3±2	10.3±1	-	1.16	4	1.15	$P2_12_12_1$ (D_2^4)	47
Testosterone valerate	$C_{24}H_{36}O_3$	Ortho.	12.3±1	16.7±2	10.3±1	-	1.18	4	1.17	$P2_12_12_1$ (D_2^4)	47
4-Androsten-17β-ol-3-one 17-bromoacetate	$C_{21}H_{29}O_3Br$	Mono.	12.413±9	15.645±9	10.164±9	96.03	1.380	4	1.385	$P2_1$ (C_2^2)	85
4-Androsten-6-bromo-17β-ol-3-one 17-acetate	$C_{21}H_{29}O_3Br$	Mono.	29.243±9	6.186±9	22.307±9	100.21	1.320	8	1.369	$C2$ (C_2^3)	85
5β-Androstan-4β-bromo-3,17-dione	$C_{19}H_{27}O_2Br$	Ortho.	16.442±9	29.364±9	7.455±9	-	1.373	8	1.353	$P2_12_12_1$ (D_2^4)	85
3β-Chloro-5-cholestene	$C_{27}H_{45}Cl$	Mono.	16.333±9	7.553±9	10.691±9	102.95	1.040	2	1.047	$P2_1$ (C_2^2)	85
5-Pregnen-17α-bromo-3β-ol-20-one 3-acetate, heptane	$C_{23}H_{33}O_3Br \cdot C_7H_{16}$	Ortho.	11.419±9	25.654±9	7.895±9	-	1.519	4	1.544	$P2_12_12_1$ (D_2^4)	85

Compound	Formula	Class	a α	b β	c γ	D_m	Z	D_x	S.G.	Ref.
5α-Pregnan-3,20-dione	$C_{21}H_{32}O_2$	Ortho.	12.233±8 / -	22.485±8 / -	6.440±8 / -	1.182	4	1.212	$P2_12_12_1$ (D_2^4)	86
4-Pregnen-17α-ol-3,20-dione	$C_{21}H_{30}O_3$	Ortho.	9.836±8 / -	23.466±8 / -	7.819±8 / -	1.156	4	1.216	$P2_12_12_1$ (D_2^4)	86
5α-Pregnan-3β,17α-diol-20-one	$C_{21}H_{34}O_3$	Tric.	6.510±8 / 93.37	12.492±8 / 107.72	6.365±8 / 103.64	1.135	1	1.166	$P1$ (C_1^1)	86
5α-Pregnan-3β,17α,20β-triol	$C_{21}H_{36}O_3$	Mono.	13.587±8 / -	10.522±8 / 103.25	7.319±8 / -	1.194	2	1.097	$P2_1$ (C_2^2)	86
5α-Pregnan-3β,11β,20β,21-tetrol	$C_{21}H_{36}O_4$	Ortho.	14.347±8 / -	23.000±8 / -	11.860±8 / -	1.156	8	1.196	$P2_12_12_1$ (D_2^4)	86
5α-Pregnan-3β-ol-20-one 3-acetate	$C_{23}H_{36}O_3$	Mono.	27.347±8 / -	7.682±8 / 91.61	9.938±8 / -	1.128	4	1.147	$P2_1$ (C_2^2)	86
5α-Pregnan-21-ol-3,20-dione acetate (form "a")	$C_{23}H_{34}O_4$	Mono.	14.096±8 / -	7.601±8 / 99.82	9.719±8 / -	1.190	2	1.212	$P2_1$ (C_2^2)	86
5α-Pregnan-21-ol-3,20-dione acetate (form "b")	$C_{23}H_{34}O_4$	Mono.	21.709±8 / -	7.608±8 / 116.42	13.923±8 / -	1.165	4	1.208	$C2$ (C_2^3)	86
5α-Pregnan-17α,21-diol-3,11,20-trione 21-acetate	$C_{23}H_{32}O_6$	Mono.	16.722±8 / -	7.663±8 / 94.70	9.787±8 / -	1.041	2	1.074	$P2_1$ (C_2^2)	86

Compound	Formula	System	a	b	c	β		Z		Space group	Ref
5α-Pregnan-3β,20α-diol diacetate	$C_{25}H_{40}O_4$	Ortho.	10.009±8	61.486±8	7.675±8	–	1.145	8	1.139	$P2_12_12_1$ (D_2^4)	86
5α-Pregnan-3β,20β-diol diacetate	$C_{25}H_{40}O_4$	Ortho.	11.665±8	29.071±8	6.839±8	–	1.166	4	1.158	$P2_12_12_1$ (D_2^4)	86
5α-Pregnan-3β,11β,17α,20β,21-pentol 3,20,21-triacetate	$C_{27}H_{42}O_8$	Mono.	13.572±8	13.755±8	7.535±8	108.81	1.222	2	1.233	$P2_1$ (C_2^2)	86
22,23-Bisnor-5-cholenic acid-3β-ol	$C_{22}H_{33}O_3$	Ortho.	11.794±8	28.039±8	6.019±8	–	1.149	4	1.154	$P2_12_12_1$ (D_2^4)	82
5β-Cholanic acid-3α,12α-diol.ethanol	$C_{24}H_{40}O_4 \cdot C_2H_5OH$	Ortho.	13.589±8	25.765±8	7.225±8	–	1.160	4	1.152	$P2_12_12_1$ (D_2^4)	82
5β-Cholanic acid-3α,6α-diol	$C_{24}H_{40}O_4$	Ortho.	11.577±8	29.924±8	6.433±8	–	1.154	4	1.170	$P2_12_12_1$ (D_2^4)	82
5β-Cholanic acid-3α,7α-diol	$C_{24}H_{40}O_4$	Ortho.	13.302±8	26.737±8	12.377±8	–	1.174	8	1.185	$P22_12_1$ (D_2^3)	82
5β-Cholanic acid-3α,7α,12α-triol tetrahydrate	$C_{24}H_{40}O_5 \cdot 4H_2O$	Mono.	14.043±8	7.849±8	13.697±8	113.53	1.156	2	1.153	$P2_1$ (C_2^2)	82
5β-Cholan-24-ol	$C_{24}H_{42}O$	Mono.	40.689±8	6.907±8	29.046±8	129.61	1.114	12	1.098	$C2$ (C_2^3)	82
5β-Cholanic acid-3α,7α,12α-triol sodium salt	$C_{24}H_{39}O_5Na$	Mono.	12.593±8	8.215±8	12.196±8	107.86	1.167	2	1.191	$P2_1$ (C_2^2)	82

442

Compound	Formula	Class	a α	b β	c γ	D_m	Z	D_x	S.G.	Ref.
5β-Cholanic acid-3α,7α,12α-triol methyl ester, ethanol	$C_{25}H_{42}O_5 \cdot C_2H_5OH$	Mono.	25.489±8 –	8.011±8 121.59	15.337±8 –	1.163	4	1.167	$C2$ (C_2^3)	82
5β-Cholanic acid-3α-ol methyl ester	$C_{25}H_{42}O_3$	Ortho.	11.337±8 –	26.796±8 –	7.013±8 –	1.132	4	1.217	$P2_12_12_1$ (D_2^4)	82
5β-Cholanic acid-3α-ol-12-one acetate methyl ester	$C_{27}H_{42}O_5$	Ortho.	11.448±8 –	59.335±8 –	7.468±8 –	1.146	8	1.170	$P2_12_12_1$ (D_2^4)	82
5β-Cholanic acid-3α-ol-acetate methyl ester	$C_{27}H_{44}O_4$	Mono.	14.989±8 –	7.740±8 107.43	11.355±8 –	1.168	2	1.144	$P2_1$ (C_2^2)	82
5β-Cholanic acid-3α,7α-diol-12-one diacetate	$C_{28}H_{42}O_7$	Tric.	14.557±8 93.47	16.671±8 98.18	6.343±8 106.51	1.123	2	1.141	$P1$ (C_1^1)	82
5β-Cholanic acid-3α,7α,12α-triol 3,7-diacetate methyl ester hemihydrate	$C_{29}H_{44}O_7 \cdot \frac{1}{2}H_2O$	Mono.	47.900±8 –	7.995±8 102.06	22.922±8 –	1.197	12	1.192	$C2$ (C_2^3)	82

REFERENCES

1. Zur Stabilität des Tris-1,10-phenanthrolin-Eisen(II) Ions. G. ALBRECHT,
    J. TSCHIRNICH and K. MADEJA, 1965. *Z. Chem.*, 5, 312-313.

2. Comparison of crystals of *l*-ephedrine copper(II) chelate obtained from
    several different solvents. Y. AMANO, K. OSAKI and T. UNO, 1965. *J. Chem.
    Soc. Japan, Pure Chem. Sect.*, 86, 1109-1111.

3. An X-ray study of 2,5-dichloronitrobenzene. L. ANNAPOORNI and B.V.R. MURTY,
    1965. *Z. Kristallogr.*, 122, 317.

4. [Cristallographic data on the lithium and cesium salts of 5,5-dimethyl-2-
    sulfo-1,3-cyclohexanedione (dimedonesulfonic acid).] S.K. APINITIS and A.F.
    IEVIN'S, 1965. *Latv. PSR Zinat Akad. Vest., Kim. Ser.*, 653-658.

5. Structure of calcium oxalate monohydrate. H.J. ARNOTT, F.G.E. PAUTARD and H.
    STEINFINK, 1965. *Nature, Lond.*, 208, 1197-1198.

6. [X-ray study of some organic silicon compounds.] R.L. AVOJAN, G.N. SAKHARIVA.
    Z.A. AKOPJAN and JU.T. STRUČKOV, 1965. *Ž. Strukt. Khim. SSSR*, 6, 792-793.

7. Partial determination of the crystal structure of methyl-2-chloro-mercuri-2-
    deoxy-α-D-talopyranoside. J. BAIN and M.M. HARDING, 1965. *J. Chem. Soc.*,
    4025-4027. [A partial structure analysis is described.]

8. A preliminary investigation of the structure of mercury dibenzyl. V.A. BAIN,
    D. CALVERT and R.C.G. KILLEAN, 1965. *Z. Kristallogr.*, 122, 476.

9. [X-ray diffraction study of the complex of 1-methyl-5-bromouracil and 9-
    methyladenine.] JU.G. BAKLAGINA, M.V. VOL'KENSTEJN and JU.D. KONDRAŠOV, 1965.
    Biofizika, SSSR, 10, 165-166. [*Biophysics, USSR*, 10, 181-182.]

10. The crystal structure of calcium 5-keto-D-gluconate (calcium D-xylo-5-hexulo-
    sonate). A.A. BALCHIN and C.H. CARLISLE, 1965. *Acta Cryst.*, 19, 103-111.

11. [Cycloveratryl clathrates.] V.M. BHATNAGAR, 1965. *Ž. Strukt. Khim. SSSR*, 6,
    794-795. [*J. Struct. Chem. USSR*, 6, 760-761.]

12. The structure of *trans*-tetrachlorobis(tetrahydrothiophen)tin(IV) in the solid
    state and in solution. I.R. BEATTIE, R. HULME and L. RULE, 1965. *J. Chem.
    Soc.*, 1581-1583.

13. [Complexes of divalent ions with asymmetric dimethylurea.] M.C. BIAGINI, L.
    COGHI and C. GUASTINI, 1965. *Gazz. Chim. Ital.*, 95, 368-374.

14. [Chlorides of bivalent metals with 2,6-dimethyl-γ-pyrone.] M.C. BIAGINI and
    C. GUASTINI, 1965. *Gazz. Chim. Ital.*, 95, 425-431.

15. Molecular compounds and complexes. II. Exploratory crystallographic study
    of some donor-acceptor molecular compounds. J.C.A. BOEYENS and F.H. HERBSTEIN,
    1965. *J. Phys. Chem.*, 69, 2153-2159.

16. [X-ray study of some heteroorganic compounds.] N.G. BOKIJ, R.L. AVOJAN, G.N.
    ZAKHAROVA, M.KH. MINASJAN, Z.A. AKOPJAN and JU.T. STRUČKOV, 1965. *Ž. Strukt.
    Khim. SSSR*, 6, 795-796. [*J. Struct. Chem. USSR*, 6, 762-763.]

17. [Unit cells and space groups of reaction products of acetylene and tolan with
    divalent germanium and silicon compounds.] N.G. BOKIJ, G.N. ZAKHAROVA and
    JU.T. STRUČKOV, 1965. *Ž. Strukt. Khim. SSSR*, 6, 476-477. [*J. Struct. Chem.
    USSR*, 6, 458-459.]

18. The crystal structure of triketoindane (anhydrous ninhydrin). A structure
    showing close C=O...C interactions. W. BOLTON, 1965. *Acta Cryst.*, 18, 5-10.

19. Optical properties and preliminary x-ray investigation of retamine and its bromide and chloride. J.M. BOSCH-FIGUEROA, L.M. MILLE and M. FONT-ALTABA, 1965. *Acta Cryst.*, 18, 921-923.

20. [Salts of hydrazinecarboxylic acid, M(II)(N$_2$H$_3$COO)$_2$.(N$_2$H$_5$)$_2$CO$_3$.] A. BRAI-BANTI, G. BIGLIARDI and R. CANALI-PADOVANI, 1965. *Ateneo Parmense, Sez.II*, 1, 75-80.

21. [Cadmium and manganese hydrazinecarboxylates.] A. BRAIBANTI, G. BIGLIARDI and A.M. MANOTTI-LANFREDI, 1965. *Ateneo Parmense, Sez. II*, 1, 81-86.

22. The growth of crystals in silica gel, and the unit-cell dimensions of cadmium oxalate and copper tartrate. C. BRIDLE and T.R. LOMER, 1965. *Acta Cryst.*, 19, 483-484.

23. [Crystal-structure studies of copper formate hemidioxane.] M. BUKOWSKA-STRZYEWSKA, 1965. *Roczn. Chem.*, 39, 507-508.

24. Crystallographic data for some ditertiary arsine-metal carbonyl complexes of group VI elements. G.J. BULLEN, 1965. *Acta Cryst.*, 18, 974.

25. The structure of basic zinc OO-dialkyl phosphorodithioates. A.J. BURN and G.W. SMITH, 1965. *Chem. Communic.*, GB, 394-396.

26. Crystal structure for sodium thymidylyl-(5'→3')-thymidylate-(5'). N. CAMERMAN and J. TROTTER, 1965. *Acta Cryst.*, 19, 867-868.

27. Stereochemistry of arsenic. XIII. 10-Chloro-5,10-dihydrophenarsazine. A. CAMERMAN and J. TROTTER, 1965. *J. Chem. Soc.*, 730-738.

28. Preliminary x-ray crystallographic data for 3,7-dideoxy-7-iodo-heptulosonate methyl ester. C.C.H. CHEN, D.B. LEVINE and B.W. LOW, 1965. *Z. Kristallogr.*, 122, 159-160.

29. Crystallographic data for some idose derivatives. S.S.C. CHU and G.A. JEFFREY, 1965. *Acta Cryst.*, 18, 820.

30. Structural relationship between two polymorphic forms of β-naphthol. P. COPPENS and I. HEAIRFIELD, 1965. *Israel J. Chem.*, 3, 25-28.

31. Crystal data for monorden (radicicol). F.W. CROMER and J. TROTTER, 1965. *Acta Cryst.*, 19, 681.

32. Raumgruppe und Gitterparameter einiger Phenylhydrazonderivate von Monosacchariden. L. CSORDAS and GY. MENCZEL, 1965. *Acta Chim. Acad. Sci. Hungar.*, 46, 191-193.

33. The structure of phosphobenzene. J.J. DALY and L. MAIER, 1965. *Nature, Lond.*, 208, 383-384.

34. Crystal data for two complexes of ethyl picrate with alkali ethylates. R. DESTRO, C.M. GRAMACCIOLI, A. MUGNOLI and M. SIMONETTA, 1965. *Tetrahedron Letters*, 2611-2615.

35. Crystal data on o-phenanthroline hydrate. G. DONNAY, J.D.H. DONNAY and M.J.C. HARDING, 1965. *Acta Cryst.*, 19, 688-689.

36. The unit-cell dimensions of potassium myristate and 1:1 acid potassium myristate. J.H. DUMBLETON, 1965. *Acta Cryst.*, 19, 279-280.

37. Androgenic steroid complexes with p-bromophenol. C. EGER and D.A. NORTON, 1965. *Nature, Lond.*, 208, 997-999.

38. Étude preliminaire de l'iodométhylate de pseudo-akuammigine. J. ETIENNE, J. LEMEN and J. LEVY, 1965. *Acta Cryst.*, 18, 130.

39. Preliminary x-ray data on phenothiazine and certain of its derivatives. D. FEIL, M.H. LINCK and J.J.H. McDOWELL, 1965. *Nature, Lond.*, 207, 285-286.

40. Crystal data of 'three-pyraphene' and of some derivatives of 'dipyraphene'. A. FERRARI, A. BRAIBANTI and A. TIRIPICCHIO, 1965. *Acta Cryst.*, 18, 979.

41. Crystallographic constants for some high-density β-phthalocyanines. P.E. FIELDING and N.C. STEPHENSON, 1965. *Australian J. Chem.*, 18, 1691-1693.

42. Crystallographic data for nickel and cobalt biuret complexes. S.K. GHOSH. R.M. SANYAL and B.K. BANERJEE, 1965. *Indian J. Phys.*, 39, 170-175.

43. Elements structuraux des diméthylphenols. H. GILLIER-PANDRAUD, 1965. *C.R. Acad. Sci. Paris*, 260, 1960-1962.

44. An x-ray study of *p*-chlorophenoxyacetic acid. K.N. GOSWAMI, 1965. *Z. Kristallogr.*, 121, 400.

45. An x-ray study of metal complexes of biguanide. K.N. GOSWAMI, 1965. *Z. Kristallogr.*, 122, 473-475.

46. Crystallographic data for the 2'-halogenobiphenyl-4-carboxylic acids and 4-acetyl-2-halogenobiphenyls. G.W. GRAY, H.H. SUTHERLAND and D.W. YOUNG, 1965. *J. Chem. Soc.*, 4208.

47. Crystallographic data for some testosterone esters. P.J.F. GRIFFITHS, K.C. JAMES and M. REES, 1965. *Acta Cryst.*, 19, 149-150.

48. Salze von Halogenophosphorsäuren. I. Über die Verbindungen $Ca(PO_2Br_2)_2 \cdot 2CH_3COOC_2H_5$ und $Mg(PO_2Br_2)_2 \cdot 2CH_3COOC_2H_5$. H. GRUNZE and K.-H. JOST, 1965. *Z. Naturf.*, B20, 268.

49. Die Konstitution und Konfiguration des **cycloneosamandions** G. HABERMEHL and S. GOETTLICHER, 1965. *Chem. Ber.*, 98, 1-10. (An approximate crystal structure is presented also.)

50. Über die Elementarzelle des Hexahydrats von **mellithsäuren** Aluminium. B. HAJEK. E. KALALOVA and F. PETRU, 1965. *Z. Chem.*, 5, 230.

51. Base adducts of β-ketoenolates. III. Complexes of cobalt(II) and nickel(II). J.T. HASHAGEN and J.P. FACKLER, 1965. *J. Amer. Chem. Soc.*, 87, 2821-2824.

52. Twinned crystals. II. α-1,2:4,5-tetrachlorobenzene. F.H. HERBSTEIN, 1965. *Acta Cryst.*, 18, 997-1000.

53. Twinned crystals. III. γ-*o*-nitroaniline. F.H. HERBSTEIN, 1965. *Acta Cryst.*, 19, 590-595.

54. Clathrates of sodium warfarin. C.F. HISKEY and V. MELNITCHENKO, 1965. *J. Pharm. Sci., USA*, 54, 1298-1302.

55. Données cristallographiques sur quelques diamides organiques. M. HOSPITAL and J. HOUSTY, 1965. *Acta Cryst.*, 18, 820.

56. Crystallographic data on some organic compounds of lithium. K. HUML, 1965. *Czech. J. Phys.*, 15, 699-701.

57. The crystal structure of securinine hydrobromide dihydrate and the molecular structure of securinine. S. IMADO, M. SHIRO and Z. HORII, 1965. *Chem. Pharm. Bull., Tokyo*, 13, 643-651.

58. X-ray powder crystallographic data on plutonium and other oxalates. I. The
     oxalates of plutonium(III) and plutonium(VI) and their isomorphs. II.
     Plutonium(VI) oxalate dihydrate, uranium(IV) oxalate hexahydrate, uranium(IV)
     oxalate dihydrate and thorium oxalate dihydrate. I.L. JENKINS, F.H. MOORE
     and M.J. WATERMAN, 1965. *J. Inorg. Nucl. Chem.*, 27, 77-80. *Ibid.*, 27,
     81-87.

59. The optical and crystallographic properties of cycloalliin hydrochloride
     monohydrates. F.T. JONES, K.S. LEE, D.R. BLACK and K.J. PALMER, 1965.
     Microscope, GB, 14, 379-383.

60. The conformation of non-aromatic ring compounds. XIV. The crystal struc-
     ture of *trans*-2,3-dichloro-1,4-dithiane at -180°C. H.T. KALFF and C. ROMERS,
     1965. *Acta Cryst.*, 18, 164-168.

61. The unit cell and space group of sodium acetate trihydrate, $NaC_2H_3O_2.3H_2O$.
     A. KALMAN, 1965. *Acta Cryst.*, 19, 853.

62. X-ray study on di- and trichloroquinaldine. Z. KALUSKI and K. GOLANKIEWICZ,
     1965. *Bull. Acad. Polon. Sci., Ser. Sci. Chim.*, 13, 93-95.

63. The crystal structure of trimethyltin hydroxide. N. KASAI, K. YASUDA and
     R. OKAWARA, 1965. *J. Organometal. Chem.*, 3, 172-173.

64. Crystallographic data on dimethyl sulphoxide complexes of zirconyl and
     thorium perchlorates. V. KRISHNAN and C.C. PATEL, 1965. *J. Inorg. Nucl.
     Chem.*, 27, 244-247.

65. The crystal structure of the β-form of trilaurin. K. LARSSON, 1965. *Ark.
     Kemi*, 23, 1-15.

66. On the crystal structure of 2-monolaurin. K. LARSSON, 1965. *Ark. Kemi*, 23,
     23-27.

67. The conformation of bicyclo[3.3.1]nonane. I. LASZLO, 1965. *Rec. trav.
     chim. Pays-Bas*, 84, 251-254.

69. Topochemistry. XI. The crystal structure of methyl *m*- and *p*-bromocinnamates.
     L. LEISEROWITZ and G.M.J. SCHMIDT, 1965. *Acta Cryst.*, 19, 311-313.

70. Kristalldaten von Hexacyanbenzol. W. LITTKE and K. WALLENFELS, 1965.
     Tetrahedron Letters, 3365-3366.

71. Crystal data for monobenzoylosmocene. A.C. MACDONALD and J. TROTTER, 1965.
     Acta Cryst., 19, 1046.

72. A test of the β-isomorphous synthesis. S.K. MAZUMDAR, 1965. *Indian J.
     Pure and Applied Physics*, 3, 411-413.

73. X-ray crystallography of 2-chloroquinoline. S. MERLINO, 1965. *Z.
     Kristallogr.*, 18, 151-157.

74. Crystal data for some dicyclopentadienyldinickel alkyne compounds. O.S.
     MILLS and B.W. SHAW, 1965. *Acta Cryst.*, 18, 562.

75. The unit cell dimensions and space group of bis-*o*-phenylene disulphide. D.J.
     MITCHELL and E.L. LIPPERT, 1965. *Acta Cryst.*, 18, 559-560.

76. [Notes on the structure of β-chloronaphthalene crystals.] R.M. MJASNIKOVA,
     V.I. ROBAS and G.K. SEMIN, 1965. *Ž. Strukt. Khim. SSSR*, 6, 474-475. [*J.
     Struct. Chem.*, 6, 455.]

77. Crystallographic data for triclinic 5-formylvanillic acid. H. MORITA and H. KODAMA, 1965. *Acta Cryst.*, 19, 687.

78. Dimorphism of DL-aspartic acid. G.S.R.K. MURTI, R. NATARAJAN and R. DEB, 1965. *Indian J. Phys.*, 39, 199-202.

79. The crystal structure of semicarbazide complexes of copper(II) and zinc chlorides. M. NARDELLI, G.F. GASPARRI. P. BOLDRINI and G.G. BATTISTINI, 1965. *Acta Cryst.*, 19, 491-500.

80. The unit cell and some properties of bis(1,3-diaminopropane)copper(II) chloride monohydrate. R. NASANEN, I. VIRTAMO and H. ERIKKILA, 1965. *Suomen Kemist.*, B, 38, 278-280.

81. An x-ray study of 3-nitro-4-toluidine. G.D. NIGAM and B.V.R. MURTY, 1965. *Z. Kristallogr.*, 122, 318.

82. Crystal data (I) for some bile acid derivatives. D.A. NORTON and B. HANER, 1965. *Acta Cryst.*, 19, 477-478.

83. The x-ray study of monobromoduclauxin. Y. OGIHARA, Y. IITAKA and S. SHIBATA, 1965. *Tetrahedron Letters*, 1289-1290. (Some details of the molecular structure are given.)

84. Crystal data (II) for some androstanes. J.M. OHRT, B.A. HANER and D.A. NORTON, 1965. *Acta Cryst.*, 19, 479.

85. Crystal data (I) for some halogenated steroids. J.M. OHRT, B.A. HANER and D.A. NORTON, 1965. *Acta Cryst.*, 19, 280.

86. Crystal data (I) for some pregnane-related compounds. J.M. OHRT, B.A. HANER and D.A. NORTON, 1965. *Acta Cryst.*, 19, 869.

87. [Crystals of palladium 8-mercaptoquinolate, $(C_9H_6SN)_2Pd$.] JA.K. OZOL, L.JA. PECH, JA.A. BANKOVSKIJ, A.F. IEVIN'S and E.A. LUKSA, 1965. *Latv. PSR Zinat. Akad. Vest., Khim. Ser.*, 251-252.

88. Unit cell and space group of cesium oxalate monohydrate, $Cs_2C_2O_4.H_2O$. B.F. PEDERSEN and B. PEDERSEN, 1965. *Acta Chem. Scand.*, 19, 1498-1499.

89. The space group of anhydrous copper propionate. A. PODDER, 1965. *Z. Kristallogr.*, 122, 317.

90. Space group and unit cell dimensions of copper propionate monohydrate. A. PODDER and S.N. GIRI, 1965. *Indian J. Phys.*, 39, 502-503.

91. [Structures of double diazonium salt crystals 1. Structure of the double salt of ferric chloride and o-methoxyphenyl diazonium chloride.] T.N. POLYNOVA, N.G. BOKIJ and M.A. PORAJ-KOŠIC, 1965. *Ž. Strukt. Khim. SSSR*, 6, 878-887. [*J. Struct. Chem.*, 6, 841-849.]

92. [Symmetry of quasi-aromatic rings.] E.G. POPOVA, D.N. SIGORIN, N.N. SAPET'KO, A.P. SKOLDINOV and G.A. GOLD'ER, 1965. *Ž. Fiz. Khim. SSSR*, 39, 2726-2729. [*Russ. J. Phys. Chem.*, 39, 1456-1458.]

93. Kristallographische Daten von Quecksilber(I)-acetat. H. PUFF, G. LORBACHER and R. SKRABS, 1965. *Z. Kristallogr.*, 122, 156-158.

94. Die Elementarzelle des Quecksilber(II)-acetate. H. PUFF, G. LORBACHER and R. SKRABS, 1965. *Acta Cryst.*, 19, 870.

95. Space group and unit-cell dimensions of 7-chloro-4-methylcoumarin. K.V.K. RAO and P.V. RAO, 1965. *Z. Kristallogr.*, 122, 318.

96. Crystal data for robustic acid methyl ether. K.V.K. RAO and P.V. RAO, 1965.
     Acta Cryst., 18, 572.

97. Optical and crystallographic data of uric acid and its dihydrate. H. RINGERTZ,
     1965. *Acta Cryst.*, 19, 286-287.

98. Unit cell and space group of sublimed fumaric acid. L.G. ROLDAN, F.J. RAHL
     and A.R. PATERSON, 1965. *Acta Cryst.*, 19, 1055.

99. Crystallographic data on ammonium dioxovanadium(V) bisoxalate dihydrate by
     the x-ray powder method. D.N. SATHYANARAYA and C.C. PATEL, 1965. *Bull.
     Chem. Soc. Japan*, 38, 1404-1405.

100. Crystallographic data for two complex oxalates of oxovanadium(IV). D.N.
     SATHYANARAYA and C.C. PATEL, 1965. *J. Inorg. Nucl. Chem.*, 27, 2549-2552.

101. [X-ray study of the addition compound of $AlCl_3$ with dimethyl ether.] K.N.
     SEMENENKO and N.JA. TUROVA, 1965. *Ž. Neorg. Khim. SSSR*, 10, 2830-2831.
     [*Russ. J. Inorg. Chem.*, 10, 1535-1536.]

102. An x-ray investigation of the sterochemistry of $Zn(NCS)_2(C_6H_5NH_2)_2$. T.M.
     SHEPHERD and I. WOODWARD, 1965. *Acta Cryst.*, 19, 479-482.

103. [Crystal chemical data on complex compounds of N-substituted derivatives of
     salicylalimine.] L.M. SKOL'NIKOVA, V.V. ZELENCOV and L.G. MAKAREVIČ, 1965.
     Ž. Strukt. Khim. SSSR, 10, 653. [*J. Struct. Chem. USSR*, 10, 626.]

104. [X-ray data for derivatives of barene and neobarene.] JU. T. STRUČKOV.
     V.I. STANKO, A.I. KLIMOVA and G.S. KON'KOVA, 1965. *Ž. Strukt. Khim. SSSR*, 6,
     923-925. [*J. Struct. Chem. USSR*, 6, 888-890.]

105. [Physical properties of guanidine perchlorate.] K.V. TITOVA and V.JA.
     ROSOLOVSKIJ, 1965. *Ž. Neorg. Khim. SSSR*, 10, 446-450. [*Russ. J. Inorg.
     Chem.*, 10, 239-244.]

106. The crystal structure of 1,1'-dibenzoylruthenocene. J. TROTTER and S.H.
     WHITLOW, 1965. *Acta Cryst.*, 19, 868.

107. The unit cell and space group of benzaldehyde phenylhydrazone, $C_6H_5.CH:N.NH.$
     C_6H_5. R.H. DE VERE, 1965. *Acta Cryst.*, 19, 681.

108. Morphology and space group of hexaantipyrine perchlorate complexes of Mg^{++},
     Ca^{++} and Zn^{++}, and pentaantipyrine perchlorate of Cu^{++}. M. VIJAYAN and M.A.
     VISWAMITRA, 1965. *Z. Kristallogr.*, 122, 153-155.

109. Crystallographic data for aldrin, $C_{12}H_8Cl_6$. H.S. VILLARROEL, 1965. *Acta
     Cryst.*, 18, 558.

110. On the preparation and crystallography of tetramethylammonium chloride
     hydrogen chloride. J.M. WILLIAMS and S.W. PETERSON, 1965. *Acta Cryst.*, 19,
     1058.

111. Crystallographic data for 2,2-diphenyl-1-picrylhydrazyl. D.E. WILLIAMS,
     1965. *J. Chem. Soc.*, 7535-7536.

449

SUBJECT INDEX

This index contains the names of substances printed at the heads of the reports, and the names of substances referred to therein. The numerical prefixes *mono* (not di, tri, bis...), *o*, *m*, *p*, D, L, etc. are disregarded in fixing the alphabetical order, although sometimes double entries are made.

FORMULA INDEX

In this index compounds containing carbon and hydrogen are listed primarily according to the number of carbon atoms in the formula; within groups with the same number of carbon atoms the order is that of increasing number of hydrogen atoms.

A few compounds that can be regarded as organic even though they do not contain hydrogen are also included. They are arranged as if the formula were written C_nH_0ABC. Polymers with indefinite suffixes have been collected at the end.

AUTHOR INDEX

Modified letters [ä, å, č, è, é, š, ž, etc.] are placed in, or immediately following the positions where the unmodified letters would occur in the normal English alphabetical order. Names beginning with Mc or Mac are collected at the beginning of the M's. Names beginning with a separated *de, van* or *von* are listed according to the main component, thus de Vries occurs with the V's.

INDEX CORRECTIONS

FORMULA INDEX

AUTHOR INDEX